VOLUME ONE HUNDRED AND SIXTEEN

PROGRESS IN MOLECULAR BIOLOGY AND TRANSLATIONAL SCIENCE

The Molecular Biology of Cadherins

VOLUME ONE HUNDRED AND SIXTEEN

PROGRESS IN MOLECULAR BIOLOGY AND TRANSLATIONAL SCIENCE

The Molecular Biology of Cadherins

Edited by

FRANS VAN ROY, Ph.D.

Department for Molecular Biomedical Research, VIB & Department of Biomedical Molecular Biology, Ghent University, Ghent, Belgium

AMSTERDAM • BOSTON • HEIDELBERG • LONDON
NEW YORK • OXFORD • PARIS • SAN DIEGO
SAN FRANCISCO • SINGAPORE • SYDNEY • TOKYO
Academic Press is an imprint of Elsevier

Academic Press is an imprint of Elsevier
The Boulevard, Langford Lane, Kidlington, Oxford, OX5 1GB, UK
32 Jamestown Road, London NW1 7BY, UK
Radarweg 29, PO Box 211, 1000 AE Amsterdam, The Netherlands
225 Wyman Street, Waltham, MA 02451, USA
525 B Street, Suite 1800, San Diego, CA 92101-4495, USA

First edition 2013

Library of Congress Cataloging-in-Publication Data
A catalog record for this book is available from the Library of Congress

British Library Cataloguing in Publication Data
A catalogue record for this book is available from the British Library

ISBN: 978-0-12-394311-8
ISSN: 1877-1173

For information on all Academic Press publications
visit our website at store.elsevier.com

13 14 15 12 11 10 9 8 7 6 5 4 3 2 1

CONTENTS

Section B
Versatility of the Cadherin Superfamily

Section D
Cadherin-Associated Proteins

CONTRIBUTORS

Panos Z. Anastasiadis
Department of Cancer Biology, Mayo Clinic, Jacksonville, Florida, USA

Geert Berx
Department for Molecular Biomedical Research, Unit of Molecular and Cellular Oncology, VIB & Department of Biomedical Molecular Biology, Ghent University, Ghent, Belgium

Fátima Carneiro
Faculty of Medicine, University of Porto; Cancer Genetics Group, Institute of Molecular Pathology and Immunology of the University of Porto (IPATIMUP), and Department of Pathology, Centro Hospitalar de São João, Porto, Portugal

Claudio Collinet
Institut de Biologie du Développement de Marseille Luminy (IBDML), Unite Mixte de Recherche 6216, Case 907, Marseille, France

Elisabetta Dejana
IFOM, FIRC Institute of Molecular Oncology, and Department of Biosciences, School of Sciences, University of Milan, Milan, Italy

Daniel J. Dickinson[1]
Cancer Biology Program, Stanford University, Stanford, California, USA

Aziz El-Amraoui
Institut Pasteur, Unité de Génétique et Physiologie de l'Audition; Inserm UMRS 1120, and UPMC, Paris, France

Joana Figueiredo
Cancer Genetics Group, Institute of Molecular Pathology and Immunology of the University of Porto (IPATIMUP), Porto, Portugal

Alexander Gheldof
Department for Molecular Biomedical Research, Unit of Molecular and Cellular Oncology, VIB & Department of Biomedical Molecular Biology, Ghent University, Ghent, Belgium

André M. Goffinet
Developmental Neurobiology, University of Louvain, Institute of Neuroscience, Box B1.73.16, Brussels, Belgium

Kathleen J. Green
Department of Pathology; Department of Dermatology, and The Robert H. Lurie Comprehensive Cancer Center, Northwestern University Feinberg School of Medicine, Chicago, Illinois, USA

[1]Present address: Department of Biology, University of North Carolina, Chapel Hill, North Carolina, USA

Ismail Sahin Gul
Department for Molecular Biomedical Research, VIB & Department of Biomedical Molecular Biology, Ghent University, Ghent, Belgium

Jeff Hardin
Department of Zoology, and Program in Cellular and Molecular Biology, University of Wisconsin, Madison, Wisconsin, USA

Teruyoshi Hirayama
KOKORO Biology Group and JST-CREST, Laboratories for Integrated Biology, Graduate School of Frontier Biosciences, Osaka University, Yamadaoka, Suita, Osaka, Japan

Ji Yeon Hong
The Catholic University of Korea, Catholic High Performance Cell Therapy Center, 505 Banpo-dong, Seocho-gu, Seoul, South Korea

Paco Hulpiau
Department for Molecular Biomedical Research, VIB & Department of Biomedical Molecular Biology, Ghent University, Ghent, Belgium

Irene Kahr
Department for Molecular Biomedical Research, VIB & Department of Biomedical Molecular Biology, Ghent University, Ghent, Belgium

Antonis Kourtidis
Department of Cancer Biology, Mayo Clinic, Jacksonville, Florida, USA

Andrew P. Kowalczyk
Department of Cell Biology; Department of Dermatology, and The Winship Cancer Institute, Emory University School of Medicine, Atlanta, Georgia, USA

Thomas Lecuit
Institut de Biologie du Développement de Marseille Luminy (IBDML), Unité Mixte de Recherche 6216, Case 907, Marseille, France

Timothy Loveless
Program in Cellular and Molecular Biology, University of Wisconsin, Madison, Wisconsin, USA

Allison Lynch
Department of Zoology, University of Wisconsin, Madison, Wisconsin, USA

Kenji Mandai
Division of Molecular and Cellular Biology, Kobe University Graduate School of Medicine, and CREST, Japan Science and Technology Agency, Kobe, Hyogo, Japan

Pierre D. McCrea
Department of Biochemistry and Molecular Biology, The University of Texas MD Anderson Cancer Center, and Program in Genes and Development, The University of Texas Graduate School of Biomedical Sciences, Houston, Texas, USA

Helen McNeill
Samuel Lunenfeld Research Institute, and Department of Molecular Genetics, University of Toronto, Toronto, Canada

Rachel K. Miller
Department of Biochemistry and Molecular Biology, The University of Texas MD Anderson Cancer Center, Houston, Texas, USA

William A. Muñoz
Department of Biochemistry and Molecular Biology, The University of Texas MD Anderson Cancer Center, and Program in Genes and Development, The University of Texas Graduate School of Biomedical Sciences, Houston, Texas, USA

W. James Nelson
Department of Biology; Department of Molecular and Cellular Physiology, and Cancer Biology Program, Stanford University, Stanford, California, USA

Siu P. Ngok
Department of Cancer Biology, Mayo Clinic, Jacksonville, Florida, USA

Carla Oliveira
Expression Regulation in Cancer Group, Institute of Molecular Pathology and Immunology of the University of Porto (IPATIMUP), and Faculty of Medicine, University of Porto, Porto, Portugal

Christine Petit
Institut Pasteur, Unité de Génétique et Physiologie de l'Audition; Inserm UMRS 1120; UPMC, and Collège de France, Paris, France

Jonathan Pettitt
Department of Molecular and Cell Biology, University of Aberdeen Institute of Medical Sciences, Aberdeen, United Kingdom

Hugo Pinheiro
Expression Regulation in Cancer Group, Institute of Molecular Pathology and Immunology of the University of Porto (IPATIMUP), Porto, Portugal

Rashmi Priya
Division of Molecular Cell Biology, Institute for Molecular Bioscience, The University of Queensland, St. Lucia, Brisbane, Australia

Glenn L. Radice
Department of Medicine, Center for Translational Medicine, Jefferson Medical College, Philadelphia, Pennsylvania, USA

Aparna Ratheesh
Division of Molecular Cell Biology, Institute for Molecular Bioscience, The University of Queensland, St. Lucia, Brisbane, Australia

Yoshiyuki Rikitake
Division of Molecular and Cellular Biology, Division of Cardiovascular Medicine, and Division of Signal Transduction, Kobe University Graduate School of Medicine, Kobe, Hyogo, Japan

Andrew T. Schiffmacher
Department of Animal and Avian Sciences, University of Maryland, 1405 Animal Sciences Center, College Park, Maryland, USA

Raquel Seruca
Faculty of Medicine, University of Porto, and Cancer Genetics Group, Institute of Molecular Pathology and Immunology of the University of Porto (IPATIMUP), Porto, Portugal

Praveer Sharma
Samuel Lunenfeld Research Institute, and Department of Molecular Genetics, University of Toronto, Toronto, Canada

Yohei Shimono
Division of Molecular and Cellular Biology, Kobe University Graduate School of Medicine, Kobe, Hyogo, Japan

Yoshimi Takai
Division of Molecular and Cellular Biology, Kobe University Graduate School of Medicine, and CREST, Japan Science and Technology Agency, Kobe, Hyogo, Japan

Lisa A. Taneyhill
Department of Animal and Avian Sciences, University of Maryland, 1405 Animal Sciences Center, College Park, Maryland, USA

Fadel Tissir
Developmental Neurobiology, University of Louvain, Institute of Neuroscience, Box B1.73.16, Brussels, Belgium

Frans van Roy
Department for Molecular Biomedical Research, VIB & Department of Biomedical Molecular Biology, Ghent University, Ghent, Belgium

Karl Vandepoele[1]
Department for Molecular Biomedical Research, VIB & Department of Biomedical Molecular Biology, Ghent University, Ghent, Belgium

Dietmar Vestweber
Max-Planck-Institute of Molecular Biomedicine, Muenster, Germany

William I. Weis
Department of Molecular and Cellular Physiology; Cancer Biology Program, and Department of Structural Biology, Stanford University, Stanford, California, USA

Takeshi Yagi
KOKORO Biology Group and JST-CREST, Laboratories for Integrated Biology, Graduate School of Frontier Biosciences, Osaka University, Yamadaoka, Suita, Osaka, Japan

Alpha S. Yap
Division of Molecular Cell Biology, Institute for Molecular Bioscience, The University of Queensland, St. Lucia, Brisbane, Australia

[1]Current address: Laboratory for Clinical Biology, Ghent University Hospital, Ghent, Belgium

PREFACE

It is with pleasure and great satisfaction that I have been editing this volume on the Molecular Biology of Cadherins. The cadherin research field is fairly young. In 1987, the Takeichi group was the first to clone a cadherin cDNA and to demonstrate the cell–cell adhesion properties of the encoded protein (Nagafuchi *et al.*, *Nature* 1987;**329**:341–3). Now, some 25 years later, the field has grown to full maturity. The cadherin superfamily consists of more than 100 members in mammals. The roles of these cadherins and cadherin-like molecules are many. They involve the development and homeostasis of numerous tissue types in all multicellular metazoans. The hallmark of cadherins is the presence of a single extracellular domain composed largely of calcium-binding "cadherin repeats," which are typically involved in specific intercellular recognition and junction formation. This basic structure has remained fairly constant over millions of years, whereas the cytoplasmic domains of several cadherin family members have evolved to different extents. This has allowed the different cadherins to interact with a wide range of cytoplasmic proteins. These interactions contribute to structural and regulatory associations with plasma membrane components, cytoskeletal structures, and various intracellular signaling pathways. While cadherins suffered for long from the connotation "molecular glue," it turned out that this "glue" is highly sophisticated, not only allowing cells in tissues to recognize each other and to form coherent functional structures but also to drive cellular behavior, selective differentiation, regenerative remodeling, and many more processes. So it is not surprising that defects in cadherin superfamily members are responsible for a wide variety of human pathologies. Cadherins and their many interaction partners could therefore serve as therapeutic targets.

This volume provides the reader with a comprehensive and timely overview of the numerous structures and functions of many representative members of the large cadherin superfamily and of the proteins interacting with these cadherins. Several of the cadherin types are presented for the first time next to each other in a single volume, and all the contributions are by expert researchers in the field. In part, this volume serves as a solid basis for translational science to appreciate and exploit the unique properties of cadherins and their interaction partners.

The first section of this volume (A) is on the molecular biology of classic cadherins and their associated proteins known as catenins. The founder of the cadherin superfamily and also its best studied representative is the epithelial E-cadherin. The assembly of cadherin-dependent cell–cell junctions is discussed in detail as well as their role in the establishment of epithelial cell polarity. Novel insights into the field have emphasized the role of cadherin interactions with contractile cytoskeletal actomyosin networks in mediating both stability and plasticity of the junctions. Further, cadherin-driven signaling influences the delicate balances of Rho GTPase family members. In particular, α-catenin and armadillo catenins are key players in these regulatory networks, as exemplified not only by intriguing results obtained from studying the archaic slime mold *Dictyostelium discoideum* but also by a diverse range of original approaches *in vitro* and *in vivo* in both invertebrate and vertebrate systems.

The next section (B) deals with the versatility of the cadherin and catenin superfamilies, which is approached by comprehensive phylogenetic analyses. The data summarized here emphasize the presence of prototypic members in organisms as basic as placozoans and sea anemones, and also the remarkable diversification of the superfamily in more recently evolved metazoans. This relates to functional specializations as exemplified by the roles of desmosomal cadherins and VE-cadherin. Desmosomal cadherins differ from classic cadherins in many respects and play indispensable tightening roles in tissues under physical tension, such as the epidermis and cardiac muscle, as is clear in several disease conditions. On the other hand, VE-cadherin (cadherin-5) is important for the formation and regulation of endothelial cell junctions, which affects vasculogenesis, angiogenesis, and vascular permeability under various conditions.

Protocadherins make up a large part of the cadherin superfamily and exist in two flavors, the more ancient ones, encoded by discrete genes, and the more modern ones, encoded by clustered genes. Although the biology of protocadherins is still largely enigmatic, recent studies indicate important roles in the establishment of the countless and complex neuronal interactions in vertebrates, in regulatory cross-talk with classic cadherins, and in an increasing number of human pathologies. Other atypical cadherin families include the Celsr proteins (flamingo in *Drosophila*), which show intriguing features in terms of planar cell polarity (PCP) phenotypes, for instance, during mouse brain development. Further, Fat and Dachsous cadherins stand out by their extended ectodomains and by their exclusive signaling roles in PCP and growth control pathways. They have been studied mainly in *Drosophila*, but data on vertebrates are emerging.

The following section (C) focuses on specific cadherin functions and cadherin type switches during the embryonic development of representative model organisms and compares that knowledge to the effect of cadherin abnormalities in human disease models and in patients. The nematode worm *Caenorhabditis elegans* is an elegant and powerful model for elucidating the developmental roles of both cadherins and their cytoplasmic interaction partners. More recently, genome-wide functional screens have further extended the usefulness of this animal model in order to identify and study novel interaction partners of cadherin–catenin complexes. Genetic engineering of mice has likewise turned out to be invaluable for elucidating the developmental and morphogenetic roles of cadherins, as demonstrated by extensive studies of N-cadherin in the heart and other organs. In view of their close homology, it is remarkable that N-cadherin and E-cadherin behave so differently in both normal and cancerous conditions. Complex and consecutive switches in cadherin type expression are frequent and relevant during neural crest cell generation, migration, and terminal differentiation. Both the underlying signaling and the resulting morphogenetic processes are rich topics for further research. Epithelial-to-mesenchymal transition (EMT) occurs both in normal development and during progression of tumor cells to higher malignancy. It turned out that an intricate network of transcriptional repressors is involved in the induction of EMT-causing gene signatures. Another way by which human cancers affect E-cadherin functionality is germline mutation, which has been shown to drive familial cancers with diffuse tumor cell distributions, such as hereditary diffuse gastric cancer and inherited lobular breast cancer. Cadherin defects in human pathologies are not restricted to cancerous conditions, as defects in more than 20 cadherin types have been linked to inherited human diseases, including skin and hair disorders, cardiomyopathies, deafness, blindness, and psychiatric disorders.

In the last section (D), full attention is given to a selection of cytoplasmic interaction partners of classic cadherins. These proteins have turned out to be also often involved in key signaling processes at both cytoplasmic and nuclear locations. For example, β-catenin is famous for its role in the canonical Wnt signaling pathway. Less well understood are the roles of the various members of the p120 armadillo protein subfamily, including p120-catenin itself and several homologs, as well as plakophilins. Besides their important influences on junctional stability and on activation of Rho family GTPases, evidence is growing for a crosstalk between these proteins and nuclear β-catenin activity. The final contribution is devoted to afadin/AF6, a cytoplasmic protein interacting with numerous proteins at or nearby the plasma membrane. This important protein plays pleiotropic functions, including regulation of the formation of cadherin-mediated cell junctions and of several other cellular processes and their key signaling cascades.

In summary, this volume covers the exciting field of the molecular biology of cadherins in a wide and comprehensive way, including bioinformatic analyses, advanced *in vitro* assays, genetics of small model organisms, generation of relevant mouse models, and discussion of many relevant and important human diseases. I am grateful to the many authors who made expert and valuable contributions to this reference work, which will be of much interest to many researchers in various disciplines. Finally, I am particularly grateful to Amin Bredan, Ph.D., who contributed significantly to this volume by meticulously editing and optimizing the originally submitted manuscripts.

Frans van Roy
Ghent University and VIB

Cadherins: Molecules and Junctions

CHAPTER ONE

Roles of Cadherins and Catenins in Cell–Cell Adhesion and Epithelial Cell Polarity

W. James Nelson[*,†,‡], **Daniel J. Dickinson**[‡,1], **William I. Weis**[†,‡,§]
[*]Department of Biology, Stanford University, Stanford, California, USA
[†]Department of Molecular and Cellular Physiology, Stanford University, Stanford, California, USA
[‡]Cancer Biology Program, Stanford University, Stanford, California, USA
[§]Department of Structural Biology, Stanford University, Stanford, California, USA
[1]Present address: Department of Biology, University of North Carolina, Chapel Hill, North Carolina, USA

Contents

Abstract

A simple epithelium is the building block of all metazoans and a multicellular stage of a nonmetazoan. It comprises a closed monolayer of quiescent cells that surround a luminal space. Cells are held together by cell–cell adhesion complexes and surrounded by extracellular matrix. These extracellular contacts are required for the formation of a polarized organization of plasma membrane proteins that regulate the directional absorption and secretion of ions, proteins, and other solutes. While advances have been made in understanding how proteins are sorted to different plasma membrane domains, less is known about how cell–cell adhesion is regulated and linked to the development of epithelial cell polarity and regulation of homeostasis.

Progress in Molecular Biology and Translational Science, Volume 116
ISSN 1877-1173
http://dx.doi.org/10.1016/B978-0-12-394311-8.00001-7

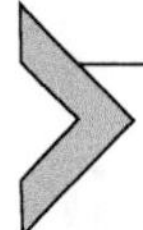

1. INTRODUCTION: THE BASIC DESIGN OF POLARIZED EPITHELIAL CELLS

A fundamental characteristic of simple epithelia is that they form a two-dimensional sheet of structurally and functionally polarized cells that is often organized into a three-dimensional tube (Fig. 1.1). The cells are surrounded by extracellular matrix (ECM) and held together by adhesive structures that include tight junctions (TJ, called septate junctions in *Drosophila*), the adherens junction (AJ), desmosomes, and several additional adhesion complexes (JAM, nectins).[1] A simple epithelial tube separates the organism into discrete compartments—generally, a specialized internal compartment separated and protected from the outside environment by the epithelium[2]—and regulates the exchange of nutrients, solutes, and waste between these two compartments thereby maintaining tissue architecture (reviewed in Ref. 3).

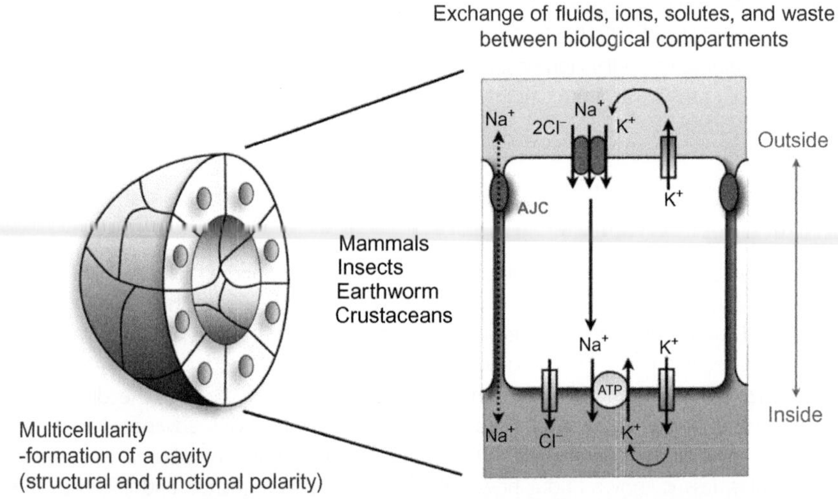

Figure 1.1 Functional Organization of a Generic Polarized Epithelium. Epithelial cells in most metazoans are organized into a closed monolayer that surrounds a luminal space. Different plasma membrane surfaces (apical and basolateral) face different biological compartments (depicted as "inside" and "outside") separated by the epithelium. Each plasma membrane domain contains different ion channels (e.g., K^+- and Cl^--channels), transporters (e.g., Na-K-2Cl-cotransporter), and pumps (e.g., Na^+/K^+-ATPase) that regulate ion and solute exchange between the two biological compartments. The cells are held together by an Apical Junctional Complex (AJC) that includes the tight junction, which regulates the paracellular flow of ions (e.g., Na^+) between the cells (dotted line). (See Color Insert.)

Each cell within a simple epithelium is polarized, with structurally and functionally distinct plasma membrane domains that face the two compartments separated by the epithelium (Fig. 1.1). The apical membrane domain faces the luminal surface—topologically the outside of the organism—and the basal and lateral (basolateral) membranes face the inner surface comprising the serosa and interstitium.[4] Each membrane domain has different membrane proteins that enable vectorial transport of ions and solutes across the epithelium (Fig. 1.1).[5,6]

The distribution of membrane proteins between the apical and basolateral membrane domains relies upon correct sorting of proteins to each plasma membrane domain (Fig. 1.2). A major site for sorting newly synthesized

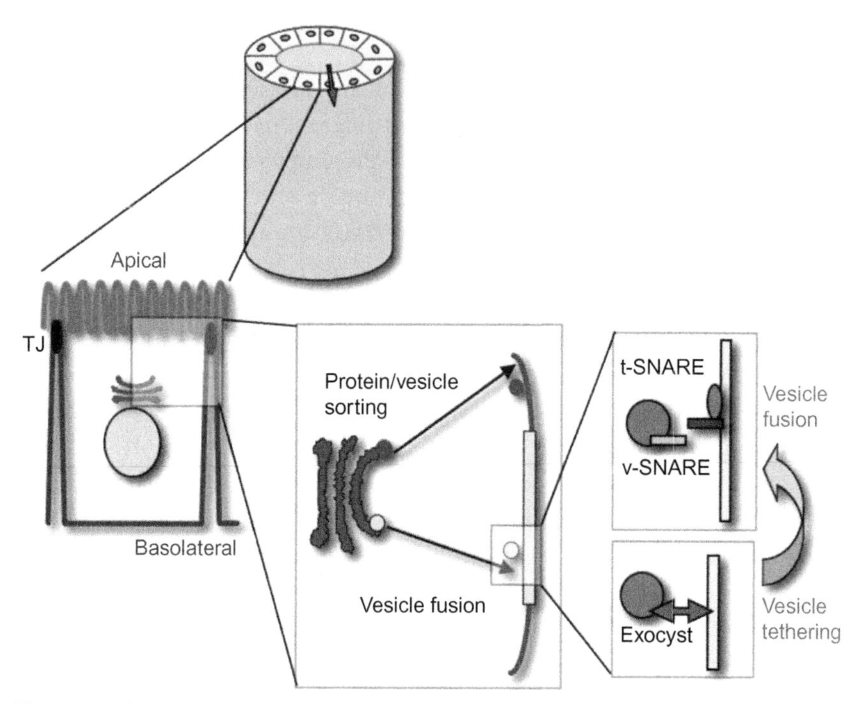

Figure 1.2 Protein sorting in a generic polarized epithelium. Each cell within an epithelial tube has apical and basolateral plasma membrane domains, which face different biological compartments and have different membrane proteins (see Fig. 1.1). Newly synthesized apical and basolateral proteins are sorted from each other on the exocytic pathway, primarily in the late Golgi complex (*trans*-Golgi network), and targeted in vesicles to the correct plasma membrane domain. At the plasma membrane, vesicle fusion is regulated by the Exocyst (a vesicle tethering complex) and the SNARE complex (vesicle fusion complex). Proteins may also be sorted in the endocytic pathway (not shown). (See Color Insert.)

apical and basolateral membrane proteins is the Trans-Golgi Complex (TGN) (reviewed in Refs. 5–7), but additional sorting events between the TGN and plasma membrane domains[8] and different membrane domains (transcytosis)[9] occur in the endocytic pathway. The docking and fusion of vesicles carrying sorted cargo proteins with the correct membrane domain require specific vesicle tethering complexes (e.g., the exocyst), Rab GTPases, and SNARE complexes localized to the apical and basolateral membrane (Fig. 1.2) (reviewed in Refs. 5,6). The distributions of proteins at the apical or basolateral membrane domain are maintained by the TJ, which provides a molecular fence at the boundary between the apical and basal–lateral membrane domains (reviewed in Ref. 10). A cytoplasmic scaffold complex of ankyrin–spectrin binds classes of membrane proteins that include ion transporters and channels and retains them on the basal–lateral membrane (reviewed in Ref. 11).

The actin, microtubule, and intermediate filament cytoskeletons in polarized epithelial cells have organizations different from those in other cell types. Actin filaments are organized as bundles within microvilli present on the apical membrane, as a ring of filament bundles circumscribing the apex of the lateral membrane often in association with the apical junctional complexes (AJC, comprising TJ and the AJ), and as dense networks lining the lateral and basal membranes (cortical cytoskeleton). In many epithelia, microtubules are prominently organized in bundles parallel to the lateral membrane with their minus-ends oriented toward the apical membrane and plus-ends toward the basal membrane, and as networks of filaments of mixed polarity underneath the apical and basal membranes. Intermediate filaments (usually members of the cytokeratin family) are organized as a structural continuum interlinking desmosomes across the cell.[12]

The structural integrity and functional polarity of epithelial tubes require both cell–cell adhesion and cell–ECM adhesion, which provide important physical and signaling cues for the initiation of cell polarization. Of the cell–cell adhesion complexes, evidence indicates that the cadherin superfamily plays an important role. For example, genetic deletion or mutations in E-cadherin disrupt the formation of the first organized tissue structures in the early mammalian embryo—the trophectoderm of the preimplantation mouse embryo[13]—and the epidermis of the *Drosophila* embryo following cellularization.[14] siRNA-mediated knockdown of E-cadherin in MDCK cells in tissue culture, which forms polarized monolayers of cells in two- and three-dimensional configurations, also inhibits the establishment of cell polarity.[15] Other proteins are involved in cell–cell adhesion, including members of the Ig superfamily (JAM1-3 and nectin), and it has been suggested that nectins initiate adhesion between cells upstream of cadherin-mediated cell–cell adhesion, and that both nectins and cadherins combine to sort cells into different arrangements[16,17] (see also Chapter 19).

Cadherins form extracellular contacts with cadherins on opposing cells through their N-terminal EC1 domain.[18] The cytoplasmic domain interacts with several proteins that either control cadherin turnover (p120 catenin, hakai) or link the complex, directly or indirectly, to the actin cytoskeleton (β-catenin and α-catenin).[19] Studies in epithelial tissues have shown that mutations of β-catenin or α-catenin, like that of E-cadherin, result in disruption of tissue organization[20]: for example, mutations of β-catenin or α-catenin in the *Drosophila* follicle cell epithelium result in defects in overall polarized cell morphology and the organization of the apical and basolateral plasma membrane domains.[21,22] Although downstream interaction of the cadherin–catenin complex with the actin cytoskeleton is complicated and not completely understood,[19,23] cadherin engagement locally regulates the activity and localization of Rho family GTPases,[24] which regulate actin organization and function.[25]

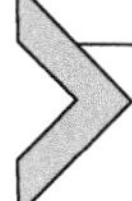

2. CADHERIN-MEDIATED CELL–CELL ADHESION AND THE FUNCTION OF POLARITY PROTEINS IN APICAL–BASAL CELL POLARITY

How cells respond to cadherin-mediated cell–cell contacts to initiate the formation of functionally different plasma membrane domains is not well understood. The initial establishment of cell surface polarity may be controlled at the plasma membrane by the position of the adhesive complex formed at cell–cell contacts and by the rapid organization of microtubules and the exocytic machinery that specifies the delivery of the correct set of transport vesicles to that site (Fig. 1.3). Basal–lateral protein accumulation at nascent cell–cell contacts in MDCK tissue culture cells requires microtubules,[26,27] the exocyst vesicle tethering complex, and the vesicle fusion t-SNARE syntaxin 4 (Figs. 1.2 and 1.3).[27] Microtubules may attach indirectly to the cadherin–catenin complex through interactions with the dynactin complex[28,29] and through interactions between p120catenin and kinesin 1.[30]

Mechanisms involved in localizing t-SNAREs to initial cell–cell contacts remain poorly understood. However, the exocyst is in a complex with E-cadherin and nectin, suggesting a direct recruitment of the exocyst to membrane sites of E-cadherin adhesion.[31] Indeed, the exocyst is recruited

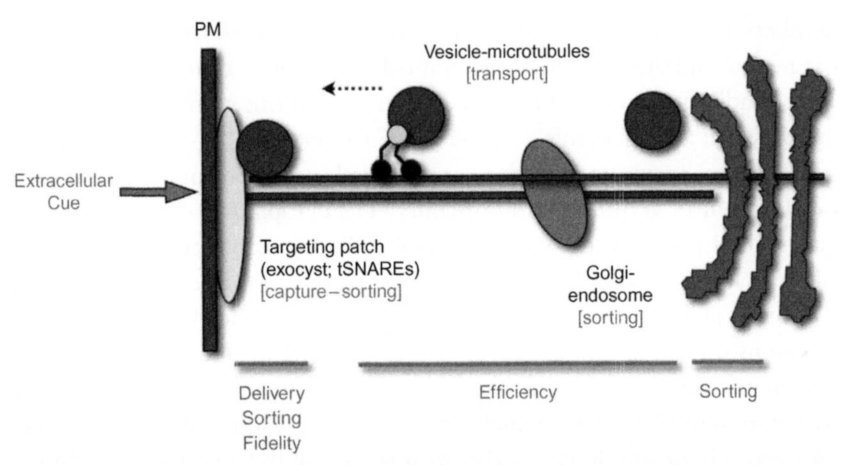

Figure 1.3 Mechanisms Involved in Protein Trafficking to Cell–Cell Contacts. An extracellular cue (cadherin-mediated cell–cell adhesion) initiates the assembly of a Targeting Patch (the exocyst, t-SNAREs) at sites of cell–cell contact, and the orientation of the microtubule cytoskeleton. In the Golgi complex and endosomes, basolateral membrane proteins are sorted into vesicles that are efficiently delivered along microtubules to the plasma membrane. At the plasma membrane, the targeting patch specifies the correct delivery and fusion of vesicles containing basolateral membrane proteins. (See Color Insert.)

to the plasma membrane and sites of cell-cell contact upon cadherin-mediated cell–cell adhesion,[31,32] and knockdown of Ral GTPases, which regulate exocyst function, affect TJ assembly and function.[33] The exocyst also regulates the distribution of polarity protein complexes (see Section 3 for details). Loss-of-function mutations in the Exo84 exocyst subunit in *Drosophila* result in loss of Crumbs and the PAR complex from the AJC and their localization along the lateral membrane, and an overall decrease in the columnar morphology of cells.[34]

A distinctive feature of polarized epithelia is the spatial distribution of evolutionarily conserved polarity protein complexes around the AJC (Fig. 1.4).[35,36] These complexes can have subtly different localizations in different contexts, but in general the Crumbs complex is located at the apical side of the AJC[37] and at the primary cilium[38]; the PAR complex is localized close to the AJC[39,40] and at the primary cilium (Fig. 1.4)[41]; and the Scribble complex is localized on the lateral membrane below the AJC.[37,40] Together, these polarity proteins regulate the organization of the apical (Crumbs [Crb], Stardust [Std], PAR complex) and basolateral membrane domain (Lethal giant larvae [Lgl], Discs large [Dlg], Scribble), respectively, in many polarized epithelial cells (Fig. 1.4).[35,37] For example, loss of expression of Crumbs

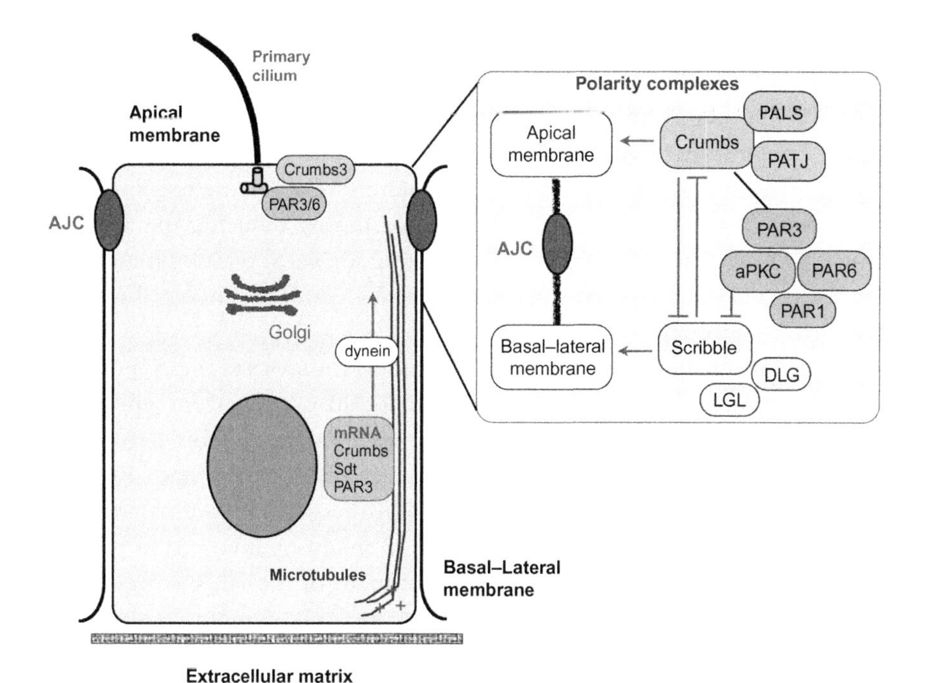

Figure 1.4 Distribution of "polarity protein" complexes in a generic polarized epithelium. Polarized epithelial cells are attached to each other by an apical junctional complex (AJC) and adhere to the underlying extracellular matrix. Each cell has a single primary cilium on the apical plasma membrane. Different "polarity protein" complexes (Crumbs, PAR, Scribble) are localized around the AJC, and regulate the organization of the apical (Crumbs) and basolateral (Scribble) plasma membrane domains. The Crumbs and Scribble complexes antagonize each other's activities. The localization of some components of the "polarity protein" complexes is regulated by cell–cell adhesion (see text) and by microtubule-dependent trafficking of mRNAs to the apical cytoplasm. (See Color Insert.)

results in disruption of the apical membrane domain and cell polarity, whereas overexpression of Crumbs leads to the expansion of the apical membrane at the expense of the (baso)lateral membrane domain; expansion of the apical membrane domain in wild-type conditions is inhibited by the (baso)lateral complex of Dlg, Scribble, Lgl, and PAR-1. Thus, the functions of apical and (baso)lateral polarity protein complexes are mutually antagonistic (Fig. 1.4).[37,40]

Studies of the ovariole follicle cell epithelium in *Drosophila* reveal that the DE-cadherin–catenin complex is required for the maintenance of follicle cell polarity.[42] Loss-of-function mutations of *DE-cadherin*,[43] *arm/β-catenin*,[37] or *α-catenin*[22] disrupt follicle cell morphology, the (baso)lateral organization of the α-spectrin/actin cytoskeleton and Dlg, and the apical organization of Crumbs and the cortical cytoskeleton of β_H-spectrin. Some of these defects appear to be due to inhibition of vesicle docking with the plasma membrane,[22] supporting a role for the cadherin–catenin complex and downstream effectors in vesicle trafficking.[27] Interestingly, the defects induced by *α-catenin* mutants can be partially ameliorated by reducing the activity of the WAVE-Arp2/3 complex,[22] supporting a role of α-catenin and the Arp2/3 complex in regulating actin assembly[44,45] and, possibly, cell surface levels of E-cadherin.[46,47]

Several mechanisms regulate the different distributions of these "polarity protein" complexes. The PAR complex localizes to nascent epithelial cell–cell contacts[48] through binding to the adhesion proteins JAM-A,[49] nectin[50], and β-catenin.[51] Analysis of cadherin–catenin and PAR-3 (bazooka) localization during initial adhesions between epidermal cells in the *Drosophila* embryo[52] indicate that PAR-3 might be involved in proper positioning of the nascent AJs. Early recruitment of the PAR complex to cell–cell contacts establishes the location of the PAR complex at the AJC between the forming apical (Crumbs complex) and basal–lateral (Scribble complex) membrane domains. PAR4/LKB1 also closely colocalizes with E-cadherin.[53] PAR4/LKB1 may locally regulate activation of PAR1 and MARKs, and thereby the organization of microtubules and the delivery of vesicles to the plasma membrane.[54–56]

Positioning the Crumbs complex close to the PAR complex in the region of the Apical junctional complex (Fig. 1.4) may be mediated by binding between different PDZ domain-containing proteins in the PAR complex (PAR3 and PAR6) and Crumbs complex (PALS1 and PATJ).[57,58] In addition, studies in *Drosophila* indicate that Crumbs mRNA is restricted to the apical domain of epithelial cells by a dynein-dependent basal to apical transport along microtubules,[59] which could result in its localized translation in the apical cytoplasm; a similar mechanism may be important in apical localization of PAR3/bazooka[60] and Stardust (Sdt, the *Drosophila* homolog of mammalian PALS1) (Fig. 1.4).[61]

Despite strong genetic evidence of the importance of these polarity protein complexes in apical–basal polarity of epithelial cells, their functions remain poorly understood. The Crumbs complex (Crumbs, PALS1, and PATJ) is required for the formation and maintenance of the TJ.[10,62] A recent study in mammalian tissue culture cells showed that PALS1 and PATJ bind a Cdc42-GAP called Rich1 (RhoGap-interacting with CIP4 homologues protein-1) and the scaffold protein AMOT (angiomotin), which together control Cdc42-dependent endocytosis of PALS1, PAR3 and overall TJ permeability.[63] Significantly, a separate study in *Drosophila* reported that deletion of either aPKC or PAR6 resulted in discontinuities in the AJC, and the appearance of E-cadherin further down the lateral membrane.[47] Similar phenotypes were observed by deleting Wasp, Arp2/3, or dynamin.[47] One possible mechanism linking these phenotypes could be a role for actin-mediated endocytosis in controlling the proper level and localization of E-cadherin and signaling activities of polarity protein complexes at the AJC.[63] This is further supported by the observation that defective endocytosis of Crumbs leads to expansion of the apical domain, a phenotype associated with "excess" Crumbs activity.[64]

Scribble is localized basal to the AJC along the lateral membrane domain. Its function is poorly understood. One possibility is that Scribble locally regulates Ca^{2+}-dependent exocytosis in a complex with βPIX and GIT1.[65] Scribble also binds PAR-1 and might regulate the localization of the PAR-1 kinase activity, and thereby exclude the PAR-3 complex.[66] Lgl is in a complex with the basal–lateral t-SNARE syntaxin 4 indicating that it too might be involved in regulating vesicle trafficking.[67] This is supported by the observation that depletion of Lgl results in inhibition of some protein delivery to the plasma membrane in neurons, including N-cadherin.[68] However, Lgl also binds myosin II[69] and the PAR-6/aPKC complex[70] and may be involved in the localization of these proteins to the lateral membrane.

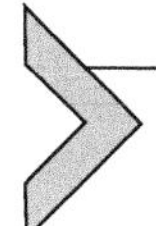

3. OTHER FUNCTIONS OF E-CADHERIN, CATENINS, AND "POLARITY PROTEIN" COMPLEXES IN MAINTAINING EPITHELIAL HOMEOSTASIS

Core components of the cadherin–catenin cell–cell adhesion complex are tumor suppressors (E-cadherin,[71,72] α-catenin[73]) and an oncogene (β-catenin[74]) and loss of function of the cadherin–catenin complex leads to increased cell proliferation, cell migration, and a general disruption of epithelial homeostasis.[20,75–77] Normally, cell growth in epithelial sheets is suppressed by contact inhibition[78], which involves cadherin-mediated cell–cell adhesion.[79,80] In addition, β-catenin and α-catenin regulate cell cycle progression: β-catenin is a Wnt pathway coactivator of genes that induce cell cycle progression (cyclin D1, c-myc),[77,81] and α-catenin regulates the Hippo pathway by sequestering in the cytoplasm the transcriptional coactivator Yki/Yap-Taz, which also drives the expression of genes that promote cell proliferation.[77,80,82,83] Another cell–cell junction associated protein, ZONAB, a Y-box transcription factor, binds to the TJ scaffold protein ZO-1[84] and interacts with the cell cycle regulator CDK4,[85] which in turn controls expression of cyclin D1 and PCNA.

Polarity protein complexes not only regulate the polarized organization of cells but also control cell proliferation, which likely contributes to the loss of both cell polarity and cell proliferation control when these proteins are mutated or deleted. For example, Scribble[86] and Dlg[87] contain a domain that is required for cell polarity (LLR domain), and another (PDZ domain) that controls cell proliferation. Significantly, the cell proliferation defect induced by deletion of the PDZ domain of Scribble can be rescued by overexpression of the LRR domain.[86] Thus, Scribble regulation of cell proliferation and cell polarity may be linked, although more work is needed to fully define the role of Scribble in these pathways.

The effects of depletion of Scribble on cell polarity and cell proliferation have also been tested in three-dimensional acini formed by MCF10A mammary epithelial cells. Depletion of Scribble has a modest effect on apical–basal polarity, but the luminal space of the acini fills with cells.[88] Luminal filling is due to a decrease in apoptosis, rather than an increase in cell proliferation, and apoptosis is indeed required to clear the luminal space of cells in this model system.[89] As noted above, Scribble forms a complex with βPIX/GIT1, which can activate Rac1; in turn, activated Rac1 induces apoptosis by activating JNK, and in the absence of Scribble (or βPIX) this does not occur, resulting in overgrowth of cells.[88]

A genetic screen in *Drosophila* for mutations that enhanced tumorigenesis caused by loss of Dlg expression identified the serine/threonine kinase Warts.[90] In *Drosophila*, Warts is a downstream component of the Hippo signaling pathway that regulates cell proliferation (see above). Interestingly, the mammalian homologs of Hippo (Mst1 and Mst2) and Warts (Lats1 and Lats2) are also tumor suppressors and loss of their expression is linked to highly aggressive breast tumors.[91] As noted above, downstream targets of Warts regulate cell cycle progression and apoptosis.[90,91] These results indicate that Dlg may also be involved in controlling cell cycle progression and apoptosis, but further studies are needed to define how it is linked to the Hippo–Warts signaling pathway.

In conclusion, the cadherin–catenin complex and polarity protein complexes not only regulate cell–cell adhesion and cell polarity, but also the control of cell proliferation in maintaining epithelial homeostasis. Thus, disruption of cell–cell interactions, cellular organization and cell proliferation that is often found in cancers and other diseases may be directly related to loss of function of protein components of these complexes.

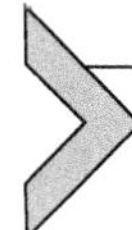

4. CADHERIN, CATENINS, AND THE ORIGINS OF EPITHELIAL POLARITY

Given the requirement of the cadherin–catenin complex in the organization of polarized epithelial cells from mammals to *Drosophila* raises the question of whether the appearance of cadherins and catenins in evolution coincide with the development of multicellularity, and with the organization of cells into functionally polarized epithelia[92]. Metazoans belong to the Unikonta, a group that includes fungi and social amoebae that can also have multicellular stages. However, it is generally thought that animals, fungi, and social amoebae evolved multicellularity independently, and that formation of an epithelium is a unique feature of animals.[93–95] Bilateria evolved additional cell types, such as mesenchymal cells that migrate to different sites in the body cavity where they contribute to the formation of secondary epithelia. This enabled a further diversification in the organization and distribution of functionally different epithelial tissues and organs.[96]

The evolution of cadherins is discussed in detail in Chapter 4, but it is notable that cadherins have only been found in metazoans and a close single-celled relative, the Choanoflagellates,[92] although Choanoflagellate cadherin lacks the catenin-binding domain found in higher metazoan cadherins and has not yet been shown to regulate cell–cell adhesion. However, a similar analysis of catenins, particularly α-catenin, reveals that homologues are found in the genomes of some nonmetazoans.[97] α-Catenin is structurally similar to vinculin, and all metazoans examined have at least one α-catenin orthologue and one clear vinculin orthologue.[97] These metazoans include sponge (*Amphimedon queenslandica*), cniderians (*Hydra magnipapillate*, *Nematostella vectensis*), and bilaterians (*Drosophila melanogaster*, *Homo sapiens*). The sequence identity of the α-catenin orthologues compared to the human αE-catenin ranges from ~60% (*D. melanogaster*) to <20% (*H. magnipapillate*). However, none of the nonmetazoans examined, including unicellular Opisthokonts (*Caspsaspora owczarzaki*, *Monosiga brevicolis*), Chytrid fungi (e.g., *Batrachytrium dendrobatidis*) and cellular slime molds (e.g., *Dictyostelium discoideum*), have more than one α-catenin family member. The sequence identity of these α-catenin orthologues compared to the human protein is <20%.[97]

D. discoideum is representative of the most basally branched clade that has an α-catenin/vinculin family member.[95,97] Biochemical and functional characterization of *Dd*α-catenin/vinculin show that it is a monomer that binds and bundles F-actin constitutively, properties that are different from those of mouse α-catenin or vinculin: mouse α-catenin is either a monomer that is conformationally inhibited or an activated homodimer that binds F-actin strongly), and vinculin is a monomer that is conformationally inhibited, and must also be activated, for example by talin, to bind F-actin. Interestingly, *Dd*α-catenin/vinculin also differs from the *Caenorhabditis elegans* α-catenin, called HMP-1 (see also Chapter 11). HMP-1 is a monomer that is conformationally inactive and does not bind F-actin *in vitro*, although studies *in vivo* indicate that HMP-1 becomes activated and that both the β-catenin- and actin-binding sites are required for normal development.[98]

Unlike vinculin, the head and tail domains of *Dd*α-catenin/vinculin do not interact and the protein does not colocalize with talin at cell–substrate adhesion sites in single cells. In addition, depletion of *Dd*α-catenin does not affect cell–substrate adhesion. *Dd*α-catenin/vinculin binds *Dd*β-catenin (also called Aardvark[99]) and, remarkably, is able to bind mouse β-catenin, indicating evolutionary conservation of binding sites and structure. Based on these properties the *Dd*α-catenin/vinculin likely encodes an α-catenin orthologue, indicating that vinculin arose later during evolution.[97]

D. discoideum expresses a β-catenin-related protein called Aardvark.[99] Both Aardvark and mammalian β-catenin contain Armadillo (Arm) repeats. Aardvark has 9 Arm repeats, compared to 12 in metazoan β-catenins. The Arm repeats of β-catenin fold to form a positively-charged groove and a similar feature is predicted in Aardvark. E-cadherin and TCF/LEF transcription factors bind along this groove in mammalian β-catenin, but none of these proteins are expressed in *Dictyostelium* and it is unknown whether there are additional binding partners. Aardvark and mammalian β-catenin share a highly conserved α-catenin-binding helix at the N-terminus of the Arm repeat domain. In contrast to these similarities, Aardvark lacks sequences in the C-terminal tail of β-catenin that are important for the transcriptional activity of β-catenin in Wnt signaling, and in the N-terminal region of β-catenin that contains GSK-3β phosphorylation sites important for targeting β-catenin for degradation. Thus, Aardvark does not have sequence features necessary for Wnt signal transduction pathway, and several other key Wnt signaling proteins are absent in *D. discoideum*.[97]

Although *D. discoideum* does not express a cadherin homologue, the α-catenin and β-catenin orthologues play critical roles in the muticellular stage of the *D. discoideum* life cycle. In the presence of plentiful food (bacteria), *D. discoideum* is a single-cell amoeba. However, when food becomes scarce the amoebae aggregate, form a motile multicellular slug, undergo culmination, and develop into a fruiting body.[100] The fruiting body comprises a rigid stalk that supports a collection of spores. Cells at the tip of the culminant form a ring surrounding the stalk tube, which contains cellulose and the ECM proteins EcmA/B that provide mechanical integrity to the stalk.[101]

Confocal microscopy showed that the tip cells comprise an organized monolayer of columnar cells surrounding the stalk (Fig. 1.5A). This organization is reminiscent of a simple epithelium surrounding a luminal space found in metazoans. Indeed, similar to metazoan epithelia, the centrosomes and Golgi localize to the stalk side of the nuclei, and the transmembrane protein cellulose synthase is localized to a plasma membrane domain adjacent to the stalk. Protein trafficking pathways that specify polarized protein distributions are unknown in *D. discoideum*. However, Sec15, a component of the Exocyst complex involved in polarized exocytosis in diverse systems, localizes at the plasma membrane adjacent to the stalk tube; this is reminiscent of exocyst localization in polarized mammalian epithelial cells. Actin is localized at all plasma membrane domains but is enriched apically with myosin II, and the microtubules have a polarized organization. These characteristics indicate that the cells form a *bona fide* epithelium (Fig. 1.5A and B).[97]

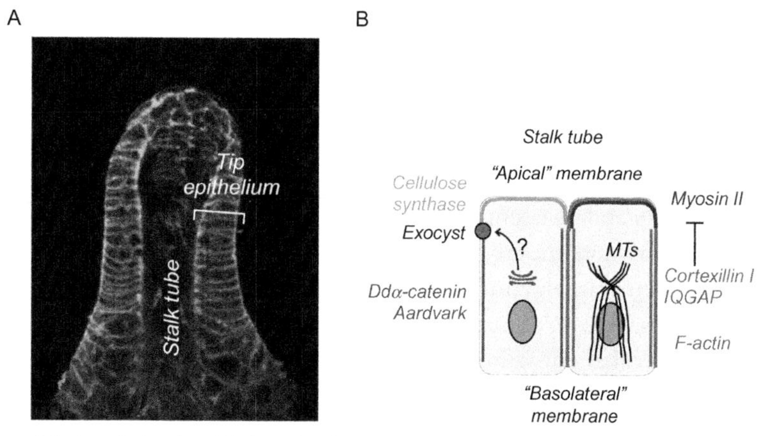

Figure 1.5 A polarized epithelium in the nonmetazoan *Dictyostelium discoideum*. (A) A confocal immunofluorescence section through the tip of the fruiting body stained for: F-actin (green), *Dd*α-catenin, (red), nuclei (blue), and microtubules (magenta). The cells are organized into a single layer (the Tip Epithelium) surrounding the stalk tube. (B) Schematic representation of the distribution of cytoskeleton (*Dd*α-catenin, *Dd*β-catenin/Aardvark, F-actin, myosin II, IQGPA, cortexillin I, micotubules/MTs), a plasma membrane protein (cellulose synthase), and components of the secretory pathway (Golgi complex, exocyst). For details about the staining and organization of proteins in the tip epithelium, see Refs. 97,102. (See Color Insert.)

Expression of Aardvark/β-catenin and α-catenin is up-regulated shortly after aggregation of the amoebae, and both proteins are present during all stages of culmination. Significantly, α-catenin colocalizes at cell–cell contacts between tip epithelial cells and is generally excluded from the membranes facing the stalk tube and the base (Fig. 1.5A and B); this distribution is similar to that in animal polarized epithelial cells in which cell–cell adhesion involves cadherins. As Aardvark/β-catenin and α-catenin form a complex, it is assumed that Aardvark/β-catenin colocalizes on the lateral membrane as well. Knockout of Aardvark/β-catenin results in loss of lateral membrane localization of α-catenin. Although the membrane protein binding site(s) for Aardvark/β-catenin and α-catenin are unknown, these results indicate that the binding hierarchy of membrane protein X—Aardvark/β-catenin—α-catenin in *Dictyostelium* may be similar to that of the cadherin–catenin complex in metazoan epithelial cells.

Knockout of Aardvark/β-catenin or knockdown of α-catenin appears to have no effect on the migration or aggregation of amoebae, or on slug formation and migration. However, multicellular development is arrested at those early stages. Cells that would normally contribute to the formation of the tip epithelium are disorganized. The stalk and tip epithelium are absent in severe *Dd*α-catenin knockdowns, but are present, although disorganized, in Aardvark knockouts and mild *Dd*α-catenin knockdowns.[102] Furthermore, the distributions of the Golgi and centrosomes are not polarized, and the transmembrane protein cellulose synthase is mislocalized intracellularly, which explains the absence of extracellular cellulose.[107] Therefore, the polarized organization of cells in the tip epithelium requires both Aardvark/β-catenin and α-catenin, similar to the requirement of catenins for the organization of epithelia in animals.

The molecular interactions required for organizing the tip epithelium are poorly understood. Proteomic analysis of β-catenin/Aardvark–α-catenin complexes revealed a downstream complex of IQGAP1 and cortexillin I that spatially regulates the distribution to myosin II (Fig. 1.5B).[102] IQGAP1 and cortexillin I form a complex.[103,104] Unlike mammalian IQGAP1, the *Dd*IQGAP1 lacks an actin-binding domain, suggesting that the actin-binding module of cortexillin I confers this activity in the complex.[102] IQGAP1 and cortexillin I colocalize with Aardvark/β-catenin and α-catenin on the lateral membrane between cells, but both proteins are absent from the apical membrane domain adjacent to the stalk tube (Fig. 1.5B). Myosin II colocalizes with actin at the membrane facing the stalk tube, and forms a continuous ring around the cell, very similar to the actomyosin ring found in metazoan epithelial tubes; myosin II is less concentrated on the basolateral plasma membrane surface that contains Aardvark/β-catenin, α-catenin, IQGAP1, and cortexillin I (Fig. 1.5B).

Significantly, deletion of Aardvark/β-catenin or α-catenin results in disorganized localizations of IQGAP1, cortexillin I and myosin II, as tip epithelium cells lose polarity and monolayer organization. Knockouts of either IQGAP1 or cortexillin I result in a multilayered epithelium in which cells are still polarized, and loss of myosin II and the actomyosin ring from the plasma membrane adjacent to the stalk tube. Disruption of the actomyosin ring results in a stalk tube that is ~2 × wider than in wild-type culminants.[102] Together these results indicate that Aardvark/β-catenin and α-catenin interact with and regulate the distribution of the IQGAP1/cortexillin I complex. In turn, the IQGAP1/cortexillin I complex excludes myosin II from the basolateral cortex and promotes accumulation of actomyosin on the membrane facing the stack tube.[102] The mechanisms involved in the exclusion of myosin II from the lateral membrane are unknown. Nevertheless, these results reveal that apical localization of myosin II is a conserved morphogenetic mechanism from nonmetazoans to vertebrates, and identify a hierarchy of proteins that regulate the polarity and organization of an epithelial tube in a simple model organism (Fig. 1.6).

The presence of a polarized epithelium in the nonmetazoan *D. discoideum* demonstrates that an epithelial tissue is not a unique feature of metazoans, and that catenins are an ancient protein module involved in the formation of membrane domains and epithelial polarity (Fig. 1.6). Thus, metazoans and social amoebae may have evolved from an ancestor that spent a portion of

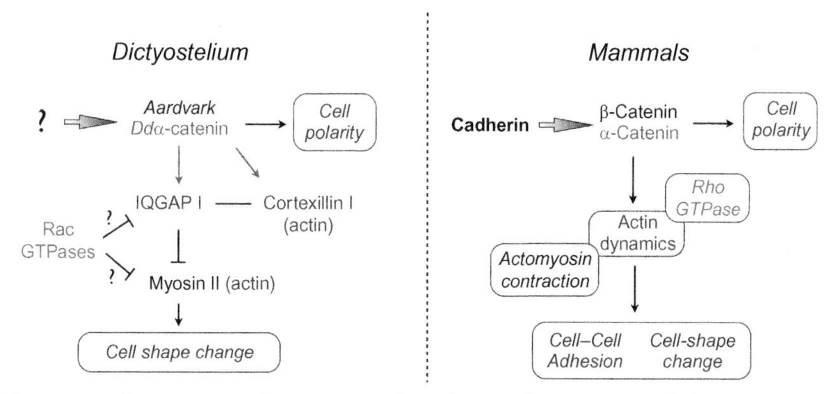

Figure 1.6 Comparison of proteins and pathways downstream of the (cadherin)–catenin complex, and their functional outcomes. For details, see text. (See Color Insert.)

its life cycle in a multicellular state and possessed the molecular machinery necessary for converting cell–cell interactions into the formation of a functionally and structurally polarized epithelial tissue.[95] Interestingly, none of the genes encoding polarity proteins are present in the *Dictyostelium* genome.[97] This raises the possibility that polarity proteins may have evolved separately, perhaps with the appearance of cadherins, to regulate the increased complexity of tissue organization in metazoans.[95]

ACKNOWLEDGMENTS

Supported by NIH grants GM35527 (W. J. N.) and GM56169 (W. I. W.).

REFERENCES

1. Niessen CM, Leckband D, Yap AS. Tissue organization by cadherin adhesion molecules: dynamic molecular and cellular mechanisms of morphogenetic regulation. *Physiol Rev* 2011;**91**:691–731.
2. Marchiando AM, Graham WV, Turner JR. Epithelial barriers in homeostasis and disease. *Annu Rev Pathol* 2010;**5**:119–44.
3. Cereijido M, Contreras RG, Shoshani L. Cell adhesion, polarity, and epithelia in the dawn of metazoans. *Physiol Rev* 2004;**84**:1229–62.
4. Bryant DM, Mostov KE. From cells to organs: building polarized tissue. *Nat Rev Mol Cell Biol* 2008;**9**:887–901.
5. Mellman I, Nelson WJ. Coordinated protein sorting, targeting and distribution in polarized cells. *Nat Rev Mol Cell Biol* 2008;**9**:833–45.
6. Tanos B, Rodriguez-Boulan E. The epithelial polarity program: machineries involved and their hijacking by cancer. *Oncogene* 2008;**27**:6939–57.
7. Folsch H. Regulation of membrane trafficking in polarized epithelial cells. *Curr Opin Cell Biol* 2008;**20**:208–13.
8. Gravotta D, Deora A, Perret E, Oyanadel C, Soza A, Schreiner R, et al. AP1B sorts basolateral proteins in recycling and biosynthetic routes of MDCK cells. *Proc Natl Acad Sci USA* 2007;**104**:1564–9.
9. Casanova JE, Breitfeld PP, Ross SA, Mostov KE. Phosphorylation of the polymeric immunoglobulin receptor required for its efficient transcytosis. *Science* 1990;**248**:742–5.
10. Shin K, Fogg VC, Margolis B. Tight junctions and cell polarity. *Annu Rev Cell Dev Biol* 2006;**22**:207–35.
11. Bennett V, Healy J. Organizing the fluid membrane bilayer: diseases linked to spectrin and ankyrin. *Trends Mol Med* 2008;**14**:28–36.
12. Nelson WJ. Adaptation of core mechanisms to generate cell polarity. *Nature* 2003;**422**:766–74.
13. Larue L, Ohsugi M, Hirchenhain J, Kemler R. E-cadherin null mutant embryos fail to form a trophectoderm epithelium. *Proc Natl Acad Sci USA* 1994;**91**:8263–7.
14. Wang F, Dumstrei K, Haag T, Hartenstein V. The role of DE-cadherin during cellularization, germ layer formation and early neurogenesis in the Drosophila embryo. *Dev Biol* 2004;**270**:350–63.
15. Capaldo CT, Macara IG. Depletion of E-cadherin disrupts establishment but not maintenance of cell junctions in Madin–Darby canine kidney epithelial cells. *Mol Biol Cell* 2007;**18**:189–200.

16. Takai Y, Ikeda W, Ogita H, Rikitake Y. The immunoglobulin-like cell adhesion molecule nectin and its associated protein afadin. *Annu Rev Cell Dev Biol* 2008;**24**:309–42.
17. Togashi H, Kominami K, Waseda M, Komura H, Miyoshi J, Takeichi M, et al. Nectins establish a checkerboard-like cellular pattern in the auditory epithelium. *Science* 2011;**333**: 1144–7.
18. Shapiro L, Weis WI. Structure and biochemistry of cadherins and catenins. *Cold Spring Harb Perspect Biol* 2009;**1**:a003053.
19. Weis WI, Nelson WJ. Re-solving the cadherin–catenin–actin conundrum. *J Biol Chem* 2006;**281**:35593–7.
20. Benjamin JM, Nelson WJ. Bench to bedside and back again: molecular mechanisms of alpha-catenin function and roles in tumorigenesis. *Semin Cancer Biol* 2008;**18**:53–64.
21. Niewiadomska P, Godt D, Tepass U. DE-Cadherin is required for intercellular motility during Drosophila oogenesis. *J Cell Biol* 1999;**144**:533–47.
22. Sarpal R, Pellikka M, Patel RR, Hui FY, Godt D, Tepass U. Mutational analysis supports a core role for Drosophila alpha-catenin in adherens junction function. *J Cell Sci* 2012;**125**:233–45.
23. Yonemura S, Wada Y, Watanabe T, Nagafuchi A, Shibata M. Alpha-catenin as a tension transducer that induces adherens junction development. *Nat Cell Biol* 2010;**12**: 533–42.
24. Yamada S, Nelson WJ. Synapses: sites of cell recognition, adhesion, and functional specification. *Annu Rev Biochem* 2007;**76**:267–94.
25. Fukata M, Kaibuchi K. Rho-family GTPases in cadherin-mediated cell–cell adhesion. *Nat Rev Mol Cell Biol* 2001;**2**:887–97.
26. Shaw RM, Fay AJ, Puthenveedu MA, von Zastrow M, Jan YN, Jan LY. Microtubule plus-end-tracking proteins target gap junctions directly from the cell interior to adherens junctions. *Cell* 2007;**128**:547–60.
27. Nejsum LN, Nelson WJ. A molecular mechanism directly linking E-cadherin adhesion to initiation of epithelial cell surface polarity. *J Cell Biol* 2007;**178**:323–35.
28. Ligon LA, Holzbaur EL. Microtubules tethered at epithelial cell junctions by dynein facilitate efficient junction assembly. *Traffic* 2007;**8**:808–19.
29. Lien WH, Gelfand VI, Vasioukhin V. Alpha-E-catenin binds to dynamitin and regulates dynactin-mediated intracellular traffic. *J Cell Biol* 2008;**183**:989–97.
30. Chen X, Kojima S, Borisy GG, Green KJ. p120 catenin associates with kinesin and facilitates the transport of cadherin–catenin complexes to intercellular junctions. *J Cell Biol* 2003;**163**:547–57.
31. Yeaman C, Grindstaff KK, Nelson WJ. Mechanism of recruiting Sec6/8 (exocyst) complex to the apical junctional complex during polarization of epithelial cells. *J Cell Sci* 2004;**117**:559–70.
32. Grindstaff KK, Yeaman C, Anandasabapathy N, Hsu SC, Rodriguez-Boulan E, Scheller RH, et al. Sec6/8 complex is recruited to cell-cell contacts and specifies transport vesicle delivery to the basal-lateral membrane in epithelial cells. *Cell* 1998;**93**:731–40.
33. Hazelett CC, Sheff D, Yeaman C. RalA and RalB differentially regulate development of epithelial tight junctions. *Mol Biol Cell* 2011;**22**:4787–800.
34. Blankenship JT, Fuller MT, Zallen JA. The Drosophila homolog of the Exo84 exocyst subunit promotes apical epithelial identity. *J Cell Sci* 2007;**120**:3099–110.
35. St Johnston D, Ahringer J. Cell polarity in eggs and epithelia: parallels and diversity. *Cell* 2010;**141**:757–74.
36. Laprise P, Tepass U. Novel insights into epithelial polarity proteins in Drosophila. *Trends Cell Biol* 2011;**21**:401–8.
37. Tanentzapf G, Tepass U. Interactions between the crumbs, lethal giant larvae and bazooka pathways in epithelial polarization. *Nat Cell Biol* 2003;**5**:46–52.

38. Fan S, Hurd TW, Liu CJ, Straight SW, Weimbs T, Hurd EA, et al. Polarity proteins control ciliogenesis via kinesin motor interactions. *Curr Biol* 2004;**14**:1451–61.
39. Izumi Y, Hirose T, Tamai Y, Hirai S, Nagashima Y, Fujimoto T, et al. An atypical PKC directly associates and colocalizes at the epithelial tight junction with ASIP, a mammalian homologue of Caenorhabditis elegans polarity protein PAR-3. *J Cell Biol* 1998;**143**:95–106.
40. Bilder D, Schober M, Perrimon N. Integrated activity of PDZ protein complexes regulates epithelial polarity. *Nat Cell Biol* 2003;**5**:53–8.
41. Sfakianos J, Togawa A, Maday S, Hull M, Pypaert M, Cantley L, et al. Par3 functions in the biogenesis of the primary cilium in polarized epithelial cells. *J Cell Biol* 2007;**179**:1133–40.
42. Tepass U, Tanentzapf G, Ward R, Fehon R. Epithelial cell polarity and cell junctions in Drosophila. *Annu Rev Genet* 2001;**35**:747–84.
43. Godt D, Tepass U. Drosophila oocyte localization is mediated by differential cadherin-based adhesion. *Nature* 1998;**395**:387–91.
44. Drees F, Pokutta S, Yamada S, Nelson WJ, Weis WI. Alpha-catenin is a molecular switch that binds E-cadherin-beta-catenin and regulates actin-filament assembly. *Cell* 2005;**123**:903–15.
45. Benjamin JM, Kwiatkowski AV, Yang C, Korobova F, Pokutta S, Svitkina T, et al. AlphaE-catenin regulates actin dynamics independently of cadherin-mediated cell–cell adhesion. *J Cell Biol* 2010;**189**:339–52.
46. Leibfried A, Fricke R, Morgan MJ, Bogdan S, Bellaiche Y. Drosophila Cip4 and WASp define a branch of the Cdc42-Par6-aPKC pathway regulating E-cadherin endocytosis. *Curr Biol* 2008;**18**:1639–48.
47. Georgiou M, Marinari E, Burden J, Baum B. Cdc42, Par6, and aPKC regulate Arp2/3-mediated endocytosis to control local adherens junction stability. *Curr Biol* 2008;**18**:1631–8.
48. Suzuki A, Ishiyama C, Hashiba K, Shimizu M, Ebnet K, Ohno S. aPKC kinase activity is required for the asymmetric differentiation of the premature junctional complex during epithelial cell polarization. *J Cell Sci* 2002;**115**:3565–73.
49. Ebnet K, Aurrand-Lions M, Kuhn A, Kiefer F, Butz S, Zander K, et al. The junctional adhesion molecule (JAM) family members JAM-2 and JAM-3 associate with the cell polarity protein PAR-3: a possible role for JAMs in endothelial cell polarity. *J Cell Sci* 2003;**116**:3879–91.
50. Takekuni K, Ikeda W, Fujito T, Morimoto K, Takeuchi M, Monden M, et al. Direct binding of cell polarity protein PAR-3 to cell–cell adhesion molecule nectin at neuroepithelial cells of developing mouse. *J Biol Chem* 2003;**278**:5497–500.
51. Wei SY, Escudero LM, Yu F, Chang LH, Chen LY, Ho YH, et al. Echinoid is a component of adherens junctions that cooperates with DE-cadherin to mediate cell adhesion. *Dev Cell* 2005;**8**:493–504.
52. McGill MA, McKinley RF, Harris TJ. Independent cadherin–catenin and Bazooka clusters interact to assemble adherens junctions. *J Cell Biol* 2009;**185**:787–96.
53. Sebbagh M, Santoni MJ, Hall B, Borg JP, Schwartz MA. Regulation of LKB1/STRAD localization and function by E-cadherin. *Curr Biol* 2009;**19**:37–42.
54. Cohen D, Brennwald PJ, Rodriguez-Boulan E, Musch A. Mammalian PAR-1 determines epithelial lumen polarity by organizing the microtubule cytoskeleton. *J Cell Biol* 2004;**164**:717–27.
55. Cohen D, Rodriguez-Boulan E, Musch A. Par-1 promotes a hepatic mode of apical protein trafficking in MDCK cells. *Proc Natl Acad Sci USA* 2004;**101**:13792–7.
56. Elbert M, Rossi G, Brennwald P. The yeast par-1 homologs kin1 and kin2 show genetic and physical interactions with components of the exocytic machinery. *Mol Biol Cell* 2005;**16**:532–49.

57. Roh MH, Margolis B. Composition and function of PDZ protein complexes during cell polarization. *Am J Physiol Renal Physiol* 2003;**285**:F377–F387.
58. Lemmers C, Michel D, Lane-Guermonprez L, Delgrossi MH, Medina E, Arsanto JP, et al. CRB3 binds directly to Par6 and regulates the morphogenesis of the tight junctions in mammalian epithelial cells. *Mol Biol Cell* 2004;**15**:1324–33.
59. Li Z, Wang L, Hays TS, Cai Y. Dynein-mediated apical localization of crumbs transcripts is required for Crumbs activity in epithelial polarity. *J Cell Biol* 2008;**180**:31–8.
60. Harris TJ, Peifer M. The positioning and segregation of apical cues during epithelial polarity establishment in Drosophila. *J Cell Biol* 2005;**170**:813–23.
61. Horne-Badovinac S, Bilder D. Dynein regulates epithelial polarity and the apical localization of stardust A mRNA. *PLoS Genet* 2008;**4**:e8.
62. Fogg VC, Liu CJ, Margolis B. Multiple regions of Crumbs3 are required for tight junction formation in MCF10A cells. *J Cell Sci* 2005;**118**:2859–69.
63. Wells CD, Fawcett JP, Traweger A, Yamanaka Y, Goudreault M, Elder K, et al. A Rich1/Amot complex regulates the Cdc42 GTPase and apical-polarity proteins in epithelial cells. *Cell* 2006;**125**:535–48.
64. Lu H, Bilder D. Endocytic control of epithelial polarity and proliferation in Drosophila. *Nat Cell Biol* 2005;**7**:1232–9.
65. Audebert S, Navarro C, Nourry C, Chasserot-Golaz S, Lecine P, Bellaiche Y, et al. Mammalian Scribble forms a tight complex with the betaPIX exchange factor. *Curr Biol* 2004;**14**:987–95.
66. Tian AG, Deng WM. Lgl and its phosphorylation by aPKC regulate oocyte polarity formation in Drosophila. *Development* 2008;**135**:463–71.
67. Musch A, Cohen D, Yeaman C, Nelson WJ, Rodriguez-Boulan E, Brennwald PJ. Mammalian homolog of Drosophila tumor suppressor lethal (2) giant larvae interacts with basolateral exocytic machinery in Madin–Darby canine kidney cells. *Mol Biol Cell* 2002;**13**:158–68.
68. Klezovitch O, Fernandez TE, Tapscott SJ, Vasioukhin V. Loss of cell polarity causes severe brain dysplasia in Lgl1 knockout mice. *Genes Dev* 2004;**18**:559–71.
69. Betschinger J, Eisenhaber F, Knoblich JA. Phosphorylation-induced autoinhibition regulates the cytoskeletal protein Lethal (2) giant larvae. *Curr Biol* 2005;**15**:276–82.
70. Betschinger J, Mechtler K, Knoblich JA. The Par complex directs asymmetric cell division by phosphorylating the cytoskeletal protein Lgl. *Nature* 2003;**422**:326–30.
71. Behrens J, Birchmeier W. Cell–cell adhesion in invasion and metastasis of carcinomas. *Cancer Treat Res* 1994;**71**:251–66.
72. Berx G, Becker KF, Hofler H, van Roy F. Mutations of the human E-cadherin (CDH1) gene. *Hum Mutat* 1998;**12**:226–37.
73. Silvis MR, Kreger BT, Lien WH, Klezovitch O, Rudakova GM, Camargo FD, et al. Alpha-catenin is a tumor suppressor that controls cell accumulation by regulating the localization and activity of the transcriptional coactivator Yap1. *Sci Signal* 2011;**4**:ra33.
74. Fodde R, Brabletz T. Wnt/beta-catenin signaling in cancer stemness and malignant behavior. *Curr Opin Cell Biol* 2007;**19**:150–8.
75. Vasioukhin V, Bauer C, Degenstein L, Wise B, Fuchs E. Hyperproliferation and defects in epithelial polarity upon conditional ablation of alpha-catenin in skin. *Cell* 2001;**104**:605–17.
76. Berx G, van Roy F. Involvement of members of the cadherin superfamily in cancer. *Cold Spring Harb Perspect Biol* 2009;**1**:a003129.
77. White BD, Chien AJ, Dawson DW. Dysregulation of Wnt/beta-catenin signaling in gastrointestinal cancers. *Gastroenterology* 2012;**142**:219–32.
78. Eagle H, Levine EM, Boone CW. Cellular growth, contact inhibition, and macromolecular synthesis. *Science* 1965;**148**:665.

79. Fagotto F, Gumbiner BM. Cell contact-dependent signaling. *Dev Biol* 1996;**180**: 445–54.
80. Lien WH, Klezovitch O, Fernandez TE, Delrow J, Vasioukhin V. alphaE-catenin controls cerebral cortical size by regulating the hedgehog signaling pathway. *Science* 2006;**311**:1609–12.
81. Nelson WJ, Nusse R. Convergence of Wnt, beta-catenin, and cadherin pathways. *Science* 2004;**303**:1483–7.
82. Boggiano JC, Fehon RG. Growth control by committee: intercellular junctions, cell polarity, and the cytoskeleton regulate Hippo signaling. *Dev Cell* 2012;**22**:695–702.
83. Schlegelmilch K, Mohseni M, Kirak O, Pruszak J, Rodriguez JR, Zhou D, et al. Yap1 acts downstream of alpha-catenin to control epidermal proliferation. *Cell* 2011;**144**: 782–95.
84. Balda MS, Matter K. Tight junctions and the regulation of gene expression. *Biochim Biophys Acta* 2009;**1788**:761–7.
85. Balda MS, Garrett MD, Matter K. The ZO-1-associated Y-box factor ZONAB regulates epithelial cell proliferation and cell density. *J Cell Biol* 2003;**160**:423–32.
86. Zeitler J, Hsu CP, Dionne H, Bilder D. Domains controlling cell polarity and proliferation in the Drosophila tumor suppressor Scribble. *J Cell Biol* 2004;**167**:1137–46.
87. Hough CD, Woods DF, Park S, Bryant PJ. Organizing a functional junctional complex requires specific domains of the Drosophila MAGUK Discs large. *Genes Dev* 1997;**11**: 3242–53.
88. Zhan L, Rosenberg A, Bergami KC, Yu M, Xuan Z, Jaffe AB, et al. Deregulation of scribble promotes mammary tumorigenesis and reveals a role for cell polarity in carcinoma. *Cell* 2008;**135**:865–78.
89. Mailleux AA, Overholtzer M, Brugge JS. Lumen formation during mammary epithelial morphogenesis: insights from in vitro and in vivo models. *Cell Cycle* 2008;**7**:57–62.
90. Zhao M, Szafranski P, Hall CA, Goode S. Basolateral junctions utilize warts signaling to control epithelial–mesenchymal transition and proliferation crucial for migration and invasion of Drosophila ovarian epithelial cells. *Genetics* 2008;**178**:1947–71.
91. Reddy BV, Irvine KD. The Fat and Warts signaling pathways: new insights into their regulation, mechanism and conservation. *Development* 2008;**135**:2827–38.
92. Nichols SA, Roberts BW, Richter DJ, Fairclough SR, King N. Origin of metazoan cadherin diversity and the antiquity of the classical cadherin/beta-catenin complex. *Proc Natl Acad Sci USA* 2012;**109**:13046–51.
93. Nichols SA, Dirks W, Pearse JS, King N. Early evolution of animal cell signaling and adhesion genes. *Proc Natl Acad Sci USA* 2006;**103**:12451–6.
94. Fahey B, Degnan BM. Origin of animal epithelia: insights from the sponge genome. *Evol Dev* 2010;**12**:601–17.
95. Dickinson DJ, Nelson WJ, Weis WI. An epithelial tissue in Dictyostelium and the origin of metazoan multicellularity. *BioEssays* 2012;**34**:833–40.
96. Baum B, Settleman J, Quinlan MP. Transitions between epithelial and mesenchymal states in development and disease. *Semin Cell Dev Biol* 2008;**19**:294–308.
97. Dickinson DJ, Nelson WJ, Weis WI. A polarized epithelium organized by beta- and alpha-catenin predates cadherin and metazoan origins. *Science* 2011;**331**:1336–9.
98. Kwiatkowski AV, Maiden SL, Pokutta S, Choi HJ, Benjamin JM, Lynch AM, et al. In vitro and in vivo reconstitution of the cadherin–catenin–actin complex from *Caenorhabditis elegans*. *Proc Natl Acad Sci USA* 2010;**107**:14591–6.
99. Grimson MJ, Coates JC, Reynolds JP, Shipman M, Blanton RL, Harwood AJ. Adherens junctions and beta-catenin-mediated cell signalling in a non-metazoan organism. *Nature* 2000;**408**:727–31.
100. Urushihara H. Developmental biology of the social amoeba: history, current knowledge and prospects. *Dev Growth Differ* 2008;**50**(Suppl. 1):S277–S281.

101. Raper KB, Fennell DI. Stalk formation in Dictyostelium. *Bull Torrey Bot Club* 1952;**19**: 1421–8.
102. Dickinson DJ, Nelson WJ, Weis WI. α-Catenin and IQGAP regulate myosin localization to control epithelial tube morphogenesis in Dictyostelium. *Dev Cell* 2012;**23**: 533–46.
103. Faix J, Dittrich W. DGAP1, a homologue of rasGTPase activating proteins that controls growth, cytokinesis, and development in Dictyostelium discoideum. *FEBS Lett* 1996;**394**:251–7.
104. Faix J, Steinmetz M, Boves H, Kammerer RA, Lottspeich F, Mintert U, et al. Cortexillins, major determinants of cell shape and size, are actin-bundling proteins with a parallel coiled-coil tail. *Cell* 1996;**86**:631–42.

CHAPTER TWO

Stability and Dynamics of Cell–Cell Junctions

Claudio Collinet, Thomas Lecuit
Institut de Biologie du Développement de Marseille Luminy (IBDML), Unite Mixte de Recherche 6216, Case 907, Marseille, France

Contents

Abstract

Adherens junctions display dual properties of robustness and plasticity. In multicellular organisms, they support both strong cell–cell adhesion and rapid cell–cell contact remodeling during development and wound healing. The core components of adherens junctions are clusters of cadherin molecules, which mediate cell–cell adhesion through homophilic interactions in *trans*. Interactions of cadherins with the actin cytoskeleton are essential for providing both stability and plasticity to adherens junctions. Cadherins regulate the turnover of actin by regulating its polymerization and anchor tensile actomyosin networks at the cell cortex. In turn, actin regulates cadherin turnover by regulating its endocytosis and actomyosin networks exert forces driving remodeling of

Progress in Molecular Biology and Translational Science, Volume 116
ISSN 1877-1173
http://dx.doi.org/10.1016/B978-0-12-394311-8.00002-9

cell–cell contacts. The interplay between adherens junctions and contractile actomyosin networks has striking outcomes during epithelial morphogenesis. Their integrated dynamics result in different morphogenetic patterns shaping tissues and organs.

1. INTRODUCTION

In multicellular organisms, intercellular adhesion underlies embryonic development and function of adult organs by providing mechanical stability to cell–cell interactions. Cell–cell adhesion is mediated by adherens junctions (AJs), which were first described as stable cellular structures of close membrane apposition between neighboring epithelial cells[1] where extracellular molecular bonds established by cadherins are stabilized by intracellular connections to a coat of actin filaments.[2] However, a more dynamic view of AJs has recently emerged and describes them as also strikingly plastic. In epithelia stable adhesion is required to establish and maintain robust and polarized monolayers, but at the same time, constant cell–cell contact remodeling takes place during development and wound healing.

Classical cadherins are Ca^{2+}-dependent transmembrane proteins that mediate the physical binding between adjacent cells by engaging in homophilic interactions in *trans* between facing membranes.[3] Cadherin homophilic complexes mechanically connect cells and are further stabilized by intracellular interactions with actin filaments through α- and β-catenin. β-catenin binds the conserved intracellular tail of cadherins and recruits α-catenin, which, in turn, directly or indirectly interacts with actin filaments.[4–7] Plasticity of AJs is not only the result of adhesion molecule turnover but also the result of active forces pulling on them. The interaction of cadherins with the actin cytoskeleton provides anchorage to contractile actomyosin networks, exerting forces responsible for cell–cell contact remodeling during development.[8–10] Thus, cadherins play a pivotal role in junction robustness and plasticity by mediating both adhesion (cohesion) and transmission of actomyosin-driven forces (remodeling).

Although the exact molecular nature of cadherin–actin dynamic interactions is still unclear[5,11] and might depend on cell types or organisms, association with actin is essential for many properties of AJs. In this chapter, we review how cadherin–actin interactions critically regulate both the stability and the dynamics of AJs. We first review the role of the actin cytoskeleton in stabilizing junctions. Then, we illustrate how both actin dynamics and mechanical forces regulate cadherin and AJ dynamics. Finally, we discuss how morphogenetic patterns emerge during development via the dynamic interplay between AJs and contractile systems. This will illustrate the relevance of AJ turnover and stability during development.

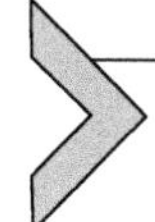

2. STABILITY OF AJs

2.1. Cadherins are organized in molecular clusters at AJs

AJs are composed of well-circumscribed clusters of cadherin molecules. These structures have been clearly identified in invertebrate epithelia[12,13] and in mammalian cells[14–16] and are thought to represent local concentrations of homophilic *trans* dimers of cadherins.[12,16] Therefore, cadherin clusters directly mediate adhesion and their number and size define the adhesive properties of AJs.

The mechanisms of cadherin clustering are still debated. Recent studies have suggested that the ectodomains of classical cadherins are sufficient to drive clustering both *in vitro*[17] and *in vivo*.[18,19] This involves both strong homophilic *trans* interactions and weaker lateral *cis* interactions, as illustrated by a thorough structure–function study of mouse E- and N-cadherin.[17] Nevertheless, *in vivo* cadherin clustering through their ectodomains is not sufficient to mediate stable adhesion. The interaction of cadherin molecules with the actin cytoskeleton regulates their membrane organization and influences both the stability and the lateral mobility of cadherin clusters (see below, Section 2.3). These are two key elements for the stability of cell–cell contacts.

2.2. Molecular nature of cadherin–actin interaction

The binding of cadherins to the actin cytoskeleton is mediated by α- and β-catenin (Fig. 2.1). β-Catenin directly binds the intracellular tail of cadherins and recruits α-catenin,[4,5] which is essential for epithelial integrity in many organisms.[20–24] α-Catenin operates at the interface with the actin cytoskeleton even though the exact mechanism of the molecular interactions remains poorly understood.[5,7,11] Indeed, α-catenin can bind both the cadherin–β-catenin complex and actin, but a quaternary complex has never been isolated.[7]

Many actin-binding proteins have been shown to interact also with α-catenin,[5] suggesting that the interaction with actin could be indirect. Noteworthy among these proteins are epithelial protein lost in neoplasm (EPLIN)[25] and vinculin.[26] EPLIN promotes actin bundling by inhibiting actin depolymerization.[27] It binds α-catenin when the latter is associated

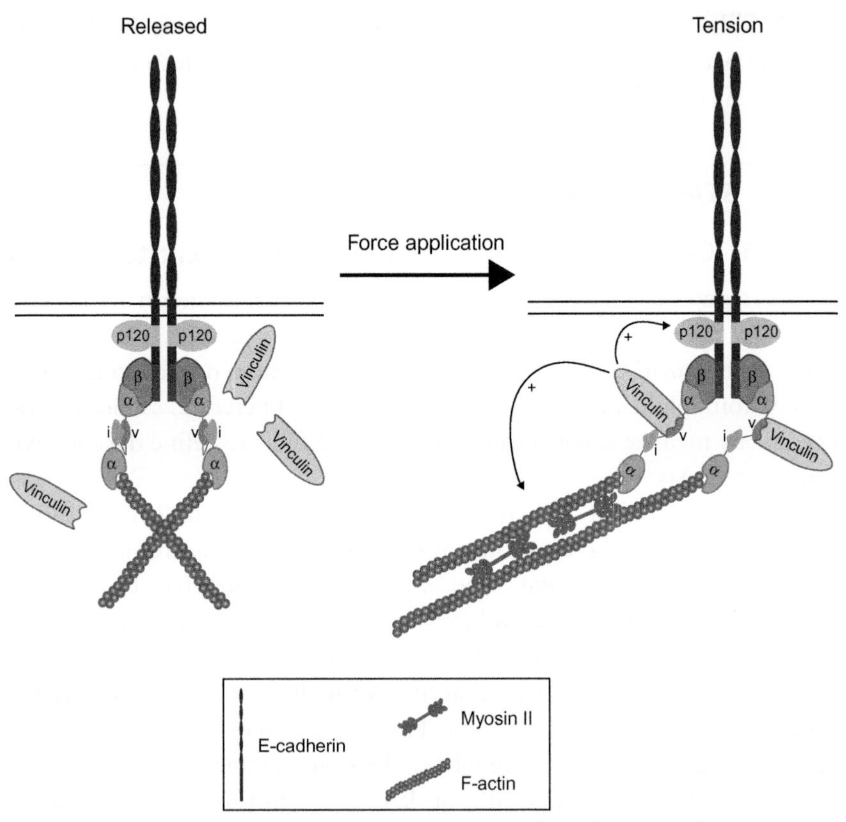

Figure 2.1 The cadherin–actin interaction and the effect of force. Cadherins interact with actin filaments through the binding of β- and α-catenin. β-Catenin binds directly to a C-terminal domain in the intracellular tail of cadherins and recruits α-catenin. Then, α-catenin binds directly or indirectly to actin filaments. α-Catenin is a stretch-activated protein and recruits vinculin in a force-dependent manner. In the absence of force (left situation), an inhibitory domain (i) masks the vinculin-binding domain (v) and prevents vinculin binding. In the presence of force (right situation), a conformational change is induced and unmasks the vinculin-binding domain (v). Vinculin can now be recruited and will stabilize α-catenin and actin recruitment. (See Color Insert.)

with cadherin and β-catenin and links this entire complex to actin.[25] Loss of EPLIN converts the circumferential actin belt (typical of mature junctions) into radially oriented actin filaments and fragments the cadherin belt into smaller spots such as those in immature junctions[25,25]. Vinculin is an actin-binding protein initially characterized in association with focal adhesions[28] and required for their stabilization under tension.[29] Recent data showed that vinculin is also recruited to AJs by α-catenin in a force-dependent manner[26] and that this stabilizes adhesion contacts (Fig. 2.1).[30]

An interesting alternative is that the cadherin–catenin complex might interact with actin directly but with very low affinity (not detectable in biochemical studies). Then, individual cadherin molecules would diffuse without a strong link to actin, but clusters (comprising a few tens to thousands of molecules) would be more efficiently tethered as the high density of cadherins would increase the resultant affinity.[31]

2.3. Regulation of cadherin cluster stability and mobility

Independently of their exact molecular nature, the interactions of cadherins with the actin cytoskeleton are important for regulating the organization of cadherin molecules in AJs at two levels: (1) they can regulate the assembly/disassembly of cadherin monomers into clusters and (2) they can regulate the lateral mobility of these clusters.

Examples of this have been provided both in mammals and in invertebrates. In mammalian cells, myosin II, and more specifically its isoform myosin IIA, which is regulated by Rho and ROCK, promotes the concentration of E-cadherin at cell–cell contacts and favors the formation of cadherin clusters.[32,33] The assembly of cadherin clusters might be favored by promoting association of monomers in *cis* as a result of the increased local concentration of E-cadherin.

Anchorage of the clusters to the cortical actin cytoskeleton influences also their lateral mobility. When dynamic epithelia reorganize their cell–cell contacts, a basal to apical flow of cadherin clusters has been observed at the cell junctions, leading to accumulation of those clusters at the apical zonula adherens.[16] This movement is dependent on actin and α-catenin.[16]

Another example in which the actin cytoskeleton influences both stability and lateral mobility of cadherin clusters is provided by the epithelial cells of the *Drosophila* embryo. Here, cadherin clusters are associated with two intermixed populations of actin filaments with different turnover rates.[31] A stable pool of actin filaments is closely associated with the cadherin clusters and underlies their stability, and a more dynamic pool constitutes a cortical actomyosin network that regulates their lateral mobility. Depolymerization of actin does not dissolve cadherin clusters (even if it reduces their fluorescent intensity) but affects their lateral mobility (not only within the junctional plane but also in the apicobasal direction). This, in turn, disrupts the integrity of the epithelium, a consequence often attributed to the destabilization of cadherin homophilic complexes.[31]

In conclusion, the actin cytoskeleton regulates the stability of AJs by two mechanisms: (1) it reinforces the stability of cadherin clusters, the basic unit of adhesion and (2) it restricts their localization at the proper sites of cell–cell contacts.

3. DYNAMICS OF AJs MEDIATED BY ACTIN TURNOVER

AJs are very dynamic structures. They continuously shrink and regrow to accommodate movements of cells during tissue and organ development. This is an active process dependent on both external and internal factors and is resisted by adhesion.

The plasticity of AJs involves a constant and regulated turnover of cadherins that depends on turnover of the actin cytoskeleton. In particular, actin polymerization is a remarkable feature of all AJs and is regulated by cadherins. In turn, actin polymerization regulates the turnover of cadherins by regulating their endocytosis. This mutual regulation is one key element of AJ plasticity.

In this section, we illustrate the modes and the pathways of actin assembly and their regulation in maturing and stable junctions. Then, we review the role of actin polymerization in the regulation of cadherin endocytosis.

3.1. Actin polymerization sustains the establishment and maturation of cell–cell contacts

During the process of formation and stabilization of cell–cell junctions, extensive actin cytoskeleton remodeling takes place and is promoted by actin polymerization (Fig. 2.2).

The process of AJ formation begins with an opportunistic contact between two cells, promoted by an exploratory activity of actin-based cell protrusions (e.g., lamellipodia and membrane ruffles in MDCK cells and filopodia in keratinocytes).[34–36] Upon contact, diffusing cadherin monomers become rapidly engaged in homophilic *trans* interactions and form small and less mobile clusters.[15] Cadherin clusters grow by recruitment of new molecules and rapidly become connected to actin.[36] At sites of cell–cell contact in MDCK cells, the lamellipodial actin array is lost and cadherin clusters form connections with actin filaments radiating outward from a central actin ring.[34] In keratinocytes, the adhesive interface develops in an "adhesion zipper" structure in which protrusions from one cell push and invaginate into the neighboring cell. This adhesion zipper structure is then resolved by actin polymerization from the tip of the invaginating protrusions. This produces forces pushing the invaginating protrusion back toward the surface of contact between the two cells.[35] In both cases, the radial actin

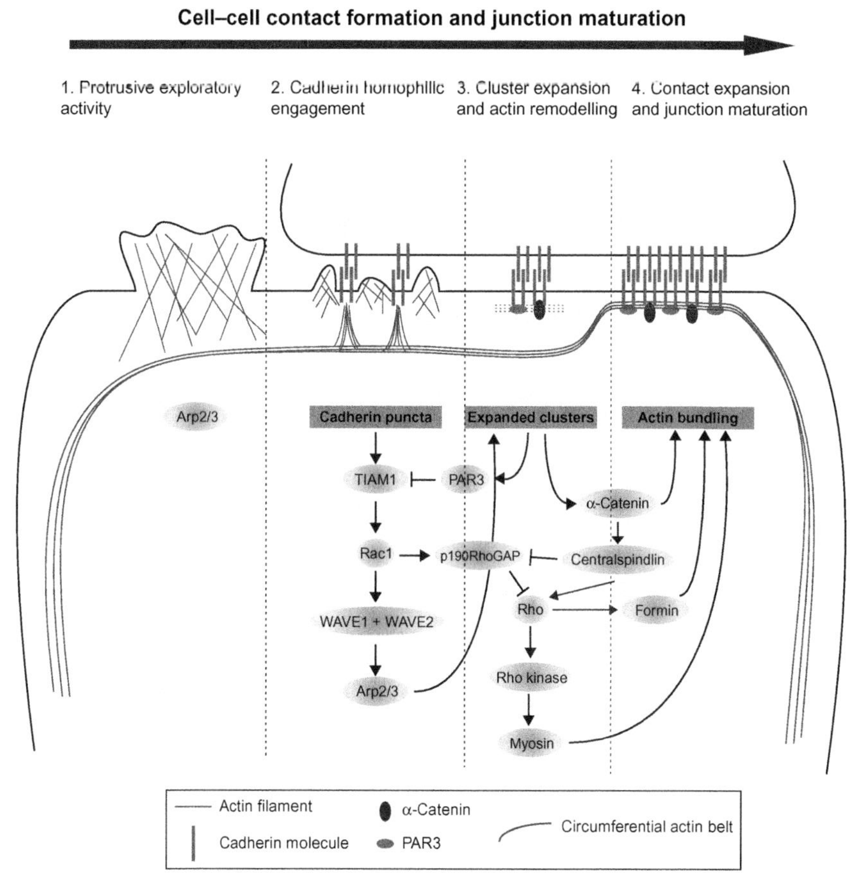

Figure 2.2 Cell–cell contact formation and signaling pathways controlling actin polymerization. Schematic representation of the formation of cell–cell contacts and maturation of adherens junctions. (1) The exploratory activity of actin-based cell protrusion promotes an opportunistic contact between cells. (2) Cadherin homophilic binding in *trans* induces branched actin polymerization via activation of Rac and ARP2/3. This promotes new membrane apposition and expansion of cadherin clusters by recruitment of new cadherin monomers. (3) The growth of the cadherin clusters recruits α-catenin and PAR3 promoting a switch from the Rac–ARP2/3-dependent polymerization of branched actin to a Rho–formin-based polymerization of unbranched actin. (4) Bundling of linear actin filaments and activation of myosin II promote the formation of the circumferential actin belt typical of mature epithelia. (See Color Insert.)

bundles are subsequently replaced by bundles running parallel to the contact. These parallel actin bundles progressively mature to the circumferential actin belt typical of mature epithelia.

This massive extent of actin remodeling requires actin polymerization by at least two types of actin nucleators: (1) the ARP2/3 complex, which promotes assembly of branched actin and (2) formins, which enhance the linear growth of actin filaments. Indeed, ARP2/3, formins, and members of the Ena/VASP family have been localized specifically to cadherin clusters, and their inhibition disrupts AJ organization as cell–cell contacts are forming.[35,37–39] Cadherins control the activity of these two types of actin nucleators by activating a cascade of signaling involving the small GTPases Rac and Rho, which activate ARP2/3 and formins, respectively (Fig. 2.2). Rac and ARP2/3 activities are typical of "young" junctions and allow cells to establish and expand contacts, whereas Rho and formin activities are typical of "older" junctions and are required at later steps of AJ maturation. Deregulation of this pattern (e.g., high ARP2/3 activity in mature junctions or high Rho activity in young ones) impairs proper junction establishment and maturation.[40,41] Therefore, during the process of cell–cell contact formation, crosstalk between Rac and Rho activity is required to control the successive activation of two different actin polymerization waves (Fig. 2.2).

Homophilic *trans* interactions between cadherin clusters activate Rac by recruiting the RacGEF, TIAM1.[42] Rac activates the ARP2/3 complex through recruitment of the ARP2/3 activator WAVE2[43] and promotes the formation of membrane protrusions to favor new membrane apposition. At the same time, Rac activity holds off the next wave of actin polymerization by recruiting the RhoGAP p190RhoGAP to p120 catenin, a component of the cadherin–catenin complex, and thereby inhibits Rho activity in young junctions.[44] Once AJs have expanded, α-catenin and PAR3 are recruited to the sites of contact and inhibit ARP2/3 and TIAM1, respectively.[6,45] Complete maturation and stabilization of AJs require the activation of Rho. Although specific activators have been implicated in Rho activation in other circumstances (e.g., during apical constriction and cytokinesis), only recently has centralspindlin, a cytokinetic regulator, been reported as a Rho activator at AJs.[46] Centralspindlin is recruited at junctions by α-catenin, where it activates Rho by recruiting the RhoGEF, ECT2, and inhibits p190RhoGAP[46] (see also Chapter 3). Therefore, at expanded contacts, α-catenin promotes Rho activation via centralspindlin. Rho activates formins to promote the polymerization of unbranched actin filaments and acts through Rho kinase to activate myosin II contractility. Myosin II and unbranched actin filaments assemble in the contractile actin belt typical of mature AJs.

Collectively, these results show that cadherins control actin polymerization and turnover, thus sustaining junction establishment.

3.2. Actin polymerization regulates cadherin endocytosis

Actin polymerization controls the turnover of cadherins by regulating their endocytosis. Cadherin endocytosis is a major mechanism of AJ remodeling[47] and trafficking has been proposed to be one of the main mechanisms of cadherin junctional turnover.[48] Several studies have established the role of dynamin- and clathrin-dependent endocytosis[47,49–51] and of clathrin adaptors, such as β-arrestin[52] and the adaptor complex 2 (AP2),[53] in cadherin endocytosis. In mammalian cells, a potent regulator of cadherin endocytosis is p120 catenin, which binds the intracellular domain of cadherins directly and inhibits their endocytosis[54–56] (see also Chapter 18). Another molecule playing a role in cadherin endocytosis is the E3 ubiquitin ligase Hakai, which binds to and ubiquitylates the cytoplasmic tail of E-cadherin, targeting it for internalization and trafficking to late endosomes/lysosomes.[57]

However, in other organisms, such as *Drosophila* and *Caenorhabditis elegans*, disruption of both p120-catenin and Hakai function has been reported to have a weak effect on cadherin endocytosis.[58–61] Therefore, other mechanisms regulating cadherin endocytosis must exist. Among others, actin polymerization has been reported to play an important role. Branched actin polymerization by ARP2/3 has an established role in endocytosis and is thought to promote membrane invagination and vesicle scission.[62] Interestingly, at AJs, actin polymerization and cadherin endocytosis are coupled by at least two different mechanisms. One involves regulation of branched actin polymerization by Cdc42 and its coupling to E-cadherin endocytosis. This mechanism has been observed in the *Drosophila* pupal notum[63,64] and involves two effectors of Cdc42, CIP4 and PAR6. CIP4, together with its downstream targets, WASP and ARP2/3, nucleates branched actin at sites of internalization, and PAR6 recruits CIP4 to AJs, where it also binds dynamin.[63,64] However, Cdc42 is also a regulator of cell polarity, as its mutation disrupts the localization[63] and induces the internalization[65] of the Crumbs complex, which is required for AJ stability. Therefore, the effect of Cdc42 on E-cadherin endocytosis could be indirect.

A second mechanism recently identified in *Drosophila* embryos involves the polymerization of unbranched actin filaments by the formin diaphanous and the activation of myosin II.[51] The activities of both diaphanous and myosin II were shown to promote cortical recruitment of AP2 and clathrin and to induce the initiation of clathrin-mediated endocytosis (CME). Initiation of CME is thought to be promoted by local *cis* clustering of E-cadherin molecules.[51]

Therefore, both branched and unbranched actin polymerization promote cadherin endocytosis via two different mechanisms controlling different steps of endocytic vesicle formation: initiation and scission. The two mechanisms might operate at the same time and cooperate for efficient cadherin endocytosis.

4. MECHANICAL REGULATION OF AJs

An established paradigm for AJs is that they anchor contractile actomyosin networks at the cell cortex and resist their mechanical tension. Recent evidence indicates that AJs not only resist tension but also dynamically adapt to it by changing their organization and modulating their mechanical properties. This uncovers a new way by which AJs regulate their dynamics, which could have important implications in cell and tissue morphogenesis. In this respect, AJs appear more similar to another cellular adhesive structure than previously thought, namely, the focal adhesions that anchor cells to the extracellular matrix. To date, properties of focal adhesions are much better characterized than those of AJs, and we believe that many regulatory principles of the former could be applied to the latter.

In this section, we present evidence suggesting that the organization and the mechanical properties of AJs respond to applied mechanical forces, invoking mechanosensation by cadherins. We then discuss parallels with focal adhesions and illustrate a potential molecular mechanism at the basis of AJ mechanosensitivity.

4.1. AJs respond to mechanical stimuli

Although the mechanosensitivity of AJs was strongly anticipated from a large number of reports showing that their remodeling during development depends largely on actomyosin contractility,[9] it was not possible to prove it purely on the basis of studies of their morphology, biochemical composition, and dynamics. Recently, more specific studies aiming at directly measuring mechanical forces applied to and resisted by AJs have shown that they can sense and respond to mechanical forces and that these effects are mediated by cadherins.

Precise measurements of forces applied by the actomyosin cytoskeleton to junctions were performed in cell duplets using elastomeric micropillars[66] or traction force microscopy[67] and have estimated the traction forces are applied to AJs. AJs, in turn, responded proportionally to the magnitude of such forces by changing their length and width.[66] Similar results were obtained also by direct mechanical stimulation with a micropipette,[66] suggesting that AJs can sense forces and adapt to them to preserve, or even enhance, junctional integrity.

Mechanosensation involves cadherins because AJs respond to mechanical stimuli specifically when they are conveyed via cadherin binding. This has been shown in two studies. In the first one, cells responded to changes of stiffness of cadherin-coated substrates,[68] and in the second one, shear forces applied with cadherin-coated ferromagnetic beads on the surface of cells rapidly induced increases of stiffness of the cell–bead junction.[30] The relative stiffness change was proportional to the magnitude of the bond shear, a typical feature of mechanosensation.

Together, these results show that cadherins can sense forces and that AJs adapt to them by reinforcing their connections.

4.2. Molecular mechanism of mechanosensation by cadherins

Mechanosensation is typical of focal adhesions. The assembly of focal adhesions is also force dependent[69,70] and depends on a mechanism of reinforcement involving characteristics of catch-bond by integrins and a force-dependent recruitment of vinculin.[9,71]

Although catch-bond-type binding of cadherins has not yet been demonstrated, a similar mechanism for vinculin recruitment was recently shown also for AJs.[26] Detailed structure–function analyses of α-catenin have shown that this is an autoinhibited protein in which a putative inhibitory domain masks an adjacent vinculin-binding site (Fig. 2.1). Myosin II-induced tension and α-catenin binding to actin are required to overcome this inhibition and recruit vinculin to AJs. Notably, in direct force measurements at AJs, vinculin contributes to the force-induced reinforcement of adhesion.[30] This is thought to happen by a positive feedback of vinculin on the stabilization of α-catenin as well as actin recruitment to adhesion sites.[26,30] These findings together suggest a model in which α-catenin acts as a stretch-activated tension-sensing protein acting between cadherins and actomyosin networks, such that increased tension relieves the autoinhibition to expose the cryptic vinculin-binding site with subsequent vinculin recruitment and junction reinforcement (Fig. 2.1).

In conclusion, AJs are plastic structures able to organize tensile actomyosin networks and at the same time able to respond to mechanical forces. This mechanosensitivity is important in the regulation of AJ dynamics and might play an important role in cell and tissue physiology.

5. DYNAMICS OF AJs DURING DEVELOPMENT

The development of multicellular organisms involves dramatic changes in the shapes of tissues and organs. Epithelial morphogenesis involves constant remodeling of cell–cell junctions and changes in the shapes of cells. These events depend largely on contractile forces generated by actomyosin networks.[8–10] AJs, however, play an important role in spatially organizing such forces to produce specific changes of shapes of cells and tissues.

Tissue invagination by apical constriction (Fig. 2.3C) and convergence–extension movements by cell intercalation (Fig. 2.3F) represent two widespread morphogenetic processes that illustrate how the interplay between AJs and the dynamics of actomyosin networks can result in different morphogenetic patterns. In the following sections, we review these two processes in the light of their differences with respect to two questions: (1) the subcellular distribution of tension and how this influences and/or is influenced by AJs and (2) how different dynamic behaviors of actomyosin networks emerge from their interactions with AJs. Finally, we discuss the role of AJs as mechanical couplers between cells integrating local cellular forces into tissue-wide morphogenetic ones.

5.1. Spatial control of tension

The spatial control of actomyosin-mediated tension is important for determining the type of deformations induced in a cell. This is nicely illustrated by epithelial cells that remodel their apical cell–cell contacts. Here, if tension is isotropically distributed, that is, equal in all directions, it results in an isotropic contraction of the apical surface and of all cell–cell contacts (Fig. 2.3A), whereas if it is anisotropically distributed (e.g., planar polarized), it leads to polarized shrinkage of junctions (Fig. 2.3D). These two cellular behaviors, integrated in a cohesive sheet of epithelial cells, produce two different morphogenetic processes: tissue bending and invagination in the first case (Fig. 2.3C) and tissue extension by cell–cell intercalation in the second (Fig. 2.3F). In these processes, AJs are not passively remodeled by actomyosin tension but are important active players providing spatial cues to actomyosin contractility (e.g., by providing cortical anchorage or by recruiting myosin activators).

Tissue invagination by apical constriction has been observed in many morphogenetic processes during development. A classic example is the closure of the neural tube in mouse and *Xenopus* embryos. Here, the bending of

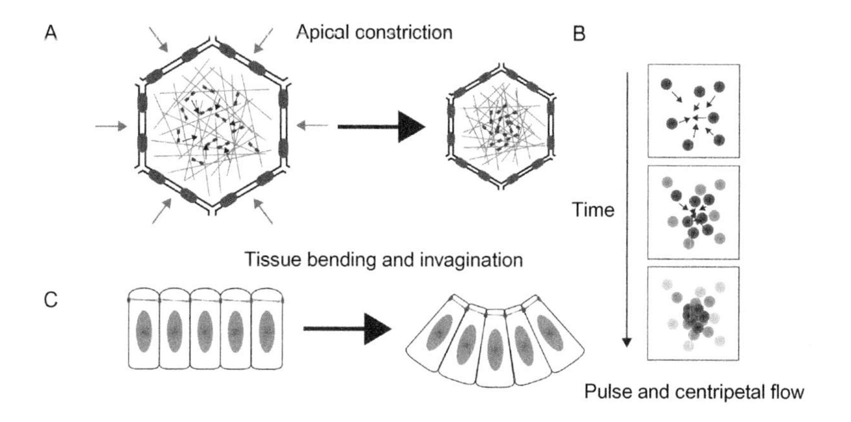

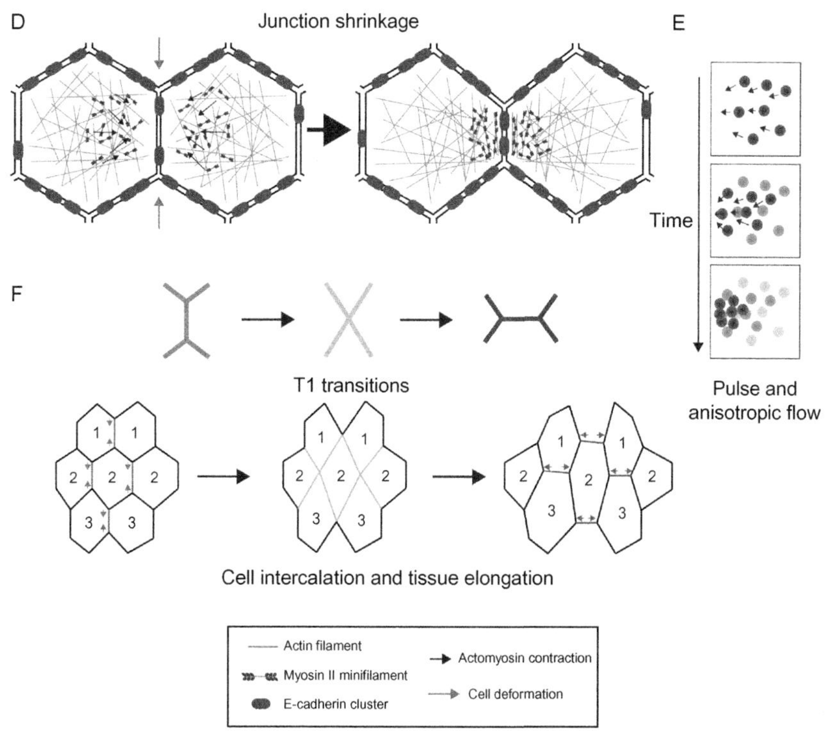

Figure 2.3 Different effects of isotropic and anisotropic tension and actomyosin network anchorage on epithelial tissue morphogenesis. (A–C) In *Drosophila* mesodermal cells, apical constriction is mediated by an isotropic tension that constricts the apical cell surface and all cell–cell contacts equally (A). This isotropic tension is exerted by an apical-medial contractile actomyosin network that is anchored at the cell cortex

(Continued)

the neuroepithelium is powered by cortical tension at apical cell junctions and is triggered by shroom, a protein that is necessary and sufficient for the recruitment and activation of myosin II there.[72–74]

In other systems, such as the mesoderms of *C. elegans*[75] and *Drosophila*,[76] apical constriction is associated with enrichment and activation of myosin II in the medial apical region of the cell rather than at junctions. In *Drosophila*, this depends on apical accumulation of RhoGEF2[77] and the combined activity of fog,[78,79] the small Gα protein concertina ($G_{\alpha q12q13}$),[80] a still unknown G-protein-coupled receptor, and the transmembrane protein T48, which binds the PDZ-binding domain of RhoGEF2.[81] Importantly, in both systems, the invagination of the mesoderm begins only with the maturation of a stable anchorage of an already contractile apical actomyosin mesh at AJs.[82] This shows the importance of the anchorage of contractile forces in this process.

Cell–cell intercalation is one of the most prominent mechanisms at the basis of the elongation of tissues and requires anisotropic tension and polarized junction remodeling. A striking example of this is provided by cell intercalation of the ventrolateral ectoderm (the germ band) of *Drosophila* embryos.[83] Here, cell intercalation is driven by a planar polarized remodeling of junctions, which comprises two steps: first, "vertical" junctions, oriented along the dorsoventral axis, shrink and then new junctions oriented along the anteroposterior axis form and extend[84] (Fig. 2.3F). This process is driven by anisotropic distribution of cortical tension,[85] due to the specific enrichment of myosin II in vertical junctions.[84,86] Rosettes form when

by cadherin clusters (A). Contraction is pulsatile and is due to the centripetal flow of myosin II molecules (red dots) that accumulate in foci (B). Shaded dots represent the previous position of myosin II molecules during the emergence of the pulse (b). The effect of apical constriction in a cohesive sheet of epithelial cells is the bending and invagination of the entire tissue (C). (D–F) In *Drosophila* ventrolateral cells, the anisotropic distribution of tension causes the selective shrinkage of vertical junctions and is associated with anisotropic distribution of cadherin clusters (and presumably with anisotropic anchorage of actomyosin networks). Here, the medial apical pulses of myosin II flow toward vertical junctions and shrink them (D). Myosin II molecules (red dots) during emergence of the pulse flow anisotropically in the anterior–posterior direction (E). Shaded dots represent the previous position of myosin II molecules during the emergence of the pulse (E). Vertical junction shrinkage is then followed by the formation and extension of a new junction in the perpendicular direction (T1 transition) (F top). In a cohesive sheet of epithelial cells, this allows an exchange of neighbors and the intercalation of cells (F bottom).Panels A, B, D, and E are modified from Ref. 9. (See Color Insert.)

myosin II accumulation in vertical junctions spans several junctions in supracellular cables, which, by shrinking, bring together five or more cells.[87,88] Remarkably, E-cadherin, α-catenin,[51,87,89] and the polarity protein PAR3[86] present a pattern of planar polarity opposite to that of myosin II, that is, are present at lower levels in shrinking junctions, suggesting weaker adhesion at these junctions. Interestingly, the complementary polarity of E-cadherin is controlled both by upregulation of endocytosis in vertical junctions[51] and by the restricted polarity of PAR3, in turn controlled by recruitment of ROCK at vertical junctions.[90] Therefore, both anisotropic tension at junctions and polarized active mechanisms to decrease adhesion cooperate to drive junction remodeling. Junctions promote polarity by recruiting upstream activators of myosin II, such as RhoGEF2, which simultaneously promote the upregulation of tension and the downregulation of adhesion.[51]

Anisotropic tension drives planar polarized junction remodeling also in other systems. One example is constituted by the elongation and bending of the vertebrate neuroepithelium during neural tube closure in chicken embryos.[91] Here, activated myosin II was detected only in a subset of AJs inducing the presence of intercalation-like cell rearrangement (rosettes) and the anisotropic bending of the neural plate.[91] Very recent work[92] identified Celsr1, a PCP-regulating cadherin, as the symmetry breaking element of the system. This junctional molecule is planar polarized and activates a signaling cascade leading to activation of Rho kinase and actomyosin contractility specifically in junctions oriented toward the mediolateral axis of the neural plate.[92]

Altogether, these results illustrate how the spatial control of tension determines patterns of tissue morphogenesis and how AJs contribute to this control by providing localized activation of myosin II and mediating cortical transmission of tension.

5.2. Dynamics of actomyosin networks: Pulses and flows

Recent observations made possible by fast live-cell imaging revealed two important properties of contractile dynamics involved in epithelial morphogenesis, that is, pulsatile contractility and actomyosin flow patterns. After their first observation in the *C. elegans* zygote,[93] these properties of actomyosin networks have been described in numerous other tissues undergoing morphogenesis, including the mesoderm,[76] the ventrolateral ectoderm[94] and the amnioserosa of *Drosophila* embryos,[95] the follicular epithelium of the *Drosophila* egg chamber,[96] the mesoderm of *C. elegans* embryos,[82] and the *Xenopus* gastrula.[97]

With respect to pulsatile contractility and actomyosin flows, the mesoderm and the ventrolateral ectoderm of the *Drosophila* embryo provide an interesting comparison. In the mesoderm, pulses of myosin II in the medial apical region induce discrete steps of apical constriction followed by phases of stabilization.[76] Similarly, in the ventrolateral ectoderm, pulses of myosin II originate in the medial apical region but then flow towards vertical junctions and shrink them. A subsequent increase in junctional myosin II stabilizes individual steps of shrinkage. The directionality of the flows determines the spatial orientation of cell deformation and is dependent on the anisotropic distribution of E-cadherin, β-catenin, and α-catenin at cell junctions.[94] AJs may play an important regulatory role by providing anchorage to medial contractile actomyosin networks. For example, pulsatile contractions could be induced by load-dependent detachment of myosin II motors, in which contractile forces are progressively resisted by the rigidity of the anchoring points. This initially determines an increase of tension by progressive recruitment of motors (inhibition of the power stroke) until deformation of the anchorage point, which releases the tension and determines the detachment of the motors. Similarly, anchorage might explain actomyosin flows in that anisotropic anchorage might underlie the formation of gradients of tension and thereby induce flows of actomyosin networks. Therefore, spatiotemporal control of anchorage might influence dynamics of actomyosin networks by defining spatial patterns of flowing behaviors. Isotropic anchorage defines centripetal actomyosin flows (pulses), and anisotropic anchorage determines anisotropic flows (Fig. 2.3B and E).

5.3. Force transmission, cell–cell coordination, and tissue integration

Tissue morphogenesis results from coordinated cell shape changes and movements that collectively deform tissues. As discussed above, individual cell shape changes and movements are driven by actomyosin-dependent subcellular forces. However, forces must be transmitted between cells and integrated across the whole tissue to result in coherent changes in tissue architecture. AJs by their nature mechanically couple cells and might play a role in the integration of mechanical forces at the tissue level. Recent studies provide examples of cell–cell coordination and tissue level integration of mechanical forces mediated by AJs.

Rosettes form during cell–cell intercalation in the *Drosophila* embryo when myosin II enrichment in vertical junctions spans multiple junctions in so-called myosin II cables.[85,87,88] Cable formation could result from feedback amplification of tension between adjacent vertical junctions mediated by their mechanical coupling. Indeed, tensile vertical junctions might guide the reorientation of neighboring junctions and stabilize myosin II there.[88]

During mesoderm invagination in *Drosophila* embryos, isotropically distributed contractile forces in epithelial cells drive apical constriction. However, the constriction of apical surfaces is not perfectly symmetrical (isotropic) and cells during invagination appear elongated along the anterior–posterior embryonic axis. This is due to a higher global tissue tension in this direction, which is transmitted to individual cells by supracellular actomyosin cables connected between cells by AJs.[98] Therefore, global tissue tension can be transmitted to individual cells via AJs and influence the patterns of their deformations.

Mechanical coordination also exists between different tissues. An example of this is provided by the process of dorsal closure in *Drosophila* embryos. Here, the lateral ectoderm migrates toward the dorsal midline and completely encloses the more dorsal amnioserosa. Recent studies indicate that pulsatile contractions of amnioserosa cells are required for their invagination and for transient dorsal displacements of the ectoderm. These contractions are coordinated with contractions of a supracellular actin cable spanning the entire leading edge of the lateral ectoderm[95] and stabilizing amnioserosa cell deformations. The mechanisms of this coordination are unclear but involve E-cadherin[99] and integrins,[100] most likely because they ensure tension transmission within and between the lateral ectoderm and the amnioserosa.

These data together illustrate the roles of AJs in tissue morphogenesis: they control the spatial patterns of tension, influence the dynamics of actomyosin networks, and integrate contractility of single cells in tissue-wide morphogenetic forces.

6. CONCLUSIONS

In this chapter, we have illustrated how interactions between cadherins and the actin cytoskeleton underlie both the stability and dynamics of AJs.

Cadherin molecules assemble into clusters that are the basic units of adhesion. Regulation of this assembly and of the cluster spatial organization at AJs is the basis of AJ stability and plasticity and is controlled by the actin cytoskeleton. Cadherins regulate the activity and the turnover of the actin cytoskeleton by regulating actin polymerization. In turn, the actin cytoskeleton controls cadherin membrane turnover by endocytosis and exerts forces to modify the organization and the mechanical properties of AJs. The behavior and the dynamics of AJs are therefore intimately linked to those of the actin cytoskeleton and of the contractile actomyosin networks exerting forces on them. This is important particularly during development, when patterns of junction remodeling determine the modes of tissue morphogenesis. Junction remodeling is powered by contractile dynamics of the actomyosin cytoskeleton but anchorage to AJs is a critical factor determining spatial patterns of cells and tissue deformations. This anchorage depends on α-catenin but its exact molecular nature and its dynamics remain elusive. Further work will be necessary to clarify the entire set of molecules involved, their dynamic molecular interactions, and their regulation during the remodeling of cell–cell contacts when forces are applied to junctions.

Cadherin clusters anchor contractile actomyosin networks involved in morphogenesis. A thorough quantitative description of their number, sizes (number of cadherin monomers in a cluster), and spatial localization in a junction and across multiple junctions in a cell will be a significant step forward in understanding how adhesion is regulated in cells. Furthermore, systematic studies comparing different systems and/or conditions (e.g., mutants) will be helpful to understand what regulates the number and size of cadherin clusters in junctions and their spatial distribution in the cell. Integration of the dynamics of anchoring molecules with detailed quantitative morphological description of cadherin clusters at junctions will help us to understand how force patterns originate in cells undergoing morphogenesis.

Finally, it will be important to be able to measure morphogenetic forces directly in developing tissues and relate them to the dynamics of both contractile actomyosin networks and AJs. This will provide insight into the nature of the forces sculpting tissues and their mode of integration at the level of the whole tissue.

ACKNOWLEDGMENTS

We thank members of Lecuit lab for critical discussions. We are particularly grateful to Romain Levayer and Stephen Kerridge for critical reading of this chapter. C. C. is supported by a Long-Term Fellowship from the Human Frontiers Science Program. T. L. is supported by the CNRS and the ANR.

REFERENCES

1. Farquhar MG, Palade GE. Junctional complexes in various epithelia. *J Cell Biol* 1963;**17**:375–412.
2. Rodewald R, Newman SB, Karnovsky MJ. Contraction of isolated brush borders from the intestinal epithelium. *J Cell Biol* 1976;**70**:541–54.
3. Takeichi M. Cadherin cell adhesion receptors as a morphogenetic regulator. *Science* 1991;**251**:1451–5.
4. Kobielak A, Fuchs E. Alpha-catenin: at the junction of intercellular adhesion and actin dynamics. *Nat Rev Mol Cell Biol* 2004;**5**:614–25.
5. Yonemura S. Cadherin-actin interactions at adherens junctions. *Curr Opin Cell Biol* 2011;**23**:515–22.
6. Drees F, Pokutta S, Yamada S, Nelson WJ, Weis WI. Alpha-catenin is a molecular switch that binds E-cadherin-beta-catenin and regulates actin-filament assembly. *Cell* 2005;**123**:903–15.
7. a Yamada S, Pokutta S, Drees F, Weis WI, Nelson WJ. Deconstructing the cadherin-catenin-actin complex. *Cell* 2005;**123**:889–901. b Maiden SL, Hardin J. The secret life of alpha-catenin: moonlighting in morphogenesis. *J Cell Biol* 2011;**195**:543–52.
8. Lecuit T, Lenne PF. Cell surface mechanics and the control of cell shape, tissue patterns and morphogenesis. *Nat Rev Mol Cell Biol* 2007;**8**:633–44.
9. Lecuit T, Lenne PF, Munro E. Force generation, transmission, and integration during cell and tissue morphogenesis. *Annu Rev Cell Dev Biol* 2011;**27**:157–84.
10. Martin AC. Pulsation and stabilization: contractile forces that underlie morphogenesis. *Dev Biol* 2010;**341**:114–25.
11. Meng W, Takeichi M. Adherens junction: molecular architecture and regulation. *Cold Spring Harb Perspect Biol* 2009;**1**:a002899.
12. Tepass U, Hartenstein V. The development of cellular junctions in the Drosophila embryo. *Dev Biol* 1994;**161**:563–96.
13. Muller HA, Wieschaus E. Armadillo, bazooka, and stardust are critical for early stages in formation of the zonula adherens and maintenance of the polarized blastoderm epithelium in Drosophila. *J Cell Biol* 1996;**134**:149–63.
14. Angres B, Barth A, Nelson WJ. Mechanism for transition from initial to stable cell-cell adhesion: kinetic analysis of E-cadherin-mediated adhesion using a quantitative adhesion assay. *J Cell Biol* 1996;**134**:549–57.
15. Adams CL, Chen YT, Smith SJ, Nelson WJ. Mechanisms of epithelial cell-cell adhesion and cell compaction revealed by high-resolution tracking of E-cadherin-green fluorescent protein. *J Cell Biol* 1998;**142**:1105–19.
16. Kametani Y, Takeichi M. Basal-to-apical cadherin flow at cell junctions. *Nat Cell Biol* 2007;**9**:92–8.
17. Harrison OJ, Jin X, Hong S, Bahna F, Ahlsen G, Brasch J, et al. The extracellular architecture of adherens junctions revealed by crystal structures of type I cadherins. *Structure* 2011;**19**:244–56.
18. Hong S, Troyanovsky RB, Troyanovsky SM. Spontaneous assembly and active disassembly balance adherens junction homeostasis. *Proc Natl Acad Sci USA* 2010;**107**:3528–33.
19. Ozaki C, Obata S, Yamanaka H, Tominaga S, Suzuki ST. The extracellular domains of E- and N-cadherin determine the scattered punctate localization in epithelial cells and the cytoplasmic domains modulate the localization. *J Biochem* 2010;**147**:415–25.
20. Kofron M, Spagnuolo A, Klymkowsky M, Wylie C, Heasman J. The roles of maternal alpha-catenin and plakoglobin in the early Xenopus embryo. *Development* 1997;**124**:1553–60.

21. Torres M, Stoykova A, Huber O, Chowdhury K, Bonaldo P, Mansouri A, et al. An alpha-E-catenin gene trap mutation defines its function in preimplantation development. *Proc Natl Acad Sci USA* 1997;**94**:901–6.
22. Vasioukhin V, Bauer C, Degenstein L, Wise B, Fuchs E. Hyperproliferation and defects in epithelial polarity upon conditional ablation of alpha-catenin in skin. *Cell* 2001;**104**: 605–17.
23. Schepis A, Sepich D, Nelson WJ. alphaE-catenin regulates cell-cell adhesion and membrane blebbing during zebrafish epiboly. *Development* 2012;**139**:537–46.
24. Kwiatkowski AV, Maiden SL, Pokutta S, Choi HJ, Benjamin JM, Lynch AM, et al. In vitro and in vivo reconstitution of the cadherin-catenin-actin complex from Caenorhabditis elegans. *Proc Natl Acad Sci USA* 2010;**107**:14591–6.
25. a Abe K, Takeichi M. EPLIN mediates linkage of the cadherin catenin complex to F-actin and stabilizes the circumferential actin belt. *Proc Natl Acad Sci USA* 2008;**105**:13–9. b Taguchi K, Ishiuchi T, Takeichi M. Mechanosensitive EPLIN-dependent remodeling of adherens junctions regulates epithelial reshaping. *J Cell Biol* 2011;**194**:643–56.
26. Yonemura S, Wada Y, Watanabe T, Nagafuchi A, Shibata M. alpha-Catenin as a tension transducer that induces adherens junction development. *Nat Cell Biol* 2010;**12**: 533–42.
27. Maul RS, Song Y, Amann KJ, Gerbin SC, Pollard TD, Chang DD. EPLIN regulates actin dynamics by cross-linking and stabilizing filaments. *J Cell Biol* 2003;**160**: 399–407.
28. Critchley DR. Cytoskeletal proteins talin and vinculin in integrin-mediated adhesion. *Biochem Soc Trans* 2004;**32**:831–6.
29. Grashoff C, Hoffman BD, Brenner MD, Zhou R, Parsons M, Yang MT, et al. Measuring mechanical tension across vinculin reveals regulation of focal adhesion dynamics. *Nature* 2010;**466**:263–6.
30. le Duc Q, Shi Q, Blonk I, Sonnenberg A, Wang N, Leckband D, et al. Vinculin potentiates E-cadherin mechanosensing and is recruited to actin-anchored sites within adherens junctions in a myosin II-dependent manner. *J Cell Biol* 2010;**189**:1107–15.
31. Cavey M, Rauzi M, Lenne PF, Lecuit T. A two-tiered mechanism for stabilization and immobilization of E-cadherin. *Nature* 2008;**453**:751–6.
32. Smutny M, Cox HL, Leerberg JM, Kovacs EM, Conti MA, Ferguson C, et al. Myosin II isoforms identify distinct functional modules that support integrity of the epithelial zonula adherens. *Nat Cell Biol* 2010;**12**:696–702.
33. Shewan AM, Maddugoda M, Kraemer A, Stehbens SJ, Verma S, Kovacs EM, et al. Myosin 2 is a key Rho kinase target necessary for the local concentration of E-cadherin at cell-cell contacts. *Mol Biol Cell* 2005;**16**:4531–42.
34. Adams CL, Nelson WJ, Smith SJ. Quantitative analysis of cadherin-catenin-actin reorganization during development of cell-cell adhesion. *J Cell Biol* 1996;**135**:1899–911.
35. Vasioukhin V, Bauer C, Yin M, Fuchs E. Directed actin polymerization is the driving force for epithelial cell-cell adhesion. *Cell* 2000;**100**:209–19.
36. Vasioukhin V, Fuchs E. Actin dynamics and cell-cell adhesion in epithelia. *Curr Opin Cell Biol* 2001;**13**:76–84.
37. Kobielak A, Pasolli HA, Fuchs E. Mammalian formin-1 participates in adherens junctions and polymerization of linear actin cables. *Nat Cell Biol* 2004;**6**:21–30.
38. Kovacs EM, Goodwin M, Ali RG, Paterson AD, Yap AS. Cadherin-directed actin assembly: E-cadherin physically associates with the Arp2/3 complex to direct actin assembly in nascent adhesive contacts. *Curr Biol* 2002;**12**:379–82.
39. Verma S, Shewan AM, Scott JA, Helwani FM, den Elzen NR, Miki H, et al. Arp2/3 activity is necessary for efficient formation of E-cadherin adhesive contacts. *J Biol Chem* 2004;**279**:34062–70.

40. Braga VM, Betson M, Li X, Lamarche-Vane N. Activation of the small GTPase Rac is sufficient to disrupt cadherin-dependent cell-cell adhesion in normal human keratinocytes. *Mol Biol Cell* 2000;**11**:3703–21.
41. Zandy NL, Playford M, Pendergast AM. Abl tyrosine kinases regulate cell-cell adhesion through Rho GTPases. *Proc Natl Acad Sci USA* 2007;**104**:17686–91.
42. Lampugnani MG, Zanetti A, Breviario F, Balconi G, Orsenigo F, Corada M, et al. VE-cadherin regulates endothelial actin activating Rac and increasing membrane association of Tiam. *Mol Biol Cell* 2002;**13**:1175–89.
43. Yamazaki D, Oikawa T, Takenawa T. Rac-WAVE-mediated actin reorganization is required for organization and maintenance of cell-cell adhesion. *J Cell Sci* 2007;**120**: 86–100.
44. Wildenberg GA, Dohn MR, Carnahan RH, Davis MA, Lobdell NA, Settleman J, et al. p120-catenin and p190RhoGAP regulate cell-cell adhesion by coordinating antagonism between Rac and Rho. *Cell* 2006;**127**:1027–39.
45. Chen X, Macara IG. Par-3 controls tight junction assembly through the Rac exchange factor Tiam1. *Nat Cell Biol* 2005;**7**:262–9.
46. Ratheesh A, Gomez GA, Priya R, Verma S, Kovacs EM, Jiang K, et al. Centralspindlin and alpha-catenin regulate Rho signalling at the epithelial zonula adherens. *Nat Cell Biol* 2012;**14**:818–28.
47. Troyanovsky SM. Regulation of cadherin-based epithelial cell adhesion by endocytosis. *Front Biosci (Schol Ed)* 2009;**1**:61–7.
48. de Beco S, Gueudry C, Amblard F, Coscoy S. Endocytosis is required for E-cadherin redistribution at mature adherens junctions. *Proc Natl Acad Sci USA* 2009;**106**:7010–5.
49. Delva E, Kowalczyk AP. Regulation of cadherin trafficking. *Traffic* 2009;**10**:259–67.
50. Troyanovsky RB, Sokolov EP, Troyanovsky SM. Endocytosis of cadherin from intracellular junctions is the driving force for cadherin adhesive dimer disassembly. *Mol Biol Cell* 2006;**17**:3484–93.
51. Levayer R, Pelissier-Monier A, Lecuit T. Spatial regulation of Dia and Myosin-II by RhoGEF2 controls initiation of E-cadherin endocytosis during epithelial morphogenesis. *Nat Cell Biol* 2011;**13**:529–40.
52. Gavard J, Gutkind JS. VEGF controls endothelial-cell permeability by promoting the beta-arrestin-dependent endocytosis of VE-cadherin. *Nat Cell Biol* 2006;**8**:1223–34.
53. Miyashita Y, Ozawa M. Increased internalization of p120-uncoupled E-cadherin and a requirement for a dileucine motif in the cytoplasmic domain for endocytosis of the protein. *J Biol Chem* 2007;**282**:11540–8.
54. Davis MA, Reynolds AB. Blocked acinar development, E-cadherin reduction, and intraepithelial neoplasia upon ablation of p120-catenin in the mouse salivary gland. *Dev Cell* 2006;**10**:21–31.
55. Ireton RC, Davis MA, van Hengel J, Mariner DJ, Barnes K, Thoreson MA, et al. A novel role for p120 catenin in E-cadherin function. *J Cell Biol* 2002;**159**:465–76.
56. Xiao K, Allison DF, Buckley KM, Kottke MD, Vincent PA, Faundez V, et al. Cellular levels of p120 catenin function as a set point for cadherin expression levels in microvascular endothelial cells. *J Cell Biol* 2003;**163**:535–45.
57. Fujita Y, Krause G, Scheffner M, Zechner D, Leddy HE, Behrens J, et al. Hakai, a c-Cbl-like protein, ubiquitinates and induces endocytosis of the E-cadherin complex. *Nat Cell Biol* 2002;**4**:222–31.
58. Myster SH, Cavallo R, Anderson CT, Fox DT, Peifer M. Drosophila p120catenin plays a supporting role in cell adhesion but is not an essential adherens junction component. *J Cell Biol* 2003;**160**:433–49.
59. Pacquelet A, Lin L, Rorth P. Binding site for p120/delta-catenin is not required for Drosophila E-cadherin function in vivo. *J Cell Biol* 2003;**160**:313–9.

60. Pettitt J, Cox EA, Broadbent ID, Flett A, Hardin J. The Caenorhabditis elegans p120 catenin homologue, JAC-1, modulates cadherin-catenin function during epidermal morphogenesis. *J Cell Biol* 2003;**162**:15–22.
61. Kaido M, Wada H, Shindo M, Hayashi S. Essential requirement for RING finger E3 ubiquitin ligase Hakai in early embryonic development of Drosophila. *Genes Cells* 2009;**14**:1067–77.
62. Kaksonen M, Toret CP, Drubin DG. Harnessing actin dynamics for clathrin-mediated endocytosis. *Nat Rev Mol Cell Biol* 2006;**7**:404–14.
63. Leibfried A, Fricke R, Morgan MJ, Bogdan S, Bellaiche Y. Drosophila Cip4 and WASp define a branch of the Cdc42-Par6-aPKC pathway regulating E-cadherin endocytosis. *Curr Biol* 2008;**18**:1639–48.
64. Georgiou M, Marinari E, Burden J, Baum B. Cdc42, Par6, and aPKC regulate Arp2/3-mediated endocytosis to control local adherens junction stability. *Curr Biol* 2008;**18**: 1631–8.
65. Harris KP, Tepass U. Cdc42 and Par proteins stabilize dynamic adherens junctions in the Drosophila neuroectoderm through regulation of apical endocytosis. *J Cell Biol* 2008;**183**:1129–43.
66. Liu Z, Tan JL, Cohen DM, Yang MT, Sniadecki NJ, Ruiz SA, et al. Mechanical tugging force regulates the size of cell-cell junctions. *Proc Natl Acad Sci USA* 2010;**107**:9944–9.
67. Maruthamuthu V, Sabass B, Schwarz US, Gardel ML. Cell-ECM traction force modulates endogenous tension at cell-cell contacts. *Proc Natl Acad Sci USA* 2011;**108**:4708–13.
68. Ladoux B, Anon E, Lambert M, Rabodzey A, Hersen P, Buguin A, et al. Strength dependence of cadherin-mediated adhesions. *Biophys J* 2010;**98**(4):534–42.
69. Balaban NQ, Schwarz US, Riveline D, Goichberg P, Tzur G, Sabanay I, et al. Force and focal adhesion assembly: a close relationship studied using elastic micropatterned substrates. *Nat Cell Biol* 2001;**3**:466–72.
70. Riveline D, Zamir E, Balaban NQ, Schwarz US, Ishizaki T, Narumiya S, et al. Focal contacts as mechanosensors: externally applied local mechanical force induces growth of focal contacts by an mDia1-dependent and ROCK-independent mechanism. *J Cell Biol* 2001;**153**:1175–86.
71. Geiger B, Spatz JP, Bershadsky AD. Environmental sensing through focal adhesions. *Nat Rev Mol Cell Biol* 2009;**10**:21–33.
72. Haigo SL, Hildebrand JD, Harland RM, Wallingford JB. Shroom induces apical constriction and is required for hingepoint formation during neural tube closure. *Curr Biol* 2003;**13**:2125–37.
73. Hildebrand JD, Soriano P. Shroom, a PDZ domain-containing actin-binding protein, is required for neural tube morphogenesis in mice. *Cell* 1999;**99**:485–97.
74. Hildebrand JD. Shroom regulates epithelial cell shape via the apical positioning of an actomyosin network. *J Cell Sci* 2005;**118**:5191–203.
75. Lee JY, Goldstein B. Mechanisms of cell positioning during C. elegans gastrulation. *Development* 2003;**130**:307–20.
76. Martin AC, Kaschube M, Wieschaus EF. Pulsed contractions of an actin-myosin network drive apical constriction. *Nature* 2009;**457**:495–9.
77. Barrett K, Leptin M, Settleman J. The Rho GTPase and a putative RhoGEF mediate a signaling pathway for the cell shape changes in Drosophila gastrulation. *Cell* 1997;**91**: 905–15.
78. Costa M, Wilson ET, Wieschaus E. A putative cell signal encoded by the folded gastrulation gene coordinates cell shape changes during Drosophila gastrulation. *Cell* 1994;**76**:1075–89.
79. Morize P, Christiansen AE, Costa M, Parks S, Wieschaus E. Hyperactivation of the folded gastrulation pathway induces specific cell shape changes. *Development* 1998; **125**:589–97.

80. Parks S, Wieschaus E. The Drosophila gastrulation gene concertina encodes a G alpha-like protein. *Cell* 1991;**64**:447–58.
81. Kolsch V, Seher T, Fernandez-Ballester GJ, Serrano L, Leptin M. Control of Drosophila gastrulation by apical localization of adherens junctions and RhoGEF2. *Science* 2007;**315**:384–6.
82. Roh-Johnson M, Shemer G, Higgins CD, McClellan JH, Werts AD, Tulu US, et al. Triggering a cell shape change by exploiting preexisting actomyosin contractions. *Science* 2012;**335**:1232–5.
83. Irvine KD, Wieschaus E. Cell intercalation during Drosophila germband extension and its regulation by pair-rule segmentation genes. *Development* 1994;**120**:827–41.
84. Bertet C, Sulak L, Lecuit T. Myosin-dependent junction remodelling controls planar cell intercalation and axis elongation. *Nature* 2004;**429**:667–71.
85. Rauzi M, Verant P, Lecuit T, Lenne PF. Nature and anisotropy of cortical forces orienting Drosophila tissue morphogenesis. *Nat Cell Biol* 2008;**10**:1401–10.
86. Zallen JA, Wieschaus E. Patterned gene expression directs bipolar planar polarity in Drosophila. *Dev Cell* 2004;**6**:343–55.
87. Blankenship JT, Backovic ST, Sanny JS, Weitz O, Zallen JA. Multicellular rosette formation links planar cell polarity to tissue morphogenesis. *Dev Cell* 2006;**11**:459–70.
88. Fernandez-Gonzalez R, Simoes Sde M, Roper JC, Eaton S, Zallen JA. Myosin II dynamics are regulated by tension in intercalating cells. *Dev Cell* 2009;**17**:736–43.
89. Tamada M, Farrell DL, Zallen JA. Abl regulates planar polarized junctional dynamics through beta-catenin tyrosine phosphorylation. *Dev Cell* 2012;**22**:309–19.
90. Simoes Sde M, Blankenship JT, Weitz O, Farrell DL, Tamada M, Fernandez-Gonzalez R, et al. Rho-kinase directs Bazooka/Par-3 planar polarity during Drosophila axis elongation. *Dev Cell* 2010;**19**:377–88.
91. Nishimura T, Takeichi M. Shroom3-mediated recruitment of Rho kinases to the apical cell junctions regulates epithelial and neuroepithelial planar remodeling. *Development* 2008;**135**:1493–502.
92. Nishimura T, Honda H, Takeichi M. Planar cell polarity links axes of spatial dynamics in neural-tube closure. *Cell* 2012;**149**:1084–97.
93. Munro E, Nance J, Priess JR. Cortical flows powered by asymmetrical contraction transport PAR proteins to establish and maintain anterior-posterior polarity in the early C. elegans embryo. *Dev Cell* 2004;**7**:413–24.
94. Rauzi M, Lenne PF, Lecuit T. Planar polarized actomyosin contractile flows control epithelial junction remodelling. *Nature* 2010;**468**:1110–4.
95. Solon J, Kaya-Copur A, Colombelli J, Brunner D. Pulsed forces timed by a ratchet-like mechanism drive directed tissue movement during dorsal closure. *Cell* 2009;**137**:1331–42.
96. He L, Wang X, Tang HL, Montell DJ. Tissue elongation requires oscillating contractions of a basal actomyosin network. *Nat Cell Biol* 2010;**12**:1133–42.
97. Kim HY, Davidson LA. Punctuated actin contractions during convergent extension and their permissive regulation by the non-canonical Wnt-signaling pathway. *J Cell Sci* 2011;**124**:635–46.
98. Martin AC, Gelbart M, Fernandez-Gonzalez R, Kaschube M, Wieschaus EF. Integration of contractile forces during tissue invagination. *J Cell Biol* 2010;**188**:735–49.
99. Gorfinkiel N, Arias AM. Requirements for adherens junction components in the interaction between epithelial tissues during dorsal closure in Drosophila. *J Cell Sci* 2007;**120**:3289–98.
100. Gorfinkiel N, Blanchard GB, Adams RJ, Martinez Arias A. Mechanical control of global cell behaviour during dorsal closure in Drosophila. *Development* 2009;**136**:1889–98.

CHAPTER THREE

Coordinating Rho and Rac: The Regulation of Rho GTPase Signaling and Cadherin Junctions

Aparna Ratheesh, Rashmi Priya, Alpha S. Yap
Division of Molecular Cell Biology, Institute for Molecular Bioscience, The University of Queensland, St. Lucia, Brisbane, Australia

Contents

Abstract

Cadherin-based cell–cell adhesions are dynamic structures that mediate tissue organization and morphogenesis. They link cells together, mediate cell–cell recognition, and influence cell shape, motility, proliferation, and differentiation. At the cellular level, operation of classical cadherin adhesion systems is coordinated with cytoskeletal dynamics, contractility, and membrane trafficking to support productive interactions. Cadherin-based cell signaling is critical for the coordination of these many cellular processes. Here, we discuss the role of Rho family GTPases in cadherin signaling. We focus on understanding the pathways that utilize Rac and Rho in junctional biology, aiming to identify the mechanisms of upstream regulation and define how the effects of these activated GTPases might regulate the actin cytoskeleton to modulate the cellular processes involved in cadherin-based cell–cell interactions.

Progress in Molecular Biology and Translational Science, Volume 116
ISSN 1877-1173
http://dx.doi.org/10.1016/B978-0-12-394311-8.00003-0

1. INTRODUCTION

Cadherin-based cell–cell junctions are sites where multiple processes—adhesion, cytoskeletal dynamics, contractility, and membrane trafficking—cooperate to mediate tissue recognition, cohesion, and patterning.[1] It has long been thought likely that cell signaling helps coordinate these processes.[2,3] Indeed, the past 15 years have seen major progress in identifying a range of junctional signals. These signals include the Rho GTPase family, which are particularly interesting for their capacity to regulate the cytoskeleton and thereby influence many other cellular processes.[4] Indeed, all the prototypical members of this family—Rho, Rac, and Cdc42[1]—have been identified at cadherin adhesions. Here, we review some salient developments in understanding the role of Rho family GTPases in cadherin biology. For clarity, we focus our discussion on understanding the roles of Rho and Rac in cadherin junctions.

2. SOME THINGS TO CONSIDER ABOUT Rho GTPases

Rho family GTPases are guanine nucleotide-binding proteins whose biochemical and biological function depends on their nucleotide status (reviewed in Ref. 4). Loading with GTP induces conformational changes in the GTPase that allow it to interact with downstream effector molecules, whereas the intrinsic GTPase activity of the molecule converts GTP to GDP, which leads to its inactivation. This intrinsic GTPase cycle is, in turn, modulated by activating and inactivating proteins[5] (Fig. 3.1A). Guanine nucleotide exchange factors (GEFs) catalyze GTP loading, thereby activating Rho GTPases, whereas GTPase-activating proteins (GAPs) stimulate the intrinsic GTPase activity of the Rho family proteins, leading to their inactivation. The kinetics of the GTPase cycle is determined by the balance between GEF and GAP activity. A third class of proteins, Rho guanine nucleotide dissociation inhibitors (GDIs), provides a further level of regulation.[6] GDIs bind to the inactive, GDP-bound form of Rho GTPases and also promote the dissociation of Rho proteins from the membrane. Both mechanisms tend to inactivate Rho GTPase signaling.

[1] A point of nomenclature: For clarity, we will refer to the class of Rho family GTPases as "Rho family GTPases," "Rho proteins," or "GTPases." Where we discuss a specific molecule, we will refer to it as "Rho," "Rac," or "Cdc42."

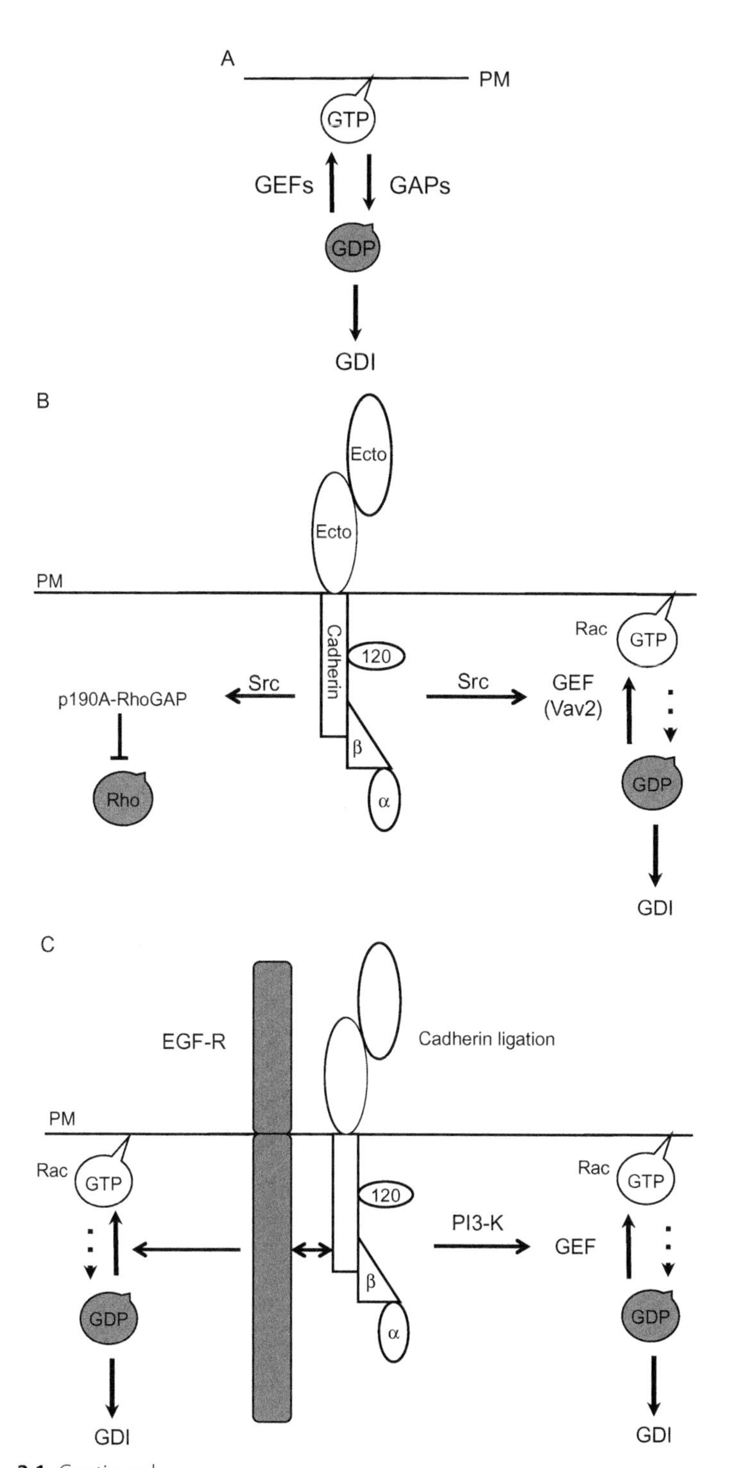

Figure 3.1 Continued

In addition to controlling activity status, regulators of the GTPase cycle can also influence the subcellular localization of Rho GTPases. The subcellular localization of GEFs will determine where Rho GTPases are activated. As well, GAPs can influence the spatial extent of GTPase signaling by limiting the zone of active protein.[7] This implies that the spatio-temporal balance between GEF and GAP activity becomes important not only for regulating the biochemical activity of the GTPase but also for shaping its precise subcellular expression.[7]

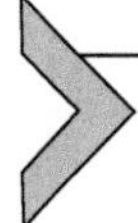

3. REGULATION OF Rho GTPase SIGNALING BY CLASSICAL CADHERINS

The potential involvement of Rho family GTPases in cadherin biology was first suggested when it was shown that manipulating GTPase function perturbed junctional integrity.[8–12] Further, immunostaining for endogenous proteins and expression of epitope-tagged transgenes revealed that Rho, Rac, and Cdc42 accumulate at cell–cell junctions.[13–15] More recently, the use of FRET-based biosensors has demonstrated that the GTP-loaded, active forms of these molecules are found at contacts, implying that they are engaged in signaling.[13–15] However, the presence of signals at junctions does not necessarily mean that they are engaged in cadherin-based cellular processes, as many other junctions[16] and signaling pathways exist at cell–cell contacts.[2] With these considerations in mind, recent studies have identified at least two broad scenarios for cadherin regulation of GTPase signaling at junctions.

Figure 3.1 Regulation of Rho family GTPase activity by classical cadherins. (A) Basics of Rho GTPase regulation. The nucleotide status of Rho family GTPases is regulated by three classes of molecules: guanine nucleotide exchange factors (GEFs), which promote the active, GTP-loaded state; GTPase-activating proteins (GAPs) that promote the inactive, GDP-loaded state; and guanine nucleotide dissociation inhibitors (GDI), which extract GTPases from membranes and sequester them in the inactive state. PM, plasma membrane. (B) A role for Src in Rac activation and cross talk with Rho. One pathway by which cadherin ligation induces Rac activation involves the Src-dependent Rac GEF, Vav2. As cadherin adhesion can activate Src, this may then mediate the downstream activation of Rac. Additionally, Src can inhibit Rho by stimulating p190A-RhoGAP. Ecto, cadherin ectodomains (involved in homophilic ligation); β, β-catenin; α, α-catenin; 120, p120-catenin. (C) Additional mechanisms for Rac activation by cadherins. E-cadherin ligation might activate Rac in two more ways. (Left) A pathway mediated by the EGF receptor (EGF-R) that is independent of EGF itself and may arise from lateral interactions between E-cadherin and EGF-R. (Right) PI3-kinase signaling induced downstream of cadherin ligation activates Rac potentially via a PI3-kinase/PI(3,4,5)P_3-dependent GEF.

3.1. Scenario 1: Activation of signaling in response to cadherin ligation

In this model, adhesive ligation of cadherin initiates a molecular cascade leading to alteration in the activity of the target signaling molecule. This is analogous to established models of signal transduction for hormones, growth factors, and cytokines. Formally, such ligation-activated signaling could lead to increased or decreased activity of the target signal. Indeed, examples of both have been identified for Rho GTPase regulation by cadherins.

3.1.1 Ligation-induced activation of Rac signaling

The concept that cadherin adhesion receptors can activate Rho family GTPase signals is best established for Rac. Biochemical studies demonstrated that cellular GTP-Rac levels rapidly increase when cells are brought into contact with one another.[16–20] Further, imaging studies using FRET biosensors showed that Rac is active in newly forming cadherin contacts.[13] Importantly, contact-induced Rac signaling was blocked when cadherin function was perturbed. For example, adding E-cadherin blocking antibodies prevented the activation of Rac when epithelial cells made contact with one another,[16,17,20] and GTP-Rac levels were lower in endothelial cells lacking VE-cadherin than in cells that possessed the cadherin.[18]

Second, ligation of cellular cadherins by recombinant adhesive ligands, immobilized on microbeads or substrata, was sufficient to activate Rac signaling.[16,19,21] Thus, cadherin ligation could activate GTPase signaling independent of the other events that occur when native cell surfaces make contact with one another. Interestingly, the activation of Rac by cell–cell contact or engagement of cadherin ligands typically occurred as an acute response, reaching a peak minutes (sometimes seconds) after adhesive engagement and subsiding thereafter.[16,21,22] Ligation-activated Rac signaling has been reported for E-cadherin,[16,21,22] *Xenopus* C-cadherin,[19] M-cadherin,[23] and N-cadherin,[24] studied in a variety of cell types. This suggests that it may be a common function of classical cadherins.

Activation of Rac implies that a GEF responds to cadherin adhesion. To date, a number of such candidate Rac GEFs have been identified. These include Trio, a Rac GEF that is necessary for the cadherin-dependent process of myoblast fusion; indeed, Trio is necessary for activation of Rac by muscle cadherin (M-cadherin) in response to adhesive ligation.[23] Similarly, Vav2, which can activate all the canonical Rho GTPases, has been implicated in E-cadherin-dependent Rac signaling in cell culture models.[25] The Rac GEF Tiam-1 has also been implicated in E-cadherin-based signaling because it localizes to cell–cell junctions in a cadherin-dependent fashion[18,26] and also regulates epithelial organization and adherens junctions through control of Rac signaling.[27,28] Overall, it is likely that multiple GEFs can mediate cadherin-dependent Rac activation, depending on the cell type and cadherin involved.

Most GEFs are controlled by upstream factors that influence their subcellular localization and/or their GEF activity.[5] The spatial specificity of junctional Rac signaling implies that mechanisms may exist to localize the activity of its upstream GEFs to cadherin adhesions. One factor may be the cadherin itself. Indeed, Vav2 is recruited to the cortex at sites where E-cadherin molecules are ligated by recombinant ligands,[25] suggesting that cadherin adhesion exerts an instructive signal to recruit Vav2 to the plasma membrane. Further, Trio was identified in a molecular complex with M-cadherin and Rac.[23]

A variety of lipid and protein kinases are also implicated in regulating GEF activity during junctional signaling. For example, E-cadherin can acutely stimulate Src,[29] the activity of which is necessary for Vav2 to activate Rac downstream of cadherin[25] (Fig. 3.1B), perhaps by disrupting an autoinhibited molecular conformation.[5] PI3-kinase signaling (Fig. 3.1C) can also be activated by E-cadherin ligation[21,30] and contributes to Rac stimulation by E-cadherin,[21] perhaps through a PI3-kinase-dependent GEF. In another example, the EGF receptor (EGF-R), a receptor tyrosine kinase, was implicated in mediating Rac activation by cadherin.[17] Interestingly, this effect did not require EGF ligand itself[17]; instead, lateral interactions between E-cadherin and EGF-R might activate EGF-R signaling[31] (Fig. 3.1C). Rac activation in response to cadherin thus involves a network of protein–protein interactions and posttranslational modifications that influence its spatial localization and activity.

In contrast to GEFs, the GAPs responsible for inactivating Rac at junctions have not been extensively investigated. We recently identified MgcRacGAP, which possesses GAP activity for Rac, Cdc42, and Rho,[32] at epithelial cadherin junctions.[14] MgcRacGAP RNAi increased the amount of active, GTP-loaded Rac at the junctions, suggesting that MgcRacGAP might serve to inactivate junctional Rac signaling (Fig. 3.3).

Finally, cadherin-activated Rac signaling can also be modulated by associated proteins other than GEFs or GAPs. This is exemplified by the actin-binding protein Ajuba, which interacts with α-catenin and localizes to cadherin junctions.[33] Ajuba also binds to Rac independent of nucleotide status, implying that it is not a simple Rac effector.[34] Instead, Ajuba RNAi decreased the degree of Rac activation that occurred when cells assembled cadherin adhesions,[34] especially in later phases of adhesion assembly. This suggests that Ajuba serves to fine-tune cadherin-induced Rac signaling during contact formation, consistent with the concept that precise levels of Rac signaling are important for cadherin biology.

3.1.2 Cadherin-inhibited cell signaling: A key role for GAPs in Rac–Rho cross talk

In contrast to Rac, Rho signaling has been reported to be inhibited by cadherin ligation. This was first studied with *Xenopus* C-cadherin. GTP-Rho levels were reduced when cells assembled contacts with one another or when cellular C-cadherin was engaged by immobilized C-cadherin ligands.[19] Consistent with this, reexpression of VE-cadherin in endothelial cells from VE-cadherin knockout mice was associated with reduced steady-state GTP-Rho levels.[18] Together, these observations suggest that cadherin adhesion can dampen Rho signaling, potentially by acting through a RhoGAP.

Indeed, several studies point to a role for p190RhoGAP in this process. Noren *et al.*[35] identified activated p190RhoGAP in lysates from cells stimulated by cadherin adhesion. Further, p190RhoGAP and its associated protein, p120RasGAP, were tyrosine phosphorylated,[35] a posttranslational modification associated with p190RhoGAP activation. As this tyrosine phosphorylation involved Src,[35] one possibility is that cadherin-activated Src[29] might stimulate p190RhoGAP to inhibit Rho signaling (Fig. 3.1B). Here, it is noteworthy that adhesion-induced reduction in GTP-Rho levels coincided with increased Rac activity,[18,19] implying that mechanisms exist to inversely coordinate these two signals at cadherin junctions, as they do in other biological contexts.[36] Taken with the observation that cadherin-activated Src signaling can also stimulate Rac through Vav2,[25] this suggests a model in which Src plays a central role in coordinating the cross talk between Rac and Rho at junctions (Fig. 3.1B).

Other recent findings also place p190RhoGAP at a strategic point in the cross talk between junctional Rac and Rho signaling (Fig. 3.2). These findings show that p190RhoGAP can be activated by Rac signaling by a number of alternative mechanisms. In one pathway, Rac-induced generation of reactive oxygen species (ROS) can inhibit protein tyrosine phosphatases, such as LMW-PTP,[37] leading to increased p190RhoGAP phosphorylation

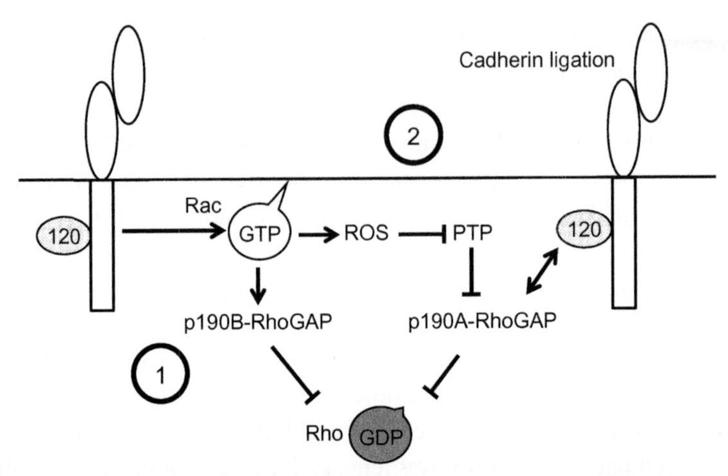

Figure 3.2 Cross-inhibition of Rho signaling by Rac through p190RhoGAP proteins. Pathways for Rac to inhibit Rho via p190RhoGAP. (1) Rac may directly activate p190B-RhoGAP to inhibit Rho. (2) Rac may inhibit p190A-RhoGAP through a pathway that involves reactive oxygen species (ROS) that inhibit the protein tyrosine phosphatases (PTP) responsible for inactivating p190A-RhoGAP. The cadherin-associated protein, p120-catenin (120), may contribute to recruiting p190A-RhoGAP to junctions.

and consequently to inhibition of Rho. Interestingly, overexpression of the cadherin-associated protein p120-catenin (p120-ctn) in fibroblasts can inhibit Rho signaling,[38] which occurs downstream of Rac, through the ROS-dependent activation of p190RhoGAP.[39] These findings led to a model in which p120-ctn coordinates the localization of Rac-activated p190RhoGAP to inhibit Rho at junctions (Fig. 3.2).

A final mechanism is revealed by the fact that there are two p190RhoGAP proteins expressed in mammalian cells. Whereas p190A-RhoGAP is predominantly regulated by protein phosphorylation and is likely to be indirectly responsive to Rac, p190B-RhoGAP can directly bind Rac,[40] a process thought to mediate its recruitment to the cell cortex. Indeed, p190B-RhoGAP can inhibit Rho signaling at cell–cell junctions in a Rac-dependent manner,[14] yielding a third model, in which Rac inhibits Rho more directly by recruiting p190B-RhoGAP to junctions (Fig. 3.2). However, p190RhoGAPs are unlikely to be the only proteins that regulate Rho inactivation at cell–cell junctions. Indeed, myosin IXA, an atypical myosin that possess Rho GAP activity,[41] modulates Rho signaling at newly forming, dynamic cell–cell junctions to influence collective cell migration.[42] These findings emphasize the richness of signaling networks that regulate Rho GTPase activity at cell–cell junctions.

3.2. Scenario 2: Cadherins as platforms for coordinating cell signaling

Acute cadherin-dependent inhibition of Rho does not, however, explain the frequent presence of active GTP-Rho at cell–cell contacts, particularly in mature junctions,[13,14] such as the epithelial zonula adherens (ZA).[14] Further, cadherin adhesion and junctional integrity in established monolayers is perturbed when Rho signaling is blocked.[14,43,44] Recent efforts to elucidate the mechanisms that support Rho signaling at cadherin junctions suggest a model for cadherin-based signaling where cadherin adhesions serve as platforms that coordinate a network of signaling molecules to regulate Rho (Fig. 3.3). This is exemplified by the effect of microtubules on epithelial cadherin junctions. Blocking the dynamic activity of microtubule plus ends without depolymerizing the whole microtubule lattices disrupted junctional integrity[45] and diminished Rho signaling at the junctions.[14] This suggests that dynamic microtubules participate in supporting Rho signaling at cadherin junctions, notably at the epithelial ZA.

Analysis of the underlying mechanism revealed that the centralspindlin complex plays a key role in junctional Rho signaling (Fig. 3.3). Centralspindlin is a heterotetramer consisting of two copies each of a kinesin family protein, MKLP1 (KIF23) and MgcRacGAP. It is best known as a regulator of cytokinesis where it controls Rho signaling by localizing the Rho GEF, Ect2, to the division furrow. However, during interphase, Ect2 is also

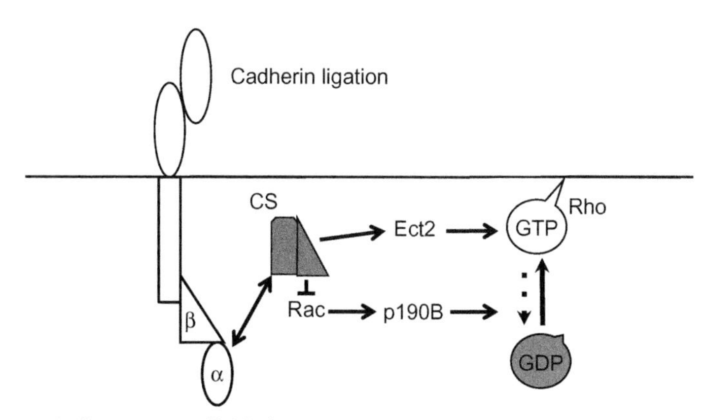

Figure 3.3 Cadherins as scaffolds for coordinating Rho signaling: a model for active Rho signaling at cadherin junctions. Centralspindlin (CS) is recruited to cadherin junctions by an interaction with α-catenin. Centralspindlin recruits the Rho GEF, Ect2, to activate Rho. Centralspindlin coordinately inhibits Rac signaling to prevent the junctional localization of p190B-RhoGAP (p190B) thereby preventing Rho inactivation. (See Color Insert.)

found at epithelial cell–cell junctions,[46] where it concentrates at the ZA and can, indeed, support Rho signaling at the ZA.[14] Centralspindlin turned out to be necessary for Ect2 localization to the ZA. Further, centralspindlin also inhibited the junctional localization of p190B RhoGAP to further promote junctional Rho signaling.[14] As noted above, this involved the inhibition of Rac at the junctions. Thus, centralspindlin controlled Rho signaling at the junctions by coordinating the precise localization of both a GEF (Ect2) and a GAP (p190B-RhoGAP). Interestingly, centralspindlin also bound to α-catenin, which is necessary for it to localize to the ZA.

Together, these findings imply that the cadherin–catenin complex serves as a cortical scaffold to specify the junctional site for centralspindlin to act. In this model, the cadherin–catenin complex may be necessary to provide the platform for signaling to occur, but Rho may not necessarily be acutely activated in response to cadherin ligation. Instead, this model may represent a form of coincidence detection, where multiple upstream regulatory molecules are spatially and temporally coordinated for robust Rho signaling to occur.[47,48] Evidence that in *Drosophila* both α-catenin and p120-ctn can bind to Rho[49] suggests that multiple intermolecular interactions can contribute to the localization and activation of Rho at cadherin adhesions.

The concept that the cadherin molecular complex supports a scaffold for signaling also extends to other elements of the Rho signaling pathway. One key effector for Rho is Rho kinase (ROCK), which is necessary for cadherin adhesion, junctional integrity, and for generation of contractile tension at junctions.[14,43,44,50–53] ROCK is present in two isoforms in mammalian cells,[54] and ROCK1 was recently identified as a binding partner for p120-ctn,[55] which was required to incorporate ROCK1 into cadherin junctions.[55] This suggests that components of the cadherin molecular complex may spatially coordinate Rho activation with downstream elements in its signaling pathway.

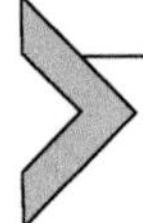

4. FUNCTIONS OF Rho AND Rac IN CADHERIN CELL–CELL INTERACTIONS

Now we turn our attention to examine how Rac and Rho signaling may contribute to cadherin biology. Of note, expression of either dominantly inhibitory or constitutively active Rho GTPase mutants can disrupt cadherin-based cell–cell contacts in cultured epithelial cells.[8–12] This implies that the precise control of signaling intensity and perhaps signaling duration by Rac and Rho is important. These cell culture observations were complemented by organismal analyses that demonstrated that signaling by both of these GTPases contributed to morphogenetic processes involving cadherin-based cell–cell interactions.[56–58] One common denominator is the capacity for Rho GTPases to regulate the actin cytoskeleton. Accordingly, in this section, we focus on discussing how GTPase regulation of the cytoskeleton may ramify to affect the cellular bases of cadherin biology.

4.1. Cadherins junctions, their actin cytoskeleton and regulation by Rac and Rho

It has long been thought that many aspects of classical cadherin biology entail close cooperation between the cadherin adhesion system and the actin cytoskeleton. Moreover, cadherin adhesions recruit many regulators of actin dynamics, organization, and contractility, often in a cadherin-dependent fashion that also involves Rho GTPases.

4.1.1 Junctional actin dynamics

The relationship between cadherins and the cytoskeleton has been most thoroughly examined in epithelia. The mature junctions in epithelia are associated with a well-developed cytoskeleton that contains a large proportion of dynamic actin filaments,[59,60] which undergo continuous turnover.[61,62] Thus, actin must be continuously assembled to replenish the junctional pool of actin filaments. The rate-limiting step in actin filament assembly is nucleation, which has been found to occur at cadherin adhesions.[60–64] Nucleation is mediated by specific molecules, including the highly conserved multiprotein Arp2/3 complex[65] and members of the formin family,[66] many of which can be regulated by Rho family GTPases. Arp2/3 activity is typically activated indirectly in response to either Rac or Cdc42, commonly mediated by the WASP/WAVE family of proteins.[67] In contrast, many members of the formin family possess GTPase-binding domains that allow their activation by GTP-Rho.[66] Thus, Rac can stimulate actin nucleation via Arp2/3, while Rho may activate formin-mediated nucleation. Both of these nucleators have been implicated in cadherin function.

The Arp2/3 complex is found at cadherin junctions in epithelial cells,[53,61,63,64] and inhibiting Arp2/3 substantially reduced actin nucleation and steady-state F-actin at junctions.[53,61,68] Further, dominant-negative Rac mutants reduce steady-state F-actin levels[12,69] and inhibit actin assembly at epithelial cell–cell junctions.[10] While both N-WASP and WAVE are found at cadherin junctions,[61,69] the principal mediator of Rac signaling appears to be WAVE2.[69] The responsiveness of WAVE proteins to Rac is mediated by the multimeric WAVE-regulatory complex,[67] and components of this complex were also found at cadherin junctions, as might be predicted for junctional WAVE to respond to GTP-Rac.[69] Thus, Rac-activated actin assembly mediated by WAVE and Arp2/3 appears to be a common feature of cadherin-based junctions in a variety of epithelial cells.

In contrast to Arp2/3, the formin family consists of many members,[66] two of which—formin-1 and mDia—have been identified at epithelial cell–cell junctions.[70,71] Both were found to contribute to junctional integrity, but their precise effects on junctional actin dynamics were not tested, and in other studies, small molecule inhibition of formin activity did not substantively affect junctional actin assembly.[64] Similarly, the reported effect of Rho inhibition on the junctional actin cytoskeleton is variable, being reduced in some studies[10] but not in others.[14] Thus, in contrast to its effect on myosin II (see below), the role of Rho and its potential effectors on junctional actin assembly is less clear. It is very likely that cell type and context may influence the contribution of Rho-dependent actin regulators to junctional filament dynamics.

4.1.2 Contractility at the junctions

The junctional actin cytoskeleton also accumulates nonmuscle myosin II (which we will refer to as myosin, for short). This is most readily evident at the ZA of polarized epithelial cells,[14,43,44,53,72] but myosins are also found at the cadherin junctions of cells that do not necessarily form a ZA.[73]

Nonmuscle myosin II is regulated by a number of signaling pathways, the best understood being serine–threonine phosphorylation of the regulatory myosin light chain (RMLC), which activates the motor function of the protein. Rho signaling activates myosin II through its effector ROCK, which can directly phosphorylate the RMLC and can also promote its phosphorylation indirectly by inhibiting myosin phosphatase.[74] Indeed, Rho signaling regulates junctional myosin. Immunofluorescence analysis using antibodies directed against the phosphorylated RMLC demonstrated that junctional myosin is active.[43,53] When either Rho or ROCK is inhibited, RMLC phosphorylation is diminished and myosin levels are reduced at cadherin junctions.[43,44] Homophilic ligation of E-cadherin by immobilized cadherin ligands is sufficient to activate myosin II and recruit it to adhesion sites, and this process is blocked by inhibitors of ROCK.[43] Further, inhibition of GEFs, such as Ect2, that support junctional Rho signaling also reduces junctional myosin II.[14] Thus, myosin II is a downstream effector of Rho signaling at cadherin junctions.

4.2. Effects of GTPase-regulated cytoskeleton on cadherin function

How, then, could cytoskeletal regulation contribute to cadherin biology? In this final section, we focus on three aspects of cadherin cell biology that illustrate the influence of the GTPase-regulated actomyosin cytoskeleton (Fig. 3.4).

4.2.1 Cortical organization of cadherins at the cell surface

Classical cadherins do not distribute uniformly at cell–cell contacts, organizing instead into clusters at adhesion zones.[75,76] Further, the apical ZA can be regarded as a chain of clusters where the surface mobility of cadherins is further stabilized.[14] Clustering is a mechanism to strengthen adhesion,[77] and the higher-order accumulation of clusters into a ZA may further amplify adhesion at cortical sites subjected to strong contractile forces during morphogenetic processes such as apical constriction.[78]

The ability of cells to organize cadherins into clusters probably involves multiple molecular mechanisms, including interactions between cadherin ectodomains. However, stable clusters in cells appear to require the cadherin cytoplasmic domain,[77] pointing to a possible contribution from the junctional cytoskeleton. Indeed, components of the cortical actin cytoskeleton coaccumulate with cadherin clusters,[79] and blocking either ROCK or myosin-II activity reduces clustering and decreases cadherin adhesion.[43,44] The actomyosin cytoskeleton also influences the higher-order organization of cadherin into the ZA in epithelial cells. Arp2/3[61] and WAVE2[69] are

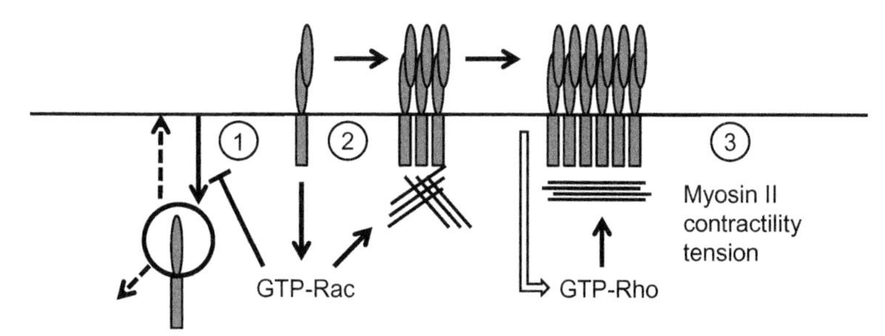

Figure 3.4 Potential effect of Rac and Rho signaling on the molecular mechanisms of cadherin junctions. Cadherin-activated Rac signaling may inhibit cadherin endocytosis (1) and induce junctional actin assembly (2). Rho activated at the junctional membrane (open arrow) signaling activates myosin II and promotes contractility and junctional tension (3).

necessary for ZA integrity, suggesting a role for regulated actin dynamics. This is likely coordinated with Rho-dependent myosin activation, as blocking Rho, its upstream activator Ect2, or myosin IIA disrupted ZA integrity and perturbed the surface stability of E-cadherin in a similar fashion.[14] GTPase-regulated actomyosin may then contribute both to small-scale cadherin clustering and to its larger scale organization into specialized junctions, like the ZA.

4.2.2 Regulation of cadherin trafficking

Like other integral membrane proteins, cadherin adhesion molecules pass through a complex network of membrane trafficking pathways, including biosynthetic exocytosis and endocytosis for recycling or degradation.[80] For cadherins, an important regulatory step occurs at endocytosis, the initial point of entry of cadherin into the cytoplasm.[80] Adhesive ligation can inhibit E-cadherin endocytosis[81,82] through a process that requires Rac and Cdc42 signaling activated by E-cadherin adhesive ligation.[82] Further, actin filament integrity and IQGAP, a Rac effector and actin-binding protein,[83] were implicated in endocytic inhibition.[82] These findings suggest that acute activation of Rac upon cadherin adhesion may stabilize E-cadherin expression on the cell surface by inhibiting its endocytosis. This is one potential explanation for the disruption of junctional integrity observed when Rac signaling is inhibited.[12]

Analysis of how Rac may influence cadherin trafficking also highlights the complexity that arises when GTPases act at multiple points in a trafficking network. Cadherin junctions can be disrupted when Rac is overactive in cells,[8] as well as when it is inhibited. Frasa *et al.*[84] recently identified a novel Rac effector, Armus, which is a GAP for the endocytic regulator Rab7. Armus supported the Rac-induced downregulation of E-cadherin by inhibiting Rab7, thereby promoting the lysosomal degradation of cadherin. Interestingly, Armus was present most prominently on intracellular vesicles, a localization that was supported by Rac.[84] This suggests that its principal site of action may be on endosomes, where it would support a Rac signaling pathway independent of a cadherin-activated pathway. Indeed, Armus appeared to participate in a Rac pathway downstream of the ARF6 GTPase,[84] which, when activated, negatively regulates junctions.[85] This reinforces the notion that the effect of Rac signaling on cadherin depends critically on the pathway in which it participates and where within the cell that pathway is active.

4.2.3 Controlling mechanical forces at cell–cell junctions

At cell–cell junctions, cells receive, generate, and integrate contractile forces.[86] The ability to resist force is essential for maintaining tissue integrity, and there is increasing evidence that cells actively respond to tensile force by reinforcing adhesion.[87,88] One manifestation of contractility is the generation of tension in junctions, as revealed by the retraction that occurs when junctions in live cells are cut with high-power lasers.[14,52] Junctional tension is reduced when myosin II or ROCK are inhibited, consistent with the role of myosin as a major contractile generator within cells.[14,52] Further, tension was reduced in cells where the Rho signaling apparatus was perturbed, either by directly inhibiting Rho or where the junctional Rho activator, Ect2, was depleted.[14] This implies that a local Rho signaling pathway supports junctional tension in the ZA, which may serve to stabilize the movement of junctions themselves.[89] By virtue of its capacity to support clustering, Rho signaling may then coordinate adhesive strengthening with contractility at cadherin junctions.

5. CONCLUDING REMARKS

In this chapter, we have focused on the pathways that might utilize Rho and Rac signaling at cadherin junctions. We have reviewed upstream signaling pathways that likely activate signaling by these GTPases at cadherin adhesions and considered how local regulation of the cytoskeleton may influence cellular processes that determine cadherin biology. Clearly, there is much still to do. In particular, there is a need to develop and apply tools that locally manipulate GTPase function in order to analyze the effect of its local action at cadherin adhesions. These signaling molecules are also embedded in complex patterns of cross talk that make analysis and interpretation challenging. Nonetheless, it is possible to discern functional modules (such as Rac-induced actin assembly and Rho-dependent contractility) within this network of interactions. We suggest that these functional modules provide a framework to elucidate the systems-level function of Rho GTPase signaling in cadherin biology.

ACKNOWLEDGMENTS

We thank our laboratory colleagues for many thoughtful discussions. The work of the authors was funded by the Human Frontiers Science Program and the National Health and Medical Research Council of Australia (APP1010489). A. S. Y. is a Research Fellow of the NHMRC (631383).

REFERENCES

1. Niessen C, Leckband D, Yap AS. Tissue organization by classical cadherin adhesion molecules: dynamic molecular and cellular mechanisms of morphogenetic regulation. *Physiol Rev* 2011;**91**:691–731.
2. Yap AS, Kovacs EM. Direct cadherin-activated cell signaling: a view from the plasma membrane. *J Cell Biol* 2003;**160**:11–6.
3. Wheelock MJ, Johnson KR. Cadherin-mediated cellular signaling. *Curr Opin Cell Biol* 2003;**15**:509–14.
4. Jaffe AB, Hall A. Rho GTPases: biochemistry and biology. *Annu Rev Cell Dev Biol* 2005;**21**:247–69.
5. Bos JL, Rehmann H, Wittinghofer A. GEFs and GAPs: critical elements in the control of small G proteins. *Cell* 2007;**129**:865–77.
6. DerMardirossian C, Bokoch GM. GDIs: central regulatory molecules in Rho GTPase activation. *Trends Cell Biol* 2005;**15**:356–63.
7. Miller AL, Bement WM. Regulation of cytokinesis by Rho GTPase flux. *Nat Cell Biol* 2009;**11**:71–7.
8. Braga VM, Betson M, Li X, Lamarche-Vane N. Activation of the small GTPase Rac is sufficient to disrupt cadherin-dependent cell-cell adhesion in normal human keratinocytes. *Mol Biol Cell* 2000;**11**:3703–21.
9. Braga VM, Del Maschio A, Machesky L, Dejana E. Regulation of cadherin function by Rho and Rac: modulation of junction maturation and cellular context. *Mol Biol Cell* 1999;**10**:9–22.
10. Braga VMM, Machesky LM, Hall A, Hotchin NA. The small GPTases rho and rac are required for the formation of cadherin-dependent cell-cell contacts. *J Cell Biol* 1997;**137**:1421–31.
11. Jou TS, Nelson WJ. Effects of regulated expression of mutant RhoA and Rac1 small GTPases on the development of epithelial (MDCK) cell polarity. *J Cell Biol* 1998;**142**:85–100.
12. Takaishi K, Sasaki T, Kotani H, Nishioka H, Takai Y. Regulation of cell-cell adhesion by Rac and Rho small G proteins in MDCK cells. *J Cell Biol* 1997;**139**:1047–59.
13. Yamada S, Nelson WJ. Localized zones of Rho and Rac activities drive initiation and expansion of epithelial cell-cell adhesion. *J Cell Biol* 2007;**178**:517–27.
14. Ratheesh A, Gomez GA, Priya R, Verma S, Kovacs EM, Jiang K, et al. Centralspindlin and alpha-catenin regulate Rho signalling at the epithelial zonula adherens. *Nat Cell Biol* 2012;**14**:818–28.
15. Terry SJ, Zihni C, Elbediwy A, Vitiello E, Leefa Chong San IV, Balda MS, et al. Spatially restricted activation of RhoA signalling at epithelial junctions by p114RhoGEF drives junction formation and morphogenesis. *Nat Cell Biol* 2011;**13**:159–66.
16. Kitt KN, Nelson WJ. Rapid suppression of activated Rac1 by cadherins and nectins during de novo cell-cell adhesion. *PLoS One* 2011;**6**:e17841.
17. Betson M, Lozano E, Zhang J, Braga VM. Rac activation upon cell-cell contact formation is dependent on signaling from the epidermal growth factor receptor. *J Biol Chem* 2002;**277**:36962–9.
18. Lampugnani MG, Zanetti A, Breviario F, Balconi G, Orsenigo F, Corada M, et al. VE-cadherin regulates endothelial actin activating Rac and increasing membrane association of Tiam. *Mol Biol Cell* 2002;**13**:1175–89.
19. Noren NK, Niessen CM, Gumbiner BM, Burridge K. Cadherin engagement regulates Rho family GTPases. *J Biol Chem* 2001;**276**:33305–8.
20. Nakagawa M, Fukata M, Yamagawa M, Itoh M, Kaibuchi K. Recruitment and activation of Rac1 by the formation of E-cadherin-mediated cell-cell adhesion sites. *J Cell Sci* 2001;**114**:1829–38.

21. Kovacs EM, Ali RG, McCormack AJ, Yap AS. E-cadherin homophilic ligation directly signals through Rac and PI3-kinase to regulate adhesive contacts. *J Biol Chem* 2002;**277**:6708–18.
22. Perez TD, Tamada M, Sheetz MP, Nelson WJ. Immediate-early signaling induced by E-cadherin engagement and adhesion. *J Biol Chem* 2008;**283**:5014–22.
23. Charrasse S, Comunale F, Fortier M, Portales-Casamar E, Debant A, Gauthier-Rouviere C. M-cadherin activates Rac1 GTPase through the Rho-GEF trio during myoblast fusion. *Mol Biol Cell* 2007;**18**:1734–43.
24. Lambert M, Padilla F, Mege RM. Immobilized dimers of N-cadherin-Fc chimera mimic cadherin-mediated cell contact formation: contribution of both outside-in and inside-out signals. *J Cell Sci* 2000;**113**:2207–19.
25. Fukuyama T, Ogita H, Kawakatsu T, Inagaki M, Takai Y. Activation of Rac by cadherin through the c-Src-Rap1-phosphatidylinositol 3-kinase-Vav2 pathway. *Oncogene* 2006; **25**:8–19.
26. Kraemer A, Goodwin M, Verma S, Yap AS, Ali RG. Rac is a dominant regulator of cadherin-directed actin assembly that is activated by adhesive ligation independently of Tiam1. *Am J Physiol Cell Physiol* 2007;**292**:C1061–C1069.
27. Hordijk PL, ten Klooster JP, van der Kammen RA, Michiels F, Oomen LC, Collard JG. Inhibition of invasion of epithelial cells by Tiam1-Rac signaling. *Science* 1997;**278**: 1464–6.
28. Malliri A, van Es S, Huveneers S, Collard JG. The Rac exchange factor Tiam1 is required for the establishment and maintenance of cadherin-based adhesions. *J Biol Chem* 2004;**279**:30092–8.
29. McLachlan RW, Kraemer A, Helwani FM, Kovacs EM, Yap AS. E-cadherin adhesion activates c-Src signaling at cell-cell contacts. *Mol Biol Cell* 2007;**18**:3214–23.
30. Pece S, Chiariello M, Murga C, Gutkind JS. Activation of the protein kinase Akt/PKB by the formation of E-cadherin-mediated cell-cell junctions. *J Biol Chem* 1999;**274**: 19347–51.
31. Pece S, Gutkind JS. Signaling from E-cadherins to the MAPK pathway by the recruitment and activation of epidermal growth factor receptors upon cell-cell contact formation. *J Biol Chem* 2000;**275**:41227–33.
32. Jantsch-Plunger V, Gonczy P, Romano A, Schnabel H, Hamill D, Schnabel R, et al. CYK-4: a Rho family GTPase activating protein (GAP) required for central spindle formation and cytokinesis. *J Cell Biol* 2000;**149**:1391–404.
33. Marie H, Pratt SJ, Betson M, Epple H, Kittler JT, Meek L, et al. The LIM protein Ajuba is recruited to cadherin-dependent cell junctions through an association with alpha-catenin. *J Biol Chem* 2003;**278**:1220–8.
34. Nola S, Daigaku R, Smolarczyk K, Carstens M, Martin-Martin B, Longmore G, et al. Ajuba is required for Rac activation and maintenance of E-cadherin adhesion. *J Cell Biol* 2011;**195**:855–71.
35. Noren NK, Arthur WT, Burridge K. Cadherin engagement inhibits RhoA via p190RhoGAP. *J Biol Chem* 2003;**278**:13615–8.
36. Sanz-Moreno V, Gadea G, Ahn J, Paterson H, Marra P, Pinner S, et al. Rac activation and inactivation control plasticity of tumor cell movement. *Cell* 2008;**135**:510–23.
37. Nimnual AS, Taylor LJ, Bar-Sagi D. Redox-dependent downregulation of Rho by Rac. *Nat Cell Biol* 2003;**5**:236–41.
38. Anastasiadis PZ, Moon SY, Thoreson MA, Mariner DJ, Crawford HC, Zheng Y, et al. Inhibition of RhoA by p120 catenin. *Nat Cell Biol* 2000;**2**:637–44.
39. Wildenberg GA, Dohn MR, Carnahan RH, Davis MA, Lobdell NA, Settleman J, et al. p120-catenin and p190RhoGAP regulate cell-cell adhesion by coordinating antagonism between Rac and Rho. *Cell* 2006;**127**:1027–39.
40. Bustos RI, Forget MA, Settleman JE, Hansen SH. Coordination of Rho and Rac GTPase function via p190B RhoGAP. *Curr Biol* 2008;**18**:1606–11.

41. Abouhamed M, Grobe K, San IV, Thelen S, Honnert U, Balda MS, et al. Myosin IXa regulates epithelial differentiation and its deficiency results in hydrocephalus. *Mol Biol Cell* 2009;**20**:5074–85.
42. Omelchenko T, Hall A. Myosin-IXA regulates collective epithelial cell migration by targeting RhoGAP activity to cell-cell junctions. *Curr Biol* 2012;**22**:278–88.
43. Shewan AM, Maddugoda M, Kraemer A, Stehbens SJ, Verma S, Kovacs EM, et al. Myosin 2 is a key Rho kinase target necessary for the local concentration of E-Cadherin at cell-cell contacts. *Mol Biol Cell* 2005;**16**:4531–2.
44. Smutny M, Cox HL, Leerberg JM, Kovacs EM, Conti MA, Ferguson C, et al. Myosin II isoforms identify distinct functional modules that support integrity of the epithelial zonula adherens. *Nat Cell Biol* 2010;**12**:696–702.
45. Stehbens SJ, Paterson AD, Crampton MS, Shewan AM, Ferguson C, Akhmanova A, et al. Dynamic microtubules regulate the local concentration of E-cadherin at cell-cell contacts. *J Cell Sci* 2006;**119**:1801–11.
46. Liu XF, Ishida H, Raziuddin R, Miki T. Nucleotide exchange factor ECT2 interacts with the polarity protein complex Par6/Par3/protein kinase Czeta (PKCzeta) and regulates PKCzeta activity. *Mol Cell Biol* 2004;**24**:6665–75.
47. Carlton JG, Cullen PJ. Coincidence detection in phosphoinositide signaling. *Trends Cell Biol* 2005;**15**:540–7.
48. Di Paolo G, De Camilli P. Phosphoinositides in cell regulation and membrane dynamics. *Nature* 2006;**443**:651–7.
49. Magie CR, Pinto-Santini D, Parkhurst SM. Rho1 interacts with p120ctn and alpha-catenin, and regulates cadherin-based adherens junction components in Drosophila. *Development* 2002;**129**:3771–82.
50. Nishimura T, Takeichi M. Shroom3-mediated recruitment of Rho kinases to the apical cell junctions regulates epithelial and neuroepithelial planar remodeling. *Development* 2008;**135**:1493–502.
51. Hildebrand JD. Shroom regulates epithelial cell shape via the apical positioning of an actomyosin network. *J Cell Sci* 2005;**118**:5191–203.
52. Fernandez-Gonzalez R, Simoes Sde M, Roper JC, Eaton S, Zallen JA. Myosin II dynamics are regulated by tension in intercalating cells. *Dev Cell* 2009;**17**:736–43.
53. Ivanov AI, Hunt D, Utech M, Nusrat A, Parkos CA. Differential roles for actin polymerization and a myosin II motor in assembly of the epithelial apical junctional complex. *Mol Biol Cell* 2005;**16**:2636–50.
54. Riento K, Ridley AJ. Rocks: multifunctional kinases in cell behaviour. *Nat Rev Mol Cell Biol* 2003;**4**:446–56.
55. Smith AL, Dohn MR, Brown MV, Reynolds AB. Association of Rho-associated protein kinase 1 with E-cadherin complexes is mediated by p120-catenin. *Mol Biol Cell* 2012;**23**:99–110.
56. Woolner S, Jacinto A, Martin P. The small GTPase Rac plays multiple roles in epithelial sheet fusion—dynamic studies of Drosophila dorsal closure. *Dev Biol* 2005; **282**:163–73.
57. Chihara T, Kato K, Taniguchi M, Ng J, Hayashi S. Rac promotes epithelial cell rearrangement during tracheal tubulogenesis in Drosophila. *Development* 2003;**130**: 1419–28.
58. Xu N, Keung B, Myat MM. Rho GTPase controls invagination and cohesive migration of the Drosophila salivary gland through Crumbs and Rho-kinase. *Dev Biol* 2008; **321**:88–100.
59. Cavey M, Rauzi M, Lenne PF, Lecuit T. A two-tiered mechanism for stabilization and immobilization of E-cadherin. *Nature* 2008;**453**:751–6.
60. Zhang J, Betson M, Erasmus J, Zeikos K, Bailly M, Cramer LP, et al. Actin at cell-cell junctions is composed of two dynamic and functional populations. *J Cell Sci* 2005;**118**: 5549–62.

61. Kovacs EM, Verma S, Ali RG, Ratheesh A, Hamilton NA, Akhmanova A, et al. N-WASP regulates the epithelial junctional actin cytoskeleton through a non-canonical post-nucleation pathway. *Nat Cell Biol* 2011;**13**:934–43.
62. Yamada S, Pokutta S, Drees F, Weis WI, Nelson WJ. Deconstructing the cadherin-catenin-actin complex. *Cell* 2005;**123**:889–901.
63. Kovacs EM, Goodwin M, Ali RG, Paterson AD, Yap AS. Cadherin-directed actin assembly: E-cadherin physically associates with the Arp2/3 complex to direct actin assembly in nascent adhesive contacts. *Curr Biol* 2002;**12**:379–82.
64. Tang VW, Brieher WM. alpha-Actinin-4/FSGS1 is required for Arp2/3-dependent actin assembly at the adherens junction. *J Cell Biol* 2012;**196**:115–30.
65. Pollard TD, Blanchoin L, Mullins RD. Molecular mechanisms controlling actin filament dynamics in nonmuscle cells. *Annu Rev Biophys Biomol Struct* 2000;**29**:545–76.
66. Goode BL, Eck MJ. Mechanism and function of formins in the control of actin assembly. *Annu Rev Biochem* 2007;**76**:593–627.
67. Insall RH, Machesky LM. Actin dynamics at the leading edge: from simple machinery to complex networks. *Dev Cell* 2009;**17**:310–22.
68. Verma S, Shewan AM, Scott JA, Helwani FM, Elzen NR, Miki H, et al. Arp2/3 activity is necessary for efficient formation of E-cadherin adhesive contacts. *J Biol Chem* 2004;**279**:34062–70.
69. Yamazaki D, Oikawa T, Takenawa T. Rac-WAVE-mediated actin reorganization is required for organization and maintenance of cell-cell adhesion. *J Cell Sci* 2007; **120**:86–100.
70. Carramusa L, Ballestrem C, Zilberman Y, Bershadsky AD. Mammalian diaphanous-related formin Dia1 controls the organization of E-cadherin-mediated cell-cell junctions. *J Cell Sci* 2007;**120**:3870–82.
71. Kobielak A, Pasolli HA, Fuchs E. Mammalian formin-1 participates in adherens junctions and polymerization of linear actin cables. *Nat Cell Biol* 2004;**6**:21–30.
72. Ivanov AI, Bachar M, Babbin BA, Adelstein RS, Nusrat A, Parkos CA. A unique role for nonmuscle myosin heavy chain IIA in regulation of epithelial apical junctions. *PLoS One* 2007;**2**:e658.
73. Krendel M, Gloushankova NA, Bonder EM, Feder HH, Vasiliev JM, Gelfand IM. Myosin-dependent contractile activity of the actin cytoskeleton modulates the spatial organization of cell-cell contacts in cultured epitheliocytes. *Proc Natl Acad Sci USA* 1999;**96**:9666–70.
74. Vicente-Manzanares M, Ma X, Adelstein RS, Horwitz AR. Non-muscle myosin II takes centre stage in cell adhesion and migration. *Nat Rev Mol Cell Biol* 2009; **10**:778–90.
75. Kametani Y, Takeichi M. Basal-to-apical cadherin flow at cell junctions. *Nat Cell Biol* 2007;**9**:92–8.
76. Harrison OJ, Jin X, Hong S, Bahna F, Ahlsen G, Brasch J, et al. The extracellular architecture of adherens junctions revealed by crystal structures of type I cadherins. *Structure* 2011;**19**:244–56.
77. Yap AS, Brieher WM, Pruschy M, Gumbiner BM. Lateral clustering of the adhesive ectodomain: a fundamental determinant of cadherin function. *Curr Biol* 1997;**7**:308–15.
78. Sawyer JM, Harrell JR, Shemer G, Sullivan-Brown J, Roh-Johnson M, Goldstein B. Apical constriction: a cell shape change that can drive morphogenesis. *Dev Biol* 2010;**341**:5–19.
79. Angres B, Barth A, Nelson WJ. Mechanism for transition from initial to stable cell-cell adhesion: kinetic analysis of E-cadherin-mediated adhesion using a quantitative adhesion assay. *J Cell Biol* 1996;**134**:549–57.
80. Bryant DM, Stow JL. The ins and outs of E-cadherin trafficking. *Trends Cell Biol* 2004;**14**:427–34.
81. Le TL, Yap AS, Stow JL. Recycling of E-cadherin: a potential mechanism for regulating cadherin dynamics. *J Cell Biol* 1999;**146**:219–32.

82. Izumi G, Sakisaka T, Baba T, Tanaka S, Morimoto K, Takai Y. Endocytosis of E-cadherin regulated by Rac and Cdc42 small G proteins through IQGAP1 and actin filaments. *J Cell Biol* 2004;**166**:237–48.
83. White CD, Erdemir HH, Sacks DB. IQGAP1 and its binding proteins control diverse biological functions. *Cell Signal* 2012;**24**:826–34.
84. Frasa MA, Maximiano FC, Smolarczyk K, Francis RE, Betson ME, Lozano E, et al. Armus is a Rac1 effector that inactivates Rab7 and regulates E-cadherin degradation. *Curr Biol* 2010;**20**:198–208.
85. Palacios F, Tushir JS, Fujita Y, D'Souza-Schorey C. Lysosomal targeting of E-cadherin: a unique mechanism for the down-regulation of cell-cell adhesion during epithelial to mesenchymal transitions. *Mol Cell Biol* 2005;**25**:389–402.
86. Gomez GA, McLachlan RW, Yap AS. Productive tension: force-sensing and homeostasis of cell-cell junctions. *Trends Cell Biol* 2011;**21**:499–505.
87. Liu Z, Tan JL, Cohen DM, Yang MT, Sniadecki NJ, Ruiz SA, et al. Mechanical tugging force regulates the size of cell-cell junctions. *Proc Natl Acad Sci USA* 2010;**107**:9944–9.
88. Maruthamuthu V, Sabass B, Schwarz US, Gardel ML. Cell-ECM traction force modulates endogenous tension at cell-cell contacts. *Proc Natl Acad Sci USA* 2011;**108**:4708–13.
89. Smutny M, Wu SK, Gomez GA, Mangold S, Yap AS, Hamilton NA. Multicomponent analysis of junctional movements regulated by myosin II isoforms at the epithelial zonula adherens. *PLoS One* 2011;**6**:e22458.

Versatility of the Cadherin Superfamily

CHAPTER FOUR

New Insights into the Evolution of Metazoan Cadherins and Catenins

Paco Hulpiau, Ismail Sahin Gul, Frans van Roy
Department for Molecular Biomedical Research, VIB & Department of Biomedical Molecular Biology, Ghent University, Ghent, Belgium

Contents

Abstract

E-Cadherin and β-catenin are the best studied representatives of the superfamilies of transmembrane cadherins and intracellular armadillo catenins, respectively. However, in over 600 million years of multicellular animal evolution, these two superfamilies have diversified remarkably both structurally and functionally. Although their basic building blocks, respectively, the cadherin repeat domain and the armadillo repeat domain, predate metazoans, the specific and complex domain compositions of the different family members and their functional roles in cell adhesion and signaling appear to be key features for the emergence of multicellular animal life. Basal animals such as placozoans and sponges have a limited number of distinct cadherins and catenins. The origin of vertebrates, in particular, coincided with a large increase in the number of cadherins and armadillo proteins, including modern "classical" cadherins, protocadherins, and plakophilins. Also, α-catenins increased. This chapter introduces the many different family members and describes the putative evolutionary relationships between them.

Progress in Molecular Biology and Translational Science, Volume 116
ISSN 1877-1173
http://dx.doi.org/10.1016/B978-0-12-394311-8.00004-2

1. CADHERINS IN MULTICELLULAR ANIMALS

1.1. General structure and nomenclature of cadherins

Since the discovery of E-cadherin (CDH1), by far the most extensively studied cadherin and considered the prototype, over a hundred different cadherin and cadherin-related members have been identified. By definition, cadherins are calcium-dependent membrane proteins with an ectodomain containing at least two consecutive extracellular cadherin (EC) repeats. Each EC repeat is about 110 amino acids (AA) long and contains an immunoglobulin-like fold consisting of seven β strands forming two β sheets. Some cadherins have additional domains in their extracellular part, such as laminin-G (lamG) and epidermal growth factor (EGF)-like domains, which are located between the aminoterminal EC repeats and the membrane-spanning region. The N-terminal end of the ectodomain is typically responsible for homophilic or heterophilic interactions. The intracellular regions of the cadherins and cadherin-related molecules are less conserved, except for some binding motifs shared within specific cadherin families. For example, two catenin-binding motifs in classical cadherins bind the armadillo proteins β-catenin and p120ctn (see below). Classification of all types of cadherins is always based on one or more of the above-mentioned features.

Phylogenetic analysis of the N-terminal EC repeats has been the most widely used method for identifying families and subfamilies within the cadherin superfamily. An example of such an analysis is shown in Fig. 4.1, which displays a tree of most of the 114 human and 17 fruit fly cadherins (modified after Ref. 1). Two major branches can be distinguished: the cadherin major branch (CMB) and the cadherin-related major branch (CrMB).

The cadherin nomenclature has not always been straightforward (see Fig. 4.1 for examples). For instance, cadherin-1, officially named CDH1 and best known as E-cadherin (epithelial cadherin), was first described as L-CAM (liver cell adhesion molecule) in chicken and as uvomorulin in mouse. The cadherin-related family members 1 and 2 (CDHR1 and CDHR2) were originally named, respectively, protocadherin 21 (PCDH21) and protocadherin 24 (PCDH24), yet they are not protocadherins but cadherin-related family members. For clarity, we refer to cadherins and cadherin-related molecules by their official nomenclature but mention common names and aliases within brackets.

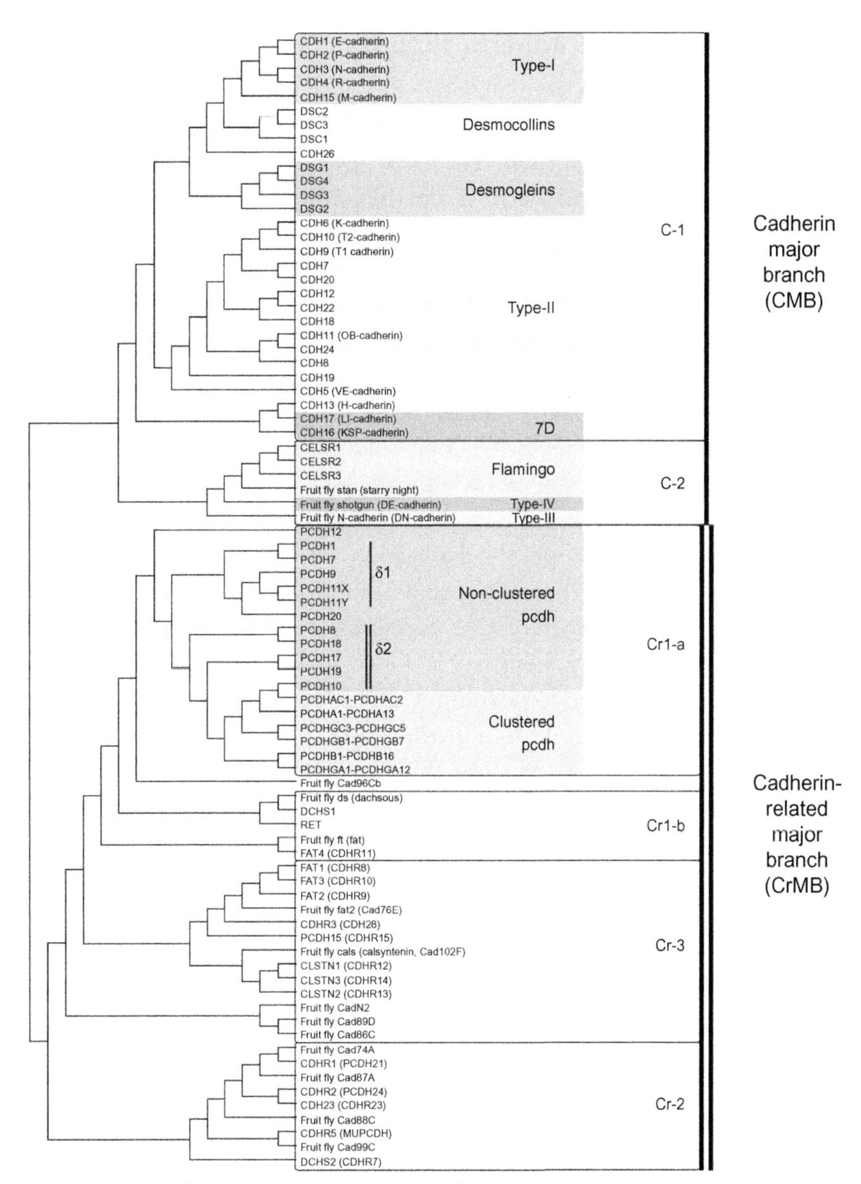

Figure 4.1 *Classification of cadherin superfamily members into six distinctive families and several branching subfamilies.*[1] The rectangular cladogram shown is based on phylogenetic analysis of the first cadherin domain of all human and fruit fly cadherin superfamily members. (See Color Insert.)

1.2. The CMB of the cadherin superfamily

The CMB consists of two large sub-branches, C-1 and C-2, and unites eight subfamilies and a few individual members (Fig. 4.1). Sub-branch C-1 comprises the type-I or classical cadherins, the type-II or atypical cadherins, desmocollins, desmogleins, and the 7D-cadherin family. All C-1 family members have five EC repeats, except for the 7D cadherins CDH16 (KSP-cadherin) and CDH17 (LI-cadherin). The latter have two additional EC repeats originating from duplication of the first two EC repeats. There are five type-I cadherins: CDH1 (E-cadherin, epithelial), CDH2 (N-cadherin, neuronal), CDH3 (P-cadherin, placental), CDH4 (R-cadherin, retinal), and CDH15 (M-cadherin, myotubule). Desmocollins, and desmogleins, together with the armadillo repeat proteins plakoglobin and plakophilins (see below), are important molecular components of desmosomes. In man, there are three desmocollins (DSC1–3) and four desmogleins (DSG1–4). Mouse has two more desmoglein types because specific duplications of desmoglein 1 led to the emergence of Dsg1a, Dsg1b, and Dsg1c. Cadherin 26 (CDH26) is a remarkable, rather isolated member of C-1 cadherins, and its specific function is unknown. The type-II cadherin subfamily contains 13 members: CDH5 (VE-cadherin, vascular endothelium), CDH6 (K-cadherin, fetal kidney), CDH7, CDH8, CDH9 (T1-cadherin, testis), CDH10 (T2-cadherin, testis), CDH11 (OB-cadherin, osteoblast), CDH12, CDH18, CDH19, CDH20, CDH22, and CDH24.

The three first subfamilies of the C-1 sub-branch are positioned close to each other while the type-II cadherins are in a separate sub-branch. One feature that can help explain this evolutionary difference is the adhesion mechanism. Desmocollins, desmogleins, type-I cadherins, and CDH26 have one tryptophan residue (Trp2 or W2) in the N-terminal adhesion arm of the first EC repeat (EC1). This Trp2 is involved in *trans* interaction with the EC1 of a cadherin on the opposite cell surface.[2] In contrast, all type-II cadherins have two conserved tryptophans (W2 and W4) responsible for formation of a strand swap dimer.[3] Consequently, the hydrophobic pocket on the opposite molecule occupied by these two tryptophans is larger in type-II cadherins. CDH13 (T-cadherin/truncated or H-cadherin/heart) and the 7D cadherins CDH16 and CDH17 are the remaining members of the C-1 family. These three cadherins do not have a typical tryptophan in the N-terminus of the first EC. Cadherin 16 and 17, however, might use the single conserved tryptophan in the beginning of their third EC repeat because the first two domains are duplications.

A heterophilic *trans* interaction is possible between LI-cadherin (CDH17) and E-cadherin (CDH1).[4] Heterophilic interactions have also been described between the members of the type-II subfamily.[5] This suggests that, besides the typical homophilic interactions of cadherins, heterophilic interactions can occur between C-1 family members containing one tryptophan in their N-terminal adhesion arm, or between those with two tryptophans. Although T-cadherin (CDH13) is a member of the C-1 family, it is exceptional in that it lacks the typical transmembrane domain present in all other cadherins. Instead, T-cadherin is attached to the cell surface by a glycosylphosphatidylinositol (GPI) anchor. Moreover, instead of using a strand-swapping adhesion mechanism, it forms an X-shaped dimer through an interface near the first and second EC repeats (EC1-EC2).[6]

The second cadherin sub-branch forming the C-2 family contains the type-III, type-IV, and flamingo cadherins. Several features distinguish them from the C-1 family: none of the cadherins in the C-2 family have a tryptophan in their EC1 repeat, they have more than five EC repeats, and they have additional domains in the extracellular region. The absence of strand-swapping tryptophan residues suggests that adhesion is mediated by a different mechanism from the C-1 family members. The type-III cadherins are exemplified by fruit fly shotgun (shg, DN-cadherin), *Caenorhabditis elegans* HMR-1 (see Chapter 11), sea urchin G-cadherin, chicken Hz-cadherin and the cHz-like cadherin in, respectively, zebrafish and the frog *Xenopus*. They have at least 13 consecutive EC repeats followed by a "primitive classic cadherin domain" (PCCD) and the transmembrane domain. This PCCD includes cysteine-rich EGF-like and LamG domains.[7] In mammals, no type-III or type-IV family members exist because of the evolutionary events discussed later in this chapter. Ectodomains of type-IV cadherins contain seven EC repeats and a PCCD. The prime example is *Drosophila* E-cadherin (DE-cadherin). All other known type-IV subfamily members are also from arthropods. Like most cadherins from the C-1 family, type-III and IV cadherins have in their cytoplasmic domain conserved motifs typical for interaction with proteins of the armadillo family. Flamingo/CELSR cadherins (see Chapter 9) have nine EC repeats followed by a flamingo box, alternating EGF-like and LamG domains, and preceding the transmembrane region a hormone receptor domain and a latrophilin/CL-like GPS domain.[8] They have a seven-transmembrane domain analogous to G-protein-coupled receptors and unlike the single transmembrane region shared by all other cadherins. Mammals typically have three paralogs,

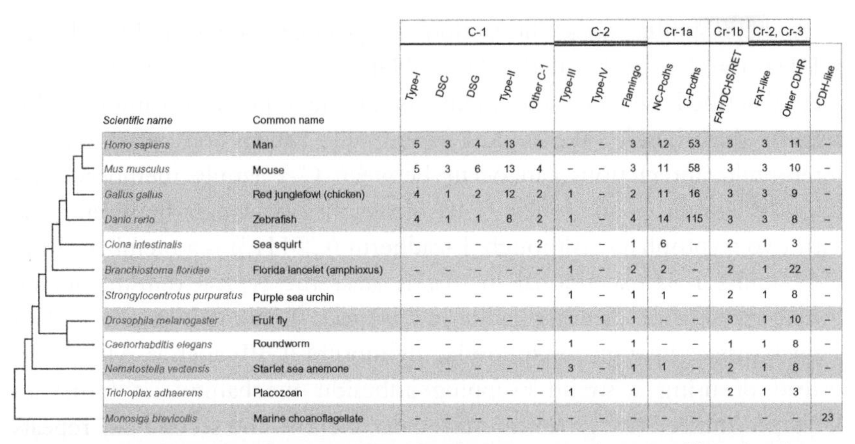

Scientific name	Common name	C-1					C-2			Cr-1a		Cr-1b	Cr-2, Cr-3		
		Type-I	DSC	DSG	Type-II	Other C-1	Type-III	Type-IV	Flamingo	NC-Pcdhs	C-Pcdhs	FAT/DCHS/RET	FAT-like	Other CDHR	CDH-like
Homo sapiens	Man	5	3	4	13	4	–	–	3	12	53	3	3	11	–
Mus musculus	Mouse	5	3	6	13	4	–	–	3	11	58	3	3	10	–
Gallus gallus	Red junglefowl (chicken)	4	1	2	12	2	1	–	2	11	16	3	3	9	–
Danio rerio	Zebrafish	4	1	1	8	2	1	–	4	14	115	3	3	8	–
Ciona intestinalis	Sea squirt	–	–	–	–	2	–	–	1	6	–	2	1	3	–
Branchiostoma floridae	Florida lancelet (amphioxus)	–	–	–	–	–	1	–	2	2	–	2	1	22	–
Strongylocentrotus purpuratus	Purple sea urchin	–	–	–	–	–	1	–	1	1	–	2	1	8	–
Drosophila melanogaster	Fruit fly	–	–	–	–	–	1	1	1	–	–	3	1	10	–
Caenorhabditis elegans	Roundworm	–	–	–	–	–	1	–	1	–	–	1	1	8	–
Nematostella vectensis	Starlet sea anemone	–	–	–	–	–	3	–	1	1	–	2	1	8	–
Trichoplax adhaerens	Placozoan	–	–	–	–	–	1	–	1	–	–	2	1	3	–
Monosiga brevicollis	Marine choanoflagellate	–	–	–	–	–	–	–	–	–	–	–	–	–	23

Figure 4.2 *Cadherin superfamily member distribution across the animal kingdom (metazoa) and the closest relatives, represented by Monosiga brevicollis.* Modern cadherins (C-1 family) and protocadherins (Cr-1a family) have expanded remarkably in vertebrates.[9] (See Color Insert.)

CELSR1, CELSR2, and CELSR3, whereas one ortholog, called starry night (stan), exists in *Drosophila* (Fig. 4.2).

1.3. The cadherin-related major branch of the cadherin superfamily

The CrMB of cadherin-related superfamily members includes four sub-branches: Cr-1a, Cr-1b, Cr-2, and Cr-3 (Fig. 4.1). With more than 60 different members, protocadherins (Pcdh) form the largest family (Cr-1a), which is subdivided in clustered and non-clustered protocadherins. Clustering refers to the genomic location of genes and not to position in the phylogenetic tree. Non-clustered protocadherins are PCDH12, PCDH20, and the delta-protocadherins (δ-Pcdh). PCDH1, PCDH7, PCDH9, and PCDH11 (X- and Y-linked) have seven EC repeats and form the δ1-protocadherin subfamily. PCDH8, PCDH10, PCDH17, PCDH18, and PCDH19 have six EC repeats like the clustered protocadherins and are called δ2-protocadherins. The δ-Pcdhs also have in their cytoplasmic domain two or three conserved motifs not present in clustered Pcdhs.

More than 50 Pcdh genes in mammalian genomes are organized in three consecutive gene clusters: Pcdhα-, Pcdhβ-, and Pcdhγ-encoding genes[10] (see Chapter 7). The protocadherins in the Pcdhα and Pcdhγ clusters have one variable exon encoding the entire ectodomain with six EC repeats, the transmembrane domain, and part of the cytoplasmic domain, whereas three constant exons encode the rest of the cytoplasmic domain, which is shared by all members of the same subcluster. Pcdhβ genes are single-exon genes. Clustered protocadherins are predominantly expressed in the brain. Mice lacking the PCDH-γ cluster manifest neurological defects indicating the involvement of PCDH-γ genes in synaptic development.[11] The organization of Pcdh gene clusters varies somewhat among the lineages in vertebrates. Duplicated protocadherin clusters in, for instance, fugu and zebrafish contain even more of these genes organized in Pcdhα and Pcdhγ subclusters (Fig. 4.2).

The cadherin-related RET, DCHS1 (Dachsous 1), and FAT4 constitute the Cr-1b sub-branch. RET is a receptor tyrosine kinase containing four EC repeats and an intracellular tyrosine kinase domain.[12] RET can phosphorylate clustered protocadherins, and both RET and DCHS1 are components of a phosphorylation-dependent signaling complex.[13] The ectodomain of DCHS1 consists of a tandem array of 27 EC repeats. *Drosophila* fat was the first identified member of the Fat subfamily[14] and is the ortholog of mammalian FAT4. Their ectodomains have 34 EC repeats followed by a few alternating EGF-like and LamG domains resembling the PCCD of C-2 family members. In *Drosophila*, the Golgi kinase Four-jointed (Fj) regulates the Fat–Dachsous interaction by phosphorylating their extracellular domains (Fat acts as receptor and Dachsous as ligand), and Fat can activate the Hippo signaling pathway[15,16] (see also Chapter 10). Mutational studies in mice have also shown that the ligand–receptor function of Dchs1 and Fat4 is required for signaling in multiple organs during murine development.[17]

Sub-branches Cr-2 and Cr-3 contain many cadherin-related molecules that fit in none of the previous groups. Cr-2 consists of CDHR1 (PCDH21), CDHR2 (PCDH24), CDHR23 (CDH23), CDHR5 (MUPCDH), and DCHS2 (CDHR7) (Fig. 4.1). Cr-3 comprises the calsyntenins, FAT-like proteins (FAT1–3), CDHR3 (CDH28), and CDHR15 (PCDH15). CDHR5 was originally named MUPCDH or μ-protocadherin because it was discovered as a mucin-like gene with an ectodomain containing four EC repeats and a triple-repeated mucin domain.[18] Although CDHR1, CDHR2, and CDHR23 possess 6, 9, and 27 EC repeats, respectively, they contain a peculiar, similarly charged N-terminus that can bind Ca^{2+}. This so-called Ca0 binding site rules out a *trans* interaction in strand-swapping mode as seen in C-1 cadherins.[19] This suggests that members of this Cr-2 subfamily have a common new interaction mechanism, but details on this are lacking. Cadherin-related 23 (CDHR23), commonly named

CDH23, forms tip links in the inner ear by interacting in *trans* with CDHR15, commonly named PCDH15 and having 11 EC repeats.[20] Both are thereby important in hearing, and many mutations in these cadherin-related molecules cause inherited deafness (see also Chapter 16). Calsyntenins (human CLSTN1–3, fruit fly cals; *C. elegans* casy-1) have only two EC repeats and are found in the postsynaptic membrane of excitatory synapses in the central nervous system.[21] They are thought to link extracellular cell adhesion to intracellular calcium signaling and to be important in learning (see Chapter 11).[22] Mammalian FAT1, FAT2, and FAT3 are orthologs of *Drosophila* fat2 (see Chapter 10) and are positioned in Cr-3. They compose the FAT-like cadherin subfamily in contrast to *Drosophila* fat and human FAT4 in Cr-1b. Nonetheless, FAT-like cadherins and FAT cadherins have a similar domain architecture consisting of 34 ECs and 1 LamG domain surrounded by several EGF-like domains.

1.4. Evolution of the cadherin superfamily in metazoans

The sequencing of many metazoan genomes has made it possible to study thoroughly the evolution of complex protein families such as the cadherin superfamily. Especially genomes of basal organisms, such as the sea anemone *Nematostella vectensis* and the primitive placozoan *Trichoplax adhaerens*, can give us clues about the genetic tools possessed by animals nearly 600 million years ago, how bilaterian body plans originated, and how genomes and gene families evolved to give rise to current higher metazoans. In the genome of *T. adhaerens* only eight cadherins were identified.[9] The cnidarian *N. vectensis* has 16 putative cadherin and cadherin-related genes, a number similar to the 17 cadherins in the *Drosophila melanogaster*, the 12 in *Caenorhabditis elegans* and the 14 in the sea urchin *Strongylocentrotus purpuratus* (Fig. 4.2).[9] Higher vertebrates contain many more cadherins, for example, 114 in human and 119 in mouse. These are generated to a large extent by two whole-genome duplications (WGD) during vertebrate evolution; this is known as Ohno's 2R hypothesis.[23] Studying cadherin repertoires in these metazoan genomes indicates that five cadherin types are present in nearly all animals: the classical cadherins, the flamingo/CELSR cadherins, and the cadherin-related Dachsous, FAT, and FAT-like types.[9] Except for classical cadherins, the other four are quite conserved among all metazoans: Dachsous has 27 ECs and no other extracellular domains, FAT, and FAT-like have 34 ECs followed by EGF-like and LamG domains, and the CELSR or flamingo cadherins have 9 ECs preceding the EGF-like and LamG domains and the 7-transmembrane domain. They probably have important and conserved basal functions in all animals.

Classical cadherins have evolved spectacularly in composition and in number. Basal metazoans, predating the bilaterians, usually have only 1 huge ancestral classical cadherin (of type-III) consisting of 32 EC repeats and LamG and EGF-like domains proximal to the transmembrane domain. This resembles the extracellular domain composition of FAT, FAT-like, and to some extent CELSR. One may assume that this kind of classical cadherin, which is not present in non-metazoans, was an essential part of the genetic toolkit of the Urmetazoan, the ancestor of all multicellular animals. Such a classical cadherin in the Urmetazoan probably acquired binding ability to ancestral armadillo proteins from the very beginning on (see also below). This is evidenced by conserved binding motifs in the cytoplasmic domain of classical cadherins ranging from placozoan to mammalians.[9] Comparative analyses of individual cadherin repeats and of consecutive blocks of EC repeats were recently done for classical cadherins from placozoan to human, including the ancestral type-III cadherin of *Trichoplax* and *Nematostella* and representative C-1 and C-2 cadherins of higher organisms. Progressive loss of the N-terminal EC repeats during evolution was proposed.[9] At the dawn of bilaterians, a first reduction in the number of EC repeats took place, as illustrated by "shortened" type-III cadherins found in protostomians and non-vertebrate deuterostomians. In arthropods (protostomians) gene duplication led to another shortened classical-like cadherin represented in the type-IV subfamily. In deuterostomians, the phylum of chordates includes cephalochordates (e.g., lancelet or amphioxus *Branchiostoma floridae*), urochordates (tunicates, e.g., the sea squirt *Ciona intestinalis*), and vertebrates (e.g., mammals, birds, and fishes). The first "short" classical cadherins with five ECs and a classical cytoplasmic domain, which belong to the C-1 family, appeared in chordates just before the origin of vertebrates. Two urochordates, *Botryllus schlosseri* and *C. intestinalis*, have one and two C-1 cadherins, respectively. The two WGD events and individual gene duplications resulted in great expansion and diversification of C-1 cadherins in vertebrates, generating type-I, type-II, and desmosomal cadherins, and in total 29 C-1 cadherins in humans. Remarkable is the loss of the type-III cadherin in almost all mammals, both marsupials and placental mammals, but not in platypus (*Ornithorhynchus anatinus*), an egg-laying mammal (monotreme). The large increase in all these different C-1 cadherins (and catenins; see below) might have contributed much to the molecular complexity of vertebrates by amplifying adhesive and signaling functions.

Protocadherins are even more prevalent in vertebrates, with over 60 Cr-1a family members in humans.[1] As protocadherins are absent in

Drosophila and *C. elegans,* they have long been considered vertebrate specific. However, one non-clustered protocadherin, with the typical seven EC repeats and two conserved motifs in the cytoplasmic domain, is encoded by the genome of the cnidarian *N. vectensis.*[9] Interestingly, also the genome of the mollusk *Aplysia califomia* contains at least one such protocadherin gene, dating the origin of protocadherins before the existence of bilaterians but also confirming its persistence in certain protostomians such as mollusks. Like classical cadherins, protocadherins increased in large numbers in vertebrates by gene and genome duplication, likely adding to the complexity of the vertebrate nervous system in which many of them are expressed.

Other cadherin-related gene products, such as CDHR15 (PCDH15) and CDHR23 (CDH23), which are involved in vertebrate hearing and vision, appeared later in metazoans or only in certain metazoan subphyla. Of the eight *Trichoplax* cadherins, five are found in nearly all other metazoans, while the other three seem to be placozoan specific. Similarly, only 9 of the 16 *Nematostella* cadherins are shared with other animals, whereas the others have not been identified in more modern metazoan lineages such as vertebrates. One such example is the cadherin Hedgling, which has a Hedge domain (Hh) and a von Willebrand factor type A domain (vWA) preceding the extracellular cadherin repeats. This is found in *Nematostella* and in the demosponge *Amphimedon queenslandica.*[24]

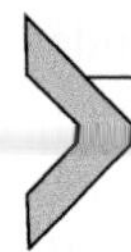

2. CATENINS WITH ARMADILLO DOMAINS

2.1. Structures and functions of armadillo catenins

Armadillo proteins share the presence of similar imperfect tandem repeats, generally composed of about 40 AA and called armadillo (ARM) repeats. ARM repeats were first identified in the *Drosophila* segment polarity protein called armadillo, which is the ortholog of mammalian β-catenin and the namesake of this large protein superfamily.[25] Armadillo proteins may differ much in the number and organization of their ARM repeats, and they are involved in a broad range of biological processes, including cell adhesion, signaling, cytoskeletal regulation, and intracellular transport. Although the sequence similarity between the individual ARM repeats of a single protein may be very low, all these repeats have conserved three-dimensional structures. Invariably, one short and two long α-helices are present per repeat. While the second and third helices are involved in hydrophobic interactions with each other and form an antiparallel motif, the first helix lies perpendicularly to the other two helices.[26] In this way, the three helices of each ARM repeat form a compact helical bundle, and individual repeats fold together as a curved superhelical structure, which allows versatile interactions with many proteins. Compared to small adapter domains such as PDZ or SH3, ARM domains have longer recognition sequences and have high affinities for their binding partners. This enables them to participate in many physiological purposes. Recently, it was shown that the conserved binding mode in ARM repeat proteins can be applied to design general peptide binding scaffolds.[27,28]

The armadillo catenin family contains the first investigated members of the ARM protein superfamily. The name catenin was derived from "catena," the Latin word for chain, and was introduced by Kemler and his colleagues.[29] They proposed that the major function of catenins might be to link proteins coding for Ca^{2+}-dependent cell adhesion molecules (CAMs) to cytoskeletal structures. In particular, they found that uvomorulin (now known as E-cadherin) associates with the catenins.

Based on their sequence homology, the mammalian catenins with ARM repeats can be divided into three subfamilies. Representative members gave the names for the corresponding subfamilies. The human β-catenin subfamily consists of β-catenin (encoded by *CTNNB1*) and plakoglobin (also called γ-catenin, encoded by *JUP*). The p120 subfamily has four members: p120ctn (p120; gene *CTNND1*), ARVCF (armadillo repeat gene deleted in velocardiofacial syndrome; gene *ARVCF*), δ-catenin (gene *CTNND2*), and p0071, also known as plakophilin-4 (gene *PKP4*). The third subfamily comprises the plakophilins: plakophilin-1 (*PKP1*), -2 (*PKP2*), and -3 (*PKP3*).[30] It should be noted that the plakophilins are sometimes considered as p120ctn subfamily members, as they show higher sequence and functional similarity to the p120ctn members than to β-catenin.

β-Catenin has broad and conserved functions. In cell adherens junctions, it bridges the cytoplasmic domain of the classical cadherins to monomeric α-catenin (see also below) and to the actin cytoskeleton.[31,32] In the nucleus, it is involved in the activation of Wnt-triggered signaling by linking TCF/Lef family transcription factors with other transcriptional regulators, such as Bcl-9.[33] In the absence of Wnt signaling, nonjunctional cytoplasmic β-catenin is targeted for degradation.[34] This process is triggered by the phosphorylation of β-catenin at its N-terminus following its association with a large, multiprotein machinery. This machinery includes proteins such as axin and APC (adenomatous polyposis coli protein), which facilitate the addition of phosphate groups to β-catenin, for instance by glycogen synthase kinase-3b (GSK3β).[35] The related subfamily member plakoglobin functions in both the adherens junctions and the desmosomes. Desmosomal plakoglobin serves as a molecular link between the cytoplasmic domains of desmosomal cadherins (desmogleins, desmocollins) and the aminoterminal ends of desmoplakin (see Chapter 5).[36] The carboxyterminal ends of desmoplakin bind to the intermediate filaments in the desmosomal plaque. Although plakoglobin has been detected in the nucleus, it does not trigger TCF/Lef-dependent transcription in response to Wnt signals.[37]

Whereas β-catenin and plakoglobin bind to a conserved motif near the C-terminus of classical cadherins, p120ctn binds to a conserved juxtamembrane domain.[31] p120ctn regulates the abundance of classical cadherins by controlling the rate of cadherin turnover.[38] Another important function of p120ctn is its regulation of Rho-GTPases, which are essential for the dynamic organization of the actin cytoskeleton.[39] Finally, p120ctn is also found in the nucleus, where it inactivates the transcriptional repressor Kaiso and consequently leads to gene activation.[40]

The other p120ctn subfamily members, ARVCF, δ-catenin, and p0071, are also reported to be involved in modulation of cadherin stability at cell–cell junctions.[41] Moreover, they directly or indirectly interact with a number of Rho-GTPases, which makes intracellular signaling and cytoskeletal control possible. The linkage of intermediate filaments to desmosomal cadherins occurs via plakoglobin (see above) and via plakophilins (PKP1–3).[36] Loss of functional PKP1 or PKP2 leads to severe pathologies, whereas ablation of PKP3 causes more minor effects (reviewed in Ref. 42). More recently, plakophilins have been implicated in cell signaling and cytoskeletal organization.

2.2. Evolution of armadillo catenins in metazoans

Human β-catenin comprises 781 AA residues and contains a central structural core consisting of 12 ARM repeats. Investigation of the β-catenin sequence and structure revealed three main domains: the ARM repeat region, an acidic N-terminal region, and a glycine-rich C-terminal region. The ARM domain of β-catenin has a typical curved structure with a positively charged groove, which creates a large binding surface for more than 20 interactors, including E-cadherin.[43,44]

Examining the evolutionary history of β-catenin by using phylogenetic analysis based on protein sequences showed that it was present far back in premetazoan species (see below). However, no β-catenin homolog has been reported in the choanoflagellate *Monosiga brevicollis*, which is a unicellular organism closely related to metazoans. The ancestral metazoan *T. adhaerens* has a β-catenin-related ARM protein.[9] Furthermore, ancestral β-catenin has gone through several lineage-specific duplications. The *C. elegans* genome contains four β-catenin genes known as *hmp-2*, *bar-1*, *wrm-1*, and *sys-1*, which are exceptionally divergent both structurally and functionally[45–47] (see also Chapter 11). In line with an additional event of whole genomic duplication in teleost fishes,[48] a second β-catenin is found in zebrafish and fugu genomes. Beyond that, several independent species-specific duplications of β-catenin have been observed, for instance in the sea anemone *N. vectensis*, the flour beetle *Tribolium castaneum* and the pea aphid *Acyrthosiphon pisum*.[49,50]

Like β-catenin, plakoglobin has 12 ARM repeats. These two closely related molecules share more than 65% global similarity at the protein level and this similarity rises to 80% in the ARM region. Phylogenetic analysis of the β-catenin subfamily showed that β-catenin underwent a duplication event in the chordate lineage, which gave arise to vertebrate β-catenin and plakoglobin (Fig. 4.3). This gene duplication allowed a neo-functionalization of plakoglobin compared to β-catenin, as described above.

Phylogenetic analysis of the complete ARM repeat region of β-catenin implies strong sequence conservation across lineages, in line with functional conservation. Comparison of individual ARM repeats of the β-catenin homologs in Cnidaria, Arthropoda, Echinodermata, and Chordata showed that each ARM repeat has an individual signature that has been conserved throughout metazoan evolution. Generally, the most conserved residues in the ARM repeat regions are involved in interactions with cadherins, axin, or APC.[45] In line with this, it has been shown that the β-catenin-binding domain in the classical cadherins of the cnidarian *N. vectensis* and the placozoan *T. adhaerens* is also well conserved.[9]

Contrary to the ARM repeat region, the N- and C-terminal regions of β-catenin orthologs are less conserved and not likely to form a stable folded structure by themselves. Nonetheless, conserved roles of N- and C-terminal regions of β-catenin are apparent in vertebrate evolution. In the bilaterian lineage, the N-terminal region is more conserved than the C-terminal region and contains the conserved binding motifs for α-catenin, Bcl-9, and GSK-3.[45,51] The C-terminal region contains the binding motif for Tax-interacting protein-1 (Tip-1), which can inhibit β-catenin transcriptional activity.[52] Both the binding site in Tip-1 and the corresponding interaction domain in β-catenin are highly conserved in vertebrates. For the duplicated β-catenins of insects, a partial subfunctionalization has been

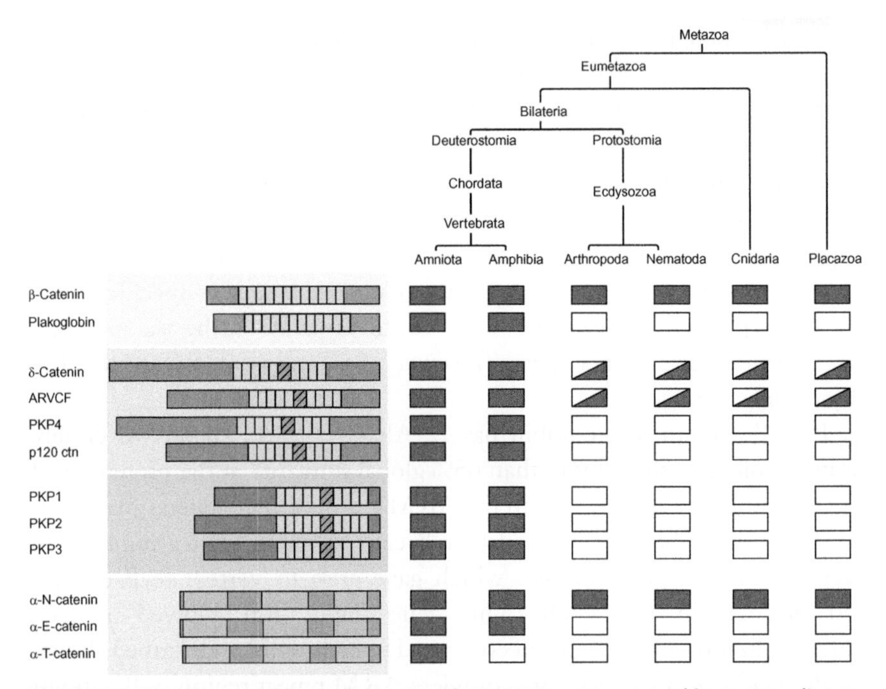

Figure 4.3 *Metazoan evolution of catenins.* The catenins are grouped here according to their subfamilies. ARM repeats are indicated as yellow boxes. Insert regions between the fifth and the sixth armadillo repeat in p120ctn subfamily members and plakophilins are indicated by hatched gray boxes. Vinculin homology domains of α-catenins are indicated by three pink boxes. The cladogram on the top is based on the widely accepted metazoan evolution with the placozoa as the outgroup. The boxes below the cladogram indicate the presence (green box) or absence (white box) of the corresponding catenins in the various lineages. The partly shaded boxes indicate that presence or absence of an ancestral p120ctn subfamily member is unresolved. (See Color Insert.)

reported, as one paralog appeared to have lost α-catenin-binding activity due to extensive changes in the N-terminal domain, whereas both paralogs remained active in Wnt signaling.[49] Although the four β-catenin orthologs in *C. elegans* are overall highly diverged, divergence is greatest in the terminal regions. This resulted in extensive subfunctionalization: HMP-2 binds to the cadherin ortholog HMR-1 and to the α-catenin ortholog HMP-1, whereas BAR-1 and WRM-1 are not present in junctional complexes[45,46] (see Chapter 11). Instead, BAR-1 is active in canonical Wnt signaling, whereas WRM-1 is active in noncanonical Wnt signaling.[46]

Members of the p120ctn and plakophilin subfamilies are structurally more closely related to each other than to the β-catenin subfamily.

Compared to the 12 repeats in β-catenin and plakoglobin, the p120ctn and plakophilin members contain nine ARM repeats and flanking N- and C-terminal regions that diverge substantially from one another. Moreover, they all have a long insert (~61 AA) between the fifth and sixth ARM repeats (Fig. 4.3). Although this region does not affect the packing of the ARM repeats, it may create a major bend in the ARM repeat region.[53]

Alike β-catenin and plakoglobin, all seven members of the p120ctn and plakophilin subfamilies are present in vertebrates (Fig. 4.3). A recent phylogenetic analysis showed that the ancestor of the p120ctn subfamily is a single δ-catenin/ARVCF-like gene present in metazoans but not in the unicellular premetazoan *M. brevicollis*.[54] Unlike plakophilins, the ancestral δ-catenin/ARVCF-like gene can be found outside the vertebrates (Fig. 4.3). The p120ctn in vertebrates probably arose by duplication of the ARVCF gene and p0071 by duplication of the δ-catenin gene. Yet it is not clear which member of the p120ctn subfamily is at the origin of the plakophilins.[50] Significantly, the appearance of plakophilins and plakoglobin coincides with the origin of desmosomes in vertebrates (Fig. 4.3).

3. ALPHA-CATENINS IN METAZOANS

Another catenin family comprises α-catenins and related proteins. Contrary to the armadillo catenins, discussed in the previous section, the α-catenins do not have ARM repeats and do not bind directly to cadherins. The family comprises αE-, αN-, and αT-catenins, which show tissue-restricted expression patterns: E stands for epithelial, N for neural, and T for testis.[55] In humans they are encoded by *CTNNA1*, *CTNNA2*, and *CTNNA3*, respectively. Together with their more distant relatives, vinculin (gene *VCL*) and α-catulin (gene *CTNNAL1*), α-catenins belong to the vinculin superfamily. α-Catenins are involved in cell–cell adhesion by binding to junctional β-catenin, and they also participate in coordinating actin dynamics in adherens junctions.[32,56] In the intercalated discs of cardiomyocytes, αT-catenin serves as a molecular bridge between classic cadherin/catenin complexes and plakophilin-2 linked to desmosomal components.[57] In that way, stress-resistant mixed-type junctions are built.

Contrary to the evolution of β-catenin and p120ctn armadillo proteins, the α-catenins have not experienced extensive duplication events in metazoans. Phylogenetic studies showed that αN-catenin is the ancestor of the other two α-catenins.[50] It gave rise to αE-catenin as a result of the vertebrate-specific whole-genome duplication, and it gave rise to αT-catenin as a result of an amniote-specific gene duplication event. In conclusion, vertebrate α-catenins share a common ancestor with the αN-catenin of non-vertebrates (Fig. 4.3).

4. CADHERINS AND CATENINS IN NON-METAZOANS

4.1. Cadherin-like molecules in non-metazoans

Signaling and cell adhesion mediated by cadherins are essential for generating and controlling multicellularity in the animal kingdom. Genuine cadherins are absent in all non-metazoan organisms such as plants and fungi. Some cadherin-like domains and molecules, however, exist outside the animal kingdom.

The choanoflagellate *M. brevicollis*, a unicellular, colony-forming organism, is one of the closest relatives of metazoans. Although *Monosiga* lacks overt cell adhesion, 23 cadherin-like genes were identified in its genome.[58] The overall domain architecture of most of these encoded proteins is quite different from those of metazoan cadherins. The EC-like domains are sometimes non-consecutive and their calcium-binding motifs are much less conserved. Based on the presence of a Ca^{2+}-free linker in the *Drosophila* N-cadherin (DN-cadherin), similar Ca^{2+}-free linkers are predictable for most choanoflagellate cadherin-like molecules.[59] The domain organizations of the *M. brevicollis* cadherin-like molecules MBCDH1 and MBCDH2 are more similar to those of metazoan cadherins, but a putative cytoplasmic catenin-binding domain could not be identified. Nonetheless, subcellular localization of these two proteins with antibodies raised against their extracellular part reveals their presence in the apical collar of *Monosiga*, where they colocalize with actin.

In the genome of *Pythium ultimum*, an oomycete plant pathogen, four cadherin-like genes were identified.[60] Most of the cadherin-like domains in *Pythium* have orthologous domains in *Phytophthora* species, another oomycete lineage. No orthologs of these genes have been found in other clades closely related to oomycetes, for example, diatoms and alveolates. Based on current knowledge, horizontal gene transfer of a cadherin-like gene from a choanoflagellate or urmetazoan ancestor to an oomycete ancestor is the most plausible hypothesis. Importantly, preservation of EC-like repeats in both *Pythium* and *Phytophthora* points to important functions. On the other hand, none of the oomycete cadherins have a catenin-binding domain.

The soil-dwelling amoeba *Dictyostelium discoideum* expresses both an α-catenin-like and a β-catenin-like protein essential for formation of a peculiar polarized epithelium without the need of any cadherin[61] (see below and Chapter 1). *Dictyostelium* does not contain cadherin or cadherin-like proteins but expresses a peculiar Ca^{2+}-dependent cell adhesion molecule, DdCAD-1, lacking a transmembrane domain and showing only limited sequence similarity to classical cadherins. Its C-terminal domain belongs to the same immunoglobulin-like fold as the extracellular domains of cadherins but the β-strand connectivity is different.[62]

It has been suggested that cadherin-like domains are found in several bacterial proteins, the yeast protein Axl2p, and human dystroglycans and sarcoglycans.[63,64] These predictions are often based on multiple sequence alignments where some similarities to a cadherin domain are found. However, important motifs shared by all true cadherin repeats are missing in these proteins. Additionally, in many cases only one such assumed cadherin domain is present in these proteins whereas most cadherin superfamily members have many consecutive cadherin repeat domains. Structure determination of part of the murine α-dystroglycan confirms that such a domain does not form a cadherin-like fold.[65]

In conclusion, only multicellular animals (metazoans) seem to possess cadherins and cadherin-related molecules. Some non-metazoans, such as choanoflagellates and oomycetes, apparently have cadherin-like proteins probably resembling those of the common ancestor of all animals (Urmetazoan). Future structural and functional studies should address whether these non-metazoan proteins have functions similar to those in animal cadherins today. Some protein domains in other organisms, for example, bacteria and yeasts, show limited sequence similarity to domains of cadherins, cadherin-related or cadherin-like proteins. These domains are often annotated as dystroglycan-type cadherin-like domains but should not be considered as *bona fide* cadherin domains.

4.2. Catenin-like molecules in non-metazoans

The transition from simple eukaryotic unicellular to multicellular organisms requires proteins that can mediate connections between cells. Even a simple polarized epithelium in ancestral metazoan species requires a cell–cell adhesion complex that contains a classical cadherin, a β-catenin homolog, a δ-catenin/ARVCF homolog, and an α-catenin homolog (similar to αN-catenin) (Fig. 4.3). To understand the early origin and evolution of specific cell–cell adhesion, it is important to determine whether cadherins, catenins, and catenin-related proteins are present in unicellular organisms and in multicellular plants.

Candida albicans and *Saccharomyces cerevisiae* are unicellular yeasts, which belong to the phylum Ascomycota. These organisms have a putative homolog of β-catenin, which is called Vac8p and is 22% identical to human β-catenin[66] (Fig. 4.4). Like the ARM domain of β-catenin, its core domain of 11 ARM repeats is involved in many protein–protein interactions.[67] Vac8p protein forms a molecular link between the vacuolar membrane and the actin cytoskeleton.[68] Similarly, metazoan β-catenin links junctional complexes in the plasma membrane to the actin filaments. Nonetheless, many differences are apparent between Vac8p and β-catenin: Vac8p is membrane-associated through acylation, and yeasts contain neither a cadherin homolog nor an α-catenin homolog. Moreover, a cell-signaling function has never been reported for Vac8p. If Vac8p is a true homolog of β-catenin, these proteins have distinctly diverged in function to a great degree.[66]

The non-metazoan but social amoeba *D. discoideum* (slime mold) undergoes multicellular development in response to starvation. Induced

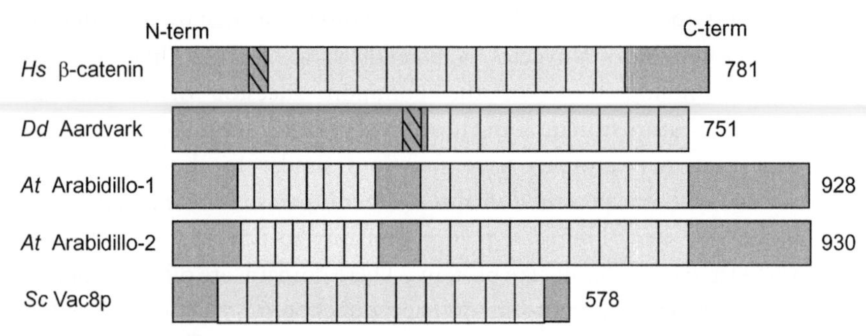

Figure 4.4 *Domain organization of β-catenin and its non-metazoan homologs.* Sequences of consecutive wider boxes indicate the ARM repeats and hatched boxes the conserved α-catenin-binding domains. β-Catenin has 12 ARM repeats; Aardvark and Arabidillos have 9 each; Vac8p has 11. Leucine-rich repeats (LRR) in Arabidillos are indicated by a sequence of more narrow boxes. While Aardvark has a conserved and functional α-catenin-binding domain, its other regions deviate much from those of mammalian β-catenins, and Aardvark is indeed nonfunctional in terms of cadherin association and Wnt signaling. Neither α-catenin homologs nor conserved α-catenin-binding motifs in ARM proteins have been reported in plants or yeast. *Hs, Homo sapiens; Dd, Dictyostelium discoideum; At, Arabidopsis thaliana; Sc, Saccharomyces cerevisiae.* Numbers indicate respective protein lengths in amino acids. N-term, aminoterminal end; C-term, carboxyterminal end. (See Color Insert.)

aggregation of single cells leads to formation of a fruiting body comprising a spore head supported by a single, vertical stalk. Grimson and colleagues showed that the appearance of a constriction near the top of the stalk tube is associated with the occurrence of close intercellular contacts, which are possibly adherens junctions and apparently coupled to actin.[69] Later, it was shown that the subcellular organization of these tip cells has the characteristics of simple polarized epithelium.[61] Identification of the proteins involved in these particular close contacts in the *D. discoideum* fruiting body was of obvious importance in the light of our understanding of the metazoan evolution of cell–cell adhesion (see also Chapter 1). BLAST searches of a *D. discoideum* cDNA database identified a β-catenin-related protein (named Aardvark) as well as an α-catenin homolog (named *Dd*α-catenin).[61,69] Aardvark is 18% identical to mammalian β-catenin. These two proteins share a highly conserved α-catenin-binding helix just in front of the ARM repeat region (Fig. 4.4). Further investigation revealed that *Dd*α-catenin is more similar to metazoan α-catenin than to vinculin. While the monomeric form of *Dd*α-catenin is able to bind F-actin filaments, the monomeric form of vertebrate α-E-catenin lacks this ability.[61] Dickinson and colleagues showed that *Dd*α-catenin can bind to both Aardvark and mouse β-catenin and that its association with cell–cell contacts *in vivo* is Aardvark-dependent.[61] Knockdown experiments showed that both Aardvark and *Dd*α-catenin are essential for the polarized organization of the tip epithelium, but not for the formation of the actin-associated cell–cell junctions. Interestingly, no cadherin-like molecule could be detected in *D. discoideum*, whereas cadherins are essential for cell–cell adhesion in metazoans. At least three other proteins, completely unrelated to cadherins, are involved in cell–cell adhesion events at various stages of *D. discoideum* development,[70] but there is no evidence for their interaction with the catenins discussed above. Despite this molecular dissimilarity to metazoan adherens junctions, it is striking that unicellular amoeboid cells of *D. discoideum* can form actin-rich junctions that are structurally and functionally reminiscent of metazoan junctions.[70]

Neither *M. brevicollis* nor oomycetes have a β-catenin-like catenin. In contrast, the presence in all metazoans and *D. discoideum* of molecular complexes composed of proteins related to armadillo catenins and α-catenins raises the possibility that common functions are mediated by these complexes in all multicellular organisms, including higher plants. In contrast to metazoan cells, plant cells have rigid cell walls located outside the cell membrane and providing strong structural and mechanical support to the cells. They lack classical cell–cell junctions that link one cell membrane directly to another. Instead, they are connected to each other by plasmodesmata, which are sophisticated intercellular transport channels, also possessing actin filaments.[71] BLAST searches based on the sequence of *D. discoideum* Aardvark identified two genes of *Arabidopsis thaliana,* named Arabidillo-1 and Arabidillo-2.[72] Like Aardvark, both of these proteins contain a core of nine ARM repeats[66] (Fig. 4.4). However, Arabidillo proteins also contain leucine-rich repeats (LRRs) in front of the ARM domain.[73] Arabidillo proteins regulate root branching and development, and they turned out to have homologs throughout the plant kingdom but are absent in algae and metazoa.[73] In metazoa, phosphorylated β-catenin is targeted to destruction by interaction with the F-box/WD40 repeat protein β-Trcp (β-transducin repeats-containing protein), which is a substrate-recognizing component of SCF E3 ubiquitin ligase.[74] Also Arabidillo proteins are sensitive to proteasome-mediated degradation, but this was found to be largely dependent on the LRR motif.[73] Interestingly, plants have recently been shown to express multiple proteins with ARM domains, most of which (64 predicted in *Arabidopsis*) are associated with a U-box and may serve as E3 ubiquitin ligases.[75] However, their evolutionary relationship to metazoan armadillo proteins and particularly to armadillo catenins is not at all clear.

ACKNOWLEDGMENTS

We thank A. Bredan for critical reading and editing of the chapter. Research supported by the Research Foundation—Flanders (FWO), by the Geconcerteerde Onderzoeksacties (GOA) of Ghent University and by the Belgian Science Policy (Interuniversity attraction poles – IAP 7/07).

REFERENCES

1. Hulpiau P, van Roy F. Molecular evolution of the cadherin superfamily. *Int J Biochem Cell Biol* 2009;**41**:343–69.
2. Harrison OJ, Jin X, Hong S, Bahna F, Ahlsen G, Brasch J, et al. The extracellular architecture of adherens junctions revealed by crystal structures of type I cadherins. *Structure* 2011;**19**:244–56.
3. Patel SD, Ciatto C, Chen CP, Bahna F, Rajebhosale M, Arkus N, et al. Type II cadherin ectodomain structures: implications for classical cadherin specificity. *Cell* 2006;**124**:1255–68.
4. Baumgartner W, Wendeler MW, Weth A, Koob R, Drenckhahn D, Gessner R. Heterotypic trans-interaction of LI- and E-cadherin and their localization in plasmalemmal microdomains. *J Mol Biol* 2008;**378**:44–54.
5. Shimoyama Y, Tsujimoto G, Kitajima M, Natori M. Identification of three human type-II classic cadherins and frequent heterophilic interactions between different subclasses of type-II classic cadherins. *Biochem J* 2000;**349**:159–67.
6. Ciatto C, Bahna F, Zampieri N, Vansteenhouse HC, Katsamba PS, Ahlsen G, et al. T-cadherin structures reveal a novel adhesive binding mechanism. *Nat Struct Mol Biol* 2010;**17**:339–47.

7. Oda H, Tsukita S. Nonchordate classic cadherins have a structurally and functionally unique domain that is absent from chordate classic cadherins. *Dev Biol* 1999;**216**:406–22.
8. Usui T, Shima Y, Shimada Y, Hirano S, Burgess RW, Schwartz TL, et al. Flamingo, a seven-pass transmembrane cadherin, regulates planar cell polarity under the control of frizzled. *Cell* 1999;**98**:585–95.
9. Hulpiau P, van Roy F. New insights into the evolution of metazoan cadherins. *Mol Biol Evol* 2011;**28**:647–57.
10. Wu Q, Maniatis T. A striking organization of a large family of human neural cadherin-like cell adhesion genes. *Cell* 1999;**97**:779–90.
11. Weiner JA, Wang XZ, Tapia JC, Sanes JR. Gamma protocadherins are required for synaptic development in the spinal cord. *Proc Natl Acad Sci USA* 2005;**102**:8–14.
12. Cabrera JR, Bouzas-Rodriguez J, Tauszig-Delamasure S, Mehlen P. RET modulates cell adhesion via its cleavage by caspase in sympathetic neurons. *J Biol Chem* 2011;**286**:14628–38.
13. Schalm SS, Ballif BA, Buchanan SM, Phillips GR, Maniatis T. Phosphorylation of protocadherin proteins by the receptor tyrosine kinase Ret. *Proc Natl Acad Sci USA* 2010;**107**:13894–9.
14. Mahoney PA, Weber U, Onofrechuk P, Biessmann H, Bryant PJ, Goodman CS. The *fat* tumor suppressor gene in Drosophila encodes a novel member of the cadherin gene superfamily. *Cell* 1991;**67**:853–68.
15. Ishikawa HO, Takeuchi H, Haltiwanger RS, Irvine KD. Four-jointed is a Golgi kinase that phosphorylates a subset of cadherin domains. *Science* 2008;**321**:401–4.
16. Badouel C, McNeill H. SnapShot: the hippo signaling pathway. *Cell* 2011;**145** (484–484):e481.
17. Mao Y, Mulvaney J, Zakaria S, Yu T, Morgan KM, Allen S, et al. Characterization of a Dchs1 mutant mouse reveals requirements for Dchs1-Fat4 signaling during mammalian development. *Development* 2011;**138**:947–57.
18. Goldberg M, Wei M, Tycko B, Falikovich I, Warburton D. Identification and expression analysis of the human mu-protocadherin gene in fetal and adult kidneys. *Am J Physiol Renal Physiol* 2002;**283**:F454–F463.
19. Elledge HM, Kazmierczak P, Clark P, Joseph JS, Kolatkar A, Kuhn P, et al. Structure of the N terminus of cadherin 23 reveals a new adhesion mechanism for a subset of cadherin superfamily members. *Proc Natl Acad Sci USA* 2010;**107**:10708–12.
20. Kazmierczak P, Sakaguchi H, Tokita J, Wilson-Kubalek EM, Milligan RA, Muller U, et al. Cadherin 23 and protocadherin 15 interact to form tip-link filaments in sensory hair cells. *Nature* 2007;**449**:87–91.
21. Hintsch G, Zurlinden A, Meskenaite V, Steuble M, Fink-Widmer K, Kinter J, et al. The calsyntenins—a family of postsynaptic membrane proteins with distinct neuronal expression patterns. *Mol Cell Neurosci* 2002;**21**:393–409.
22. Ikeda DD, Duan Y, Matsuki M, Kunitomo H, Hutter H, Hedgecock EM, et al. CASY-1, an ortholog of calsyntenins/alcadeins, is essential for learning in Caenorhabditis elegans. *Proc Natl Acad Sci USA* 2008;**105**:5260–5.
23. Kasahara M. The 2R hypothesis: an update. *Curr Opin Immunol* 2007;**19**:547–52.
24. Adamska M, Matus DQ, Adamski M, Green K, Rokhsar DS, Martindale MQ, et al. The evolutionary origin of hedgehog proteins. *Curr Biol* 2007;**17**:R836–R837.
25. Riggleman B, Wieschaus E, Schedl P. Molecular analysis of the armadillo locus: uniformly distributed transcripts and a protein with novel internal repeats are associated with Drosophila segment polarity gene. *Genes Dev* 1989;**3**:96–113.
26. Striegl H, Andrade-Navarro MA, Heinemann U. Armadillo motifs involved in vesicular transport. *PLoS One* 2010;**5**:e8991.
27. Parmeggiani F, Pellarin R, Larsen AP, Varadamsetty G, Stumpp MT, Zerbe O, et al. Designed armadillo repeat proteins as general peptide-binding scaffolds: consensus

design and computational optimization of the hydrophobic core. *J Mol Biol* 2008;**376**:1282–304.
28. Madhurantakam C, Varadamsetty G, Grutter MG, Pluckthun A, Mittl PR. Structure based optimization of designed armadillo-repeat proteins. *Protein Sci* 2012;**21**:1015–28.
29. Kemler R, Ozawa M. Uvomorulin-catenin complex: cytoplasmic achorage of a Ca(2+)-dependent cell adhesion molecule. *Bioessays* 1989;**11**:88–91.
30. McCrea PD, Gu D. The catenin family at a glance. *J Cell Sci* 2010;**123**:637–42.
31. van Roy F, Berx G. The cell-cell adhesion molecule E-cadherin. *Cell Mol Life Sci* 2008;**65**:3756–88.
32. Maiden SL, Hardin J. The secret life of alpha-catenin: moonlighting in morphogenesis. *J Cell Biol* 2011;**195**:543–52.
33. van Es JH, Barker N, Clevers H. You Wnt some, you lose some: oncogenes in the Wnt signaling pathway. *Curr Opin Genet Dev* 2003;**13**:28–33.
34. Peifer M, Polakis P. Wnt signaling in oncogenesis and embryogenesis—a look outside the nucleus. *Science* 2000;**287**:1606–9.
35. Behrens J, Jerchow BA, Wurtele M, Grimm J, Asbrand C, Wirtz R, et al. Functional interaction of an axin homolog, conductin, with beta-catenin, APC, and GSK3beta. *Science* 1998;**280**:596–9.
36. Desai BV, Harmon RM, Green KJ. Desmosomes at a glance. *J Cell Sci* 2009;**122**:4401–7.
37. Shimizu M, Fukunaga Y, Ikenouchi J, Nagafuchi A. Defining the roles of beta-catenin and plakoglobin in LEF/T-cell factor-dependent transcription using beta-catenin/plakoglobin-null F9 cells. *Mol Cell Biol* 2008;**28**:825–35.
38. Ireton RC, Davis MA, van Hengel J, Mariner DJ, Barnes K, Thoreson MA, et al. A novel role for p120 catenin in E-cadherin function. *J Cell Biol* 2002;**159**:465–76.
39. Anastasiadis PZ. p120-ctn: a nexus for contextual signaling via Rho GTPases. *Biochim Biophys Acta* 2007;**1773**:34–46.
40. Park JI, Kim SW, Lyons JP, Ji H, Nguyen TT, Cho KC, et al. Kaiso/p120-catenin and TCF/beta-catenin complexes coordinately regulate canonical Wnt gene targets. *Dev Cell* 2005;**8**:843–54 [corr. in vol. 849, p. 305].
41. McCrea PD, Park JI. Developmental functions of the p120-catenin subfamily. *Biochim Biophys Acta* 2007;**1773**:17–33.
42. Neuber S, Muhmer M, Wratten D, Koch PJ, Moll R, Schmidt A. The desmosomal plaque proteins of the plakophilin family. *Dermatol Res Pract* 2010;**2010**:101452.
43. Shapiro L, Weis WI. Structure and biochemistry of cadherins and catenins. *Cold Spring Harb Perspect Biol* 2009;**1**:a003053.
44. Xing Y, Takemaru K, Liu J, Berndt JD, Zheng JJ, Moon RT, et al. Crystal structure of a full-length beta-catenin. *Structure* 2008;**16**:478–87.
45. Schneider SQ, Finnerty JR, Martindale MQ. Protein evolution: structure-function relationships of the oncogene beta-catenin in the evolution of multicellular animals. *J Exp Zool B Mol Dev Evol* 2003;**295**:25–44.
46. Korswagen HC. Canonical and non-canonical Wnt signaling pathways in Caenorhabditis elegans: variations on a common signaling theme. *Bioessays* 2002;**24**:801–10.
47. Liu J, Phillips BT, Amaya MF, Kimble J, Xu W. The C. elegans SYS-1 protein is a bona fide beta-catenin. *Dev Cell* 2008;**14**:751–61.
48. Van de Peer Y, Maere S, Meyer A. The evolutionary significance of ancient genome duplications. *Nat Rev Genet* 2009;**10**:725–32.
49. Bao R, Fischer T, Bolognesi R, Brown SJ, Friedrich M. Parallel duplication and partial subfunctionalization of beta-catenin/armadillo during insect evolution. *Mol Biol Evol* 2012;**29**:647–62.
50. Zhao ZM, Reynolds AB, Gaucher EA. The evolutionary history of the catenin gene family during metazoan evolution. *BMC Evol Biol* 2011;**11**:198.

51. Sampietro J, Dahlberg CL, Cho US, Hinds TR, Kimelman D, Xu WQ. Crystal structure of a beta-catenin/BCL9/Tcf4 complex. *Mol Cell* 2006;**24**:293–300.
52. Zhang J, Yan X, Shi C, Yang X, Guo Y, Tian C, et al. Structural basis of beta catenin recognition by Tax-interacting protein-1. *J Mol Biol* 2008;**384**:255–63.
53. Choi HJ, Weis WI. Structure of the armadillo repeat domain of plakophilin 1. *J Mol Biol* 2005;**346**:367–76.
54. Carnahan RH, Rokas A, Gaucher EA, Reynolds AB. The molecular evolution of the p120-catenin subfamily and its functional associations. *PLoS One* 2010;**5**:e15747.
55. Janssens B, Goossens S, Staes K, Gilbert B, van Hengel J, Colpaert C, et al. alpha-T-Catenin: a novel tissue-specific beta-catenin-binding protein mediating strong cell-cell adhesion. *J Cell Sci* 2001;**114**:3177–88.
56. Kobielak A, Fuchs E. Alpha-catenin: at the junction of intercellular adhesion and actin dynamics. *Nat Rev Mol Cell Biol* 2004;**5**:614–25.
57. Goossens S, Janssens B, Bonné S, De Rycke R, Braet F, van Hengel J, et al. A unique and specific interaction between alpha-T-catenin and plakophilin-2 recruits desmosomal proteins to the adherens junctions of the heart. *J Cell Sci* 2007;**120**:2126–36.
58. Abedin M, King N. The premetazoan ancestry of cadherins. *Science* 2008;**319**:946–8.
59. Jin X, Walker MA, Felsovalyi K, Vendome J, Bahna F, Mannepalli S, et al. Crystal structures of Drosophila N-cadherin ectodomain regions reveal a widely used class of Ca(2+)-free interdomain linkers. *Proc Natl Acad Sci USA* 2011;**109**:E127–134.
60. Levesque CA, Brouwer H, Cano L, Hamilton JP, Holt C, Huitema E, et al. Genome sequence of the necrotrophic plant pathogen Pythium ultimum reveals original pathogenicity mechanisms and effector repertoire. *Genome Biol* 2010;**11**:R73.
61. Dickinson DJ, Nelson WJ, Weis WI. A polarized epithelium organized by beta- and alpha-catenin predates cadherin and metazoan origins. *Science* 2011;**331**:1336–9.
62. Lin Z, Sriskanthadevan S, Huang HB, Siu CH, Yang DW. Solution structures of the adhesion molecule DdCAD-1 reveal new insights into Ca(2+)-dependent cell-cell adhesion. *Nat Struct Mol Biol* 2006;**13**:1016–22.
63. Dickens NJ, Beatson S, Ponting CP. Cadherin-like domains in alpha-dystroglycan, alpha/epsilon sarcoglycan and yeast and bacterial proteins. *Curr Biol* 2002;**12**: R197–R199.
64. Cao LH, Yan XM, Borysenko CW, Blair HC, Wu CQ, Yu L. CHDL: a cadherin-like domain in Proteobacteria and Cyanobacteria. *FEMS Microbiol Lett* 2005; **251**:203–9.
65. Bozic D, Sciandra F, Lamba D, Brancaccio A. The structure of the N-terminal region of murine skeletal muscle alpha-dystroglycan discloses a modular architecture. *J Biol Chem* 2004;**279**:44812–6.
66. Tewari R, Bailes E, Bunting KA, Coates JC. Armadillo-repeat protein functions: questions for little creatures. *Trends Cell Biol* 2010;**20**:470–81.
67. Tang FS, Peng YT, Nau JJ, Kauffman EJ, Weisman LS. Vac8p, an armadillo repeat protein, coordinates vacuole inheritance with multiple vacuolar processes. *Traffic* 2006;**7**:1368–77.
68. Wang YX, Catlett NL, Weisman LS. Vac8p, a vacuolar protein with armadillo repeats, functions in both vacuole inheritance and protein targeting from the cytoplasm to vacuole. *J Cell Biol* 1998;**140**:1063–74.
69. Grimson MJ, Coates JC, Reynolds JP, Shipman M, Blanton RL, Harwood AJ. Adherens junctions and beta-catenin-mediated cell signalling in a non-metazoan organism. *Nature* 2000;**408**:727–31.
70. Abedin M, King N. Diverse evolutionary paths to cell adhesion. *Trends Cell Biol* 2010;**20**:734–42.
71. White RG, Barton DA. The cytoskeleton in plasmodesmata: a role in intercellular transport? *J Exp Bot* 2011;**62**:5249–66.

72. Coates JC. Armadillo repeat proteins: beyond the animal kingdom. *Trends Cell Biol* 2003;**13**:463–71.
73. Nibau C, Gibbs DJ, Bunting KA, Moody LA, Smiles EJ, Tubby JA, et al. ARABIDILLO proteins have a novel and conserved domain structure important for the regulation of their stability. *Plant Mol Biol* 2011;**75**:77–92.
74. Liu CM, Kato Y, Zhang ZH, Do VM, Yankner BA, He X. beta-Trcp couples beta-catenin phosphorylation-degradation and regulates Xenopus axis formation. *Proc Natl Acad Sci USA* 1999;**96**:6273–8.
75. Yee D, Goring DR. The diversity of plant U-box E3 ubiquitin ligases: from upstream activators to downstream target substrates. *J Exp Bot* 2009;**60**:1109–21.

CHAPTER FIVE

Structure, Function, and Regulation of Desmosomes

Andrew P. Kowalczyk[*,†,‡], **Kathleen J. Green**[§,¶,||]

[*]Department of Cell Biology, Emory University School of Medicine, Atlanta, Georgia, USA
[†]Department of Dermatology, Emory University School of Medicine, Atlanta, Georgia, USA
[‡]The Winship Cancer Institute, Emory University School of Medicine, Atlanta, Georgia, USA
[§]Department of Pathology, Northwestern University Feinberg School of Medicine, Chicago, Illinois, USA
[¶]Department of Dermatology, Northwestern University Feinberg School of Medicine, Chicago, Illinois, USA
[||]The Robert H. Lurie Comprehensive Cancer Center, Northwestern University Feinberg School of Medicine, Chicago, Illinois, USA

Contents

Abstract

Desmosomes are adhesive intercellular junctions that mechanically integrate adjacent cells by coupling adhesive interactions mediated by desmosomal cadherins to the intermediate filament cytoskeletal network. Desmosomal cadherins are connected to intermediate filaments by densely clustered cytoplasmic plaque proteins comprising members of the armadillo gene family, including plakoglobin and plakophilins, and members of the plakin family of cytolinkers, such as desmoplakin. The importance of desmosomes in tissue integrity is highlighted by human diseases caused by mutations in desmosomal genes, autoantibody attack of desmosomal cadherins, and bacterial toxins that selectively target desmosomal cadherins. In addition to reviewing the

Progress in Molecular Biology and Translational Science, Volume 116
ISSN 1877-1173
http://dx.doi.org/10.1016/B978-0-12-394311-8.00005-4

well-known roles of desmosomal proteins in tissue integrity, this chapter also highlights the growing appreciation for how desmosomal proteins are integrated with cell signaling pathways to contribute to vertebrate tissue organization and differentiation.

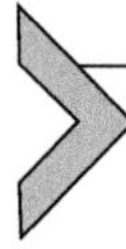

1. DESMOSOME COMPOSITION AND ARCHITECTURE

1.1. Desmosome structure and morphology

Desmosomes are specialized and highly ordered membrane domains that mediate cell–cell contact and strong adhesion. Adhesive interactions at the desmosome are coupled to the intermediate filament cytoskeleton. By mediating both cell–cell adhesion and cytoskeletal linkages, desmosomes mechanically integrate cells within tissues and thereby function to resist mechanical stress.[1–3] This essential structural and mechanical function is highlighted by the prominent distribution of desmosomes in tissues that are routinely subjected to physical forces, such as the heart and skin, and the wide range of desmosomal diseases that result from disruption of desmosome function.[4–6] At the ultrastructural level, desmosomes appear as electron-dense discs approximately 0.2–0.5 μm in diameter, which assemble into a mirror image arrangement at cell–cell interfaces[1,7,8] (Fig. 5.1). Large bundles of intermediate filaments extend from the nuclear surface and cell interior out toward the plasma membrane, where they attach to desmosomes by interweaving with the cytoplasmic plaque of the adhesive complex. The overall adhesive function of the desmosome is dependent upon the tethering of intermediate filaments to the desmosomal plaque, highlighting the integrated functions of adhesion and cytoskeletal elements. Thus, desmosomes are modular structures comprising adhesion molecules that bolt cells together, cytoskeletal cables that disperse forces, and linking molecules at the cytoplasmic plaque of the desmosome that carry mechanical load from the adhesion molecules to the intermediate filament cytoskeleton.

1.2. Desmosome molecular composition

Three major gene families encode desmosomal proteins. Desmosomal cadherins, comprising two subtypes called desmogleins and desmocollins, are a subfamily of the cadherin superfamily that mediate calcium-dependent cell–cell adhesion.[1,3] In humans, four genes encode desmogleins (Dsg1–4) and three genes encode desmocollins (Dsc1–3)[2] (Fig. 5.2). These proteins include five extracellular cadherin repeats, each of which form Ig-like

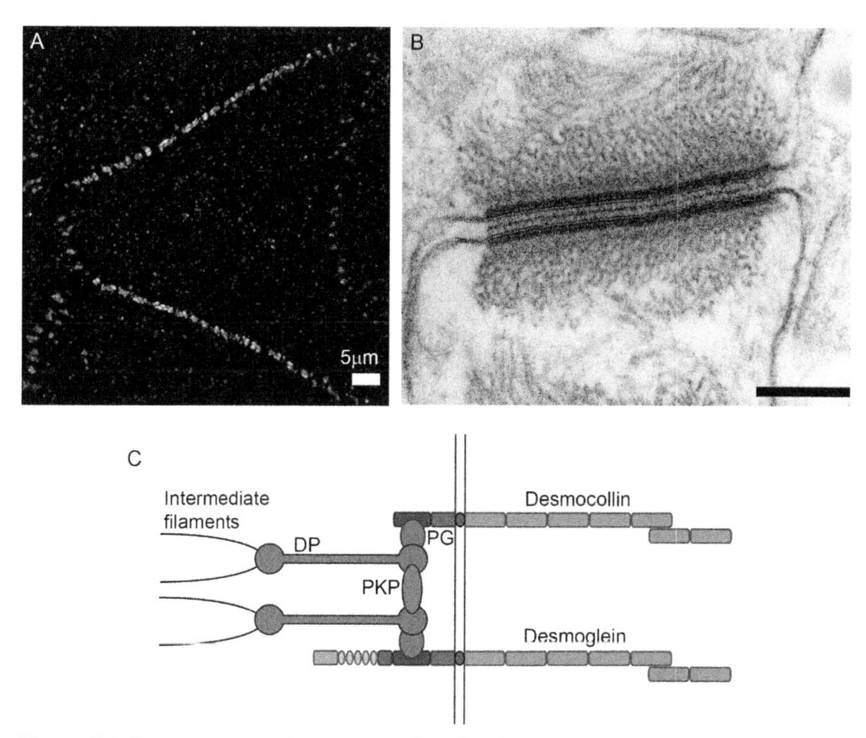

Figure 5.1 Desmosome architecture and molecular composition. (A) Superresolution immunofluorescence localization of desmoglein-3 (red) and desmoplakin (green) in cultured keratinocytes obtained using structured illumination microscopy (NIkon N-SIM). (B) Electron micrograph of a desmosome from bovine tongue epithelium. (C) Molecular components and model of desmosomal protein organization (see Fig. 5.2 for domain annotations of desmosomal cadherins). Scale bar = 0.25 μm. (See Color Insert.)

globular domains with calcium binding sites between each pair of consecutive repeats.[9] The cytoplasmic domains of desmogleins and desmocollins both contain an intracellular anchor and a cadherin-like sequence (ICS), which is conserved in classical cadherins. The desmogleins have additional unique sequences with unknown functions, including a proline-rich linker region, a repeat unit domain, and a desmoglein terminal domain[10] (Fig. 5.2). Each of the three desmocollin RNAs can be alternatively spliced to yield an "a" and a "b" isoform. In the "b" splice variant, the region encoding the ICS domain is truncated and terminates with an additional 11 amino acids in Dsc1 and 2, and eight residues in Dsc3, not found in the "a" form.[3] The desmosomal cadherin genes are expressed in a tissue- and differentiation-specific manner[1,3] (Table 5.1), and emerging evidence suggests important roles for these expression patterns in driving epithelial patterning and differentiation.

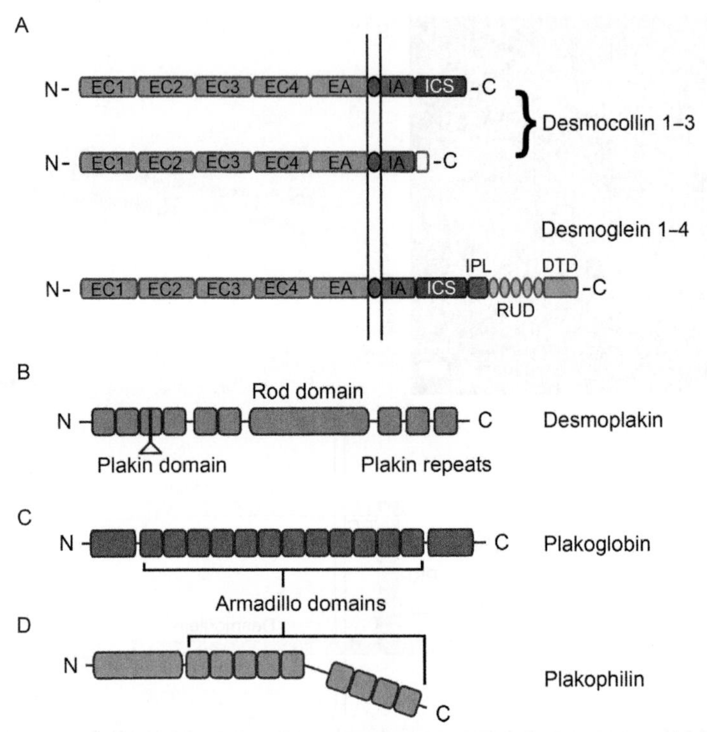

Figure 5.2 Domain structure of major desmosomal proteins. (A) Desmosomal cadherins comprise five extracellular cadherin repeats (ECs), a single pass transmembrane domain, and an intracellular domain that associates with desmosomal plaque proteins (EA, extracellular anchor; IA, intracellular anchor; ICS, intracellular cadherin-like sequence; IPL, intracellular proline-rich linker; RUD, repeat unit domain; DTD, desmoglein terminal domain). (B) Desmoplakin is a prototypical member of the plakin family with amino and carboxyl-terminal globular domains joined by a central alpha-helical coiled-coil rod domain. (C) Plakoglobin comprises amino and carboxyl terminal domains flanking 12 central armadillo repeat (Arm) domains. (D) Plakophilins are also armadillo family proteins, but contain nine Arm repeats as well as amino and carboxyl terminal domains which can vary between isoforms as a result of alternative RNA splicing. Please refer to text for details of domain organization and function. (See Color Insert.)

Structural analysis of cadherin ectodomain interactions has yielded important insights into how cadherins mediate adhesive interactions between cells. In the case of classical cadherins, the defining molecular interaction mediating adhesion is termed a strand-swapped dimer. This *trans* dimer is formed when a conserved tryptophan residue (Trp-2) on a cadherin from one cell is inserted into the hydrophobic pocket of a partnering

Table 5.1 Desmosomal cadherin gene expression patterns and disease associations

Desmosomal gene	Tissue distribution	Type of disease	Human disease
DSG1	Stratifying epithelia	Genetic, autoimmune, infectious	SPPK, pemphigus foliaceus, pemphigus vulgaris (mucocutaneous type), bullous impetigo/SSSS, paraneoplastic pemphigus, Netherton's syndrome
DSG2	Epithelia, heart	Genetic, infectious	Arrhythmogenic cardiomyopathy, dilated cardiomyopathy, adenoviral infection of respiratory and urinary tract
DSG3	Stratifying epithelia	Autoimmune	Pemphigus vulgaris (mucosal dominant type, mucocutaneous type), paraneoplastic pemphigus
DSG4	Stratifying epithelia, hair	Genetic	Localized recessive hypotrichosis, recessive monilethrix
DSC1	Stratifying epithelia	–	None known
DSC2	Epithelia, heart	Genetic	Wooly hair, keratoderma, cardiomyopathy (+/− keratoderma)
DSC3	Epithelia	–	Hypotrichosis with scalp vesicles
Plakoglobin (JUP)	Widespread	Genetic	Naxos disease (with cardiomyopathy, wooly hair, keratoderma), arrhythmogenic cardiomyopathy, keratoderma/wooly hair, lethal congenital epidermolysis bullosa
Plakophilin 1 (PKP1)	Stratifying epithelia	Genetic	Skin fragility-ectodermal dysplasia syndrome
Plakophilin 2 (PKP2)	Widespread	Genetic	Arrhythmogenic cardiomyopathy
Plakophilin 3 (PKP3)	Widespread	–	None known

Continued

Table 5.1 Desmosomal cadherin gene expression patterns and disease associations—cont'd

Desmosomal gene	Tissue distribution	Type of disease	Human disease
Desmoplakin (DPI/II)	Epithelia, heart	Genetic	Carvajal syndrome (with cardiomyopathy, wooly hair, keratoderma), SPPK, arrhythmogenic cardiomyopathy, dilated cardiomyopathy, lethal acantholytic epidermolysis bullosa
Corneodesmosin	Epidermis, hair, hard palate	Genetic	Hypotrichosis simplex, generalized peeling skin syndrome

Abbreviations: SPPK, striate palmoplantar keratoderma; SSSS, Staphylococcal scalded skin syndrome.

cadherin on an adjacent cell. This adhesive dimer requires formation of an X-dimer intermediate, which lowers the energy needed for tryptophan swapping.[9,11] Atomic level structural information on desmosomal cadherin ectodomain interactions is currently unavailable. However, the desmosomal cadherins, both desmogleins and desmocollins, possess the conserved tryptophan in their EC1 domain.[9] Furthermore, cryoelectron tomography and modeling based on classical cadherin ectodomain structures is consistent with EC1 domain interactions between desmosomal cadherins.[12–14] Also, cross-linking experiments suggest that the conserved Trp-2 residue in desmocollin is required for homophilic interactions of this desmosomal cadherin.[15] Nonetheless, we do not fully understand how this large and complex subfamily of cadherins interacts to facilitate desmosome assembly and adhesion.

Armadillo (Arm) family proteins represent cadherin binding partners that play important roles in tissue integrity and cell signaling. Two types of armadillo proteins are present in desmosomes[16,17] (Fig. 5.2). Plakoglobin is a founding member of the armadillo gene family, defined by the presence of a 42-amino acid repeat motif found in the closely related *Drosophila* armadillo protein and its vertebrate orthologue, β-catenin.[18] Plakoglobin binds to the ICS domain of desmogleins and desmocollins and functions as a bridge between the cadherin tail and intermediate filament binding proteins such as desmoplakin[2] (see below). Desmosomes also contain plakophilins, members of the p120-catenin subfamily, which have nine Arm repeats in contrast to the 12 repeats in plakoglobin and β-catenin.[17,19] Humans express three plakophilins (PKP1–3) that are expressed differentially in simple and stratified epithelia, cardiomyocytes, endothelia, and other cell types.[16] Biochemical and cryoelectron microscopy tomography suggest that desmoplakin, through interactions with both plakoglobin and plakophilins, drives clustering and lateral interactions between desmosomal cadherins.[20–23] In addition to their important structural roles, both plakoglobin and plakophilins exhibit diverse junction-independent functions in the cytoplasm and nucleus.[19,24,25]

The third major gene family encoding desmosomal proteins is the plakin family.[26] Among its members, desmoplakin is an obligate desmosomal protein that couples intermediate filaments to the desmosomal plaque[27,28] (Fig. 5.2). The desmoplakin amino-terminus[26] binds directly to plakoglobin and the plakophilins.[2,20,29,30] An extended alpha-helical coiled-coil rod links the plakin domain to a carboxyl-terminal globular domain, which binds directly to intermediate filaments.[31–33] Humans have a single desmoplakin gene, which is alternatively spliced to yield desmoplakin I and desmoplakin II, with the desmoplakin II variant lacking about two-thirds of the rod domain.[34] The desmoplakin-I-specific sequences might facilitate association with microtubule binding proteins (see below and Ref. 35), while the desmoplakin II variant appears to be particularly important in epidermal adhesion.[36]

1.3. Desmosome-associated proteins

A variety of minor and/or differentiation-specific proteins are associated with desmosomes.[37] Corneodesmosin is a glycoprotein expressed primarily in the upper layers of the epidermis and hair follicles, where it contributes to keratinocyte cohesion and is a protease target during epidermal desquamation.[38] Another recently discovered desmosome protein with restricted expression is Perp (p53 apoptosis effector related to PMP-22).[39] As a target of p53/p63 transcriptional regulation, Perp is expressed predominantly in stratifying epithelia where it plays an essential role in intercellular adhesion.[39] Other reported desmosome-associated proteins include the calmodulin binding protein keratocalmin, the adaptor protein and RhoA regulator kazrin, a keratin-binding protein pinin, and several members of the plakin family.[26,37,40,41] How these proteins contribute to desmosome function and the stoichiometry of their associations with the major desmosomal proteins is not well understood.

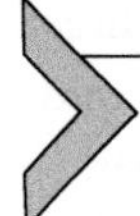

2. MAKING AND BREAKING DESMOSOMES: REGULATION OF ADHESIVE STRENGTH

2.1. Desmosomal cadherin trafficking

Desmosomes are constructed from distinct, cadherin- and plaque-associated complexes that form in the cytoplasm and are delivered to regions of cell–cell contact, where final assembly occurs. The calcium sensitive nature of desmosomal adhesion has served as tool to manipulate desmosome formation. When cells are cultured in low calcium, desmosomal proteins are recruited to the cell surface but are rapidly retrieved and degraded due to the inability of the desmosomal cadherins to mediate adhesion in the absence of calcium.[2] Upon addition of calcium, junction assembly is triggered and junctional proteins are stabilized. Early biochemical studies revealed that the desmosomal cadherins and plakoglobin cofractionate, while desmoplakin is a part of distinct multiprotein complexes that separately traffic to cell–cell contacts.[42] This early work has been advanced by the use of fluorescently tagged desmosomal components and time-lapse imaging techniques, allowing direct visualization and analysis of desmosome molecule dynamics, including movement triggered by cells coming into contact with each other in response to calcium or at the edge of a scrape wound.[43,44]

Recent studies implicate the microtubule network and kinesin motor proteins in transporting desmogleins and desmocollins to sites of cell adhesion.[45] Interestingly, desmocollin delivery to the membrane requires kinesin-2, whereas desmoglein transport is dependent upon kinesin-1. Selective inactivation of either kinesin weakens cell–cell adhesion strength, with knockdown of both motors having additive effects. This dynamic arrangement points to a potential mechanism for regulating desmosome structure and adhesive strength through the independent control of individual cadherin subunits. PKP2 silencing interferes with long-range microtubule-dependent trafficking of Dsc2 but not Dsg2, indicating that armadillo proteins might functionally couple desmosomal cadherins to the motor complex in some cases.

Advances have also been made in understanding how desmosomes are dismantled. Disruption of adhesion by calcium chelating agents was used in early studies to inactivate desmosomal cadherins, resulting in their rapid retrieval from the plasma membrane by the engulfment of entire junctions.[1] Similar mechanisms might be involved in the downregulation of desmosomal adhesion upon epidermal injury.[46] More recent studies of desmosomal cadherin endocytosis have utilized antibody-mediated disruption of adhesion, using

IgG from patients afflicted with pemphigus vulgaris, a skin disease involving the generation of IgG against desmogleins (Dsg3 and Dsg1, see below). These autoantibodies cluster desmogleins on the cell surface, leading to lipid-raft-mediated endocytosis and loss of cell surface desmogleins.[47,48] Consistent with this pathway for desmoglein endocytosis, several recent studies have reported that desmosomal proteins cofractionate with membrane raft microdomains[49,50] (SN Stahley and AP Kowalczyk, unpublished). Thus, lipid raft carriers appear to play a role in regulating desmosomal cadherin trafficking, perhaps during both assembly and disassembly of desmosomes.

While sequences in the desmoglein cytoplasmic domain that regulate its association with lipid rafts or the endocytic machinery are not well understood, recent evidence suggests that intermolecular tail–tail interactions requiring the desmoglein unique region dampen Dsg2 internalization. A mutation in this region associated with arrhythmogenic cardiomyopathy abrogates Dsg2 tail interactions and significantly enhances Dsg2 internalization.[51] Components of the trafficking machinery could be targets in inherited diseases that cannot be explained by mutations in known desmosomal genes.

2.2. Desmoplakin and plakophilins

Desmosomal plaque proteins in cultured keratinocytes exhibit a multiphasic assembly process involving the formation and dynamic recruitment of nonmembrane-bound particles to sites of junction assembly. The initial wave of desmosome assembly is characterized by the localized accumulation of PKP2 at sites of cell–cell contact, followed closely by its binding partner desmoplakin.[44,52] Desmoplakin accumulation at cell borders is followed by the *de novo* formation and translocation of desmoplakin/PKP2-containing cytoplasmic particles to the cell surface, a process requiring PKP2. The results suggest a model whereby *de novo* cell–cell contacts send signals to cytoplasmic plaque precursors, which are then translocated to regions of cell adhesion in an actin-dependent manner.[52]

To create a functional desmosome, desmoplakin needs to tightly couple desmosomal cadherins to the intermediate filament cytoskeleton. On the other hand, desmoplakin association with intermediate filaments needs to be highly dynamic in order for the protein to be transported efficiently to the cell surface during desmosome assembly. These competing requirements appear to be resolved by the ability of the plakophilins to scaffold serine/threonine kinases that modify desmoplakin association with the intermediate filament network. Supporting this idea, PKP2 associates with PKCα in a complex with desmoplakin and PKP2 silencing disrupts this complex.[53] PKCα inhibition, PKP2 silencing, or mutation of S2849 downstream of the intermediate filament binding site in desmoplakin all result in retention of desmoplakin on intermediate filament networks and impair desmoplakin accumulation at sites of cell–cell contact. Desmoplakin translocation also requires actin and is regulated by RhoA through a PKP2-dependent mechanism.[52] Thus, plakophilins play critical roles in desmoplakin trafficking to sites of desmosome assembly through the regulation of both desmoplakin carboxyl-terminal domain phosphorylation and actin dynamics. We do not fully understand the relative contributions of the three plakophilin genes, which appear to drive desmosome assembly with different efficacies.

The regulation of desmoplakin dynamics might contribute to compromised epidermal adhesion observed in human Darier's disease. This disease is caused by mutations in the ATPase that pumps calcium ions from the cytosol into the ER and is thus critical for maintaining calcium homeostasis.[54] Recent evidence suggests that silencing the Darier's calcium pump interferes with desmoplakin transport to desmosomes as a result of defects in PKCα signaling and consequent retention of desmoplakin along keratin filaments.[55] The accompanying defects in cell–cell adhesion could be reversed by activating PKC.[55] These studies highlight the importance of defining fundamental mechanisms of desmosome modulation in order to gain insights into disease pathomechanisms and to reveal new therapeutic targets.

2.3. Desmosome maturation and regulation of adhesion

During the first several days after their formation, desmosomes continue to depend on extracellular calcium for their maintenance at the plasma membrane. However, over a period of a week or so they undergo a maturation process during which they become calcium independent and achieve what has been coined a "hyperadhesive" state by Garrod and colleagues.[56] Upon wounding, desmosomes lose their hyperadhesive state in response to recruitment and activation of PKCα to the wound edge.[57] While the mechanism underlying the acquisition of hyperadhesion is unclear, one model is that desmosomal cadherin ectodomains undergo a conformational change that contributes to this state.[56] It has also been reported that keratinocyte cell sheets expressing a desmoplakin mutant with enhanced keratin-binding properties acquire strong adhesion over time and resist reduced calcium and activation of PKC.[58] Similarly, PKP1 overexpression (DK Tucker and AP Kowalczyk, unpublished) or loss of function mutations[59] alter desmosome adhesive strength and dependency on calcium. Together, these observations underscore the importance of desmosomal plaque protein composition in modulating desmosome adhesion.

Desmosomes in the epidermal stratum corneum are highly cross-linked and stabilized, creating a unique challenge for homeostatic regulation of desmosome turnover in the framework of a regenerating epithelium. In this context, desmosome turnover is thought to involve a carefully orchestrated process of degradation regulated by serine proteases and protease inhibitors expressed in the skin. Desmogleins are substrates for members of the kallikrein protease family, eight of which are expressed in the epidermis,[60] as well as ADAM (a disintegrin and metalloprotease) family proteases.[61] Upsetting the balance of desmoglein synthesis and degradation and consequent alterations in desmosome function can result in human diseases such as Netherton syndrome, discussed in the section below.

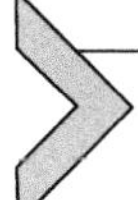

3. DESMOSOMES IN DISEASE

3.1. Epidermal autoimmune disorders

The first studies to recognize a direct role for desmosomes in disease came from dermatologists studying pemphigus, a family of autoimmune disorders associated with epidermal fragility and blistering.[4] The hallmark of pemphigus is the loss of adhesion between keratinocytes (acantholysis) due to the generation of autoantibodies against desmogleins. In pemphigus foliaceus, autoantibodies (IgG) targeting Dsg1 are generated, resulting in superficial blisters in the granular layer of the epidermis. In pemphigus vulgaris, IgG targeting both Dsg3 and Dsg1 are produced, resulting in deep epidermal blisters and oral erosions.[2,62,63] The localization of blistering generally correlates with the expression patterns of the desmoglein antigens (Table 5.1), providing important evidence that desmosomes are essential for maintaining strong cell–cell adhesion in stratifying epithelia.

Pemphigus disease activity can be transferred by introducing IgG from the patient into mice or cell culture.[64] This property has provided investigators with a tool to investigate disease pathomechanisms and basic cellular mechanisms of desmosome regulation. One proposed mechanism for antibody-mediated loss of adhesion is steric hindrance, which would be predicted to occur if antibodies occupy sites on the cadherin extracellular domain that are involved in cell–cell adhesion.[65,66] Evidence from atomic force microscopy and other approaches lend credence to this mechanism of pathogenicity.[67] However, other studies have shown that signal transduction pathways, particularly the p38 MAPK pathway, are activated downstream of antibody binding,[68] and inhibitors of p38 MAPK prevent loss of adhesion.[69,70] Furthermore, pemphigus antibodies cause desmoglein endocytosis, thereby depleting cell surface levels of the cadherin and compromising the adhesive potential of the cell surface.[47,48,71,72] Desmoglein internalization and antibody-mediated loss of adhesion can be prevented by p38 MAPK and tyrosine kinase inhibitors.[48,73,74] If pemphigus antibodies cause loss of adhesion by steric hindrance alone, it is difficult to understand how inhibition of intracellular signaling pathways prevents loss of adhesion strength. These and other data suggest that these antibodies cause a range of effects on desmogleins and their associated regulatory pathways.[70,75–77] Additional study of the pathomechanisms of pemphigus should yield new therapeutic targets and insights into how keratinocytes regulate desmoglein cell surface levels and adhesion.

3.2. Desmosomes as targets for pathogens and proteolysis

Whereas epidermal desmogleins are targets for autoimmune antibodies in skin disease, Dsg2 was recently identified as a receptor for a subclass of adenoviruses (serotypes 3, 7, 11, and 14) that cause respiratory and urinary tract infections.[78] Adenoviral particle binding to Dsg2 triggered phenotypic changes in the target epithelial cells similar to those induced during epithelial–mesenchymal transitions. These data are interesting with respect to viral pathogenicity and also provide evidence that desmosomal cadherins modulate the epithelial phenotype.

Desmogleins are also targets for both bacterially produced and endogenous proteases. The toxin exfoliative toxin A (ETA), produced by the *Staphylococcus* bacteria that causes bullous impetigo, is a serine protease that cleaves Dsg1 after residue 381 between ECs 3 and 4 (Table 5.1). This cleavage removes sequences required for cell–cell adhesion in the superficial epidermis, resulting in focal lesions that histologically resemble pemphigus foliaceus.[79] The specificity of the protease is quite remarkable; no other protein is known to be cleaved by ETA or the other closely related proteases produced by staphylococcal bacteria.[80] Dramatic evidence for this specificity is provided by staphylococcal scalded skin syndrome (SSSS), which is observed in infants and immune-compromised patients. In SSSS, the bacterial infection becomes systemic with extensive epidermal involvement. No other organ systems are affected by the protease, and the disorder can be successfully treated with antibiotic regimens to eliminate the bacterial infection. The discovery that Dsg1 is the target of ETA (Table 5.1) provides unique verification for the role of Dsg1 in pemphigus foliaceus. Indeed, injection of either pemphigus foliaceus IgG or ETA into mouse epidermis produces an identical phenotype.[62] These observations, along with the genetic disorders discussed below, firmly establish the role of desmosomal cadherins in epidermal function and integrity.

Netherton syndrome is an autosomal recessive disease caused by mutations in the serine protease inhibitor LEKT1 (lymphoepithelial Kazal-type-related inhibitor).[81] Loss of this protease inhibitor results in excessive tryptic and chymotryptic enzyme activity attributed to members of the kallikrein protease family. Reduced proteolysis of Dsg1 was proposed to be a central contributor to the aberrant desquamation and keratinization in this disorder.[82] Kallikrein-5-dependent proteolysis of Dsg1[83] and ADAM-dependent proteolysis of Dsg2 have also been suggested to promote their turnover in oral squamous cell carcinoma cells.[84] Further, retention of Dsg2 was observed in the epidermis of patients with a recessive loss of function mutation in ADAM17 resulting in neonatal-onset inflammatory skin and bowel disease.[85] Finally, Dsg2 cleavage via cysteine proteinases was also reported to contribute to stimulus-induced apoptosis in intestinal epithelial cells.[86] Together, these observations suggest that aberrant proteolysis of desmogleins contributes to human disease pathogenesis.

3.3. Desmosomes as targets of inherited diseases of skin and heart

Mutations associated with desmosomal genes are frequently manifested in heart and skin, both of which are exposed to high mechanical stress[87] (Table 5.1). Mutations in Dsg1, Dsg4, and PKP1, which are found in desmosomes of the differentiated layers of the epidermis, cause epidermal disorders without cardiac involvement. In contrast, mutations in genes such as Dsg2 and PKP2 lead to cardiac defects without epidermal disease. Other desmosomal components with more widespread expression patterns, such as desmoplakin, plakoglobin, and Dsc2, affect both heart and skin.[5,6] Interestingly, the effects of desmosome gene mutations do not appear to be clinically manifest in internal organs, suggesting that desmosome function is most critical in stratifying epithelia and myocardium.

Common to most of the disorders affecting the epidermis is a loss of tissue integrity, likely resulting from a reduction in desmosome numbers and impaired intermediate filament attachment to the plasma membrane.[5,87]

Similarly, structural changes in cardiac intercellular junctions, the intercalated discs,[88,89] are a major contributing factor in the development of arrhythmogenic cardiomyopathy, which can result in sudden death in individuals with this "disease of the desmosome."[89,90] Advanced stages of the disease are accompanied by the loss of cardiac myocytes and replacement with fibrofatty tissue. How this transition occurs and the extent to which these changes in heart structure lead to arrhythmia is poorly understood. It now seems that loss of mechanical integrity alone cannot completely explain the arrhythmias that can occur in these individuals. Indeed, the mammalian intercalated disc is now known to remodel postnatally into hybrid junctions in which desmosome components associate not only with other mechanical junction components but also with gap junction components (e.g., Cx43) and the sodium channel protein Nav1.5.[88,90] The most commonly mutated gene associated with arrhythmogenic cardiomyopathy is the one encoding PKP2. Loss of this protein leads to changes in electrical coupling and excitability associated with redistribution of these channels from their normal sites at cardiomyocyte cell–cell junctions.[91,92] It has also been reported that the redistribution of desmosomal armadillo proteins might interfere with nuclear transcription pathways driven by β-catenin, contributing to the fibrofatty replacement during disease pathogenesis.[93] This finding also prompted the suggestion that ectodermal defects such as wooly hair and changes in nails seen in cutaneous manifestations of desmosome disease involve reprogramming of cell fate through alterations in Wnt signaling.[93,94] This idea is controversial, however, as alterations in β-catenin activity associated with desmosome gene mutations are variable.[95] The finding that PKP2 deficiency in epithelial cells results in global activation of RhoA and PKCα[52,53] suggests that desmosome impairment may affect multiple signaling pathways contributing to skin and heart disease progression. A role for desmosome molecules in signaling has also been implicated in normal tissue differentiation, as discussed below.

4. DESMOSOMAL PROTEINS IN DIFFERENTIATION, DEVELOPMENT, AND CANCER

Investigators have long speculated that the highly patterned distribution of desmosome molecules in complex epithelia might have functional importance beyond adhesion[96,97] (Table 5.1). The idea that desmosomal molecules play a more active role in tissue morphogenesis gained credence from studies in which desmosomal cadherin expression was forced in epidermal layers where they are normally not expressed.[96] These studies have been complemented by animal gene knockouts as well as selective silencing of desmosomal cadherins in human organotypic model systems.

A theme emerging from these studies is that basal and suprabasal desmosomal cadherins support and/or promote cell phenotypes appropriate to the epidermal compartment in which they are normally located. For instance, forced expression of Dsg3 in suprabasal epidermis resulted in a tissue that took on properties of oral mucosal epithelium, where Dsg3 naturally predominates over Dsg1.[98] Forced suprabasal expression of Dsg2, normally expressed at low levels in basal epidermis, resulted in hyperplasia and appearance of precancerous papillomas.[99] The possible importance of Dsg2 in cell growth is also supported by the observation that this desmosomal cadherin is required for proliferation of embryonic stem cells in developing mouse embryos, and that its ablation leads to early embryonic lethality.[100] The lethality of Dsg2 ablation is manifested before the appearance of desmosomal junctions, suggesting junction-independent functions for this cadherin.[100] Likewise, mice lacking Dsc3 die embryonically before desmosomes are present.[101] This observation is particularly interesting as Dsc3 was originally thought to be exclusively expressed in complex epithelia.[102] Furthermore, mice with forced expression of Dsc3 suprabasally in the epidermis develop alopecia and changes in keratinocyte differentiation associated with increased β-catenin-dependent transcription activity.[103] Together, these observations are consistent with a role for Dsc3 that extends beyond adhesion.

The molecular mechanisms linking desmosomal cadherins to signaling pathways important for tissue morphogenesis and homeostasis are not well understood. The observation that Dsg2-dependent hyperplasia was accompanied by increased cell growth and survival cues[99] raises the possibility that, like classic cadherins, desmosomal cadherins couple with receptor tyrosine kinase signaling pathways. Indeed, evidence that desmosomal cadherins couple with ErbB family members comes from studies of Dsg1 in a human organotypic model of epidermal morphogenesis.[104] In this system, Dsg1 was found to drive inhibitory cues that promote differentiation. Interestingly, the adhesive ectodomain of this cadherin was dispensable for this function, which was attributed to attenuation of Erk1/2 activity. The identification of Erbin, an ErbB2 interacting protein, as a direct binding partner of Dsg1 might help to explain the molecular connection between Dsg1 and the MAPK pathway (RM Harmon and KJ Green, unpublished). These data support the idea that desmosomal cadherins might provide a balance between proproliferative and prodifferentiation pathways during differentiation.

Desmosomes also control cytoarchitectural cues important for tissue development and differentiation. In addition to facilitating intermediate filament attachment, Lechler and colleagues revealed for the first time that the desmosome protein desmoplakin governs the redistribution of microtubules that occurs as epidermal cells begin to stratify.[35,105] In basal cells, microtubules are radially distributed, with minus ends emanating from perinuclear centrosomes and plus ends extending out to the membrane. During differentiation, minus ends are released from the centrosome and microtubules take on a more cortical distribution, running parallel to the plasma membrane. This reorganization is accompanied by a redistribution of the minus end proteins ninein, Lis1, and Ndel1 to the plasma membrane in a desmoplakin-dependent manner. Interestingly, this function was attributed to the desmoplakin-I isoform only, suggesting a role for the region of the desmoplakin alpha-helical coiled-coil rod domain in this process. Lis1 ablation in the epidermis also resulted in destabilization/loss of desmosomes and barrier defects in the skin of mice, uncovering a bidirectional role between desmoplakin and microtubules during differentiation.[35]

Additional roles for desmoplakin were revealed from conditional knockout of this desmosome plaque protein in mouse gut epithelium.[106] Unlike its loss in the skin and heart, loss of desmoplakin in the intestinal epithelium did not affect adhesion or overall organization of simple epithelial keratins, in spite of their dissociation from desmosomes. It also did not have an obvious effect on microtubule organization, even though minus end proteins were mislocalized to the cytoplasm. Loss of desmoplakin did, however, alter microvillar structure. It was suggested that desmoplakin might be required for stability of the actin-rich terminal web and possibly for signaling that regulates actin organization.[106]

Changes in expression of desmosomal cadherins during cancer progression has, like loss of classic cadherins (see Chapter 14), been implicated in acquisition of the metastatic phenotype.[107] However, the differential patterns of expression in complex epithelia raise the possibility that they can also contribute to the balance of cell growth and differentiation during tumorigenesis. Indeed, it has been reported that certain desmosomal proteins are lost independently of classic cadherins during cancer progression, and loss of Dsg1 has been reported as a better prognostic marker than loss of E-cadherin in head and neck squamous cell carcinomas.[108] These findings raise the possibility that desmosomes play suppressive roles during cancer progression that extend beyond loss of adhesion during epithelial to mesenchymal transitions.[107,109]

For example, the p53/63 target Perp was identified as a suppressor of UVB-induced squamous cell skin carcinoma[110] as well as mammary gland tumorigenesis.[111] In the mouse skin carcinoma model, Perp-deficient skin tumors exhibited loss of desmosomal constituents while adherens junctions were still present, a scenario also observed in human squamous cell carcinoma samples. Perp loss in mouse mammary epithelia interfered with desmosome integrity and tissue homeostasis, and promoted mammary carcinogenesis in a p53-deficient tumor model. Interestingly, in both the skin and mammary gland, recruitment of immune cells involved in inflammatory responses was observed. These data are consistent with a role for Perp in suppressing tumorigenesis, and depending on the model, Perp appears to act at multiple stages of disease progression.[107] In another case, transcriptional profiling of noninvasive and invasive neuroendocrine tumors in an animal model of pancreatic cancer revealed reductions in multiple desmosome components.[109] Experimental ablation of desmoplakin resulted in micrometastases in spite of retention of E-cadherin at cell–cell junctions. Altogether these data support the idea that the tumor suppressive role of desmosomes is complex and likely to transcend desmosomal protein functions at cell–cell junctions.

5. FUTURE DIRECTIONS

Over the past decade, desmosomes have emerged as much more than simple spot welds. However, there is still much to learn about the adhesion-dependent and -independent contributions of desmosomes to tissue differentiation and homeostasis, and how these functions are coordinated. Interestingly, classical cadherins and adherens junctions have emerged as mediators of mechanochemical signaling.[112] At present, we know very little about how desmosomes, which might be ideally suited for translating mechanical information to cellular biochemical signaling pathways, participate in mechanotransduction under various pathophysiological circumstances. In addition to modulating receptors on the plasma membrane, mechanical cues might also regulate the distribution and function of desmosome molecules in different subcellular compartments within the cytosol and nucleus (see Chapter 17), to regulate signal transduction, gene expression, protein translation, and other cellular activities.

Additional insights into desmosome regulation will also be derived from higher resolution imaging of desmosome architecture at the light, electron microscopic, and atomic levels. For example, our understanding of ribosomes, the ciliary axoneme, and the nuclear pore complex, have been advanced by applying a combination of crystallography and electron tomography to yield extraordinary imagery and thus a detailed understanding of how proteins within these macromolecular complexes are organized. These approaches have been successful to some extent with the desmosome, but additional studies are needed. Furthermore, our understanding of normal desmosomal architecture would benefit by comparison with samples derived from patients and model organisms harboring genetic defects in desmosomal genes. Last, we lack an understanding of how cells spatially and temporally constrain desmosome assembly and disassembly to yield structures that can display both remarkable stability and resistance to mechanical insult, yet also remain sufficiently dynamic to allow for tissue plasticity. Continued development of imaging tools, in combination with emerging proteomic analysis of desmosomes at various states of assembly, will likely yield significant new insights into the dynamic regulation of desmosomes in the next decade.

ACKNOWLEDGMENTS

The authors are grateful to Kowalczyk lab members for helpful comments and advice on this chapter, to Dr. Rebecca Oas for assistance with artwork, and to Sara Stahley for immunofluorescence imaging shown in Fig. 5.1. A. P. K. is supported by National Institutes of Health R01AR050501 and R01AR048266. K. J. G. is supported by National Institutes of Health RO1 AR041836, R37 AR043380, and R01 CA122151, by a grant from the Leducq Foundation, and by the Joseph L. Mayberry Senior Endowment.

REFERENCES

1. Thomason HA, Scothern A, McHarg S, Garrod DR. Desmosomes: adhesive strength and signalling in health and disease. *Biochem J* 2010;**429**:419–33.
2. Delva E, Tucker DK, Kowalczyk AP. The desmosome. *Cold Spring Harb Perspect Biol* 2009;**1**:a002543.
3. Saito MT, Tucker DK, Kholhorst D, Niessen CM, Kowalczyk AP. Classical and desmosomal cadherins at a glance. *J Cell Sci* 2012;**125**:2547–52.
4. Amagai M, Stanley JR. Desmoglein as a target in skin disease and beyond. *J Invest Dermatol* 2012;**132**:776–84.
5. Petrof G, Mellerio JE, McGrath JA. Desmosomal genodermatoses. *Br J Dermatol* 2012;**166**: 36–45.
6. Delmar M, McKenna WJ. The cardiac desmosome and arrhythmogenic cardiomyopathies: from gene to disease. *Circ Res* 2010;**107**:700–14.
7. Franke WW. Discovering the molecular components of intercellular junctions—a historical view. *Cold Spring Harb Perspect Biol* 2009;**1**:a003061.
8. North AJ, Bardsley WG, Hyam J, Bornslaeger EA, Cordingley HC, Trinnaman B, et al. Molecular map of the desmosomal plaque. *J Cell Sci* 1999;**112**:4325–36.
9. Brasch J, Harrison OJ, Honig B, Shapiro L. Thinking outside the cell: how cadherins drive adhesion. *Trends Cell Biol* 2012;**22**:299–310.

10. Koch PJ, Walsh MJ, Schmelz M, Goldschmidt MD, Zimbelmann R, Franke WW. Identification of desmoglein, a constitutive desmosomal glycoprotein, as a member of the cadherin family of cell adhesion molecules. *Eur J Cell Biol* 1990;**53**:1–12.
11. Ishiyama N, Ikura M. The three-dimensional structure of the cadherin-catenin complex. *Subcell Biochem* 2012;**60**:39–62.
12. Owen GR, Stokes DL. Exploring the nature of desmosomal cadherin associations in 3D. *Dermatol Res Pract* 2010;**2010**:930401.
13. He W, Cowin P, Stokes DL. Untangling desmosomal knots with electron tomography. *Science* 2003;**302**:109–13.
14. Al-Amoudi A, Diez DC, Betts MJ, Frangakis AS. The molecular architecture of cadherins in native epidermal desmosomes. *Nature* 2007;**450**:832–7.
15. Nie Z, Merritt A, Rouhi-Parkouhi M, Tabernero L, Garrod D. Membrane-impermeable cross-linking provides evidence for homophilic, isoform-specific binding of desmosomal cadherins in epithelial cells. *J Biol Chem* 2011;**286**:2143–54.
16. Hatzfeld M. Plakophilins: multifunctional proteins or just regulators of desmosomal adhesion? *Biochim Biophys Acta* 2007;**1773**:69–77.
17. Carnahan RH, Rokas A, Gaucher EA, Reynolds AB. The molecular evolution of the p120-catenin subfamily and its functional associations. *PLoS One* 2010;**5**:e15747.
18. Peifer M, McCrea PD, Green KJ, Wieschaus E, Gumbiner BM. The vertebrate adhesive junction proteins beta-catenin and plakoglobin and the Drosophila segment polarity gene armadillo form a multigene family with similar properties. *J Cell Biol* 1992;**118**:681–91.
19. Bass-Zubek AE, Godsel LM, Delmar M, Green KJ. Plakophilins: multifunctional scaffolds for adhesion and signaling. *Curr Opin Cell Biol* 2009;**21**:708–16.
20. Kowalczyk AP, Hatzfeld M, Bornslaeger EA, Kopp DS, Borgwardt JE, Corcoran CM, et al. The head domain of plakophilin-1 binds to desmoplakin and enhances its recruitment to desmosomes. Implications for cutaneous disease. *J Biol Chem* 1999;**274**:18145–8.
21. Hatzfeld M, Haffner C, Schulze K, Vinzens U. The function of plakophilin 1 in desmosome assembly and actin filament organization. *J Cell Biol* 2000;**149**:209–22.
22. Bonne S, Gilbert B, Hatzfeld M, Chen X, Green KJ, van Roy F. Defining desmosomal plakophilin-3 interactions. *J Cell Biol* 2003;**161**:403–16.
23. Al-Amoudi A, Castano-Diez D, Devos DP, Russell RB, Johnson GT, Frangakis AS. The three-dimensional molecular structure of the desmosomal plaque. *Proc Natl Acad Sci USA* 2011;**108**:6480–5.
24. Clevers H, Nusse R. Wnt/beta-catenin signaling and disease. *Cell* 2012;**149**:1192–205.
25. Pieperhoff S, Barth M, Rickelt S, Franke WW. Desmosomal molecules in and out of adhering junctions: normal and diseased states of epidermal, cardiac and mesenchymally derived cells. *Dermatol Res Pract* 2010;**2010**:139167.
26. Sonnenberg A, Liem RK. Plakins in development and disease. *Exp Cell Res* 2007;**313**:2189–203.
27. Bornslaeger EA, Corcoran CM, Stappenbeck TS, Green KJ. Breaking the connection: displacement of the desmosomal plaque protein desmoplakin from cell-cell interfaces disrupts anchorage of intermediate filament bundles and alters intercellular junction assembly. *J Cell Biol* 1996;**134**:985–1001.
28. Gallicano GI, Kouklis P, Bauer C, Yin M, Vasioukhin V, Degenstein L, et al. Desmoplakin is required early in development for assembly of desmosomes and cytoskeletal linkage. *J Cell Biol* 1998;**143**:2009–22.
29. Smith EA, Fuchs E. Defining the interactions between intermediate filaments and desmosomes. *J Cell Biol* 1998;**141**:1229–41.
30. Kowalczyk AP, Bornslaeger EA, Borgwardt JE, Palka HL, Dhaliwal AS, Corcoran CM, et al. The amino-terminal domain of desmoplakin binds to plakoglobin

and clusters desmosomal cadherin-plakoglobin complexes. *J Cell Biol* 1997;**139**: 773–84.
31. Choi HJ, Park-Snyder S, Pascoe LT, Green KJ, Weis WI. Structures of two intermediate filament-binding fragments of desmoplakin reveal a unique repeat motif structure. *Nat Struct Biol* 2002;**9**:612–20.
32. Meng JJ, Bornslaeger EA, Green KJ, Steinert PM, Ip W. Two-hybrid analysis reveals fundamental differences in direct interactions between desmoplakin and cell type-specific intermediate filaments. *J Biol Chem* 1997;**272**:21495–503.
33. Kouklis PD, Hutton E, Fuchs E. Making a connection: direct binding between keratin intermediate filaments and desmosomal proteins. *J Cell Biol* 1994;**127**:1049–60.
34. Green KJ, Parry DA, Steinert PM, Virata ML, Wagner RM, Angst BD, et al. Structure of the human desmoplakins. Implications for function in the desmosomal plaque. *J Biol Chem* 1990;**265**:2603–12.
35. Sumigray KD, Chen H, Lechler T. Lis1 is essential for cortical microtubule organization and desmosome stability in the epidermis. *J Cell Biol* 2011;**194**:631–42.
36. Cabral RM, Tattersall D, Patel V, McPhail GD, Hatzimasoura E, Abrams DJ, et al. The DSPII splice variant is critical for desmosome-mediated HaCaT keratinocyte adhesion. *J Cell Sci* 2012;**125**:2583–861.
37. Holthofer B, Windoffer R, Troyanovsky S, Leube RE. Structure and function of desmosomes. *Int Rev Cytol* 2007;**264**:65–163.
38. Jonca N, Leclerc EA, Caubet C, Simon M, Guerrin M, Serre G. Corneodesmosomes and corneodesmosin: from the stratum corneum cohesion to the pathophysiology of genodermatoses. *Eur J Dermatol* 2011;**21**(Suppl. 2):35–42.
39. Ihrie RA, Marques MR, Nguyen BT, Horner JS, Papazoglu C, Bronson RT, et al. Perp is a p63-regulated gene essential for epithelial integrity. *Cell* 2005;**120**:843–56.
40. Shi J, Sugrue SP. Dissection of protein linkage between keratins and pinin, a protein with dual location at desmosome-intermediate filament complex and in the nucleus. *J Biol Chem* 2000;**275**:14910–5.
41. Sevilla LM, Nachat R, Groot KR, Watt FM. Kazrin regulates keratinocyte cytoskeletal networks, intercellular junctions and differentiation. *J Cell Sci* 2008;**121**:3561–9.
42. Pasdar M, Krzeminski KA, Nelson WJ. Regulation of desmosome assembly in MDCK epithelial cells: coordination of membrane core and cytoplasmic plaque domain assembly at the plasma membrane. *J Cell Biol* 1991;**113**:645–55.
43. Green KJ, Getsios S, Troyanovsky S, Godsel LM. Intercellular junction assembly, dynamics, and homeostasis. *Cold Spring Harb Perspect Biol* 2010;**2**:a000125.
44. Godsel LM, Hsieh SN, Amargo EV, Bass AE, Pascoe-McGillicuddy LT, Huen AC, et al. Desmoplakin assembly dynamics in four dimensions: multiple phases differentially regulated by intermediate filaments and actin. *J Cell Biol* 2005;**171**:1045–59.
45. Nekrasova OE, Amargo EV, Smith WO, Chen J, Kreitzer GE, Green KJ. Desmosomal cadherins utilize distinct kinesins for assembly into desmosomes. *J Cell Biol* 2011;**195**: 1185–203.
46. Garrod DR, Berika MY, Bardsley WF, Holmes D, Tabernero L. Hyper-adhesion in desmosomes: its regulation in wound healing and possible relationship to cadherin crystal structure. *J Cell Sci* 2005;**118**:5743–54.
47. Calkins CC, Setzer SV, Jennings JM, Summers S, Tsunoda K, Amagai M, et al. Desmoglein endocytosis and desmosome disassembly are coordinated responses to pemphigus autoantibodies. *J Biol Chem* 2006;**281**:7623–34.
48. Delva E, Jennings JM, Calkins CC, Kottke MD, Faundez V, Kowalczyk AP. Pemphigus vulgaris IgG-induced desmoglein-3 endocytosis and desmosomal disassembly are mediated by a clathrin- and dynamin-independent mechanism. *J Biol Chem* 2008;**283**: 18303–13.

49. Resnik N, Sepcic K, Plemenitas A, Windoffer R, Leube R, Veranic P. Desmosome assembly and cell-cell adhesion are membrane raft-dependent processes. *J Biol Chem* 2011;**286**:1499–507.
50. Brennan D, Peltonen S, Dowling A, Medhat W, Green KJ, Wahl 3rd JK, et al. A role for caveolin-1 in desmoglein binding and desmosome dynamics. *Oncogene* 2012;**31**:1636–48.
51. Chen J, Nekrasova OE, Patel DM, Koetsier J, Amargo EV, Desai BV, et al. The unique region of a desmosomal cadherin, desmoglein 2, regulates internalization via tail-tail interactions. *J Cell Biol* 2012; in press.
52. Godsel LM, Dubash AD, Bass-Zubek AE, Amargo EV, Klessner JL, Hobbs RP, et al. Plakophilin 2 couples actomyosin remodeling to desmosomal plaque assembly via RhoA. *Mol Biol Cell* 2010;**21**:2844–59.
53. Bass-Zubek AE, Hobbs RP, Amargo EV, Garcia NJ, Hsieh SN, Chen X, et al. Plakophilin 2: a critical scaffold for PKC alpha that regulates intercellular junction assembly. *J Cell Biol* 2008;**181**:605–13.
54. Savignac M, Edir A, Simon M, Hovnanian A. Darier disease: a disease model of impaired calcium homeostasis in the skin. *Biochim Biophys Acta* 2011;**1813**:1111–7.
55. Hobbs RP, Amargo EV, Somasundaram A, Simpson CL, Prakriya M, Denning MF, et al. The calcium ATPase SERCA2 regulates desmoplakin dynamics and intercellular adhesive strength through modulation of PKCα signaling. *FASEB J* 2011;**25**: 990–1001.
56. Garrod D, Kimura TE. Hyper-adhesion: a new concept in cell-cell adhesion. *Biochem Soc Trans* 2008;**36**:195–201.
57. Thomason HA, Cooper NH, Ansell DM, Chiu M, Merrit AJ, Hardman MJ, et al. Direct evidence that PKCalpha positively regulates wound re-epithelialization: correlation with changes in desmosomal adhesiveness. *J Pathol* 2012;**227**:346–56.
58. Hobbs RP, Green KJ. Desmoplakin regulates desmosome hyperadhesion. *J Invest Dermatol* 2012;**132**:482–5.
59. South AP, Wan H, Stone MG, Dopping-Hepenstal PJ, Purkis PE, Marshall JF, et al. Lack of plakophilin 1 increases keratinocyte migration and reduces desmosome stability. *J Cell Sci* 2003;**116**:3303–14.
60. Ovaere P, Lippens S, Vandenabeele P, Declercq W. The emerging roles of serine protease cascades in the epidermis. *Trends Biochem Sci* 2009;**34**:453–63.
61. Bech-Serra JJ, Santiago-Josefat B, Esselens C, Saftig P, Baselga J, Arribas J, et al. Proteomic identification of desmoglein 2 and activated leukocyte cell adhesion molecule as substrates of ADAM17 and ADAM10 by difference gel electrophoresis. *Mol Cell Biol* 2006;**26**:5086–95.
62. Payne AS, Hanakawa Y, Amagai M, Stanley JR. Desmosomes and disease: pemphigus and bullous impetigo. *Curr Opin Cell Biol* 2004;**16**:536–43.
63. Stanley JR, Amagai M. Pemphigus, bullous impetigo, and the staphylococcal scalded-skin syndrome. *N Engl J Med* 2006;**355**:1800–10.
64. Anhalt GJ, Labib RS, Voorhees JJ, Beals TF, Diaz LA. Induction of pemphigus in neonatal mice by passive transfer of IgG from patients with the disease. *N Engl J Med* 1982;**306**:1189–96.
65. Payne AS, Ishii K, Kacir S, Lin C, Li H, Hanakawa Y, et al. Genetic and functional characterization of human pemphigus vulgaris monoclonal autoantibodies isolated by phage display. *J Clin Invest* 2005;**115**:888–99.
66. Tsunoda K, Ota T, Aoki M, Yamada T, Nagai T, Nakagawa T, et al. Induction of pemphigus phenotype by a mouse monoclonal antibody against the amino-terminal adhesive interface of desmoglein 3. *J Immunol* 2003;**170**:2170–8.
67. Heupel WM, Zillikens D, Drenckhahn D, Waschke J. Pemphigus vulgaris IgG directly inhibit desmoglein 3-mediated transinteraction. *J Immunol* 2008;**181**:1825–34.

68. Mao X, Sano Y, Park JM, Payne AS. p38 MAPK activation is downstream of the loss of intercellular adhesion in pemphigus vulgaris. *J Biol Chem* 2011;**286**:1283–91.
69. Sharma P, Mao X, Payne AS. Beyond steric hindrance: the role of adhesion signaling pathways in the pathogenesis of pemphigus. *J Dermatol Sci* 2007;**48**:1–14.
70. Getsios S, Waschke J, Borradori L, Hertl M, Muller EJ. From cell signaling to novel therapeutic concepts: international pemphigus meeting on advances in pemphigus research and therapy. *J Invest Dermatol* 2010;**130**:1764–8.
71. Jennings JM, Tucker DK, Kottke MD, Saito M, Delva E, Hanakawa Y, et al. Desmosome disassembly in response to pemphigus vulgaris IgG occurs in distinct phases and can be reversed by expression of exogenous Dsg3. *J Invest Dermatol* 2011;**131**:706–18.
72. Yamamoto Y, Aoyama Y, Shu E, Tsunoda K, Amagai M, Kitajima Y. Anti-desmoglein 3 (Dsg3) monoclonal antibodies deplete desmosomes of Dsg3 and differ in their Dsg3-depleting activities related to pathogenicity. *J Biol Chem* 2007;**282**:17866–76.
73. Berkowitz P, Hu P, Warren S, Liu Z, Diaz LA, Rubenstein DS. p38MAPK inhibition prevents disease in pemphigus vulgaris mice. *Proc Natl Acad Sci USA* 2006;**103**: 12855–60.
74. Jolly PS, Berkowitz P, Bektas M, Lee HE, Chua M, Diaz LA, et al. p38MAPK signaling and desmoglein-3 internalization are linked events in pemphigus acantholysis. *J Biol Chem* 2010;**285**:8936–41.
75. Williamson L, Raess NA, Caldelari R, Zakher A, de Bruin A, Posthaus H, et al. Pemphigus vulgaris identifies plakoglobin as key suppressor of c-Myc in the skin. *EMBO J* 2006;**25**:3298–309.
76. Spindler V, Waschke J. Role of Rho GTPases in desmosomal adhesion and pemphigus pathogenesis. *Ann Anat* 2011;**193**:177–80.
77. Grando SA. Pemphigus autoimmunity: hypotheses and realities. *Autoimmunity* 2012;**45**:7–35.
78. Wang H, Li ZY, Liu Y, Persson J, Beyer I, Moller T, et al. Desmoglein 2 is a receptor for adenovirus serotypes 3, 7, 11 and 14. *Nat Med* 2011;**17**:96–104.
79. Amagai M, Matsuyoshi N, Wang ZH, Andl C, Stanley JR. Toxin in bullous impetigo and staphylococcal scalded-skin syndrome targets desmoglein 1. *Nat Med* 2000;**6**: 1275–7.
80. Hanakawa Y, Schechter NM, Lin C, Garza L, Li H, Yamaguchi T, et al. Molecular mechanisms of blister formation in bullous impetigo and staphylococcal scalded skin syndrome. *J Clin Invest* 2002;**110**:53–60.
81. Chavanas S, Bodemer C, Rochat A, Hamel-Teillac D, Ali M, Irvine AD, et al. Mutations in SPINK5, encoding a serine protease inhibitor, cause Netherton syndrome. *Nat Genet* 2000;**25**:141–2.
82. Descargues P, Deraison C, Bonnart C, Kreft M, Kishibe M, Ishida-Yamamoto A, et al. Spink5-deficient mice mimic Netherton syndrome through degradation of desmoglein 1 by epidermal protease hyperactivity. *Nat Genet* 2005;**37**:56–65.
83. Jiang R, Shi Z, Johnson JJ, Liu Y, Stack MS. Kallikrein-5 promotes cleavage of desmoglein-1 and loss of cell-cell cohesion in oral squamous cell carcinoma. *J Biol Chem* 2011;**286**:9127–35.
84. Klessner JL, Desai BV, Amargo EV, Getsios S, Green KJ. EGFR and ADAMs cooperate to regulate shedding and endocytic trafficking of the desmosomal cadherin desmoglein 2. *Mol Biol Cell* 2009;**20**:328–37.
85. Brooke MA, Nitoiu D, Kelsell DP. Cell-cell connectivity: desmosomes and disease. *J Pathol* 2012;**226**:158–71.
86. Nava P, Laukoetter MG, Hopkins AM, Laur O, Gerner-Smidt K, Green KJ, et al. Desmoglein-2: a novel regulator of apoptosis in the intestinal epithelium. *Mol Biol Cell* 2007;**18**:4565–78.

87. Lai-Cheong JE, Arita K, McGrath JA. Genetic diseases of junctions. *J Invest Dermatol* 2007;**127**:2713–25.
88. Franke WW, Borrmann CM, Grund C, Pieperhoff S. The area composita of adhering junctions connecting heart muscle cells of vertebrates. I. Molecular definition in intercalated disks of cardiomyocytes by immunoelectron microscopy of desmosomal proteins. *Eur J Cell Biol* 2006;**85**:69–82.
89. Li J, Radice GL. A new perspective on intercalated disc organization: implications for heart disease. *Dermatol Res Pract* 2010;**2010**:207835.
90. Delmar M. Desmosome-ion channel interactions and their possible role in arrhythmogenic cardiomyopathy. *Pediatr Cardiol* 2012;**33**:975–9.
91. Sato PY, Musa H, Coombs W, Guerrero-Serna G, Patino GA, Taffet SM, et al. Loss of plakophilin-2 expression leads to decreased sodium current and slower conduction velocity in cultured cardiac myocytes. *Circ Res* 2009;**105**:523–6.
92. Sato PY, Coombs W, Lin X, Nekrasova O, Green KJ, Isom LL, et al. Interactions between ankyrin-G, Plakophilin-2, and Connexin43 at the cardiac intercalated disc. *Circ Res* 2011;**109**:193–201.
93. Garcia-Gras E, Lombardi R, Giocondo MJ, Willerson JT, Schneider MD, Khoury DS, et al. Suppression of canonical Wnt/beta-catenin signaling by nuclear plakoglobin recapitulates phenotype of arrhythmogenic right ventricular cardiomyopathy. *J Clin Invest* 2006;**116**:2012–21.
94. MacRae CA, Birchmeier W, Thierfelder L. Arrhythmogenic right ventricular cardiomyopathy: moving toward mechanism. *J Clin Invest* 2006;**116**:1825–8.
95. Li D, Liu Y, Maruyama M, Zhu W, Chen H, Zhang W, et al. Restrictive loss of plakoglobin in cardiomyocytes leads to arrhythmogenic cardiomyopathy. *Hum Mol Genet* 2011;**20**:4582–96.
96. Simpson CL, Patel DM, Green KJ. Deconstructing the skin: cytoarchitectural determinants of epidermal morphogenesis. *Nat Rev Mol Cell Biol* 2011;**12**:565–80.
97. Schmidt A, Koch PJ. Desmosomes: just cell adhesion or is there more? *Cell Adh Migr* 2007;**1**:28–32.
98. Elias PM, Matsuyoshi N, Wu H, Lin C, Wang ZH, Brown BE, et al. Desmoglein isoform distribution affects stratum corneum structure and function. *J Cell Biol* 2001;**153**:243–9.
99. Brennan D, Hu Y, Joubeh S, Choi YW, Whitaker-Menezes D, O'Brien T, et al. Suprabasal Dsg2 expression in transgenic mouse skin confers a hyperproliferative and apoptosis-resistant phenotype to keratinocytes. *J Cell Sci* 2007;**120**:758–71.
100. Eshkind L, Tian Q, Schmidt A, Franke WW, Windoffer R, Leube RE. Loss of desmoglein 2 suggests essential functions for early embryonic development and proliferation of embryonal stem cells. *Eur J Cell Biol* 2002;**81**:592–8.
101. Den Z, Cheng X, Merched-Sauvage M, Koch PJ. Desmocollin 3 is required for pre-implantation development of the mouse embryo. *J Cell Sci* 2006;**119**:482–9.
102. Chidgey MA, Yue KK, Gould S, Byrne C, Garrod DR. Changing pattern of desmocollin 3 expression accompanies epidermal organisation during skin development. *Dev Dyn* 1997;**210**:315–27.
103. Hardman MJ, Liu K, Avilion AA, Merritt A, Brennan K, Garrod DR, et al. Desmosomal cadherin misexpression alters beta-catenin stability and epidermal differentiation. *Mol Cell Biol* 2005;**25**:969–78.
104. Getsios S, Simpson CL, Kojima S, Harmon R, Sheu LJ, Dusek RL, et al. Desmoglein 1-dependent suppression of EGFR signaling promotes epidermal differentiation and morphogenesis. *J Cell Biol* 2009;**185**:1243–58.
105. Lechler T, Fuchs E. Desmoplakin: an unexpected regulator of microtubule organization in the epidermis. *J Cell Biol* 2007;**176**:147–54.

106. Sumigray KD, Lechler T. Desmoplakin controls microvilli length but not cell adhesion or keratin organization in the intestinal epithelium. *Mol Biol Cell* 2012;**23**:792–9.
107. Dusek RL, Attardi LD. Desmosomes: new perpetrators in tumour suppression. *Nat Rev Cancer* 2011;**11**:317–23.
108. Wong MP, Cheang M, Yorida E, Coldman A, Gilks CB, Huntsman D, et al. Loss of desmoglein 1 expression associated with worse prognosis in head and neck squamous cell carcinoma patients. *Pathology* 2008;**40**:611–6.
109. Chun MG, Hanahan D. Genetic deletion of the desmosomal component desmoplakin promotes tumor microinvasion in a mouse model of pancreatic neuroendocrine carcinogenesis. *PLoS Genet* 2010;**6**:e1001120.
110. Beaudry VG, Jiang D, Dusek RL, Park EJ, Knezevich S, Ridd K, et al. Loss of the p53/p63 regulated desmosomal protein Perp promotes tumorigenesis. *PLoS Genet* 2010;**6**: e1001168.
111. Dusek RL, Bascom JL, Vogel H, Baron S, Borowsky AD, Bissell MJ, et al. Deficiency of the p53/p63 target Perp alters mammary gland homeostasis and promotes cancer. *Breast Cancer Res* 2012;**14**:R65.
112. Leckband DE, le Duc Q, Wang N, de Rooij J. Mechanotransduction at cadherin-mediated adhesions. *Curr Opin Cell Biol* 2011;**23**:523–30.

CHAPTER SIX

The Role of VE-Cadherin in Vascular Morphogenesis and Permeability Control

Elisabetta Dejana*,†, Dietmar Vestweber‡

*IFOM, FIRC Institute of Molecular Oncology, University of Milan, Milan, Italy
†Department of Biosciences, School of Sciences, University of Milan, Milan, Italy
‡Max-Planck-Institute of Molecular Biomedicine, Muenster, Germany

Contents

Abstract

VE-cadherin is an endothelial-specific cadherin that is essential for the formation and regulation of endothelial cell junctions. The adhesive function and expression levels of VE-cadherin at endothelial contacts are central determinants of the control of vascular permeability and leukocyte recruitment into tissue. In addition to controlling junctional integrity, VE-cadherin modulates a multitude of signaling processes that influence the behavior of endothelial cells, such as proliferation, survival, migration, polarity,

Progress in Molecular Biology and Translational Science, Volume 116
ISSN 1877-1173
http://dx.doi.org/10.1016/B978-0-12-394311-8.00006-6

expression of other junctional components, and tube and lumen formation of blood vessels. This chapter highlights recent progress in understanding how VE-cadherin modulates these various cellular processes. In addition, the current knowledge about how VE-cadhern participates in the regulation of the endothelial barrier in the adult organism is discussed.

1. INTRODUCTION

Endothelial cells are linked to each other by two major types of cell–cell junctions named adherens junctions (AJ) and tight junctions (TJ). These structures are formed by transmembrane adhesive proteins, which promote homophilic cell-to-cell adhesion. In turn, the transmembrane proteins are linked to specific intracellular partners, which mediate their anchorage to the actin cytoskeleton and, as a consequence, stabilize junctions.[1,2]

Junctions in the endothelium control different features of vascular homeostasis. For instance, permeability to plasma solutes is controlled, to a good extent, by junctions. In addition, leukocyte extravasation and infiltration into inflamed areas require regulated opening and closing of cell-to-cell contacts.[3,4] Notably, junctional proteins can also transfer intracellular signals, which modulate several endothelial-specific functions, including contact inhibition of cell growth, cell polarity, tubulogenesis, resistance to apoptotic stimuli, and others.[5]

Endothelial junctions are highly specialized along the vascular tree according to the specific needs of the organs. For instance, in the brain, where strict control of permeability between the blood and the nervous system is required, junctions are well developed and rich in TJ.[6] In contrast, in capillary lymphatics, where dynamic traffic of lymphocytes, dendritic cells, and lymph occurs, junctions have different specialized features. In these vessels, junctional adhesive proteins are not organized in zipper-like structures but, instead, form separate and well-defined clusters (buttons) along the cell membrane with membrane flaps in between (Fig. 6.1). These particular structures allow dynamic entrance of cells and solutes, while the membrane flaps form a sort of valve inhibiting their exit.[7]

AJ and TJ are formed by different molecular components, with few exceptions (for a review, see Ref. 1). At TJ, adhesion is mediated by members of the claudin family and endothelial cells express several members of this family (claudin-1, -3, -5, -12, and others) depending on the type of vessel. Claudin-5 is an endothelial-specific claudin present in the endothelium of the majority of vessels. Claudins are linked inside the cells

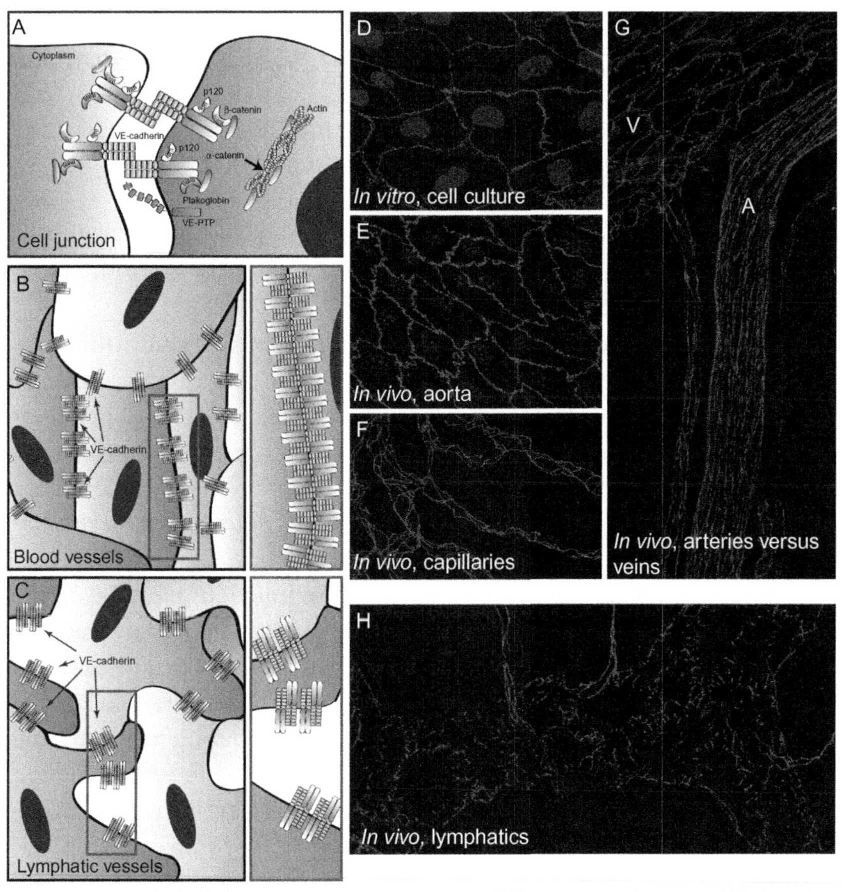

Figure 6.1 Endothelial junction organization in different vessel types. (A) Junction between endothelial cells comprising VE-cadherin and its direct molecular partners, armadillo proteins β-catenin, plakoglobin, and p120. The protein tyrosine phosphatase VE-PTP associates with VE-cadherin via extracellular domains. The α-catenin interacts as monomer with cadherin-bound β-catenin or plakoglobin, but as dimer with filamentous actin. (B) VE-cadherin forms zipper-like structures along cell-to-cell contacts in blood vessels. (C) VE-cadherin and other adherens junctional and tight junctional proteins form button-type junctions in capillary lymphatics. (D–H) Immunofluorescence staining of VE-cadherin in cultured endothelial cells (D) and *in vivo* in the aorta (E), capillaries (F), arteries and veins (A and V in panel G), and lymphatics (H). Note the similarities between VE-cadherin junctional staining in cultured endothelial cells and in blood vessels *in vivo*. In lymphatics, VE-cadherin shows a discontinuous button-like staining. See the text for details. (See Color Insert.)

to intracellular partners, which include the members of the ZO family, cingulin, ZONAB, and others. These proteins mediate the anchorage to actin but they may also transfer intracellular signals (for a review, see Ref. 8).

At AJ, on the other hand, adhesion is mediated by members of the cadherin family. In particular, endothelial cells express VE-cadherin (cadherin-5), N-cadherin (cadherin-2) and, in some vessels, also P- (cadherin-3) and T-cadherin (cadherin-13) (see below). VE- and N-cadherins can directly bind armadillo catenins such as β-catenin, plakoglobin, and p120 (for a review, see Ref. 1). As in the case of TJ, these AJ proteins promote junction stability through their interaction with the actin cytoskeleton. As discussed below in more detail, these proteins may also transfer intracellular signals that regulate different vascular functions.

In general, junctions are dynamic structures. In the adult, they can quickly open or close in response to agents that increase permeability or to the passage of leukocytes. Furthermore, endothelial cell-to-cell junctions need to adapt to changes in blood flow and hydrostatic pressure, which accompany the movements of the body. Junction organization may therefore vary not only in different types of vessels but also upon exposure to different hemodynamic conditions.

During vascular development in the embryo, junctions undergo different steps of maturation that parallel vascular remodeling and stabilization.[9] In the growing vasculature, the vessels are more permeable and adhesion at endothelial cell-to-cell junctions is relatively weak. The intracellular partners of the junctional adhesive proteins may also vary during junction maturation. In addition, even after contacts have been stably formed, adhesion proteins are still in dynamic equilibrium at junctions and recycle continuously between the plasma membrane and intracellular compartments.[10,11]

In this review, we focus on the molecular organization and functional changes of AJ and, in particular, of VE-cadherin and its molecular partners. The role of AJ in vascular morphogenesis and in the control of permeability to inflammatory cells and plasma solutes is discussed in detail. In contrast, we consider only marginally the structure and function of TJ in the endothelium. Excellent reviews on TJ organization and function are available in the literature.[8,12]

2. STRUCTURAL PROPERTIES OF VE-CADHERIN

The structural features of the *trans* interactions of cadherins have been intensively studied because they form the basis of homophilic, cadherin-specific cell recognition and adhesion. VE-cadherin belongs to the "classical" cadherins, which comprise two major subfamilies named type-I and type-II cadherins (see also Chapter 4). Both types of classical cadherins have five extracellular cadherin (EC) domains. Their elongated and substantially curved structure is stabilized by Ca^{2+} ions that bind to the boundaries between the EC domains.

Whereas the most membrane-distal, N-terminal EC1 domains of type-I cadherins contain a single highly conserved tryptophan at position 2 (W2), type-II cadherins contain in addition to W2 also a conserved tryptophan at position 4 (W4). Atomic resolution structural studies of type-I cadherins revealed that the EC1 domains form a *trans*-dimer interface based on the exchange or swapping of a β-strand between the two partner domains. This swapping of domains is stabilized by the insertion of the side chain of the conserved W2 tryptophan into a complementary hydrophobic pocket in the partner EC1 domain.[13–15] A principally similar basis for *trans* interactions of EC1 domains of different type-II cadherins was recently described.[16] These interactions rely on the insertion of the W2 and W4 tryptophan residues into a larger hydrophobic pocket of the partner EC1 domains.

In addition, lateral *cis* interactions of type-I cadherins may allow cadherin clustering and provide the interface for *trans* interactions. *Cis* interactions of type-I cadherins are mediated by lateral interactions between the EC1 and EC2 domains of two neighboring molecules, which are facilitated by the elongated curved structure of the cadherin, as was first demonstrated for the type-I C-cadherin of *Xenopus*.[14]

For type-II VE-cadherin, a different model was initially suggested. A recombinant form of VE-cadherin expressed in bacteria and containing domains EC1–EC4 was analyzed by biochemical techniques and visualized by cryoelectron microscopy. 3D reconstructions of the electron micrographs on the basis of the crystal structure of C-cadherin[14] suggested a hexameric complex formed by two helical *cis* trimers with each monomer forming a *trans* dimer.[17,18] A hexameric complex was in agreement with the recombinant protein molecular weight as determined by biochemical methods. The *cis* interactions of three parallel VE-cadherin molecules in one triple helix were reported to rely on interactions between the EC4 domains.[17,18]

However, the analysis of a recombinant form of VE-cadherin containing all five EC domains fused to a multimerizing coiled-coil domain of cartilage matrix protein suggested that VE-cadherin homophilic interaction follows the same mode described for type-I cadherins.[19] Since this recombinant protein was expressed in mammalian cells, it had the chance to become glycosylated. This fusion protein gave rise to ring-like structures and double-ring-like structures in electron micrographs, which was in line with EC1–EC2 *cis* interactions and EC1-based *trans* interactions, in full agreement with the elongated curved crystal structure of full length C-cadherin. Ahrens *et al.*[19] suggested that the hexameric structure of the EC1–EC4 protein was due to the lack of glycosylation and the absence of the EC5 domain. This assumption was indeed verified when it was shown that the natively glycosylated whole VE-cadherin ectodomain (EC1–EC5) forms dimers, but no higher order multimers.[20] Moreover, EC3 and EC4 fragments formed higher order multimers only when glycosylation was absent. Finally, imaging the overall arrangement of human recombinant EC1–EC5 VE-cadherin by atomic force microscopy[20] confirmed the results by Ahrens *et al.*[19] Thus, VE-cadherin homophilic interactions are probably not based on cylindrical hexameric structures but are rather comparable to structures as described for classical type-I cadherins.

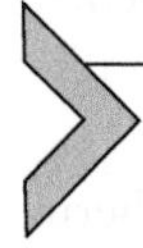

3. VE-CADHERIN PARTNERS AND INTRACELLULAR SIGNALING

3.1. General characteristics of VE-cadherin signaling

A major function of VE-cadherin is to promote homotypic cell–cell adhesion and thereby create and maintain endothelial integrity. This property is particularly important during vascular development when single endothelial cells need to preserve adhesion to one another to form new vessels (for review, see Refs. 12,21).

Besides promoting adhesion between endothelial cells, VE-cadherin and AJ, in general, can transfer intracellular signals. The signaling network transduced by VE-cadherin is quite complex and varies in different functional conditions and in growing or resting vasculature.[22,23] As a general concept, the natural state of endothelial cells in mature vessels is that of a confluent, resting monolayer, where VE-cadherin is clustered at cell–cell contacts and AJ are correctly organized. In these conditions, signaling through VE-cadherin promotes vascular stability, such as contact inhibition of growth, protection from apoptosis, and control of permeability.

However, also during angiogenesis, VE-cadherin contributes to key and specific functional properties of endothelial cells. These include the establishment of cell polarity and consequently lumen formation and vascular tubulogenesis,[24,25] rearrangement of the cytoskeleton,[4] and TJ remodeling and gene transcription.[26] The complexity of VE-cadherin signaling is underlined by the long list of signaling proteins found to be directly or indirectly associated to it or to AJ in the endothelium (Table 6.1).

Table 6.1 VE-cadherin functional partners[a]

Partners	Function	References
β-Catenin	Junction architecture and signaling	27–29
p120	VEC stabilization and signaling	30–32
Plakoglobin	Junction architecture and signaling	33,34
α-Catenin, α-actinin, Eplin	Cytoskeletal anchorage and organization	35–38
PECAM	Mechanosensor	39,40
TIAM	Activation of Rac, junction stability	41,42
Rap1, Raf -1, MAGI	Junction stability and maturation	24,43–45
PI3 kinase, Akt	Signal transduction	26
Src, FAK	Increase in permeability	46–48
DEP-1	VEGFR2 dephosphorylation	49
VE-PTP	VE-cadherin dephosphorylation, permeability control, leukocyte diapedesis	34,50,51
RPTP-μ	Permeability control	52
PTP1B, SHP2	Permeability control	53–55
Csk	Src inactivation, inhibition of cell growth	56
PAR3/PAR6/aPKC	Cell polarity, lumen formation	24,57–59
CCM1/Krit1	Cell polarity, junction stability	24,44
VEGFR2	Contact inhibition of cell growth, permeability control	34,46,49,60
FGFR	Cell motility and growth	61,62
TGF β receptor	Inhibition of cell motility	63
Caveolin-1	Junction permeability	64

[a]The table reports some of the best studied VE-cadherin partners. We apologize for not providing a more complete list due to space constraints.

VE-cadherin molecules are expressed in high numbers on the endothelial cell membrane. Therefore, it is likely that VE-cadherin forms distinct complexes with different partners also in the same cell. These complexes may transfer different types of signals to the cells depending on the type of stimulus to which the cells are exposed. Furthermore, it is likely that VE-cadherin association with one or another type of complex is reversible and both temporally and spatially regulated.

VE-cadherin modifications that might induce specific association and activation of one or another partner are still poorly understood. It has been reported that phosphorylation on tyrosine[56,65] or on serine residues[60] of VE-cadherin mediates the interaction with specific intracellular proteins. Alternatively, components of the VE-cadherin complex such as β-catenin may indirectly mediate the interaction with signaling mediators.[49]

For clarity, in the subsections below, we divide the intracellular VE-cadherin partners according to their activities in order to draw a tentative picture of the multiple roles of this protein in vascular responses.

3.2. Cytoskeletal remodeling by VE-cadherin

Interactions of VE-cadherin with the cytoskeleton enable VE-cadherin to modulate endothelial cell contacts during embryogenesis and influence vasculogenesis, angiogenesis, lumen formation and endothelial cell migration.[5] In addition, several of these cytoskeletal interaction partners in combination with kinases and phosphatases also regulate endothelial junctions in the adult organism, thereby determining vascular permeability and leukocyte trafficking, which is discussed at the end of this review. The cytoplasmic tail of VE-cadherin is highly homologous to that of other classic cadherins[27] and allows its interaction with the typical cadherin partners called catenins. In particular, VE-cadherin binds quite tightly through its cytoplasmic domain to β-catenin and plakoglobin (Fig. 6.1A). Importantly, in other cell types, such as epithelial cells, plakoglobin is a component of desmosomes, which are absent in endothelial cells (see also Chapter 5). Since plakoglobin binds with high affinity to desmosomal cadherins, such as desmogleins and desmocollins, little plakoglobin is available to interact with cadherins in the epithelium. However, in the endothelium, desmosomal cadherins are absent and therefore plakoglobin can effectively bind to VE-cadherin.[33,34] It is possible that in endothelial cells, plakoglobin forms types of junctions that are different from those containing β-catenin by preferentially interacting with desmoplakin, which in turn associates with vimentin instead of actin microfilaments.[66,67]

Both β-catenin and plakoglobin can bind α-catenin, which regulates the actin cytoskeleton, and therefore cell shape and motility. How cadherin–catenin complexes interact with the actin cytoskeleton is still a matter of debate, since direct interaction of the cadherin–β-catenin–α-catenin complex with actin filaments has been questioned.[35] In its monomeric state, α-catenin binds β-catenin coupled to cadherin, but not actin.[68] However, homodimeric α-catenin binds actin, and not β-catenin, and promotes the bundling of actin filaments.[69] Thus, the binding of α-catenin to the cadherin complex may negatively regulate its action on actin polymerization.[35] Various actin-binding proteins, such as α-actinin,[36,70] vinculin,[71,72] and Eplin,[37] can also bind α-catenin. Of these candidate linker molecules, Eplin was shown to directly link the E-cadherin–β-catenin–α-catenin complex with actin.[37] Eplin was also found in endothelial cells at junctions, where it was suggested to link the VE-cadherin–catenin complex with actin and to act as a mechanotransmitter, which enables vinculin to bind to α-catenin and to form a secondary link between the VE-cadherin complex and actin.[38] Interestingly, based on the regulated binding of vinculin, so-called focal adherens junctions between endothelial cells were recently described.[73] These junctions were defined as attachment sites for radial F-actin bundles and sites of junction remodeling. They were formed upon stimulation with junction-challenging agents such as thrombin. The binding of vinculin was dependent on actomyosin contractility and on exposure of a specific vinculin-binding α-catenin domain. As a consequence of this force-dependent binding of vinculin, mechanical stability was reinforced and endothelial junctions were prevented from opening.[73]

An additional cytoplasmic partner of cadherins is p120, which is a Src substrate and, like β-catenin and plakoglobin, contains Armadillo repeats.[30] However, p120 binds to a membrane-proximal domain of VE-cadherin and does not associate with α-catenin or other actin-binding proteins. The functional role of p120 is complex. It is known to stabilize cadherins and, in particular, VE-cadherin at the membrane by inhibiting their internalization and degradation.[30,31] Furthermore, when p120 is free in the cytoplasm, it is a powerful inhibitor of RhoA, as it acts as a RhoA guanine nucleotide dissociation inhibitor[30] (see also Chapter 18).

VE-cadherin expression and clustering might activate the small GTPase Rac through the Rac GEF (guanosine exchange factor) TIAM, which was also found at the AJ. Rac can stabilize AJ,[41,42] whereas activation of Rho tends to induce actin-based endothelial cell retraction.[74] Other small GTPases able to interact with VE-cadherin are Rap1[43,44] and Raf-1.[45] Rap1 stabilizes cell-to-cell junctions and binds to AJ indirectly, through its link to afadin or CCM1/Krit.[24,44] Raf1 is recruited to VE-cadherin-based AJ by Rap1 and, in turn, brings the Rho effector Rok-α to nascent AJs.[45] This interaction helps the timely maturation of AJ during angiogenesis.

3.3. VE-cadherin and TJ

Cadherin clustering at AJ promotes junction maturation and TJ organization.[35] This effect may be mediated simply by tethering the cells together and allowing the assembly of TJ adhesion molecules at the membrane. However, VE-cadherin might also act by upregulating the expression of TJ proteins such as claudin-5, a cell-specific adhesion protein of endothelial TJs. VE-cadherin upregulates claudin-5 expression by releasing a transcription inhibitory complex comprising FoxO1 and β-catenin from the claudin-5 promoter.[26] This pathway is mediated by VE-cadherin-triggered activation of PI3 kinase and Akt. These, in turn, inactivate FoxO1. Furthermore, ZO-1, which is commonly considered a cytoplasmic component of TJ, could be found at the AJ during the early steps of their organization.[75] ZO-1 localizes at the AJ through binding to β-catenin, but this is transient because ZO-1 subsequently moves away and concentrates at the TJ.

3.4. VE-cadherin mediates contact inhibition of cell growth

Endothelial cells grow in monolayers and strongly inhibit their growth as soon as they come in contact with neighboring cells. This process, called contact inhibition of cell growth, is mediated by the action of different signaling pathways.

VE-cadherin can interact and cocluster with growth factor receptors and control their intracellular signal. More specifically, VE-cadherin interacts with vascular endothelial growth factor receptor 2 (VEGFR2) and inhibits its downstream proliferation signal.[49] The association of VE-cadherin with the VEGFR2 complex requires the cytoplasmic domain of VE-cadherin and β-catenin. When VEGFR2 is bound to VE-cadherin, this receptor is retained at the cell membrane, where it is quickly dephosphorylated by density-enhanced phosphatase 1.[49]

VE-cadherin can also interact and promote formation of the TGF-β receptor complex by mediating the coclustering of TGFβRII with Alk-1 and Alk-5. This process increases TGF-β signaling through Smad phosphorylation and causes inhibition of endothelial cell growth and motility.[63]

Other pathways that can contribute to inhibition of cell growth have been described. These include the binding of the kinase Csk to the VE-cadherin tail on tyrosine 685. Csk is able to inactivate Src and reduce endothelial cell proliferation.[56]

3.5. VE-cadherin and transcriptional signaling

In parallel with the broad range of VE-cadherin functions at the cell surface, an increasing body of evidence shows that VE-cadherin can influence nuclear processes by regulating the nuclear trafficking of transcriptionally active partners.

β-Catenin, one of the key players in canonical Wnt signaling,[28] is constitutively bound to classical cadherins. When β-catenin is released into the cytoplasm, it is quickly inactivated by phosphorylation and ubiquitination by a protein complex that includes axin and adenomatous polyposis coli. However, when the amount of VE-cadherin expressed is reduced or when endothelial cells are activated by Wnt, β-catenin can translocate to the nucleus and modulate gene transcription.[29] β-Catenin can do so through its interaction with members of the lymphoid enhancer factor (LEF)/T-cell factor (TCF) protein family and with other transcription factors such as FoxO1, Smads, HIFs, and others (see for review, Ref. 29). In addition, cadherins may inhibit β-catenin signaling by promoting its degradation by a membrane-associated degradation complex.[76] Thus, cadherins can interfere with the transcriptional activity of β-catenin, not only by diverting it away from the nucleus but also by increasing its degradation at the membrane.

Besides being a membrane trap for β-catenin, VE-cadherin may also act as such a trap for other armadillo proteins with transcriptional activity, such as p120 and plakoglobin. At the transcriptional level, p120 binds the transcriptional repressor Kaiso and displaces it from promoters allowing their activation (for a review, see Ref. 32). The transcriptional activity of plakoglobin has not been studied as extensively as that of β-catenin. The effect of plakoglobin on gene expression can be positive, negative, or neutral, depending on the cellular context.[1,32] Although plakoglobin has been shown to bind to TCF/LEF transcription factors and exert transcriptional activity, other reports have shown that the plakoglobin–TCF complex is unable to bind DNA.[32,77]

Taken together, the data discussed above imply that cadherins, and VE-cadherin in particular, may repress or induce gene transcription. They do so in an unusual and indirect way, by binding transcriptionally active proteins and retaining them at the membrane. Notably, such binding can occur in the absence of cell-to-cell adhesion, but it becomes more stable upon clustering of cadherins.[78]

3.6. VE-cadherin mediates endothelial cell polarity

In the mouse embryo, absence of VE-cadherin results in alteration of the lumen of the aorta and cardinal vein and enlargement of the lumen of the cephalic vessels.[24] Likewise, in zebrafish, knockdown of the gene encoding VE-cadherin induces severe defects in the vascular lumen.[79]

Although different mechanisms might regulate vascular lumen formation, the establishment of endothelial apical–basal polarity is expected to be an essential requirement for all of them. The partitioning defective (PAR) polarity complex is a key determinant of cell polarity. This complex is formed by the small GTPases Cdc42 and Rac, TIAM; the adaptor and regulatory proteins Par-3 and Par-6; and the effector PKCζ (for a review, see Ref. 57). VE-cadherin was found to be able to interact with Par-3 and Par-6,[58] and Cdc 42 and Par-3 have been shown to regulate lumen formation in 3D cultures of endothelial cells.[59] More recent data show that VE-cadherin is needed for the correct localization and activation of the Par-3/Par-6/PKCζ polarity complex, which in turn determines endothelial apical–basal polarity.[24]

CCM1/Krit is needed for the correct localization of VE-cadherin and the polarity complex at junctions.[24] In the absence of Krit, apical–basal polarity of endothelial cells is lost and the vascular lumen severely affected. These data might explain why patients lacking CCM1/Krit develop severe vascular malformations in the brain microvasculature.[24]

Investigations of vascular lumen formation in the developing aorta showed that CD34 sialomucins, moesin, F-actin, and myosin II localize at cell-to-cell junctions through the action of VE-cadherin.[25] This, in combination with activation of the cells with VEGF and the consequent shape change, leads to lumen formation in the developing aorta.

3.7. VE-cadherin and junctional proteins as mechanosensors

Endothelial cells can sense and respond to mechanic stimuli. In other words, they can translate forces into biochemical signals. This is an important property of these cells, which are in continuous contact with shear and stretching forces induced by the hemodynamic conditions to which they are exposed. Several mechanosensors have been proposed, including cilia, caveolin, ion channels, integrins, and others (for a review, see Ref. 39). An endothelial-specific flow sensor has been described.[40] It includes PECAM (platelet and endothelial cell adhesion molecule), which acts as a mechanosensor, VE-cadherin, and VEGFR2. Exposure of the cells to shear stress activates members of the Src family, which, in turn, phosphorylates PECAM. VE-cadherin associates with PECAM and brings VEGFR2 to the complex. This activates VEGFR2 in a ligand-independent way and VEGFR2, in turn, activates signaling pathways such as Akt, MAP kinases, and nitric oxide synthetase.[40] These complex molecular interactions tell us that VE-cadherin and other adhesive proteins may play important signaling roles independently from their adhesive properties.

4. VE-CADHERIN AND N-CADHERIN CROSS TALK

Even if VE-cadherin is the most prominent cadherin at endothelial AJs, it is not the only cadherin expressed in endothelial cells. N-cadherin can be found at comparable levels in most of the endothelial cells examined so far.[61,80] P-cadherin expression was detected by PCR analysis and T-cadherin was also found in the vasculature in tissue sections.[81]

N-cadherin is not endothelial specific as it is also present in other cell types, including neuronal cells, lens cells, skeletal and heart muscle cells, osteoblasts, pericytes, and fibroblasts.[27] Its upregulation has been frequently associated to increased cell motility and, consistently, it is upregulated in highly invasive and poorly polarized cancer cells (see also Chapters 12 and 14).

The overall protein sequences of N- and VE-cadherin are quite homologous. Conservation is particularly high at the cytoplasmic tail, where -catenin-binding regions are located.[27] Although these two cadherins have a similar structure and bind the same intracellular partners, they seem to perform different functions in the endothelium. VE-cadherin mediates adhesion between adjacent endothelial cells and is located at AJ, while N-cadherin is mostly dispersed on the endothelial cell membrane.

It is possible that N-cadherin mediates heterotypic contacts between endothelial cells and mural cells such as pericytes or smooth muscle cells. Contacts between endothelial cells and the underlying smooth muscle cells, pericytes, or astrocytes have indeed been described (for a review, see Ref. 1), suggesting that these interactions may be important in elongation of vascular sprouts or in protection of endothelial cells from apoptosis. However, endothelial-specific loss of N-cadherin leads to embryonic lethality at mid-gestation, which occurs before pericyte investment of the vasculature.[80] This raises the possibility that N-cadherin plays a more complex functional role than that originally hypothesized.

Using endothelial cells expressing either VE- or N-cadherin, we found that both cadherins inhibit cell growth and apoptosis and limit FoxO1 and β-catenin nuclear signaling.[61] In contrast, only N-cadherin could induce endothelial cell motility, while VE-cadherin had an opposite effect. This appears to be due to the capacity of VE-cadherin to associate with FGFR (fibroblast growth factor receptor) and the phosphatase DEP-1, which, in turn, inhibits FGFR phosphorylation and signaling. In our analysis of gene expression profiles, we consistently found that a set of genes, downstream of FGF signaling and important in cell migration, were selectively upregulated by N-cadherin but not by VE-cadherin.[61]

Thus, changes in the levels and types of cadherins expressed by endothelial cells may affect endothelial cell functions and vascular development. It is possible that N-cadherin promotes endothelial migration and elongation of newly forming vessels. In contrast, VE-cadherin, when clustered at AJ, would promote vascular stability and limit cell motility and growth. It is likely that differences in signaling of cadherins are due, at least in part, to association with growth factor receptors and modulation of their downstream signaling.

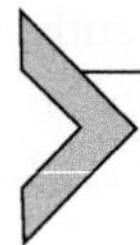

5. VE-CADHERIN IN LEUKOCYTE TRANSMIGRATION AND VASCULAR PERMEABILITY

5.1. Tyrosine phosphorylation of the VE-cadherin–catenin complex

The homophilic adhesive function of VE-cadherin was initially established by analyzing CHO cells transfected with VE-cadherin cDNA.[82] By using blocking antibodies against VE-cadherin, various laboratories demonstrated that the function of VE-cadherin is essential for the maintenance of endothelial cell contact stability. These antibodies abrogated cell contact stability *in vitro* between cultured endothelial cells as well as *in vivo* in mice, causing vascular leaks as well as enhancing extravasation of leukocytes in various tissues.[83–85]

To analyze whether posttranslational modifications of VE-cadherin play a role in the regulation of its adhesive function, various tyrosine residues within its cytoplasmic tail were mutated to either phenylalanine or glutamic acid. Analysis of these point mutants in transfected CHO cells suggested that phosphorylation of tyrosine residues 658 and 731 might lead to the destabilization of VE-cadherin-mediated cell contacts.[65] Indeed, the expression of Y658F or Y731F mutants of VE-cadherin in endothelial cells reduced the transmigration efficiency of myeloid cells through monolayers of the transfected cells.[86] Systematic testing of the ability of Y/F mutants of almost every tyrosine in VE-cadherin revealed that mutants for tyrosine residues 645, 731, and 733 are inhibitory for leukocyte transmigration, whereas all other mutants and remarkably also the Y658F mutant showed no effect.[87] Since docking of leukocytes to endothelial cells indeed stimulated tyrosine phosphorylation of VE-cadherin, these results demonstrated a role for VE-cadherin during leukocyte migration through endothelial cell monolayers. However, the identity of the relevant tyrosine residues has remained controversial.

In addition, these results suggested the importance of VE-cadherin and of the junctional, paracellular route for leukocyte diapedesis (transmigration through the endothelial barrier). This is in agreement with studies documenting that an alternative route via transcytosis through endothelial cell layers, which has been well documented *in vitro*, is only used by about 5–10% of the transmigrating leukocytes.[88] *In vivo*, both pathways have been documented by electron microscopy. However, due to technical limitations, it was never possible to analyze large numbers of cells and quantify whether the two routes are of similar importance *in vivo*.

A genetic approach based on modifying VE-cadherin in mice was recently adopted to clarify this question. In this approach, knock-in mice were generated resulting in replacement of VE-cadherin with a VE-cadherin–α-catenin fusion construct.[89] It was shown that this fusion protein interacts more efficiently with the actin cytoskeleton. Mice expressing VE-cadherin–α-catenin instead of VE-cadherin have highly stabilized endothelial cell contacts in the skin, which were resistant to VEGF- or histamine-induced vascular leakiness. Furthermore, neutrophil and T-lymphocyte recruitment into various tissues was strongly reduced.[89] This study established that VE-cadherin is an essential component of mechanisms that regulate the diapedesis of leukocytes *in vivo*. In addition, the results strongly suggest that enhancing the interaction between VE-cadherin and the actin cytoskeleton stabilizes endothelial cell contacts. A third conclusion resulting from this study is that the paracellular, junctional route is the preferred pathway for leukocyte extravasation in various tissues and for various inflammatory stimuli.[89]

5.2. Tyrosine phosphatases, kinases, and dissociation of the cadherin–catenin complex

The physiological mechanisms that regulate the adhesive function of VE-cadherin and thereby the stability of endothelial cell contacts *in vivo* are not yet very well understood. The role of various tyrosine residues in the cytoplasmic domain of VE-cadherin in the regulation of *in vitro* transmigration of leukocytes may be a hint that tyrosine phosphorylation is important in the regulation of endothelial junctions. The vascular endothelial protein tyrosine phosphatase (VE-PTP) is an endothelial-specific transmembrane protein,[90,91] which associates with VE-cadherin via extracellular domains and supports VE-cadherin adhesion in transfected cells[50] (Fig. 6.1A). Silencing VE-PTP in cultured endothelial cells attenuated the adhesive function of VE-cadherin, reduced endothelial cell contact integrity, and enhanced leukocyte diapedesis.[34] Since leukocyte binding to endothelial cells triggers the dissociation of VE-PTP from VE-cadherin, it was speculated that this dissociation is a prerequisite for diapedesis.[34] This hypothesis was verified *in vivo* in a study where inhibition of the dissociation of the VE-PTP/VE-cadherin complex led to a strong reduction of leukocyte recruitment into various tissues and also abrogated the induction of vascular permeability by VEGF.[51] This study was based on fusing two different additional protein domains to, respectively, the C-termini of VE-cadherin and VE-PTP. These additional domains contained different binding sites for a small molecular weight compound that could be used to stabilize the complex of the two proteins and prevent dissociation. Knock-in mice were generated by inserting the two gene constructs, encoding the respective fusion proteins, into the VE-cadherin locus. In these mice, it was demonstrated that blocking the dissociation of VE-cadherin/VE-PTP complex with the small chemical compound indeed blocked the induction of vascular permeability and strongly reduced the extravasation of leukocytes.[51] Thus, the dissociation of this tyrosine phosphatase turned out to be required for leukocyte extravasation and opening of endothelial contacts *in vivo*.

Several other lines of *in vitro* evidence document that tyrosine phosphorylation of components of the VE-cadherin–catenin complex or herewith associated molecules are important for the regulation of endothelial cell contacts. Besides VE-PTP, other receptor-type PTPs and cytosolic PTPs are involved in this regulation. RPTP-μ was found to bind directly to the cytosolic domain of VE-cadherin, and silencing of RPTP-μ by siRNA enhanced albumin permeability across monolayers of cultured human lung endothelial cells.[52] Overexpression of a catalytically inactive form of the cytosolic PTP1B in lung endothelial cells increased permeability, and PTP1B could be coprecipitated with the VE-cadherin–catenin complex.[53] Similar effects were reported from experiments using pharmacological inhibitors of SHP2 or overexpression of a phosphatase-dead mutant of SHP2 in lung endothelial cells.[54] SHP2 had also been reported before to associate with the

VE-cadherin–catenin complex and dissociation from this complex was seen upon stimulation with thrombin.[55]

Vascular permeability induced by VEGF was reported to require Src,[46,92] possibly affecting VE-cadherin in direct or indirect ways. VEGF was found to stimulate the phosphorylation of VE-cadherin at various tyrosine residues. Based on peptide mapping, one study reported that VEGF triggers phosphorylation at tyrosine residue 685 exclusively.[47] In addition, Src was found to coprecipitate with VE-cadherin and to phosphorylate a corresponding peptide of VE-cadherin *in vitro*.[47] However, others reported, based on phospho-tyrosine-specific antibodies, that VEGF stimulates phosphorylation of Y658 and Y731 of VE-cadherin. It was suggested that the consequence of this was the dissociation of p120 or β-catenin, respectively, since such dissociations were not seen when either VE-cadherin Y658F or Y731F mutants were analyzed.[93] Another recently published mechanism by which VEGF can trigger the dissociation of β-catenin involves the phosphorylation of β-catenin instead of VE-cadherin.[48] It was reported that VEGF induces activation and recruitment of the kinase FAK to endothelial junctions, where it binds via its FERM domain to the cytoplasmic domain of VE-cadherin, phosphorylates Y142 of β-catenin, and triggers its dissociation from VE-cadherin, thereby contributing to the VEGF-induced opening of endothelial junctions.[48] In the case of thrombin treatment, it was suggested that this triggers the association of phosphorylated caveolin-1 with β-catenin and plakoglobin, which leads to their dissociation from VE-cadherin and the opening of endothelial junctions.[64]

Due to the importance of catenins for the adhesive function of cadherins, the dissociation of the cadherin–catenin complex has always been an attractive mechanism to explain how cadherin-mediated cell contacts could be weakened or opened (Fig. 6.2A and B). An alternative mechanism could be that the interaction of the cadherin–catenin complex with the actin cytoskeleton becomes abrogated instead of the dissociation of catenins from cadherins or each other. Indeed, histamine has been reported to stimulate tyrosine phosphorylation of VE-caderin, β-catenin, and plakoglobin and to enhance extractability of the intact cadherin–catenin complex with detergent.[94,95]

5.3. Endocytosis of VE-cadherin

Besides dissociation of the cadherin–catenin complex or disconnecting the cadherin–catenin complex from the actin cytoskeleton, a third mechanism for opening endothelial cell contacts was reported to rely on endocytosis of

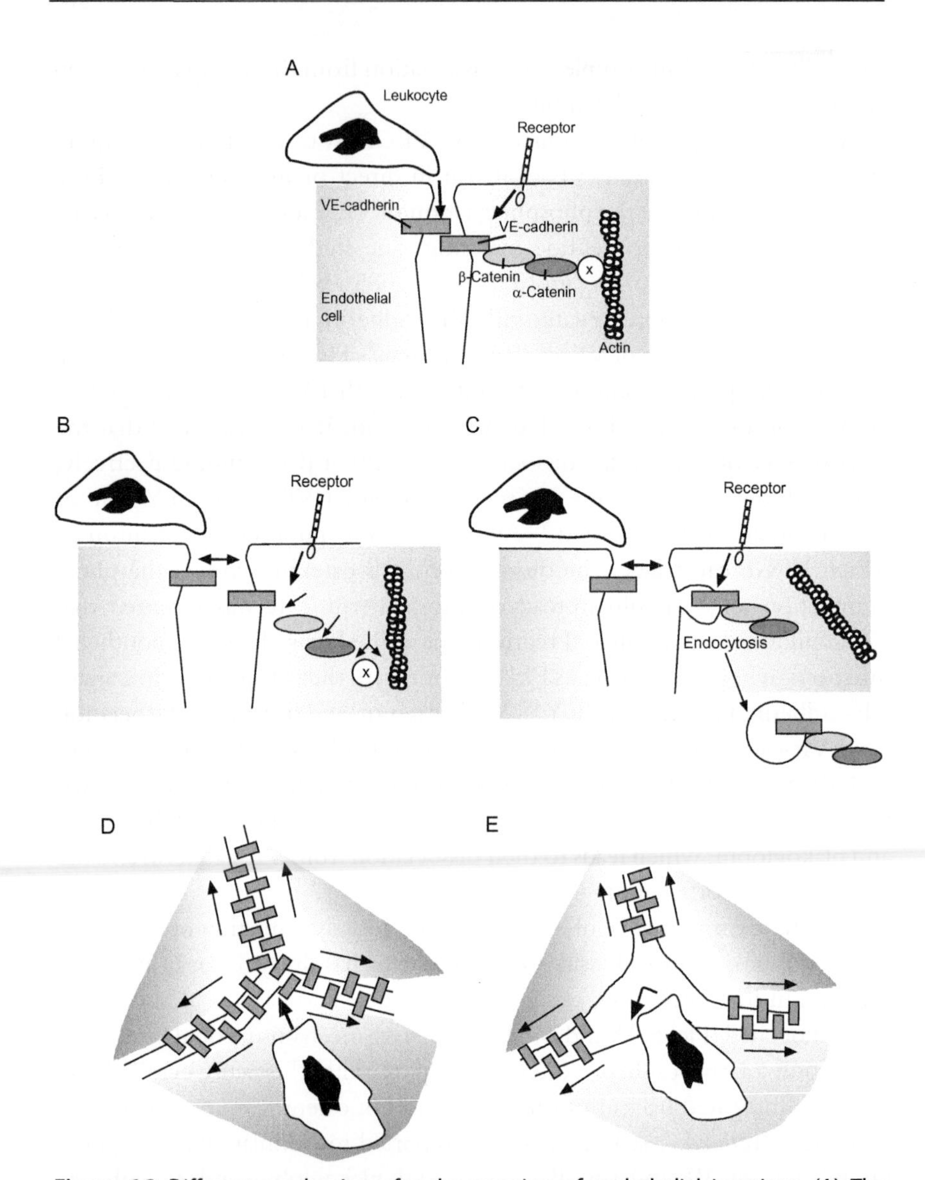

Figure 6.2 Different mechanisms for the opening of endothelial junctions. (A) The VE-cadherin–catenin complex stabilizes the endothelial junctions that form a barrier for transmigrating leukocytes and for the action of receptors inducing vascular permeability. (B) Either the dissociation of the cadherin–catenin complex or the disconnection of this complex from the actin cytoskeleton could lead to the opening of endothelial cell contacts. (C) Endocytosis could remove VE-cadherin from junctions and thereby open them. (D) View from above of an endothelial monolayer, with VE-cadherin molecules sealing the cell–cell contacts between three endothelial cells. Arrows indicate the lateral movement of VE-cadherin within the plane of the membrane. (E) Gaps for

VE-cadherin (Fig. 6.2C). VEGF was found to stimulate the phosphorylation of serine 665 in the cytoplasmic domain of VE-cadherin via Src-dependent phosphorylation of the guanine exchange factor Vav2, which activates the GTPase Rac, which, in turn, activates the p21-activated kinase (see also Chapter 3). Phosphorylation of S665 resulted in the recruitment of β-arrestin-2, which promotes the internalization of VE-cadherin into clathrin-coated vesicles and uptake into the cells.[60] Interestingly, the armadillo catenin p120 is known to bind to the membrane-proximal region of VE-cadherin, in the vicinity of S665. Binding of p120 to VE-cadherin was shown to stabilize VE-cadherin at the cell surface and control its internalization and degradation.[96,97] Binding of p120 to VE-cadherin may also be involved in the mechanism by which FGF can stabilize endothelial cell contacts. Inhibiting FGF signaling by various means was found to stimulate the dissociation of p120 from VE-cadherin and to reduce the cell surface expression of VE-cadherin.[62] This could be due to enhanced tyrosine phosphorylation of VE-cadherin, since interference with FGF signaling enhanced SHP2 levels in the cells due to inhibition of FGF-controlled SHP2 protein degradation.[98]

p120 was also reported to inhibit leukocyte transmigration through endothelial monolayers by reducing tyrosine phosphorylation of VE-cadherin,[99] what leads to reduced association with activated Src.[100] These effects were independent of p120-mediated effects on small GTPases, either inhibition of RhoA or activation of Rac1 and Cdc42.[101,102] Indeed, they were also seen with a p120 mutant lacking the RhoA binding domains.[100]

5.4. Opening of junctions by lateral movement of VE-cadherin

A VE-cadherin–GFP fusion protein was observed to "be moved laterally aside" at sites where leukocytes transmigrated through the monolayer of endothelial cells.[103] Such mechanism could destabilize endothelial junctions without dissociating VE-cadherin from the actin cytoskeleton and without the necessity to endocytose VE-cadherin (Fig. 6.2D and E). Although overexpression of p120 in endothelial cells was found to inhibit endothelial gap formation and leukocyte transmigration, no signs of internalization of VE-cadherin at sites of leukocyte transmigration were seen.[99] Overexpression of p120 reduced

transmigrating leukocytes could be formed between endothelial cells by a mechanism that would allow VE-cadherin to be moved aside within the plane of the plasma membrane. This mechanism may require neither dissociation of VE-cadherin from the actin cytoskeleton nor endocytosis of VE-cadherin. (See Color Insert.)

VE-cadherin tyrosine phosphorylation, and it was hypothesized that this might impair the lateral movement of VE-cadherin molecules.

6. CONCLUSIONS

VE-cadherin is a strictly endothelial-specific adhesion molecule that influences a broad spectrum of cellular responses of endothelial cell biology. Knowledge of the molecular processes that are influenced by VE-cadherin and that in turn modulate the adhesive function of VE-cadherin are of central importance for understanding the formation of blood vessels and the regulation of the barrier function of the blood vessel system. Despite much progress in the past, elucidating the mechanistic details that control junctional plasticity during embryonic development and in inflammatory settings of the adult organism is still a major challenge in vascular biology. Further insights into these processes will foster the development of new approaches to target the various dysfunctions of the vascular system.

ACKNOWLEDGMENTS

The work was supported in part by the Associazione Italiana per la Ricerca sul Cancro and "Special Program Molecular Clinical Oncology 5x1000" to AGIMM (AIRC-Gruppo Italiano Malattie Mieloproliferative) and by the European Research Council, the European Community Networks (EUSTROKE-contract-202213, OPTISTEM-contract-223098, ANGIOSCAFF-NMP3-LA-2008-214402 Networks; ENDOSTEM-HEALTH-2009-241440 and JUSTBRAIN HEALTH 2009 241861), the Italian Ministry of University and Research, the Cariplo Foundation, the Fondation Leducq Transatlantic Network of Excellence, the Max-Planck-Society, and the Deutsche Forschungsgemeinschaft (SFB629).

REFERENCES

1. Bazzoni G, Dejana E. Endothelial cell-to-cell junctions: molecular organization and role in vascular homeostasis. *Physiol Rev* 2004;**84**:869–901.
2. Vestweber D, Winderlich M, Cagna G, Nottebaum AF. Cell adhesion dynamics at endothelial junctions: VE-cadherin as a major player. *Trends Cell Biol* 2009;**19**:8–15.
3. Nourshargh S, Hordijk PL, Sixt M. Breaching multiple barriers: leukocyte motility through venular walls and the interstitium. *Nat Rev Mol Cell Biol* 2010;**11**:366–78.
4. Komarova Y, Malik AB. Regulation of endothelial permeability via paracellular and transcellular transport pathways. *Annu Rev Physiol* 2010;**72**:463–93.
5. Dejana E, Tournier-Lasserve E, Weinstein BM. The control of vascular integrity by endothelial cell junctions: molecular basis and pathological implications. *Dev Cell* 2009;**16**:209–21.
6. Liebner S, Czupalla CJ, Wolburg H. Current concepts of blood-brain barrier development. *Int J Dev Biol* 2011;**55**:467–76.
7. Baluk P, Fuxe J, Hashizume H, Romano T, Lashnits E, Butz S, et al. Functionally specialized junctions between endothelial cells of lymphatic vessels. *J Exp Med* 2007;**204**:2349–62.

8. Steed E, Balda MS, Matter K. Dynamics and functions of tight junctions. *Trends Cell Biol* 2010;**20**:142–9.
9. Potente M, Gerhardt H, Carmeliet P. Basic and therapeutic aspects of angiogenesis. *Cell* 2011;**146**:873–87.
10. de Beco S, Amblard F, Coscoy S. New insights into the regulation of E-cadherin distribution by endocytosis. *Int Rev Cell Mol Biol* 2012;**295**:63–108.
11. Hong S, Troyanovsky RB, Troyanovsky SM. Spontaneous assembly and active disassembly balance adherens junction homeostasis. *Proc Natl Acad Sci USA* 2010;**107**: 3528–33.
12. Wallez Y, Huber P. Endothelial adherens and tight junctions in vascular homeostasis, inflammation and angiogenesis. *Biochim Biophys Acta* 2008;**1778**:794–809.
13. Shapiro L, Fannon AM, Kwong PD, Thompson A, Lehmann MS, Grubel G, et al. Structural basis of cell-cell adhesion by cadherins. *Nature* 1995;**374**:327–37.
14. Boggon TJ, Murray J, Chappuis-Flament S, Wong E, Gumbiner BM, Shapiro L. C-cadherin ectodomain structure and implications for cell adhesion mechanisms. *Science* 2002;**296**:1308–13.
15. Haussinger D, Ahrens T, Aberle T, Engel J, Stetefeld J, Grzesiek S. Proteolytic E-cadherin activation followed by solution NMR and X-ray crystallography. *EMBO J* 2004;**23**:1699–708.
16. Patel SD, Ciatto C, Chen CP, Bahna F, Rajebhosale M, Arkus N, et al. Type II cadherin ectodomain structures: implications for classical cadherin specificity. *Cell* 2006;**124**:1255–68.
17. Legrand P, Bibert S, Jaquinod M, Ebel C, Hewat E, Vincent F, et al. Self-assembly of the vascular endothelial cadherin ectodomain in a Ca2+-dependent hexameric structure. *J Biol Chem* 2001;**276**:3581–8.
18. Hewat EA, Durmort C, Jacquamet L, Concord E, Gulino-Debrac D. Architecture of the VE-cadherin hexamer. *J Mol Biol* 2007;**365**:744–51.
19. Ahrens T, Lambert M, Pertz O, Sasaki T, Schulthess T, Mege RM, et al. Homoassociation of VE-cadherin follows a mechanism common to "classical" cadherins. *J Mol Biol* 2003;**325**:733–42.
20. Brasch J, Harrison OJ, Ahlsen G, Carnally SM, Henderson RM, Honig B, et al. Structure and binding mechanism of vascular endothelial cadherin, a divergent classical cadherin. *J Mol Biol* 2011;**408**:57–73.
21. Vestweber D, Broermann A, Schulte D. Control of endothelial barrier function by regulating vascular endothelial-cadherin. *Curr Opin Hematol* 2010;**17**:230–6.
22. Dejana E, Orsenigo F, Molendini C, Baluk P, McDonald DM. Organization and signaling of endothelial cell-to-cell junctions in various regions of the blood and lymphatic vascular trees. *Cell Tissue Res* 2009;**335**:17–25.
23. Harris ES, Nelson WJ. VE-cadherin: at the front, center, and sides of endothelial cell organization and function. *Curr Opin Cell Biol* 2010;**22**:651–8.
24. Lampugnani MG, Orsenigo F, Rudini N, Maddaluno L, Boulday G, Chapon F, et al. CCM1 regulates vascular-lumen organization by inducing endothelial polarity. *J Cell Sci* 2010;**123**:1073–80.
25. Strilic B, Kucera T, Eglinger J, Hughes MR, McNagny KM, Tsukita S, et al. The molecular basis of vascular lumen formation in the developing mouse aorta. *Dev Cell* 2009;**17**:505–15.
26. Taddei A, Giampietro C, Conti A, Orsenigo F, Breviario F, Pirazzoli V, et al. Endothelial adherens junctions control tight junctions by VE-cadherin-mediated upregulation of claudin-5. *Nat Cell Biol* 2008;**10**:923–34.
27. Takeichi M. The cadherins: cell-cell adhesion molecules controlling animal morphogenesis. *Development* 1988;**102**:639–55.

28. Clevers H. Wnt/beta-catenin signaling in development and disease. *Cell* 2006;**127**: 469–80.
29. Dejana E. The role of wnt signaling in physiological and pathological angiogenesis. *Circ Res* 2010;**107**:943–52.
30. Reynolds AB, Carnahan RH. Regulation of cadherin stability and turnover by p120ctn: implications in disease and cancer. *Semin Cell Dev Biol* 2004;**15**:657–63.
31. Xiao K, Oas RG, Chiasson CM, Kowalczyk AP. Role of p120-catenin in cadherin trafficking. *Biochim Biophys Acta* 2007;**1773**:8–16.
32. McCrea PD, Gu D, Balda MS. Junctional music that the nucleus hears: cell-cell contact signaling and the modulation of gene activity. *Cold Spring Harb Perspect Biol* 2009;**1**: a002923.
33. Lampugnani MG, Corada M, Caveda L, Breviario F, Ayalon O, Geiger B, et al. The molecular organization of endothelial cell to cell junctions: differential association of plakoglobin, beta-catenin, and alpha-catenin with vascular endothelial cadherin (VE-cadherin). *J Cell Biol* 1995;**129**:203–17.
34. Nottebaum AF, Cagna G, Winderlich M, Gamp AC, Linnepe R, Polaschegg C, et al. VE-PTP maintains the endothelial barrier via plakoglobin and becomes dissociated from VE-cadherin by leukocytes and by VEGF. *J Exp Med* 2008;**205**:2929–45.
35. Hartsock A, Nelson WJ. Adherens and tight junctions: structure, function and connections to the actin cytoskeleton. *Biochim Biophys Acta* 2008;**1778**:660–9.
36. Knudsen KA, Soler AP, Johnson KR, Wheelock MJ. Interaction of alpha-actinin with the cadherin/catenin cell-cell adhesion complex via alpha-catenin. *J Cell Biol* 1995;**130**:67–77.
37. Abe K, Takeichi M. EPLIN mediates linkage of the cadherin catenin complex to f-actin and stabilizes the circumferential actin belt. *Proc Natl Acad Sci USA* 2008;**105**:13–9.
38. Chervin-Pétinot A, Courçon M, Almagro S, Nicolas A, Grichine A, Grunwald D, et al. Epithelial protein lost in neoplasm (EPLIN) interacts with α-catenin and actin filaments in endothelial cells and stabilizes vascular capillary network in vitro. *J Biol Chem* 2012;**287**:7556–72.
39. Conway D, Schwartz MA. Lessons from the endothelial junctional mechanosensory complex. *F1000 Biol Rep* 2010;**1**:1.
40. Tzima E, Irani-Tehrani M, Kiosses WB, Dejana E, Schultz DA, Engelhardt B, et al. A mechanosensory complex that mediates the endothelial cell response to fluid shear stress. *Nature* 2005;**437**:426–31.
41. Lampugnani MG, Zanetti A, Breviario F, Balconi G, Orsenigo F, Corada M, et al. VE-cadherin regulates endothelial actin activating Rac and increasing membrane association of Tiam. *Mol Biol Cell* 2002;**13**:1175–89.
42. Cain RJ, Vanhaesebroeck B, Ridley AJ. The PI3k p110alpha isoform regulates endothelial adherens junctions via Pyk2 and Rac1. *J Cell Biol* 2010;**188**:863–76.
43. Sakurai A, Fukuhara S, Yamagishi A, Sako K, Kamioka Y, Masuda M, et al. MAGI-1 is required for Rap1 activation upon cell-cell contact and for enhancement of vascular endothelial cadherin-mediated cell adhesion. *Mol Biol Cell* 2006;**17**:966–76.
44. Glading A, Han J, Stockton RA, Ginsberg MH. KRIT-1/CCM1 is a Rap1 effector that regulates endothelial cell cell junctions. *J Cell Biol* 2007;**179**:247–54.
45. Wimmer R, Cseh B, Maier B, Scherrer K, Baccarini M. Angiogenic sprouting requires the fine tuning of endothelial cell cohesion by the Raf-1/Rok-alpha complex. *Dev Cell* 2012;**22**:158–71.
46. Weis S, Shintani S, Weber A, Kirchmair R, Wood M, Cravens A, et al. Src blockade stabilizes a Flk/cadherin complex, reducing edema and tissue injury following myocardial infarction. *J Clin Invest* 2004;**113**:885–94.
47. Wallez Y, Cand F, Cruzalegui F, Wernstedt C, Souchelnytskyi S, Vilgrain I, et al. Src kinase phosphorylates vascular endothelial-cadherin in response to vascular endothelial

growth factor: identification of tyrosine 685 as the unique target site. *Oncogene* 2007;**26**:1067–77.

48. Chen XL, Nam JO, Jean C, Lawson C, Walsh CT, Goka E, et al. VEGF-induced vascular permeability is mediated by FAK. *Dev Cell* 2012;**22**:146–57.
49. Lampugnani MG, Orsenigo F, Gagliani MC, Tacchetti C, Dejana E. Vascular endothelial cadherin controls VEGFR-2 internalization and signaling from intracellular compartments. *J Cell Biol* 2006;**174**:593–604.
50. Nawroth R, Poell G, Ranft A, Samulowitz U, Fachinger G, Golding M, et al. VE-PTP and VE-cadherin ectodomains interact to facilitate regulation of phosphorylation and cell contacts. *EMBO J* 2002;**21**:4885–95.
51. Broermann A, Winderlich M, Block H, Frye M, Rossaint J, Zarbock A, et al. Dissociation of VE-PTP from VE-cadherin is required for leukocyte extravasation and for VEGF-induced vascular permeability in vivo. *J Exp Med* 2011;**208**:2393–401.
52. Sui XF, Kiser TD, Hyun SW, Angelini DJ, Del Vecchio RL, Young BA, et al. Receptor protein tyrosine phosphatase microregulates the paracellular pathway in human lung microvascular endothelia. *Am J Pathol* 2005;**166**:1247–58.
53. Grinnell KL, Chichger H, Braza J, Duong H, Harrington EO. Protection against LPS-induced pulmonary edema through the attenuation of protein tyrosine phosphatase-1B oxidation. *Am J Respir Cell Mol Biol* 2012;**46**:623–32.
54. Grinnell KL, Casserly B, Harrington EO. Role of protein tyrosine phosphatase SHP2 in barrier function of pulmonary endothelium. *Am J Physiol Lung Cell Mol Physiol* 2010;**298**:L361–L370.
55. Ukropec JA, Hollinger MK, Salva SM, Woolkalis MJ. SHP2 association with VE-cadherin complexes in human endothelial cells is regulated by thrombin. *J Biol Chem* 2000;**275**:5983–6.
56. Baumeister U, Funke R, Ebnet K, Vorschmitt H, Koch S, Vestweber D. Association of Csk to VE-cadherin and inhibition of cell proliferation. *EMBO J* 2005;**24**:1686–95.
57. McCaffrey LM, Macara IG. Signaling pathways in cell polarity. *Cold Spring Harb Perspect Biol* 2012;**4**:a009654.
58. Iden S, Rehder D, August B, Suzuki A, Wolburg-Buchholz K, Wolburg H, et al. A distinct PAR complex associates physically with VE-cadherin in vertebrate endothelial cells. *EMBO Rep* 2006;**7**:1239–46.
59. Koh W, Mahan RD, Davis GE. Cdc42- and Rac1-mediated endothelial lumen formation requires Pak2, Pak4 and Par3, and PKC-dependent signaling. *J Cell Sci* 2008;**121**:989–1001.
60. Gavard J, Gutkind JS. VEGF controls endothelial-cell permeability by promoting the beta-arrestin-dependent endocytosis of VE-cadherin. *Nat Cell Biol* 2006;**8**:1223–34.
61. Giampietro C, Taddei A, Corada M, Sarra-Ferraris GM, Alcalay M, Cavallaro U, et al. Overlapping and divergent signaling pathways of N-cadherin and VE-cadherin in endothelial cells. *Blood* 2012;**119**:2159–70.
62. Murakami M, Nguyen LT, Zhuang ZW, Moodie KL, Carmeliet P, Stan RV, et al. The FGF system has a key role in regulating vascular integrity. *J Clin Invest* 2008;**118**:3355–66.
63. Rudini N, Felici A, Giampietro C, Lampugnani M, Corada M, Swirsding K, et al. VE-cadherin is a critical endothelial regulator of TGF-beta signalling. *EMBO J* 2008;**27**:993–1004.
64. Kronstein R, Seebach J, Grossklaus S, Minten C, Engelhardt B, Drab M, et al. Caveolin-1 opens endothelial cell junctions by targeting catenins. *Cardiovasc Res* 2012;**93**:130–40.
65. Potter MD, Barbero S, Cheresh DA. Tyrosine phosphorylation of VE-cadherin prevents binding of p120- and beta-catenin and maintains the cellular mesenchymal state. *J Biol Chem* 2005;**280**:31906–12.

66. Cattelino A, Liebner S, Gallini R, Zanetti A, Balconi G, Corsi A, et al. The conditional inactivation of the beta-catenin gene in endothelial cells causes a defective vascular pattern and increased vascular fragility. *J Cell Biol* 2003;**162**:1111–22.
67. Schmelz M, Moll R, Kuhn C, Franke WW. Complexus adhaerentes, a new group of desmoplakin-containing junctions in endothelial cells: Ii. Different types of lymphatic vessels. *Differentiation* 1994;**57**:97–117.
68. Yamada S, Pokutta S, Drees F, Weis WI, Nelson WJ. Deconstructing the cadherin-catenin-actin complex. *Cell* 2005;**123**:889–901.
69. Drees F, Pokutta S, Yamada S, Nelson WJ, Weis WI. Alpha-catenin is a molecular switch that binds E-cadherin-beta-catenin and regulates actin-filament assembly. *Cell* 2005;**123**:903–15.
70. Nieset JE, Redfield AR, Jin F, Knudsen KA, Johnson KR, Wheelock MJ. Characterization of the interactions of alpha-catenin with alpha-actinin and beta-catenin/plakoglobin. *J Cell Sci* 1997;**110**:1013–22.
71. Weiss EE, Kroemker M, Rudiger AH, Jockusch BM, Rudiger M. Vinculin is part of the cadherin-catenin junctional complex: complex formation between alpha-catenin and vinculin. *J Cell Biol* 1998;**141**:755–64.
72. Watabe-Uchida M, Uchida N, Imamura Y, Nagafuchi A, Fujimoto K, Uemura T, et al. Alpha-catenin-vinculin interaction functions to organize the apical junctional complex in epithelial cells. *J Cell Biol* 1998;**142**:847–57.
73. Huveneers S, Oldenburg J, Spanjaard E, van der Krogt G, Grigoriev I, Akhmanova A, et al. Vinculin associates with endothelial VE-cadherin junctions to control force-dependent remodeling. *J Cell Biol* 2012;**196**:641–52.
74. Stockton RA, Shenkar R, Awad IA, Ginsberg MH. Cerebral cavernous malformations proteins inhibit Rho kinase to stabilize vascular integrity. *J Exp Med* 2010;**207**:881–96.
75. Itoh M, Nagafuchi A, Moroi S, Tsukita S. Involvement of ZO-1 in cadherin-based cell adhesion through its direct binding to alpha-catenin and actin filaments. *J Cell Biol* 1997;**138**:181–92.
76. Maher MT, Flozak AS, Stocker AM, Chenn A, Gottardi CJ. Activity of the beta-[illegible] based adhesion. *J Cell Biol* 2009;**186**:219–28.
77. Zhurinsky J, Shtutman M, Ben-Ze'ev A. Differential mechanisms of LEF/TCF family-dependent transcriptional activation by beta-catenin and plakoglobin. *Mol Cell Biol* 2000;**20**:4238–52.
78. Cavallaro U, Dejana E. Adhesion molecule signalling: not always a sticky business. *Nat Rev Mol Cell Biol* 2011;**12**:189–97.
79. Montero-Balaguer M, Swirsding K, Orsenigo F, Cotelli F, Mione M, Dejana E. Stable vascular connections and remodeling require full expression of VE-cadherin in zebrafish embryos. *PLoS One* 2009;**4**:e5772.
80. Luo Y, Radice GL. N-cadherin acts upstream of VE-cadherin in controlling vascular morphogenesis. *J Cell Biol* 2005;**169**:29–34.
81. Hebbard LW, Garlatti M, Young LJ, Cardiff RD, Oshima RG, Ranscht B. T-cadherin supports angiogenesis and adiponectin association with the vasculature in a mouse mammary tumor model. *Cancer Res* 2008;**68**:1407–16.
82. Breviario F, Caveda L, Corada M, Martin Padura I, Navarro P, Golay J, et al. Functional properties of human vascular endothelial cadherin (7b4/cadherin-5), an endothelium-specific cadherin. *Arterioscler Thromb Vasc Biol* 1995;**15**:1229–39.
83. Gotsch U, Borges E, Bosse R, Böggemeyer E, Simon M, Mossmann H, et al. VE-cadherin antibody accelerates neutrophil recruitment in vivo. *J Cell Sci* 1997;**110**:583–8.

84. Matsuyoshi N, Toda K, Horiguchi Y, Tanaka T, Nakagawa S, Takeichi M, et al. In vivo evidence of the critical role of cadherin-5 in murine vascular integrity. *Proc Assoc Am Physicians* 1997;**109**:362–71.
85. Corada M, Mariotti M, Thurston G, Smith K, Kunkel R, Brockhaus M, et al. Vascular endothelial-cadherin is an important determinant of microvascular integrity in vivo. *Proc Natl Acad Sci USA* 1999;**96**:9815–20.
86. Allingham MJ, van Buul JD, Burridge K. ICAM-1-mediated, src- and pyk2-dependent vascular endothelial cadherin tyrosine phosphorylation is required for leukocyte transendothelial migration. *J Immunol* 2007;**179**:4053–64.
87. Turowski P, Martinelli R, Crawford R, Wateridge D, Papagiorgiou A-P, Lampugnani MG, et al. Phosphorylation of vascular endothelial cadherin controls lymphocyte emigration. *J Cell Sci* 2008;**121**:29–37.
88. Carman CV, Springer TA. A transmigratory cup in leukocyte diapedesis both through individual vascular endothelial cells and between them. *J Cell Biol* 2004;**167**:377–88.
89. Schulte D, Küppers V, Dartsch N, Broermann A, Li H, Zarbock A, et al. Stabilizing the VE-cadherin-catenin complex blocks leukocyte extravasation and vascular permeability. *EMBO J* 2011;**30**:4157–70.
90. Fachinger G, Deutsch U, Risau W. Functional interaction of vascular endothelial-protein-tyrosine phosphatase with the angiopoietin receptor tie-2. *Oncogene* 1999;**18**: 5948–53.
91. Baumer S, Keller L, Holtmann A, Funke R, August B, Gamp A, et al. Vascular endothelial cell specific phospho-tyrosine phosphatase (VE-PTP) activity is required for blood vessel development. *Blood* 2006;**107**:4754–62.
92. Eliceiri BP, Paul R, Schwartzberg PL, Hood JD, Leng J, Cheresh DA. Selective requirement for Src kinases during VEGF-induced angiogenesis and vascular permeability. *Mol Cell* 1999;**4**:915–24.
93. Monaghan-Benson E, Burridge K. The regulation of vascular endothelial growth factor-induced microvascular permeability requires rac and reactive oxygen species. *J Biol Chem* 2009;**284**:25602–11.
94. Andriopoulou P, Navarro P, Zanetti A, Lampugnani MG, Dejana E. Histamine induces tyrosine phosphorylation of endothelial cell-to-cell adherens junctions. *Arterioscler Thromb Vasc Biol* 1999;**19**:2286–97.
95. Shasby DM, Ries DR, Shasby SS, Winter MC. Histamine stimulates phosphorylation of adherens junction proteins and alters their link to vimentin. *Am J Physiol Lung Cell Mol Physiol* 2002;**282**:L1330–L1338.
96. Xiao K, Allison DF, Buckley KM, Kottke MD, Vincent PA, Faundez V, et al. Cellular levels of p120 catenin function as a set point for cadherin expression levels in microvascular endothelial cells. *J Cell Biol* 2003;**163**:535–45.
97. Xiao K, Garner J, Buckley KM, Vincent PA, Chiasson CM, Dejana E, et al. p120-catenin regulates clathrin-dependent endocytosis of VE-cadherin. *Mol Biol Cell* 2005;**16**:5141–51.
98. Hatanaka K, Lanahan AA, Murakami M, Simons M. Fibroblast growth factor signaling potentiates VE-cadherin stability at adherens junctions by regulating shp2. *PLoS One* 2012;**7**:e37600.
99. Alcaide P, Newton G, Auerbach S, Sehrawat S, Mayadas TN, Golan DE, et al. p120-catenin regulates leukocyte transmigration through an effect on VE-cadherin phosphorylation. *Blood* 2008;**112**:2770–9.
100. Alcaide P, Martinelli R, Newton G, Williams MR, Adams A, Vincent PA, et al. p120-catenin prevents neutrophil transmigration independently of Rho A inhibition by impairing Src dependent VE-cadherin phosphorylation. *Am J Physiol Cell Physiol* 2012;**303**:C385–C395.

101. Anastasiadis PZ, Reynolds AB. Regulation of Rho GTPases by p120-catenin. *Curr Opin Cell Biol* 2001;**13**:604–10.
102. Noren NK, Niessen CM, Gumbiner BM, Burridge K. Cadherin engagement regulates Rho family GTPases. *J Biol Chem* 2001;**276**:33305–8.
103. Shaw SK, Bamba PS, Perkins BN, Luscinskas FW. Real-time imaging of vascular endothelial-cadherin during leukocyte transmigration across endothelium. *J Immunol* 2001;**167**:2323–30.

CHAPTER SEVEN

Clustered Protocadherins and Neuronal Diversity

Teruyoshi Hirayama, Takeshi Yagi
KOKORO Biology Group and JST-CREST, Laboratories for Integrated Biology, Graduate School of Frontier Biosciences, Osaka University, Yamadaoka, Suita, Osaka, Japan

Contents

Abstract

Neuronal diversity is a fundamental requirement for complex neuronal networks and brain function. The clustered protocadherin (Pcdh) family possesses several characteristic features that are important for the molecular basis of neuronal diversity. Clustered Pcdhs are expressed predominantly in the central nervous system, in neurites, growth cones, and synapses. They consist of about 60 isoforms, and their expression is stochastically and combinatorially regulated in individual neurons. The multiple clustered Pcdhs expressed in individual neurons form heteromultimeric protein complexes that exhibit homophilic adhesion properties. Theoretically, the clustered Pcdhs could generate more than 3×10^{10} possible variations in each neuron and 12,720 types of *cis*-tetramers per neuron. The clustered Pcdhs are important for normal neuronal development. The clustered *Pcdh* genes have also attracted attention as a target for epigenetic regulation.

Progress in Molecular Biology and Translational Science, Volume 116
ISSN 1877-1173
http://dx.doi.org/10.1016/B978-0-12-394311-8.00007-8

1. INTRODUCTION

Protocadherins (Pcdhs) were initially discovered when researchers used degenerate primers to amplify the extracellular repeats of classical cadherins.[1] The sequences of the PCR fragments revealed extracellular cadherin (EC) domains strikingly different from those of classical cadherins. These new cadherins, which were isolated from various vertebrate and invertebrate species, retained many features of the primordial cadherin motif (Fig. 7.1B). Hence, Suzuki and colleagues named them Pcdhs, which comes from the Greek word *protos*, meaning "first."[3]

About 80 different Pcdhs have been identified and they are expressed predominantly in the nervous system. The Pcdh family is classified into

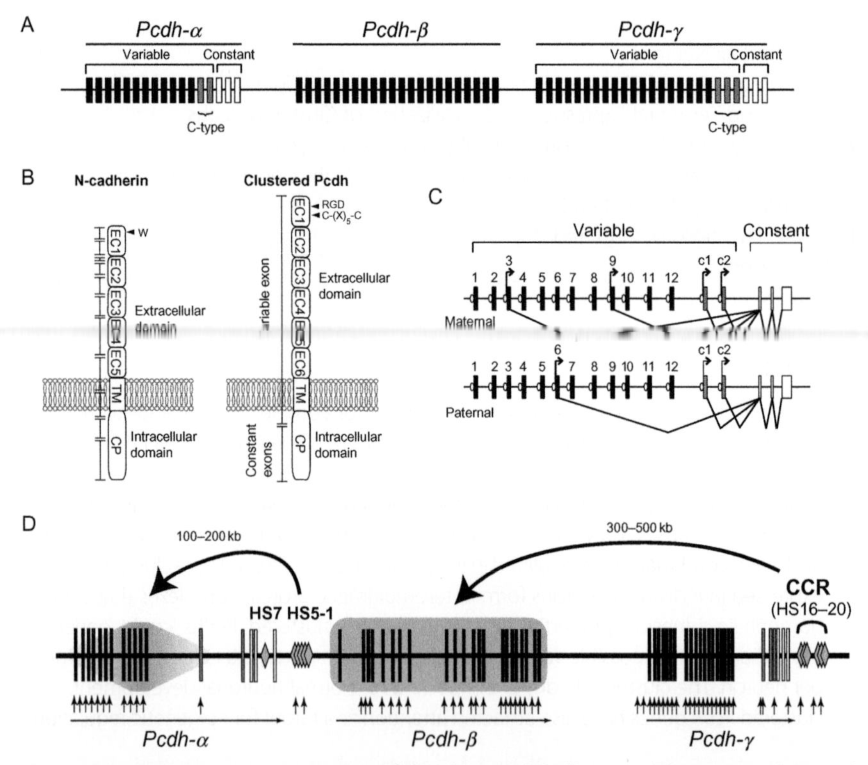

Figure 7.1 (A) Genomic organization of the clustered *Pcdh* genes. The mouse clustered *Pcdh* genes (*Pcdh-α*, *Pcdh-β*, and *Pcdh-γ*) are tandemly arranged on chromosome 18. The *Pcdh-α* and *Pcdh-γ* clusters are composed of variable exons (black boxes) and constant (white boxes) exons. *Pcdh-α* and *Pcdh-γ* also have C-type variable exons (gray boxes) with a different sequence profile from that of other variable exons. (B) Comparison

two subgroups based on genomic structure: clustered Pcdhs and non-clustered Pcdhs[4] (see also Chapter 4). In mammals, the clustered Pcdhs include about 60 members whose genes are tandemly aligned at the same chromosomal locus, while the genes of nonclustered Pcdhs are scattered across different chromosomes.

Pcdhs typically have six or seven ECs, whereas classical cadherins have five, at least in chordates (Fig. 7.1B).[5,6] Classical type-I cadherins have a conserved tryptophan at position 2 (W2) of their EC1 domain, and classical type-II cadherins have conserved tryptophan residues at position 2 (W2) and position 4 (W4) of the EC1 domain. These conserved tryptophan residues line a hydrophobic acceptor pocket, which is involved in homophilic binding.[7,8] Neither the tryptophan residues nor the hydrophobic pocket are conserved in the Pcdh family.[9–11]

The extracellular domain of the Pcdhs has other characteristic conserved sequences (Fig. 7.1B). One is an RGD motif that is conserved in clustered Pcdh-α, the nonclustered Pcdh17 and Pcdh19, and Cdhr5 (Mucdhl).[5,10] The RGD motif has a well-known role in integrin-dependent cell adhesion,

of the molecular domain and gene structures between N-cadherin and a clustered Pcdh. N-cadherin is shown as an example of a classical cadherin. It has five extracellular cadherin (EC) repeats and an essential tryptophan residue (W) within the first EC1 domain. The lines on the left represent exon–intron boundaries. Classical cadherins are encoded by many small exons. Clustered Pcdhs have six ECs. The EC1 domain has an RGD and a C-$(X)_5$-C motif. Its extracellular domain, transmembrane domain (TM), and part of the cytoplasmic domain (CP) are encoded by a single variable exon. (C) Stochastic and monoallelic expression of the *Pcdh-α* genes. The transcription of *Pcdh-α* in a neuron is shown as an example of expression from the clustered *Pcdh* locus. Each allele, paternal and maternal, is independently regulated, and from each allele some isoforms are expressed as a combination of variable and constant exons. C-type isoforms are expressed from both alleles, but all other isoforms are expressed in a monoallelic fashion. As a result, the mRNAs of multiple isoforms are differentially transcribed in individual neurons. Open circles indicate individual gene-specific promoter sequences. (D) Gene regulation of the mouse *Pcdh* cluster. Long-range *cis*-regulatory elements HS5-1 (depicted by diamonds) strongly regulate the expression of the isoforms of *Pcdh-α10*, *-α11*, and *-α12*. In addition, HS5-1 affects the expression of *Pcdh-α3* to *-α12* with a strength depending on the distance of the respective genes from HS5-1. The cluster control region (CCR), situated at the 3′ end of the *Pcdh-γ* gene cluster and containing sites HS16-20 (diamonds), has a strong effect on the expression of *Pcdh-β* genes, but only a weak effect on the *Pcdh-γ* cluster, and no effect on the expression of the *Pcdh-α* gene cluster. Small arrows at the bottom indicate CTCF-binding sites. Most of the promoter regions and HS regions of the clustered *Pcdhs* have CTCF-binding sites. Conditional knockout of CTCF in neurons results in the dramatic downregulation of stochastically expressed clustered *Pcdh* isoforms (see text).[2]

and its ligands include fibronectin, vitronectin, fibrinogen, von Willebrand factor, and many other large glycoproteins.[12] Clustered Pcdh-α 4 exhibits integrin-dependent cell adhesion in the HEK293T cell line.[13] The other conserved sequence is a Cys-$(X)_5$-Cys motif, which is completely conserved in the clustered Pcdhs and delta-2 Pcdhs.[10,14] This sequence is important for forming disulfide bonds between neighboring Pcdhs and probably contributes to *cis*-tetramerization. These conserved sequences suggest that Pcdhs mediate novel adhesion mechanisms, which are still uncharacterized.

2. CLUSTERED Pcdhs

The clustered *Pcdh* genes were first identified in 1998 as a cadherin-related neuronal receptor (*Cnr*) gene family in mouse brain (Fig. 7.1A).[15,16] CNR1 and CNR2 were first isolated in a yeast two-hybrid screen as weak interactors with the N-terminal region of the Fyn tyrosine kinase, including Fyn's SH3 and SH2 motifs. CNR1 and CNR2 are 87% similar at the amino acid level, and interestingly, the 3′ regions of their respective cDNA sequences are completely identical.

The *Cnr3* to *Cnr8* cDNAs were then isolated by reverse transcription PCR (RT-PCR) using consensus primers. Genomic DNA analysis of the mouse *Cnr* family members suggested that their diverse 5′ exons, tandemly located on the same chromosome, combined with a common set of 3′ exons. In 1999, a BLAST search for CNRs turned up 52 clustered *Pcdh* genes in the human genome project database.[17] The human genes are organized into three tandemly linked clusters, *Pcdh-α*, *Pcdh-β*, and *Pcdh-γ* (Fig. 7.1A). The human *Pcdh-α* members are homologs of the mouse *Cnrs*.[17,18]

The genomic organization of the *Pcdh-α* and *Pcdh-γ* clusters consists of, respectively, variable-region exons and a set of common constant-region exons (Fig. 7.1A). The variable-region exons encode six N-terminal EC-like domains, a single transmembrane domain, and a short part of the cytoplasmic domain (Fig. 7.1B). The constant-region exons encode the rest of the cytoplasmic domain. Each variable-region exon of the mouse *Pcdh-α* and *Pcdh-γ* clusters, which has 14 and 22 variable exons, respectively, splices to the first constant exon (Fig. 7.1A and C).

Thus, within the Pcdh-α and Pcdh-γ clusters, the isoforms share a large part of their cytoplasmic domain sequence, but have highly homologous yet distinct extracellular domains. On the other hand, the 22 exons of *Pcdh-β* are tandemly arranged, but lack a constant region with common exons; that is, the *Pcdh-β* genes are one-exon genes (Fig. 7.1A).[17] Owing to the lack of constant-region exons, each *Pcdh-β* gene encodes a unique but similar protein with a short cytoplasmic domain.

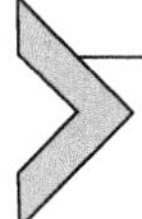

3. STOCHASTIC AND COMBINATORIAL GENE EXPRESSIONS GENERATE MOLECULAR DIVERSITY IN INDIVIDUAL NEURONS

In the case of *Pcdh-α* and *Pcdh-γ*, each variable-exon transcript is spliced to the first of three constant-region exons. This genomic organization is similar to that of the immunoglobulin (*Ig*) and T-cell receptor (*TCR*) gene clusters. *Ig* and *TCR* generate enormous variability in the antibody repertoire by somatic DNA recombination and mutation. This similarity suggested that the clustered *Pcdhs* might generate enormous diversity in the brain by a related, but novel, mechanism. In support of this idea, double *in situ* hybridization experiments showed that different neurons express different sets of *Pcdh* genes.[16] This observation suggested that the clustered *Pcdhs* provide a molecular basis for individual neurons to have a unique molecular identity that could contribute to the establishment of complex neural networks.

Wu and Maniatis[17] proposed four models for *Pcdh* gene regulation that are compatible with the genomic organization of the clustered *Pcdhs*: (1) single promoter and *cis* alternative splicing, (2) multiple promoters and *cis* alternative splicing, (3) multiple promoters and *trans* alternative splicing, and (4) DNA rearrangements. Using genetically modified clustered *Pcdh* alleles, two research groups independently demonstrated that clustered *Pcdhs* are transcribed from promoters located upstream of each variable exon and that the pre-mRNA is predominantly processed into mature mRNA by *cis*-splicing (illustrated in Fig. 7.1C).[19,20] Products of intercluster *trans* alternative splicing were also found, albeit at a much lower level.[19] These results suggested that the *Pcdh* gene clusters are regulated at least at the level of transcription and that differential promoter activation is a key factor for generating molecular diversity in individual neurons.

Single-cell RT-PCR analysis confirmed that multiple different *Pcdh* isoforms are expressed by individual neurons.[21,22] To study expression at the allelic level, polymorphisms between the C57BL/6 and MSM or JF1 mouse strains were used. In this way, the allelic expression patterns of *Pcdh-α* and *Pcdh-γ* in individual Purkinje cells in the F1 mice, obtained by interstrain crossings, could be determined. The results showed that single Purkinje cells indeed express several different clustered *Pcdh* transcripts.

Interestingly, in single neurons, some isoforms are derived from the maternal allele, while others are derived from the paternal allele (Fig. 7.1C). On the other hand, the *C-type* isoforms (*Pcdh-α C1* and *C2*, and *Pcdh-γ C3*, *C4*, and *C5*) were biallelically expressed. These findings showed that the expression of clustered *Pcdhs* is monoallelically and combinatorially regulated, with the exception of the *C-type* isoforms (Fig. 7.1C). Thus, if for instance 15 possible isoforms are randomly expressed from the clustered *Pcdh* genes in an individual neuron, approximately 3×10^{10} variations of tetramers can be generated[14] (see Section 7).

Monoallelic gene expression is mainly classified into three groups: (1) random X-chromosome inactivation in female cells, (2) imprinting genes in autosomes from only the maternal or the paternal allele, and (3) random monoallelic gene expression in autosomes, including random allelic exclusion.[23] This third group includes the genes encoding *Ig*, *TCR*, natural killer-cell receptors, and odorant and pheromone receptors. Some members of the interleukin gene family are also expressed from either the maternal or the paternal allele, although cells expressing both alleles occur as well. The expression of clustered *Pcdh* genes may be considered a subclass of this third group, because their regulation differs from the above examples. Namely, different sets of monoallelically regulated isoforms are expressed from each allele, concomitant with biallelically expressed *C-type* isoforms in individual cells.

Genome-wide analyses have demonstrated that random monoallelic gene expression is more widespread among autosomal genes in mammals than previously thought (>5% in human and 10% in mouse).[23,24] *Pcdh-γ* genes have also been described as cross-effect genes that influence the relative expression of maternally versus paternally inherited alleles.[25] These random monoallelic gene expressions are thought to function to generate diversity in individual cells and their clonal descendants, thereby creating more phenotypic variation than genotypic variation alone.[23,24]

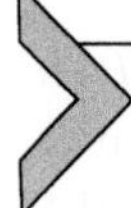

4. GENE REGULATION OF CLUSTERED Pcdhs IS ASSOCIATED WITH EPIGENETIC MODIFICATIONS

Gene expression compensation has been observed in the *Pcdh-α* cluster.[26] When the number of exons in the variable region is altered in the mouse brain, either by deletion or by duplication, the remaining or excessive exons show compensation at the transcriptional level, so that the total expression level of *Pcdh-α* isoforms is maintained independently of the number of variable exons.

As mentioned above (Fig. 7.1C), clustered *Pcdh* genes have multiple alternative promoters and therefore might use stochastic promoter choice mechanisms to generate molecular diversity in individual neurons. Comparison of the upstream sequences of clustered *Pcdh* genes revealed a highly conserved sequence element (CSE) upstream of the transcription start site (TSS) of each variable exon.[27] The CSEs consist of a common core sequence, CGCT, shared among all *Pcdh* clusters except the *C-type* isoforms, and flanking sequences that are more *Pcdh* cluster specific. *In vitro* reporter assays indicated that the CSEs are required for promoter activity.[19] The adjacent upstream region of the CSE, termed the specific sequence element (SSE), is also critical for promoter activity.[28]

DNase-I hypersensitivity (HS) assays led to the identification of long-range *cis*-regulatory elements that enhance the promoter activities in the *Pcdh-α* and *Pcdh-β* gene clusters.[29,30] In the *Pcdh-α* cluster, two elements were identified[29] (Fig. 7.1D). One is HS5-1, which is located about 30-kb downstream of the last *Pcdh-α* constant-region exon. Another is HS7, which is located between the *Pcdh-α* constant-region exons 2 and 3. *In vivo* reporter assays showed that HS5-1 is active in the central nervous system (CNS), especially in the cortex and olfactory bulb, whereas HS7 is preferentially active in the sensory organs. HS5-1 deletion in mouse causes strong downregulation of the 3′ portion of *Pcdh-α* variable exons, especially the *Pcdh-α 10* to *12* isoforms.[30,31] However, its influence on the expression of 5′ *Pcdh-α* variable exons on the other end of the *Pcdh-α* cluster is low. An exception to this gradient is the expression of *Pcdh-αC2*, which is at the 3′-most end of the cluster, but is not at all influenced by the HS5-1 deletion.

The HS7 deletion mutant mouse shows a more moderate and uniform decrease in the expression of the *Pcdh-α* cluster genes, including *Pcdh-α C2*.[31] These results indicate that multiple enhancer elements act as tissue- and development-specific enhancers, and suggest the existence of other enhancer elements that regulate the 5′ side of *Pcdh-α* variable exons.

With respect to the *Pcdh-β* gene cluster, a cluster control region (CCR) containing six HS sites (HS16–20 and 17′) was identified.[30] CCR-deleted mutant mice show a dramatic downregulation of all the *Pcdh-β* cluster genes. Surprisingly, this CCR is located downstream of the last *Pcdh-γ* constant-region exon and has a specific effect on the *Pcdh-β* cluster, even though the 320-kb region containing the *Pcdh-γ* cluster lies between them.

Chromosomal *cis* interactions between sites separated by long distances and *trans* interactions contribute to the complexity of gene regulation mechanisms. CCCTC-binding protein (CTCF) is a key molecule that mediates chromosomal interactions.[32] Genome-wide analyses using nonneuronal cells showed that the clustered *Pcdh* locus has many CTCF-binding sites (Fig. 7.1D).[33–35] The enhancer regions CCR and HS5-1 contain CTCF-binding motifs. CTCF also binds to SSE–CSE regions in the individual promoter regions of clustered Pcdhs. This CTCF binding to the promoter regions is significantly correlated with enhancer activity.[28,31,36] Using neuron-specific conditional CTCF-knockout mice, CTCF was shown to be essential for the stochastic gene expression of the clustered *Pcdhs*.[2]

The cohesin complex is another macromolecular structure with a key role in mediating chromosomal interactions. It colocalizes with CTCF. Mice with a heterozygous mutation in Nipped-B-like, which promotes cohesin loading onto DNA, show a significant decrease in *Pcdh-β* cluster gene expression.[37] Cohesin complex subunit SA1 (cohesin-SA1) binds to TSSs of the clustered Pcdhs. The ablation of cohesin-SA1 leads to the downregulation of clustered *Pcdh* genes in the brain.[38] In addition, the cohesin complex subunit Rad21 binds to the promoter and enhancer regions of clustered Pcdhs in CAD and N2a cells, and this binding is positively correlated with *Pcdh* expression in the brain.[36] These findings suggest that complex chromatin organization is involved in the expression mechanisms of clustered *Pcdhs*.

Neuron-restrictive silencer factor (NRSF) is known to inhibit neuron-specific gene expression in nonneuronal tissues. In the fugu *Pcdh* cluster, multiple NRSF-binding elements ensure the neuron-specific expression of clustered *Pcdhs*.[39] In mammals, NRSF strongly binds HS5-1 to inhibit the *Pcdh-α* gene expression in nonneuronal cells.[31]

It is well established that the CpG methylation state of promoter regions influences gene expression.[40] Different sets of clustered *Pcdhs* are expressed in discrete human and mouse cell lines. The expressed *Pcdh* genes were found to be hypomethylated, and the unexpressed *Pcdh* genes were hypermethylated.[19,41] In the brain, the stochastically expressed isoforms of *Pcdh-α* show mosaic or mixed methylation states, in contrast to the constitutively expressed isoforms, Pcdh-αC1 and -αC2, which show hypomethylation.[41] In addition, perturbation of methyl-CpG-binding protein 2 (MeCP2), the causative gene of Rett syndrome, can lead to the misregulation of clustered *Pcdh* transcription.[42] These results indicate that the expression of clustered Pcdhs is influenced by DNA methylation.

Aberrant, but nonrandom DNA methylation states have been found in cancer cells,[43] and clustered *Pcdhs* are reported to be targets of DNA methylation in cancer cells. For example, Pcdh-γA11 is reported to be a methylation target in astrocytic gliomas.[44] In breast cancer, the clustered *Pcdh* locus, comprising *Pcdh-α*, *Pcdh-β*, and *Pcdh-γ* genes, is widely methylated.[43] Long-range epigenetic silencing (LRES), in which more than 100 kb of the DNA region becomes highly methylated, is reported in cancer tissues.[45,46] LRES on the clustered *Pcdh* locus is found in Wilms' tumor, a pediatric tumor of the kidney.[47]

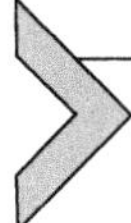

5. SPATIAL AND TEMPORAL EXPRESSION AND LOCALIZATION OF CLUSTERED Pcdhs

The spatial and temporal expression patterns of clustered *Pcdhs* have been investigated in several species, including rat, mouse, chick, and zebrafish. A general feature is that the *Pcdhs* are widely expressed in the CNS.[16,48–52] In rodents, the expression of clustered *Pcdhs* has been studied extensively by *in situ* hybridization. Both the *Pcdh-α* and *Pcdh-γ* clusters exhibit high expression levels in the pyramidal and dentate gyrus cells of the hippocampus; the Purkinje and granule cells of the cerebellum; the mitral, granular, and periglomerular cells of the olfactory bulb; cortical neurons of the cerebrum; and the motor neurons, sensory neurons, and interneurons of the spinal cord.[16,49,52,53] Furthermore, Northern blot analysis revealed that the expression of clustered *Pcdhs* in the developing mouse brain is temporally regulated. For example, the expression of *Pcdh-α4* (*CNR1*) increases from at least embryonic day (E) 15 to postnatal day (P) 0 and gradually decreases from P7 to P56.[54]

The expression of *Pcdh-γA1*, *-γA12*, *-γB6*, and *-γc3* is also observed in the embryonic brain. This expression increases until the second postnatal week (P9) and is then slightly downregulated.[49] C-type isoforms of *Pcdh-α*, *Pcdh-αC1*, and *Pcdh-αC2* are expressed throughout the developing brain and peak at P10,[54] whereas the expression of *Pcdh-γC4* and *Pcdh-γC5* changes markedly during brain development. *Pcdh-γC4* is expressed in the embryonic brain but is dramatically downregulated after the second postnatal week. In contrast, the expression of *Pcdh-γC5* is detected at about the first postnatal week and increases until the second postnatal week, after which it is maintained.[49] By RT-PCR, the expression of *Pcdh-β* is detected from E12.5 in the brain, after which it increases and peaks around the time of birth.[53]

Protein localization information can be important for predicting function, because specific protein functions usually require localization in specific areas.

Immunoelectron microscopy using antibodies to detect the cytoplasmic constant-region domain showed that Pcdh-α and Pcdh-γ are localized to synaptic regions, suggesting that the clustered Pcdhs may mediate synaptic interactions.[16,55,56] However, in mature neurons, many synapses were not labeled.

Immunoreactivity for Pcdh-γ is found in tubulovesicular structures within axonal and dendritic terminals and in nonsynaptic plasma membranes.[56] Subcellular fractionation analysis also showed that the Pcdh proteins are present in various fractions: the crude synaptosomal fraction (P2), large tubulovesicular organelles (P3), osmotically shocked synaptosomes (LP1), and synaptic vesicles (LP2).[55–57] These observations indicated that clustered Pcdhs are transferred between synapses and intracellular compartments.

Changes in the distribution of clustered Pcdhs during development have been reported. In primary cultures, Pcdh-γ exhibits strong expression in most immature neurons, especially in neurites and both axonal and dendritic growth cones.[56,58] In contrast, Pcdh-γ localizes to the somatodendritic compartment of mature neurons, with low levels of expression in the dendritic spines. Pcdh-γ is also expressed in cells of the choroid plexus in the brain.[59]

In vivo analysis revealed that myelination acts as a trigger for a dramatic decline in Pcdh-α axonal localization.[60] Initially, in the late embryonic and early postnatal stages, Pcdh-α is strongly expressed in major axonal tracts: the optic nerve, internal capsule, and corpus callosum. However, around P10 Pcdh-α is severely decreased in these axonal tracts, in parallel with increased myelin marker expression. This alteration of Pcdh-α expression is similar to that of L1 and GAP-43, which also exhibit an inverse pattern with myelination.[61–63] In the shiverer mutant mouse,[64] which carries a deletion in the gene encoding myelin basic protein, the intense localization of Pcdh-α continues at least until P19, and weaker expression is observed only in the adult. One report describes a close relationship between the formation of appropriate neuronal circuits in the visual cortex and the termination of myelination.[65] These data suggest that the axonal localization of Pcdh-α is controlled by myelin sheath compaction, which might contribute to neuronal growth and neural circuit maturation during development.

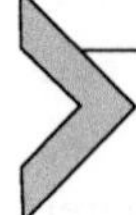

6. INTRACELLULAR SIGNALING BY CLUSTERED Pcdhs AND PROTEIN DYNAMICS

Using the yeast CytoTrap two-hybrid system, two tyrosine kinases, proline-rich tyrosine kinase 2 (PYK2) and focal adhesion kinase (FAK), were found to bind to Pcdh-γ's cytoplasmic domain[66] (Fig. 7.2). Similarly, Pcdh-α interacts with PYK2 and FAK. The kinase activities of PYK2 and

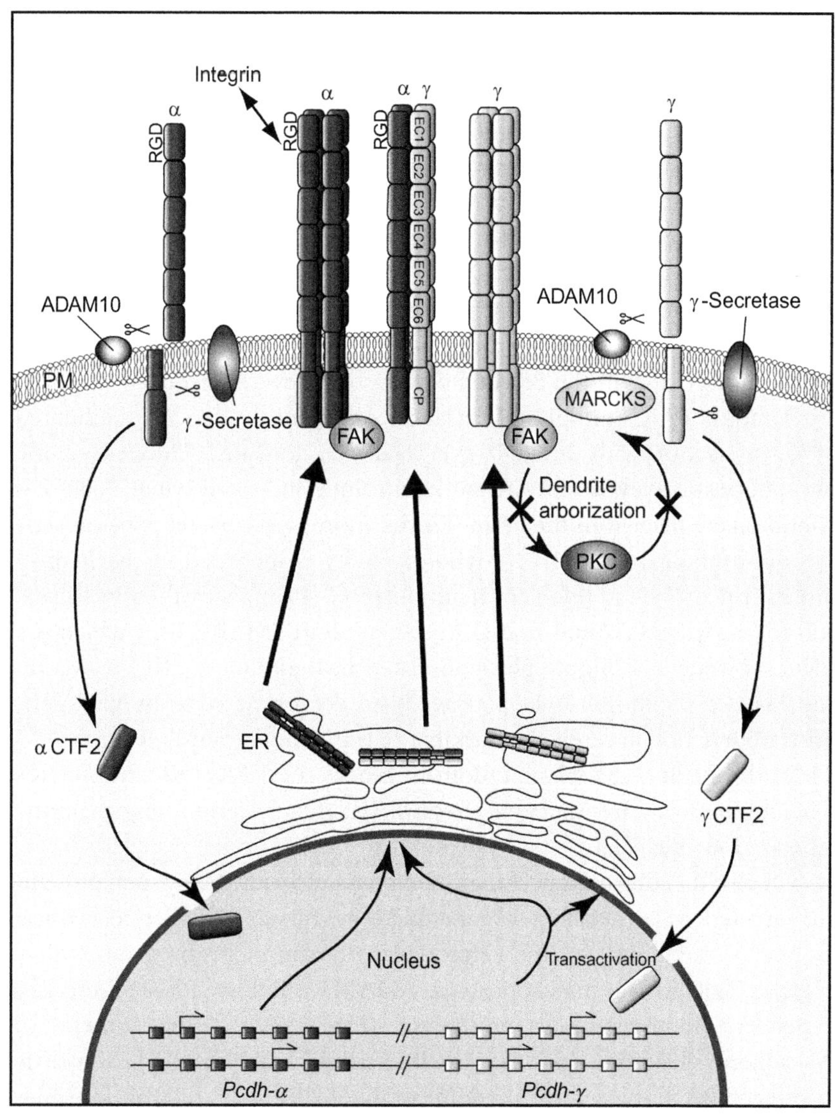

Figure 7.2 Schematic overview of the intracellular dynamics and intracellular signaling of Pcdh proteins. The Pcdh-γ proteins are known to form *cis*-homotetramers, whereas the depicted Pcdh-α *cis*-homotetramers and Pcdh-α/Pcdh-γ *cis*-heterotetramers are presumptive. At least a fraction of the Pcdh-α proteins is thought to form hetero-complexes with Pcdh-γ. β1-Integrin may interact with the RGD sequence in the EC1 of Pcdh-α. The Pcdh-α and Pcdh-γ cytoplasmic domains interact with FAK kinase, resulting in its inhibition. Suppression of the FAK–PKC–MARCKS signaling pathway is important for neuronal dendrite arborization. After the ectodomain of Pcdh-α and -γ is cleaved off by the ADAM10 metalloproteinase, their cytoplasmic domains are cleaved

(Continued)

FAK are inhibited by their interaction with Pcdh-α and Pcdh-γ, and PYK2 is activated in Pcdh-γ-deficient neurons. Overexpression of PYK2 induces apoptosis in the chicken spinal cord.[66] Pcdh-γ ablation in mouse and morpholino-mediated Pcdh-α downregulation in zebrafish induce apoptosis in subpopulations of neurons.[55,67] Thus, the negative regulation of PYK2 via its interaction with Pcdh-α and Pcdh-γ suggests that the cytoplasmic domains of Pcdh-α and Pcdh-γ contribute to the cell survival of subsets of neurons.[66]

On the other hand, Pcdh-γ regulates dendritic arborization in cortical neurons via the FAK signaling pathway (Fig. 7.2).[68] Mice with a conditional deletion of Pcdh-γ in the cerebral cortex show severely reduced dendritic arborization. FAK's autophosphorylation at tyrosine residue 397 is inhibited by its interaction with the Pcdh-γ cytoplasmic domain.[66] Conversely, the phosphorylation levels of FAK are significantly increased when Pcdh-γ is conditionally deleted in the brain. Furthermore, FAK overexpression activates phospholipase C-γ1,[69,70] whose activity is increased in the Pcdh-γ mutant mice.[68] In addition, protein kinase C (PKC) activity is increased, and myristoylated alanine-rich C-kinase substrate (MARCKS), which is a PKC substrate,[71] is highly phosphorylated in these mice. PKC activation and MARCKS phosphorylation have been demonstrated to prevent dendrite arborization in cerebellar Purkinje cells and hippocampal neurons.[72–74] Thus, the Pcdh-γ cytoplasmic domain blocks the FAK/PKC/MARCKS signaling pathway by interacting with FAK, thereby promoting dendritic arborization (Fig. 7.2).

Clustered Pcdhs are type-I transmembrane proteins. Many such proteins are targeted by a metalloproteinase and γ-secretase.[75] Similar to classical E- and N-cadherins,[76,77] the extracellular domain of Pcdh-α and Pcdh-γ is cleaved off by the metalloprotease ADAM10, followed by γ-secretase-dependent intramembrane proteolysis (Fig. 7.2).[78–81] In contrast to N-cadherin shedding is stimulated by the activation of N-methyl-D-aspartic acid receptors, however, the shedding of the Pcdh-γ extracellular domain is activated by stimulation with α-amino-3-hydroxy-5-methylisoxazole-4-propionic acid hydrate.[80] The intracellular Pcdh-α and Pcdh-γ fragments then translocate to the nucleus, probably mediated by the putative nuclear

by γ-secretase. The resulting C-terminal fragments, αCTF2 and γCTF2, are transported into the nucleus. γCTF2 directly or indirectly transactivates the *Pcdh-γ* genes. α, Pcdh-α; γ, Pcdh-γ; EC, extracellular cadherin domain; TM, transmembrane domain; CP, cytoplasmic domain; PM, plasma membrane; ER, endoplasmic reticulum.

transport signal present in the common cytoplasmic sequences in all isoforms of Pcdh-α and Pcdh-γ (Fig. 7.1C).[81] Luciferase reporter assays showed that the cytoplasmic fragments of Pcdh-γ transactivate all the Pcdh-γ promoters,[78] but the existence of other target genes and the *in vivo* function of such transactivations remain unclear.

In general, when classical cadherins are expressed in a cell line or neuron, they are expressed on the cell surface and exhibit cell–cell adhesion activity. In contrast, only small amounts of the Pcdh-α proteins are expressed on the cell surface by HEK293T cells. However, when Pcdh-α and Pcdh-γ are coexpressed in HEK293T cells, the expression of Pcdh-α at the cell surface increases more than fivefold.[57] Similar results are obtained in SH-SY5Y cells.[81] Nonetheless, in neurons of Pcdh-γ knockout mice, the Pcdh-α protein is delivered to the cell surface at levels similar to those of wild-type neurons.[81] These results indicate that the surface delivery of Pcdh-γ is not essential for Pcdh-α's expression at the cell surface (Fig. 7.2). Interestingly, however, the combinatorial expression of Pcdh-α and Pcdh-γ results in their decreased susceptibility to γ-secretase.[81] In fact, an interaction between Pcdh-α and Pcdh-γ was demonstrated by coimmunoprecipitation experiments using extracts of mouse brain and neuroblastoma.[57] This combinatorial expression of the Pcdh-α and Pcdh-γ proteins permits individual neurons to express even more diverse codes at the protein level.

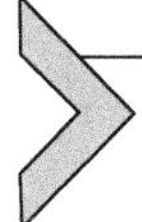

7. HETEROMULTIMERIC PROTEIN COMPLEXES OF CLUSTERED Pcdhs GENERATE MOLECULAR DIVERSITY AT THE PROTEIN LEVEL

In the conventional cell-aggregation assays used to demonstrate the binding properties of classical cadherins, the clustered Pcdhs do not show similarly strong homophilic binding activities.[10,13,82] However, more sensitive assays can detect a homophilic and Ca^{2+}-dependent adhesion property of Pcdh-γ.[49,80] These Pcdh-γ-mediated cell–cell interactions are regulated by intracellular trafficking.[83]

The existence of strictly homophilic Pcdh-γ interactions was shown using leukemia cells (K562),[84] which do not express endogenous Pcdh-γ or any classical cadherins.[85] This study also revealed several characteristic features of the Pcdh-γ adhesion activity.[84] First, this activity is only partially Ca^{2+} dependent. Second, the EC1 domain, especially a tyrosine (at position 30 in the mouse Pcdh-γA3) and the Cys-$(X)_5$-Cys motif, is important for Pcdh-γ's cell-aggregation activity and surface expression. Third, the homophilic interaction is mediated through Pcdh-γ's EC2–EC3 domains. A remarkable characteristic of Pcdh-γ isoforms is that they form promiscuous *cis*-multimers, which show no isoform specificity and are mainly predicted to be tetramers. In contrast, these tetramers exhibit strictly homophilic interactions in *trans*. Pcdh-γ also forms complexes with Pcdh-α and Pcdh-β isoforms.[86]

These findings together suggest that the heteromultimerization of clustered Pcdhs generates high molecular diversity at the protein level in individual neurons. Theoretically, assuming that 15 possible isoforms can be expressed in an individual neuron and that each isoform of clustered Pcdhs contributes equally to form tetramers, 12,720 types of possible *cis*-tetramers could be generated in an individual neuron (Fig. 7.3).[14] In fact, the variety of *cis*-tetramers may contribute much to the diversity of individual synapses

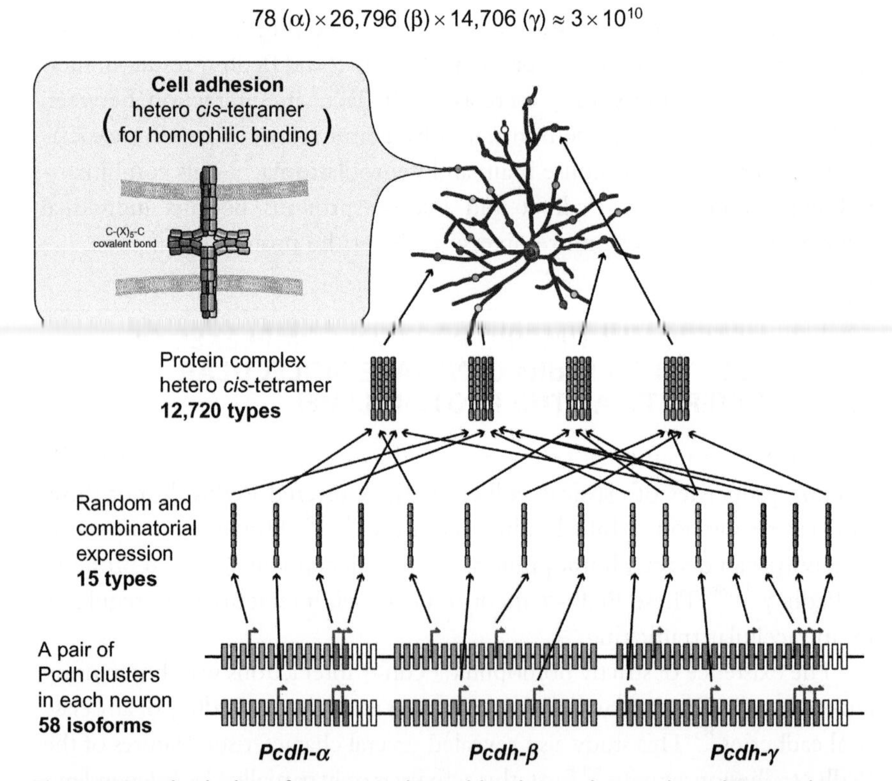

Figure 7.3 Individuality of neurons achieved by the random and combinatorial transcription of clustered *Pcdh* isoforms and by the random and combinatorial production of *cis*-heterotetramers of Pcdh proteins. In this example, for each allele of clustered *Pcdh* genes, a combination of one α, two β, and two γ isoforms is randomly transcribed.

formed by a neuron. Interestingly, the matching probability between two neurons is rapidly decreased by the expression of a small number of different isoforms. For example, if out of 15 expressed isoforms, only three differ between two neurons, the matching probability becomes 41.1% (below 50%). Thus, the stochastic expression of clustered Pcdh isoforms may provide a molecular code for stamping a high degree of individuality on every neuron in the brain.[14]

8. *IN VIVO* FUNCTIONS OF CLUSTERED Pcdhs

The *in vivo* functions of clustered Pcdhs have been investigated by gene targeting in mice. A mouse with a deletion of the *Pcdh-α* constant-region exons (ΔCR) in the olfactory system was generated and investigated.[87] The olfactory system has very interesting features: more than 1200 odorant receptor (OR) genes are encoded in the mouse genome, and each olfactory sensory neuron (OSN) expresses only one OR gene.[88–90] Axons of OSNs expressing the same OR are sorted into specific glomeruli of the olfactory bulb in a neural-activity-dependent manner.[55,91] The Pcdh-αΔCR mice exhibit aberrant axonal projections from the OSNs into glomeruli. They show many, small ectopic glomeruli, without obvious synaptic abnormalities. Small ectopic glomeruli are found in the early developmental stages of wild-type mice, but they are normally eliminated in an activity-dependent manner. In contrast, ectopic glomeruli are found in the adult stages of Pcdh-αΔCR mice.[87] These findings indicated that Pcdh-α is needed for the correct projection of OSN axons into OR-specific glomeruli and that Pcdh-α may play a role in glomerular remodeling.

Pcdh-α is also required for serotonergic projections.[92] Serotonergic neurons of the raphe nuclei project their axons into various regions of the CNS,

---

Hence, from the two alleles in each neuron, a total of two α, four β, and four γ isoforms are randomly expressed. There are 78 possible combinations of two for the 12α isoforms, 26,796 combinations of four for the 22β isoforms, and 14,706 combinations of four for 19γ isoforms.[14] As a result, individual neurons could have approximately 3×10^{10} variations of variable *Pcdh* transcripts, as calculated at the top of the figure. The five C-type isoforms (αC1, αC2, γC3, γC4, and γC5) are constitutively expressed from both alleles in each neuron. Therefore, 15 isoforms are expressed in each neuron: 10 random isoforms (2α+4β+4γ) and 5 constitutive isoforms. At the protein level, 12,720 types of *cis*-tetramers are possible for each set of 15 isoforms. The C-$(X)_5$-C motifs in the EC1 domain, conserved among all clustered Pcdh isoforms, are important for forming the *cis*-tetramers.[14] *This figure is modified from Yagi.*(See Color Insert.)

and Pcdh-α is strongly expressed in the raphe nuclei. Pcdh-αΔCR mice and control mice show similar patterns of serotonergic projections until the first week after birth, but at P21, when the serotonergic axonal termini form arbors in their target areas, obvious abnormalities are found in the hippocampus, globus pallidus, and substantia nigra of the mutant mice. For instance, hippocampal serotonergic fibers are more concentrated at the stratum lacunosum-moleculare, and the stratum oriens shows a more sparse distribution of serotonergic fibers compared with controls. Similar results are found in Pcdh-α hypomorphic mice, in which 56 amino acids at the carboxyl terminus of the A-type alternative splice form are deleted.[92]

Behavior analyses were carried out on mice with isoform-specific knockouts of either Pcdh-α A or Pcdh-α B, which are alternative splice variants of the constant region.[18,93] Such alternative splice variants are not found for Pcdh-γ. The Pcdh-α A and Pcdh-α B mutant mice are born at the expected Mendelian ratio, are fertile, and have no apparent abnormality. However, Pcdh-α A isoform-specific knockout mice show a significant increase in fear-induced freezing behavior in a contextual test, but no difference in a tone-cued test. Pcdh-α A knockout mice also show differences from wild type in solving tasks in an eight-arm radial maze. In contrast, there are no behavioral differences in Pcdh-α B knockout mice. These results suggest that the Pcdh-α A isoform, which has a longer cytoplasmic domain than the Pcdh-α B isoform, plays an important role in normal contextual learning and working memory. These phenotypic differences may be caused by altered signal transduction or by other specific molecular interactions through the Pcdh-α A isoform-specific domain.

The function of Pcdh-γ has been well investigated in the spinal cord. Sanes and colleagues generated Pcdh-γ deletion mutant mice,[55] which are born at the expected Mendelian frequency but die within 12 h after birth, with a hunched posture, shallow and irregular breathing, and repetitive limb tremors.[55] However, no clear difference from wild-type mice is detected prior to E15.5, and neuronal migration, axon outgrowth, and synapse formation progress without obvious abnormality. Hence, Pcdh-γ may be dispensable for the early steps of neural development. However, in these Pcdh-γ mutant mice, the spinal cord is smaller than that of controls, and severe neurodegeneration and apoptosis are observed in the spinal cord at E18. Interestingly, the apoptosis is mainly detected in interneurons, while only limited apoptosis is seen in the sensory and motor neurons of the spinal cord; in the brain, limited apoptosis was observed in the hippocampal and cortical neurons.

To further analyze the synaptic development of Pcdh-γ-deleted neurons, Pcdh-γ/BAX double-mutant mice were generated.[94] Almost all apoptosis, including that of the interneurons in the spinal cord, is blocked in these double-mutant mice, which die soon after birth, like the single Pcdh-γ deletion mutant mice. In the double-mutant mice, the number of synapses in the spinal cord is significantly reduced in both excitatory and inhibitory neurons. Similar results are found in Pcdh-γ hypomorphic mice, in which 57 amino acids at the carboxyl terminal are deleted. These results suggest that the observed reduction in synapse number is not caused by neuronal loss but may be due to defects in neural connectivity that depend on Pcdh-γ.

Weiner and colleagues carried out a detailed analysis of Pcdh-γ function in the spinal cord.[95,96] Using various developmental and cell-type-specific markers, they found that the apoptosis of interneurons proceeds in a ventral-to-dorsal temporal gradient and is associated with the first wave of synaptogenesis.[95] Moreover, apoptosis does not occur equally in interneuron populations. For example, over 80% of engrailed-1-positive V1 neurons are lost, whereas 30% of Chx10-positive V2a neurons are lost. They also showed, using cell-type-specific Pcdh-γ-deleted mice, that Pcdh-γ is required noncell autonomously for spinal interneuron survival. Pcdh-γ-deleted neurons can survive when they are surrounded by normal neurons, but normal neurons cannot survive when they are surrounded by Pcdh-γ-deleted ones.

Similar results were reported using retina-specific Pcdh-γ mutant mice.[97] The apoptosis of interneurons and retinal ganglion cells, which normally occurs during the first 2 postnatal weeks, is increased in these mutant mice. The retina-specific Pcdh-γ mutant mice also show reduced numbers of synapses. The apoptosis of retinal ganglion cells occurs noncell autonomously. Nonetheless, the retinal structure, lamina-specific arbor, and functional visual circuits develop normally in the absence of Pcdh-γ. In contrast to the spinal cord, the reduced synapse number is rescued when apoptosis is blocked with a BAX mutant. These results suggest that the effects of Pcdh-γ loss are distinct in different neural regions and that different mechanisms may underlie synapse formation.

Pcdh-γ regulates synapse development via astrocyte–neuron interactions.[98] Astrocytes play an important role in synaptogenesis through the modulation of various secreted molecules and contacts,[99,100] and Pcdh-γ is expressed in both neurons and astrocytes. When wild-type astrocytes are cultured with Pcdh-γ-deleted neurons, synaptogenesis is significantly reduced. Similar results are observed in cocultures of Pcdh-γ-deleted astrocytes and wild-type neurons during early developmental stages, but synapses eventually form at normal levels upon further maturation. The effect of the Pcdh-γ-deleted astrocytes is not due to secreted molecules, but requires astrocyte–neuron contacts. The astrocyte-specific Pcdh-γ deletion does not result in increased apoptosis. These results showed that Pcdh-γ contributes to direct interactions between astrocytes and neurons that affect synapse formation.

Thus, although Pcdh-γ-mediated astrocyte–neuron contacts promote synaptogenesis, the disruption of these contacts is not sufficient to induce apoptosis, as observed in the above-described Pcdh-γ mutant mice. This raises the possibility that the increased apoptosis is caused by the disruption of neuron–neuron contacts involving Pcdh-γ.

9. CONCLUSIONS

The clustered *Pcdh* gene family encodes diverse cell adhesion and recognition proteins that are specifically expressed in the vertebrate brain. Their molecular diversity increased during vertebrate evolution. Regulation of the gene expression of the *Pcdh* clusters in individual cells can generate high neuronal diversity in the brain. At the protein level, the heteromultimerization of *cis*-tetramers contributes to synapse diversity in the brain. Through their specific protein functions, including cell–cell interactions and cytoplasmic signaling, the diverse clustered Pcdhs may help generate the complex neural networks required for the many demanding functions of the brain.

REFERENCES

1. Sano K, Tanihara H, Heimark RL, Obata S, Davidson M, St John T, et al. Protocadherins: a large family of cadherin-related molecules in central nervous system. *EMBO J* 1993;**12**:2249–56.
2. Hirayama T, Tarusawa E, Yoshimura Y, Galjart N, Yagi T. CTCF is required for neural development and stochastic expression of clustered Pcdh genes in neurons. *Cell Rep* 2012;**2**:345–57.
3. Frank M, Kemler R. Protocadherins. *Curr Opin Cell Biol* 2002;**14**:557–62.
4. Yagi T, Takeichi M. Cadherin superfamily genes: functions, genomic organization, and neurologic diversity. *Genes Dev* 2000;**14**:1169–80.
5. Kim SY, Yasuda S, Tanaka H, Yamagata K, Kim H. Non-clustered protocadherin. *Cell Adh Migr* 2011;**5**:97–105.
6. Hulpiau P, van Roy F. Molecular evolution of the cadherin superfamily. *Int J Biochem Cell Biol* 2009;**41**:349–69.
7. Boggon TJ, Murray J, Chappuis-Flament S, Wong E, Gumbiner BM, Shapiro L. C-cadherin ectodomain structure and implications for cell adhesion mechanisms. *Science* 2002;**296**:1308–13.
8. Patel SD, Chen CP, Bahna F, Honig B, Shapiro L. Cadherin-mediated cell-cell adhesion: sticking together as a family. *Curr Opin Struct Biol* 2003;**13**:690–8.

9. Redies C, Vanhalst K, Roy F. delta-Protocadherins: unique structures and functions. *Cell Mol Life Sci* 2005;**62**:2840–52.
10. Morishita H, Umitsu M, Murata Y, Shibata N, Udaka K, Higuchi Y, et al. Structure of the cadherin-related neuronal receptor/protocadherin-alpha first extracellular cadherin domain reveals diversity across cadherin families. *J Biol Chem* 2006;**281**:33650–63.
11. Morishita H, Yagi T. Protocadherin family: diversity, structure, and function. *Curr Opin Cell Biol* 2007;**19**:584–92.
12. Takagi J. Structural basis for ligand recognition by RGD (Arg-Gly-Asp)-dependent integrins. *Biochem Soc Trans* 2004;**32**:403–6.
13. Mutoh T, Hamada S, Senzaki K, Murata Y, Yagi T. Cadherin-related neuronal receptor 1 (CNR1) has cell adhesion activity with beta1 integrin mediated through the RGD site of CNR1. *Exp Cell Res* 2004;**294**:494–508.
14. Yagi T. Molecular codes for neuronal individuality and cell assembly in the brain. *Front Mol Neurosci* 2012;**5**:45.
15. Kai N, Mishina M, Yagi T. Molecular cloning of Fyn-associated molecules in the mouse central nervous system. *J Neurosci Res* 1997;**48**:407–24.
16. Kohmura N, Senzaki K, Hamada S, Kai N, Yasuda R, Watanabe M, et al. Diversity revealed by a novel family of cadherins expressed in neurons at a synaptic complex. *Neuron* 1998;**20**:1137–51.
17. Wu Q, Maniatis T. A striking organization of a large family of human neural cadherin-like cell adhesion genes. *Cell* 1999;**97**:779–90.
18. Sugino H, Hamada S, Yasuda R, Tuji A, Matsuda Y, Fujita M, et al. Genomic organization of the family of CNR cadherin genes in mice and humans. *Genomics* 2000;**63**:75–87.
19. Tasic B, Nabholz CE, Baldwin KK, Kim Y, Rueckert EH, Ribich SA, et al. Promoter choice determines splice site selection in protocadherin alpha and gamma pre-mRNA splicing. *Mol Cell* 2002;**10**:21–33.
20. Wang X, Su H, Bradley A. Molecular mechanisms governing Pcdh-gamma gene expression: evidence for a multiple promoter and cis-alternative splicing model. *Genes Dev* 2002;**16**:1890–905.
21. Esumi S, Kakazu N, Taguchi Y, Hirayama T, Sasaki A, Hirabayashi T, et al. Monoallelic yet combinatorial expression of variable exons of the protocadherin-alpha gene cluster in single neurons. *Nat Genet* 2005;**37**:171–6.
22. Kaneko R, Kato H, Kawamura Y, Esumi S, Hirayama T, Hirabayashi T, et al. Allelic gene regulation of Pcdh-alpha and Pcdh-gamma clusters involving both monoallelic and biallelic expression in single Purkinje cells. *J Biol Chem* 2006;**281**:30551–60.
23. Gimelbrant A, Hutchinson JN, Thompson BR, Chess A. Widespread monoallelic expression on human autosomes. *Science* 2007;**318**:1136–40.
24. Zwemer LM, Zak A, Thompson BR, Kirby A, Daly MJ, Chess A, et al. Autosomal monoallelic expression in the mouse. *Genome Biol* 2012;**13**:R10.
25. Gregg C, Zhang J, Weissbourd B, Luo S, Schroth GP, Haig D, et al. High-resolution analysis of parent-of-origin allelic expression in the mouse brain. *Science* 2010;**329**:643–8.
26. Noguchi Y, Hirabayashi T, Katori S, Kawamura Y, Sanbo M, Hirabayashi M, et al. Total expression and dual gene-regulatory mechanisms maintained in deletions and duplications of the Pcdha cluster. *J Biol Chem* 2009;**284**:32002–14.
27. Wu Q, Zhang T, Cheng JF, Kim Y, Grimwood J, Schmutz J, et al. Comparative DNA sequence analysis of mouse and human protocadherin gene clusters. *Genome Res* 2001;**11**:389–404.
28. Golan-Mashiach M, Grunspan M, Emmanuel R, Gibbs-Bar L, Dikstein R, Shapiro E. Identification of CTCF as a master regulator of the clustered protocadherin genes. *Nucleic Acids Res* 2012;**40**:3378–91.

29. Ribich S, Tasic B, Maniatis T. Identification of long-range regulatory elements in the protocadherin-alpha gene cluster. *Proc Natl Acad Sci USA* 2006;**103**:19719–24.
30. Yokota S, Hirayama T, Hirano K, Kaneko R, Toyoda S, Kawamura Y, et al. Identification of the cluster control region for the protocadherin-beta genes located beyond the protocadherin-gamma cluster. *J Biol Chem* 2011;**286**:31885–95.
31. Kehayova P, Monahan K, Chen W, Maniatis T. Regulatory elements required for the activation and repression of the protocadherin-alpha gene cluster. *Proc Natl Acad Sci USA* 2011;**108**:17195–200.
32. Phillips JE, Corces VG. CTCF: master weaver of the genome. *Cell* 2009;**137**: 1194–211.
33. Kim TH, Abdullaev ZK, Smith AD, Ching KA, Loukinov DI, Green RD, et al. Analysis of the vertebrate insulator protein CTCF-binding sites in the human genome. *Cell* 2007;**128**:1231–45.
34. Xie X, Mikkelsen TS, Gnirke A, Lindblad-Toh K, Kellis M, Lander ES. Systematic discovery of regulatory motifs in conserved regions of the human genome, including thousands of CTCF insulator sites. *Proc Natl Acad Sci USA* 2007;**104**:7145–50.
35. Handoko L, Xu H, Li G, Ngan CY, Chew E, Schnapp M, et al. CTCF-mediated functional chromatin interactome in pluripotent cells. *Nat Genet* 2011;**43**:630–8.
36. Monahan K, Rudnick ND, Kehayova PD, Pauli F, Newberry KM, Myers RM, et al. Role of CCCTC binding factor (CTCF) and cohesin in the generation of single-cell diversity of Protocadherin-alpha gene expression. *Proc Natl Acad Sci USA* 2012;**109**: 9125–30.
37. Kawauchi S, Calof AL, Santos R, Lopez-Burks ME, Young CM, Hoang MP, et al. Multiple organ system defects and transcriptional dysregulation in the Nipbl(+/-) mouse, a model of Cornelia de Lange Syndrome. *PLoS Genet* 2009;**5**:e1000650.
38. Remeseiro S, Cuadrado A, Gomez-Lopez G, Pisano DG, Losada A. A unique role of cohesin-SA1 in gene regulation and development. *EMBO J* 2012;**31**:2090–102.
39. Tan YP, Li S, Jiang XJ, Loh W, Foo YK, Loh CB, et al. Regulation of protocadherin gene expression by multiple neuron-restrictive silencer elements scattered in the gene cluster. *Nucleic Acids Res* 2010;**38**:4985–97.
40. Robertson KD, Wolffe AP. DNA methylation in health and disease. *Nat Rev Genet* 2000;**1**:11–9.
41. Kawaguchi M, Toyama T, Kaneko R, Hirayama T, Kawamura Y, Yagi T. Relationship between DNA methylation states and transcription of individual isoforms encoded by the protocadherin-alpha gene cluster. *J Biol Chem* 2008;**283**:12064–75.
42. Chahrour M, Jung SY, Shaw C, Zhou X, Wong ST, Qin J, et al. MeCP2, a key contributor to neurological disease, activates and represses transcription. *Science* 2008;**320**:1224–9.
43. Novak P, Jensen T, Oshiro MM, Watts GS, Kim CJ, Futscher BW. Agglomerative epigenetic aberrations are a common event in human breast cancer. *Cancer Res* 2008;**68**:8616–25.
44. Waha A, Guntner S, Huang TH, Yan PS, Arslan B, Pietsch T, et al. Epigenetic silencing of the protocadherin family member PCDH-gamma-A11 in astrocytomas. *Neoplasia* 2005;**7**:193–9.
45. Frigola J, Song J, Stirzaker C, Hinshelwood RA, Peinado MA, Clark SJ. Epigenetic remodeling in colorectal cancer results in coordinate gene suppression across an entire chromosome band. *Nat Genet* 2006;**38**:540–9.
46. Novak P, Jensen T, Oshiro MM, Wozniak RJ, Nouzova M, Watts GS, et al. Epigenetic inactivation of the HOXA gene cluster in breast cancer. *Cancer Res* 2006;**66**:10664–70.
47. Dallosso AR, Hancock AL, Szemes M, Moorwood K, Chilukamarri L, Tsai HH, et al. Frequent long-range epigenetic silencing of protocadherin gene clusters on chromosome 5q31 in Wilms' tumor. *PLoS Genet* 2009;**5**:e1000745.

48. Morishita H, Murata Y, Esumi S, Hamada S, Yagi T. CNR/Pcdhalpha family in subplate neurons, and developing cortical connectivity. *Neuroreport* 2004;**15**:2595–9.
49. Frank M, Ebert M, Shan W, Phillips GR, Arndt K, Colman DR, et al. Differential expression of individual gamma-protocadherins during mouse brain development. *Mol Cell Neurosci* 2005;**29**:603–16.
50. Cronin KD, Capehart AA. Gamma protocadherin expression in the embryonic chick nervous system. *Int J Biol Sci* 2007;**3**:8–11.
51. Bass T, Ebert M, Hammerschmidt M, Frank M. Differential expression of four protocadherin alpha and gamma clusters in the developing and adult zebrafish: DrPcdh2gamma but not DrPcdh1gamma is expressed in neuronal precursor cells, ependymal cells and non-neural epithelia. *Dev Genes Evol* 2007;**217**:337–51.
52. Zou C, Huang W, Ying G, Wu Q. Sequence analysis and expression mapping of the rat clustered protocadherin gene repertoires. *Neuroscience* 2007;**144**:579–603.
53. Junghans D, Heidenreich M, Hack I, Taylor V, Frotscher M, Kemler R. Postsynaptic and differential localization to neuronal subtypes of protocadherin beta16 in the mammalian central nervous system. *Eur J Neurosci* 2008;**27**:559–71.
54. Takei Y, Hamada S, Senzaki K, Mutoh T, Sugino H, Yagi T. Two novel CNRs from the CNR gene cluster have molecular features distinct from those of CNR1 to 8. *Genomics* 2001;**72**:321 30.
55. Wang X, Weiner JA, Levi S, Craig AM, Bradley A, Sanes JR. Gamma protocadherins are required for survival of spinal interneurons. *Neuron* 2002;**36**:843–54.
56. Phillips GR, Tanaka H, Frank M, Elste A, Fidler L, Benson DL, et al. Gamma-protocadherins are targeted to subsets of synapses and intracellular organelles in neurons. *J Neurosci* 2003;**23**:5096–104.
57. Murata Y, Hamada S, Morishita H, Mutoh T, Yagi T. Interaction with protocadherin-gamma regulates the cell surface expression of protocadherin-alpha. *J Biol Chem* 2004;**279**:49508–16.
58. Kallenbach S, Khantane S, Carroll P, Gayet O, Alonso S, Henderson CE, et al. Changes in subcellular distribution of protocadherin gamma proteins accompany maturation of spinal neurons. *J Neurosci Res* 2003;**72**:549–56.
59. Lobas MA, Helsper L, Vernon CG, Schreiner D, Zhang Y, Holtzman MJ, et al. Molecular heterogeneity in the choroid plexus epithelium: the 22-member gamma-protocadherin family is differentially expressed, apically localized, and implicated in CSF regulation. *J Neurochem* 2012;**120**:913–27.
60. Morishita H, Kawaguchi M, Murata Y, Seiwa C, Hamada S, Asou H, et al. Myelination triggers local loss of axonal CNR/protocadherin alpha family protein expression. *Eur J Neurosci* 2004;**20**:2843–7.
61. Stallcup WB, Beasley LL, Levine JM. Antibody against nerve growth factor-inducible large external (NILE) glycoprotein labels nerve fiber tracts in the developing rat nervous system. *J Neurosci* 1985;**5**:1090–101.
62. Bartsch U, Kirchhoff F, Schachner M. Immunohistological localization of the adhesion molecules L1, N-CAM, and MAG in the developing and adult optic nerve of mice. *J Comp Neurol* 1989;**284**:451–62.
63. Kapfhammer JP, Schwab ME. Inverse patterns of myelination and GAP-43 expression in the adult CNS: neurite growth inhibitors as regulators of neuronal plasticity? *J Comp Neurol* 1994;**340**:194–206.
64. Inoue Y, Nakamura R, Mikoshiba K, Tsukada Y. Fine structure of the central myelin sheath in the myelin deficient mutant Shiverer mouse, with special reference to the pattern of myelin formation by oligodendroglia. *Brain Res* 1981;**219**:85–94.
65. McGee AW, Yang Y, Fischer QS, Daw NW, Strittmatter SM. Experience-driven plasticity of visual cortex limited by myelin and Nogo receptor. *Science* 2005;**309**:2222–6.
66. Chen J, Lu Y, Meng S, Han MH, Lin C, Wang X. alpha- and gamma-Protocadherins negatively regulate PYK2. *J Biol Chem* 2009;**284**:2880–90.

67. Emond MR, Jontes JD. Inhibition of protocadherin-alpha function results in neuronal death in the developing zebrafish. *Dev Biol* 2008;**321**:175–87.
68. Garrett AM, Schreiner D, Lobas MA, Weiner JA. gamma-Protocadherins control cortical dendrite arborization by regulating the activity of a FAK/PKC/MARCKS signaling pathway. *Neuron* 2012;**74**:269–76.
69. Zhang X, Chattopadhyay A, Ji QS, Owen JD, Ruest PJ, Carpenter G, et al. Focal adhesion kinase promotes phospholipase C-gamma1 activity. *Proc Natl Acad Sci USA* 1999;**96**:9021–6.
70. Tvorogov D, Wang XJ, Zent R, Carpenter G. Integrin-dependent PLC-gamma1 phosphorylation mediates fibronectin-dependent adhesion. *J Cell Sci* 2005;**118**:601–10.
71. Heemskerk FM, Chen HC, Huang FL. Protein kinase C phosphorylates Ser152, Ser156 and Ser163 but not Ser160 of MARCKS in rat brain. *Biochem Biophys Res Commun* 1993;**190**:236–41.
72. Metzger F, Kapfhammer JP. Protein kinase C activity modulates dendritic differentiation of rat Purkinje cells in cerebellar slice cultures. *Eur J Neurosci* 2000;**12**:1993–2005.
73. Schrenk K, Kapfhammer JP, Metzger F. Altered dendritic development of cerebellar Purkinje cells in slice cultures from protein kinase C gamma-deficient mice. *Neuroscience* 2002;**110**:675–89.
74. Li H, Chen G, Zhou B, Duan S. Actin filament assembly by myristoylated alanine-rich C kinase substrate-phosphatidylinositol-4,5-diphosphate signaling is critical for dendrite branching. *Mol Biol Cell* 2008;**19**:4804–13.
75. Marambaud P, Robakis NK. Genetic and molecular aspects of Alzheimer's disease shed light on new mechanisms of transcriptional regulation. *Genes Brain Behav* 2005;**4**:134–46.
76. Marambaud P, Wen PH, Dutt A, Shioi J, Takashima A, Siman R, et al. A CBP binding transcriptional repressor produced by the PS1/epsilon-cleavage of N-cadherin is inhibited by PS1 FAD mutations. *Cell* 2003;**114**:635–45.
77. Maretzky T, Reiss K, Ludwig A, Buchholz J, Scholz F, Proksch E, et al. ADAM10 mediates E-cadherin shedding and regulates epithelial cell-cell adhesion, migration, and beta-catenin translocation. *Proc Natl Acad Sci USA* 2005;**102**:9182–7.
78. Hambsch B, Grinevich V, Seeburg PH, Schwarz MK. {gamma}-Protocadherins, presenilin-mediated release of C-terminal fragment promotes locus expression. *J Biol Chem* 2005;**280**:15888–97.
79. Haas IG, Frank M, Veron N, Kemler R. Presenilin-dependent processing and nuclear function of gamma-protocadherins. *J Biol Chem* 2005;**280**:9313–9.
80. Reiss K, Maretzky T, Haas IG, Schulte M, Ludwig A, Frank M, et al. Regulated ADAM10-dependent ectodomain shedding of gamma-protocadherin C3 modulates cell-cell adhesion. *J Biol Chem* 2006;**281**:21735–44.
81. Bonn S, Seeburg PH, Schwarz MK. Combinatorial expression of alpha- and gamma-protocadherins alters their presenilin-dependent processing. *Mol Cell Biol* 2007;**27**:4121–32.
82. Obata S, Sago H, Mori N, Rochelle JM, Seldin MF, Davidson M, et al. Protocadherin Pcdh2 shows properties similar to, but distinct from, those of classical cadherins. *J Cell Sci* 1995;**108**(Pt. 12):3765–73.
83. Fernandez-Monreal M, Kang S, Phillips GR. Gamma-protocadherin homophilic interaction and intracellular trafficking is controlled by the cytoplasmic domain in neurons. *Mol Cell Neurosci* 2009;**40**:344–53.
84. Schreiner D, Weiner JA. Combinatorial homophilic interaction between gamma-protocadherin multimers greatly expands the molecular diversity of cell adhesion. *Proc Natl Acad Sci USA* 2010;**107**:14893–8.
85. Ozawa M, Kemler R. Altered cell adhesion activity by pervanadate due to the dissociation of alpha-catenin from the E-cadherin.catenin complex. *J Biol Chem* 1998;**273**:6166–70.

86. Han MH, Lin C, Meng S, Wang X. Proteomics analysis reveals overlapping functions of clustered protocadherins. *Mol Cell Proteomics* 2010;**9**:71–83.
87. Hasegawa S, Hamada S, Kumode Y, Esumi S, Katori S, Fukuda E, et al. The protocadherin-alpha family is involved in axonal coalescence of olfactory sensory neurons into glomeruli of the olfactory bulb in mouse. *Mol Cell Neurosci* 2008;**38**:66–79.
88. Buck L, Axel R. A novel multigene family may encode odorant receptors: a molecular basis for odor recognition. *Cell* 1991;**65**:175–87.
89. Malnic B, Hirono J, Sato T, Buck LB. Combinatorial receptor codes for odors. *Cell* 1999;**96**:713–23.
90. Mombaerts P. Genes and ligands for odorant, vomeronasal and taste receptors. *Nat Rev Neurosci* 2004;**5**:263–78.
91. Yu CR, Power J, Barnea G, O'Donnell S, Brown HE, Osborne J, et al. Spontaneous neural activity is required for the establishment and maintenance of the olfactory sensory map. *Neuron* 2004;**42**:553–66.
92. Katori S, Hamada S, Noguchi Y, Fukuda E, Yamamoto T, Yamamoto H, et al. Protocadherin-alpha family is required for serotonergic projections to appropriately innervate target brain areas. *J Neurosci* 2009;**29**:9137–47.
93. Fukuda E, Hamada S, Hasegawa S, Katori S, Sanbo M, Miyakawa T, et al. Down-regulation of protocadherin-alpha A isoforms in mice changes contextual fear conditioning and spatial working memory. *Eur J Neurosci* 2008;**28**:1362–76.
94. Weiner JA, Wang X, Tapia JC, Sanes JR. Gamma protocadherins are required for synaptic development in the spinal cord. *Proc Natl Acad Sci USA* 2005;**102**:8–14.
95. Prasad T, Wang X, Gray PA, Weiner JA. A differential developmental pattern of spinal interneuron apoptosis during synaptogenesis: insights from genetic analyses of the protocadherin-gamma gene cluster. *Development* 2008;**135**:4153–64.
96. Prasad T, Weiner JA. Direct and indirect regulation of spinal cord Ia afferent terminal formation by the gamma-protocadherins. *Front Mol Neurosci* 2011;**4**:54.
97. Lefebvre JL, Zhang Y, Meister M, Wang X, Sanes JR. gamma-Protocadherins regulate neuronal survival but are dispensable for circuit formation in retina. *Development* 2008;**135**:4141–51.
98. Garrett AM, Weiner JA. Control of CNS synapse development by {gamma}-protocadherin-mediated astrocyte-neuron contact. *J Neurosci* 2009;**29**:11723–31.
99. Barres BA. The mystery and magic of glia: a perspective on their roles in health and disease. *Neuron* 2008;**60**:430–40.
100. Theodosis DT, Poulain DA, Oliet SH. Activity-dependent structural and functional plasticity of astrocyte-neuron interactions. *Physiol Rev* 2008;**88**:983–1008.

CHAPTER EIGHT

Delta-Protocadherins in Health and Disease

Irene Kahr, Karl Vandepoele[1], Frans van Roy
Department for Molecular Biomedical Research, VIB & Department of Biomedical Molecular Biology, Ghent University, Ghent, Belgium
[1]Current address: Laboratory for Clinical Biology, Ghent University Hospital, Ghent, Belgium

Contents

Abstract

The protocadherin family comprises clustered and nonclustered protocadherin genes. The nonclustered genes encode mainly δ-protocadherins, which deviate markedly from classical cadherins. They can be subdivided phylogenetically into δ0-protocadherins (protocadherin-20), δ1-protocadherins (protocadherin-1, -7, -9, and -11X/Y), and δ2-protocadherins (protocadherin-8, -10, -17, -18, and -19). δ-Protocadherins share a similar gene structure and are expressed as multiple alternative splice forms differing mostly in their cytoplasmic domains (CDs). Some δ-protocadherins reportedly show cell–cell adhesion properties. Individual δ-protocadherins appear to be involved in specific signaling pathways, as they interact with proteins such as TAF1/Set, TAO2β, Nap1, and the Frizzled-7 receptor. The spatiotemporally restricted expression of δ-protocadherins in various tissues and species and their functional analysis suggest that they play multiple, tightly regulated roles in vertebrate development. Furthermore, several δ-protocadherins have been implicated in neurological disorders and in cancers, highlighting the importance of scrutinizing their properties and their dysregulation in various pathologies.

Progress in Molecular Biology and Translational Science, Volume 116
ISSN 1877-1173
http://dx.doi.org/10.1016/B978-0-12-394311-8.00008-X

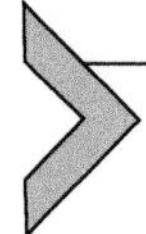

1. IDENTIFICATION OF δ-PROTOCADHERIN FAMILY MEMBERS

Protocadherins (Pcdhs) were first identified by Sano *et al.*[1] two decades ago. They cloned human protocadherin-42 and -43, which were later renamed protocadherin-1 and protocadherin-γC3, respectively. The overall structure of these proteins was found to resemble that of classical cadherins. Their ectodomains are composed of six or seven extracellular cadherin (EC) repeats, whereas classical cadherins have five. However, their cytoplasmic domains (CDs) have no significant homology with those of classical cadherins. Evidence has been provided that Pcdhs are most likely derived from a cadherin-related molecule but were not themselves the first cadherin-like molecules.[2]

The number of protocadherin family members identified has dramatically increased over the past years, making Pcdhs the largest subgroup in the cadherin superfamily.[3] More than 70 different *Pcdh* genes have been identified in mammals. Based on genomic structure, the Pcdh family can be divided into clustered protocadherins (C-Pcdhs) and nonclustered protocadherins (NC-Pcdhs).[4,5] C-Pcdhs are dealt with in Chapter 7 by T. Hirayama and T. Yagi. Except for Pcdh12, NC-Pcdhs are considered members of the δ-protocadherin subfamily (Table 8.1) and can be subdivided into three subgroups: δ0-, δ1-, and δ2-protocadherins. Pcdh1, Pcdh7, Pcdh9, and Pcdh11 constitute the δ1-Pcdh subfamily. They have an ectodomain with seven EC domains and three conserved motifs (CM1–CM3) in their CD.[3,6,21] Human *PCDH11* genes were mapped to the XY homology regions of the sex chromosomes and therefore are expressed as PCDH11X and PCDH11Y, whereas in nonhominids, Pcdh11 is encoded only by the X chromosome.[25] Pcdh20 is phylogenetically related to δ1-Pcdhs but lacks the CMs, and has therefore been classified as δ0-Pcdh.[3] δ2-Pcdhs comprise Pcdh8, Pcdh10, Pcdh17, Pcdh18 and Pcdh19, and have six EC domains and a CD bearing two conserved motifs (CM1, CM2).[6]

2. STRUCTURAL PROPERTIES OF δ-PROTOCADHERINS

A striking difference in the genomic organization of *Pcdh* genes and classical cadherin genes is that the ectodomain of Pcdhs is generally encoded by a single large exon, while the extracellular region of classical cadherins is encoded by multiple exons.[4,6] This large exon in the *Pcdh* genes usually encodes the entire ectodomain, the single transmembrane domain, and part of the CD.

Table 8.1 The δ-protocadherin family

Gene symbol	Name	Synonyms	Interaction partners	Knockout/knockdown models[a]	References
PCDH1	Protocadherin-1	Protocadherin-42 Axial protocadherin (AXPC)	RYK PP1α	KD in *Xenopus*: inhibition of morphogenesis in axial mesoderm; KD in chicken: defect in neural crest cell localization.	6,7
PCDH7	Protocadherin-7	BH-protocadherin NF-protocadherin (NFPC)	PP1α, RYK TAF1/Set	Expression of DN Pcdh7 in *Xenopus*: disruption of ectoderm integrity, and defects in axon initiation and elongation; KD in *Xenopus*: neural tube closure defects.	7–9
PCDH8	Protocadherin-8	Arcadlin	TAO2β, N-cadherin	KO in mouse: no phenotypic abnormality.	10,11
PAPC	Paraxial protocadherin	*Xenopus* PCDH8 paralog	Frizzled-7, Sprouty, ANR5, CK2β, C-cadherin, FLRT3	KD in *Xenopus* and zebrafish: defects in tissue separation, convergent extension movements during gastrulation; Expression of DN PAPC in mouse tails: disruption of epithelial organization between segmental borders of somites.	12–17
PCDH9	Protocadherin-9		PP1α, RYK		6,7
PCDH10	Protocadherin-10	OL-protocadherin (OL-pc)	Nap1, RYK	KO in mouse: defects in axon trajectories in brain.	7,18,49

Continued

Table 8.1 The δ-protocadherin family—cont'd

Gene symbol	Name	Synonyms	Interaction partners	Knockout/knockdown models	References
PCDH11X	Protocadherin-11X	Protocadherin-X	PP1α		6
PCDH11Y	Protocadherin-11Y	Protocadherin-Y Protocadherin-PC	PP1α β-catenin		6,19
PCDH17	Protocadherin-17	Protocadherin-68	N-cadherin		20
PCDH18	Protocadherin-18	Protocadherin-68-like	Dab1, RYK		7,21
PCDH19	Protocadherin-19		Nap1, RYK, N-cadherin	KD in zebrafish: disruption of early brain morphogenesis.	7,20,22,23
PCDH20	Protocadherin-20	Protocadherin-13	RYK		7

[a]DN, dominant negative; KD, knockdown; KO, knockout.

Classical cadherins have five EC domains with characteristic motifs. Pcdhs contain six or seven EC repeats differing in several aspects from those of classical cadherins.[1,2] Pcdhs also differ in lacking a prodomain and in the frequent occurrence of variants generated by alternative splicing or alternative polyadenylation.[4] Usually, the shortest transcripts of the δ-*Pcdh* genes encode a transmembrane protein with a fairly short CD lacking the conserved motifs.[6] In particular, PCDH11X and -Y are expressed as many splice variants due to the alternative use of small exons encoding either the 5′ UTR or parts of the CD.[6,26] A unique feature of PCDH11 proteins is the presence of an internal repeat at the carboxy-terminus of the long isoforms. A sequence of 11 amino acids is repeated nine times (eight times in mouse Pcdh11X) and contains on average two proline residues, suggesting a peculiar 3D structure. Most of the differences between PCDH11X and PCDH11Y have occurred in this repeat region.[25]

A cysteine at the beginning of the CD is conserved in all δ-Pcdhs.[3] This cysteine is palmitoylated in rat Pcdh1 and several other δ-Pcdhs.[27] Palmitoylation is important in protein trafficking and compartmentalization in polarized neurons.[27] When cells were treated with a palmitoylation inhibitor, PCDH1 was no longer localized at the membrane (our unpublished observations). However, PCDH1 proteins with a mutation in this cysteine still localized to the membrane, indicating that other residues contribute.

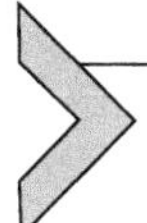

3. PHYSIOLOGICAL AND DEVELOPMENTAL FUNCTIONS OF δ-PROTOCADHERINS

3.1. Tissue morphogenesis and cell–cell adhesion

3.1.1 Homophilic cell–cell adhesion

Several δ-Pcdhs can mediate Ca^{2+}-dependent homophilic cell–cell adhesion, although binding appears to be generally weaker than that of classical cadherins. For instance, *in vitro* adhesive activity has been demonstrated for Pcdh1/AXPC,[1,28] Pcdh7/NFPC,[29] Pcdh8/Arcadlin/PAPC,[30,31] Pcdh10[32], and Pcdh19.[22] Pcdh19-mediated adhesion was strengthened when its CD was removed or replaced by that of E-cadherin,[22] suggesting that its ectodomain is adhesion-compatible, but that the CD does not efficiently stabilize or negatively regulates homophilic adhesion. Adhesive strength is likewise increased when the cytoplasmic tails of Pcdh1/AXPC or PAPC are removed,[28,31] although the results for PAPC could not be confirmed in another study.[12] In contrast to classical cadherins, Pcdhs lack a β-catenin binding site in their CD and they do not interact with any other molecules that link classical cadherins to the cytoskeleton, and thereby stabilize the adhesion complex. It seems that Pcdhs are less involved in typical cell–cell adhesion but might play a role in intercellular regulation. Some δ-Pcdhs can promote cell sorting *in vitro* and *in vivo* by preferring homophilic over heterophilic interactions. AXPC and PAPC are reciprocally expressed in axial and paraxial mesoderm and mediate cell sorting and movements during embryonic gastrulation in *Xenopus*.[28,31] Both Pcdh19[22] and PAPC[33] form oligomers via the conserved cysteine residues in their ectodomains. The PAPC CD is not required for its cell-sorting activity, whereas both the transmembrane and extracellular domains are necessary.[33] In contrast, Pcdh7/NFPC mediates cell sorting in the epidermis of *Xenopus* in a homophilic manner, but not in the absence of its CD.[29] Thus, δ-Pcdhs can mediate weak homophilic cell–cell interactions, seemingly by various mechanisms.

3.1.2 Heterophilic cis interactions

There are also reports on δ-Pcdhs mediating heterophilic cis interactions with classical cadherins. PAPC was shown to regulate C-cadherin adhesion activity in *Xenopus* embryos.[12] It has been proposed that PAPC and C-cadherin form a complex with FLRT3 (fibronectin leucine-rich domain transmembrane protein-3), thereby controlling cell sorting and morphogenetic movements in *Xenopus*.[13] Both PAPC and FLRT3 specifically interact with human E-cadherin in transfected A431 cells.[13] Moreover, the rat ortholog of Pcdh8 interacts laterally with N-cadherin in rat hippocampal neurons and this interaction is increased by neural stimulation.[10] The interaction domains were localized to their transmembrane segments. Pcdh8 was shown to downregulate the adhesive activity of N-cadherin by triggering N-cadherin endocytosis (see below). N-cadherin also forms a *cis*-complex with Pcdh19 *in vitro* and *in vivo*[23] (Fig. 8.1A). Disruption of this interaction impairs cell movements during neurulation in *Xenopus*,[23] and knockdown of Pcdh19 by morpholino oligonucleotides leads to a severe disruption in early brain morphogenesis of zebrafish embryos caused at least in part by defective cell movements in the anterior neural plate.[34] This phenotype is reminiscent of N-cadherin mutants and morphants,[35] suggesting that Pcdh19 and N-cadherin function together in brain morphogenesis. Adhesion by the Pcdh19–N-cadherin complex is not disrupted by mutations that inhibit N-cadherin homophilic binding but is impaired by a mutation in Pcdh19, suggesting that Pcdh19 plays the dominant role in this mechanism of cell adhesion.[20] Also Pcdh17 associates with N-cadherin, but this complex segregates from the Pcdh19–N-cadherin complex.[20] These data imply that

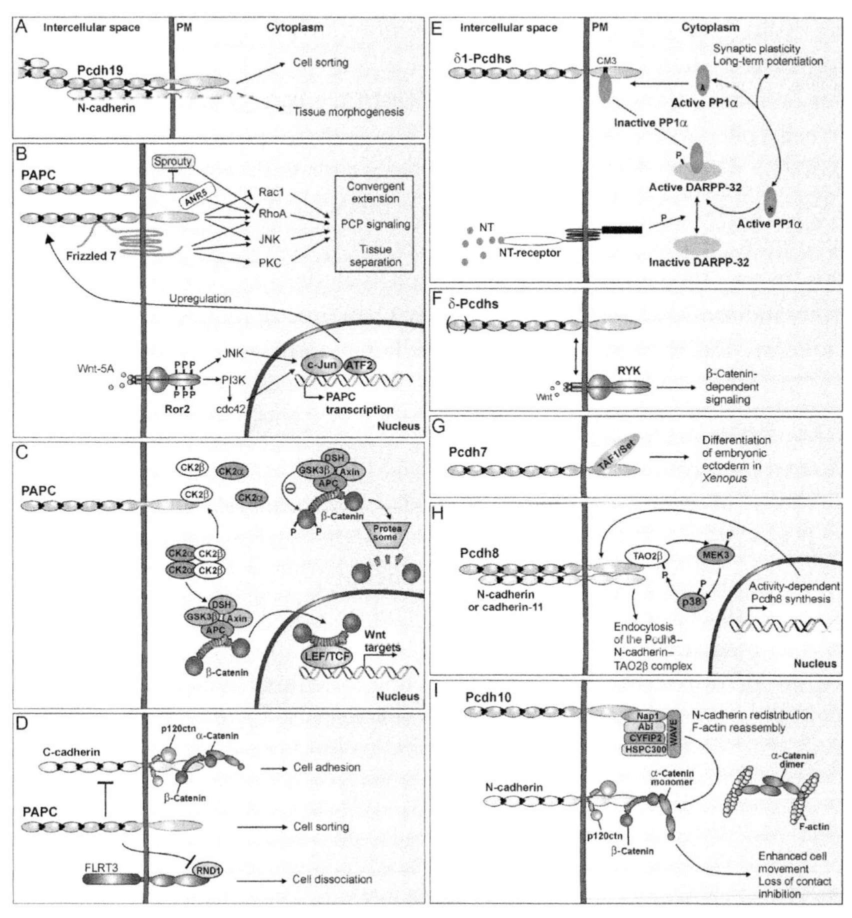

Figure 8.1 Schematic overview of intracellular signaling pathways involving δ-protocadherins. See text for references and acronyms. (A) Protocadherin-19 (Pcdh19) mediates a new mechanism of homophilic cell adhesion with N-cadherin acting as a required *cis*-cofactor. (B) Paraxial protocadherin (PAPC) interacts with Frizzled-7 to activate downstream signaling through RhoA, JNK, and PKC, which leads to tissue separation during convergent extension movements in *Xenopus* embryos. Furthermore, PAPC interacts with ANR5 and Sprouty, which, respectively, activate or inhibit RhoA signaling, and thereby promote the planar cell polarity (PCP) pathway. PAPC expression is regulated by binding of Wnt-5A to Ror2, which in turn activates JNK signaling, leading to upregulation of PAPC. P, phosphotyrosine. (C) PAPC can inhibit canonical Wnt signaling by interacting with CK2β. This interferes with CK2 complex formation, which leads to proteasomal degradation of β-catenin and subsequent inhibition of Wnt target gene transcription. (D) PAPC influences morphogenetic movements and cell sorting in *Xenopus* embryos by downregulating the adhesive activity of C-cadherin. Moreover, PAPC and C-cadherin form a functional complex with FLRT3. Unregulated FLRT3 causes excessive cell dissociation, but PAPC binding inhibits the ability of FLRT3 to recruit the

(Continued)

δ2-Pcdhs can associate with N-cadherin to form adhesive *cis*-complexes, where Pcdhs rather than N-cadherin mediate binding specificity. Furthermore, mouse Pcdh10 was shown to regulate N-cadherin assembly at cell–cell contacts by recruiting Nck-associated protein 1 (Nap1) and WAVE1 to cell–cell contacts, but a direct link with N-cadherin has not been shown.[18] Nap1 regulates WAVE-mediated actin assembly and has an important role in various morphogenetic events, such as neural tube closure and migration of endoderm and mesoderm.[36] These data collectively indicate that δ-Pcdhs can directly or indirectly interact with classical cadherins and thereby regulate their adhesive behavior.

3.1.3 Tissue morphogenesis

Studies in various model organisms have demonstrated that δ-Pcdhs are involved in regulating morphogenesis. For example, knockdown of Pcdh1/AXPC inhibits the morphogenesis of the axial mesoderm in *Xenopus* embryos, resulting in a severely bent phenotype reflective of a structural defect in the notochord.[28,37] In chick embryos, Pcdh1 plays an important role in mediating the localization and/or retention of neural crest cells within the dorsal root ganglia,[38] providing functional evidence that Pcdh-mediated cell sorting is important in the developing peripheral nervous system.

small GTPase RND1, and this reduces the magnitude of FLRT3-induced cell dissociation. (E) δ1-Pcdhs interact with PP1α, probably inactivating it. They might work synergistically with neurotransmitters (NT) to regulate synaptic plasticity. (F) Several δ-Pcdhs were shown to associate with RYK, which functions as a transmembrane receptor for ligands of the Wnt family. Both δ-Pcdhs and RYK were shown to be involved in the regulation of cell movements during convergent extension in zebrafish and frog, and of axon path finding *in vivo* and *in vitro*. (G) Protocadherin-7 (Pcdh7) interacts with the histone-regulating protein TAF1/Set; this interaction could be causally linked to differentiation of *Xenopus* embryonic ectoderm. (H) On neural activation, N-cadherin and cadherin-11 associate with protocadherin-8 (Pcdh8), which binds to the MAPKKK TAO2β. TAO2β activates p38 MAPK, which in turn feeds back to TAO2β, in the end triggering endocytosis of the Pcdh8-N-cadherin–TAO2β complex. (I) Protocadherin-10 (Pcdh10) associates with Nap1, which is a known mediator of WAVE-mediated actin assembly. By recruiting Nap1 to cell–cell contact sites, Pcdh10 might alter F-actin organization, which in turn causes N-cadherin redistribution. As the redistributed N-cadherin molecules can no longer induce contact inhibition, the result is uncooperative, accelerated cell movement. Shown is also the classical cadherin–catenin complex and the binding of α-catenin dimers to F-actin (see Chapter 1). (See Color Insert.)

Xenopus PAPC is one of the best-studied δ-Pcdhs. It is required for morphogenesis of the paraxial mesoderm during gastrulation, including convergent extension movements, tissue separation, and somitogenesis.[12,14,31,39] Surprisingly, the paralog Pcdh8 has hardly been studied in *Xenopus*. Pcdh8-null mice show no obvious segmentation defect and are viable and fertile,[11,40] probably due to redundancy of related Pcdhs. Indeed, in zebrafish, Pcdh10 has a comparable function to the *Xenopus* PAPC.[41] PAPC interacts with the Wnt receptor Frizzled-7 to activate downstream signaling through RhoA, Rac1, and JNK[14,39] (Fig. 8.1B), which are regulators or effectors of the planar cell polarity (PCP) pathway. Moreover, the intracellular domain of PAPC interacts with ANR5 (ankyrin repeats domain protein 5) and Sprouty and thereby promotes the PCP pathway[15,16] (Fig. 8.1B). Transcription of PAPC is regulated by Wnt-5A in early *Xenopus* development.[42] This requires the receptor tyrosine kinase Ror2, which activates its downstream effector JNK, which in turn leads to upregulation of PAPC[42] (Fig. 8.1B). This pathway was suggested to represent an alternative to noncanonical Wnt signaling that controls gene expression and is required for regulation of convergent extension movements in *Xenopus* gastrulation.

PAPC was shown to inhibit canonical Wnt signaling by interacting with casein kinase 2β (CK2β).[17] The CK2 complex activates Dishevelled and stabilizes β-catenin in the cytoplasm, which precedes nuclear entry and Wnt target gene regulation. Binding of PAPC to CK2β antagonizes this activity, thereby restricting the expression of Wnt target genes during *Xenopus* gastrulation[17] (Fig. 8.1C). Thus, PAPC can regulate both canonical and noncanonical Wnt signaling. Furthermore, PAPC can influence morphogenetic movements in *Xenopus* by downregulating the adhesion activity of C-cadherin[12] (Fig. 8.1D). Whether a direct interaction of PAPC with C-cadherin is involved is unknown. C-cadherin mediates cell–cell adhesion between *Xenopus* blastomeres and both its inhibition and its hyperactivity interfere with normal blastopore closure. PAPC serves as an inhibitor of C-cadherin. Interestingly, signaling through the Frizzled-7 receptor is not required for PAPC regulation of C-cadherin,[12] suggesting that Frizzled-7 signaling and C-cadherin regulation are distinct branches of the PAPC pathway inducing morphogenetic movements. PAPC might also regulate cell sorting and morphogenetic movements during *Xenopus* gastrulation by formation of a functional complex with C-cadherin and FLRT3.[13] Unregulated FLRT3 strongly induces cell dissociation, but upon PAPC binding, this effect is diminished. This results in a more physiological modulation of adhesion so that cells can sort out within the tissue[13] (Fig. 8.1D).

Thus, although PAPC can directly downregulate cadherin-mediated adhesion to some extent, it can also function as a molecular "buffer" to limit the activity of the more potent cell adhesion inhibitor FLRT3. The resulting modulation of cadherin-mediated adhesion is optimal for morphogenesis.

Ectopic expression of a dominant-negative form of Pcdh7/NFPC results in disruption of ectoderm integrity in *Xenopus* embryos, leading to cell dissociation and blister formation.[29] This defect can be rescued by overexpression of Pcdh7 or by overexpression of TAF1, a cytoplasmic interaction partner of Pcdh7.[8] Furthermore, Pcdh7 and TAF1 mediate cell adhesion within the neural folds and both are required for proper neurulation in *Xenopus*.[43] Disruption of Pcdh7 or TAF1 causes failure of neural tube closure along the entire anterior–posterior axis.

3.2. Brain development

δ-Pcdhs appear to exercise tightly regulated functions in the development, regionalization, and functional differentiation of the brain. Like classical cadherins, δ-Pcdhs exhibit diverse, spatially restricted, and temporally regulated expression patterns in the developing and mature central nervous system (CNS) of various species (for instance, Refs. 44–46). Based on their findings, Redies and colleagues proposed that the combinatorial expression of multiple cadherin genes contributes to the formation and maintenance of the complex neural circuitry of the CNS. A similar molecular adhesive code has been proposed for clustered Pcdhs in the brain[47] (see Chapter 7). In line with this model, Pcdh10-deficient mice have severe defects in axon trajectories through the ventral telencephalon.[48] Another study provided proof that Pcdh10 has an important role in axon guidance of olfactory sensory neurons (OSNs)[49] by demonstrating that Pcdh10 is regulated by neural activity in the mouse olfactory epithelium and that misexpression of Pcdh10 in OSNs affects late-stage axonal phenotypes. In *Xenopus*, inhibition of Pcdh7/NFPC severely reduces retinal axon and dendrite initiation and elongation.[50] These data suggest that δ-Pcdhs are general players in axonogenesis, dendritogenesis, and axon guidance.

Particular δ-Pcdhs are linked via their cytoplasmic interaction partners to signaling pathways implicated in neuronal development. Protein phosphatase 1α (PP1α), which interacts with all δ1-Pcdhs,[6] is implicated in long-term potentiation and depression, which are important in memory and learning.[51] The interaction of Pcdh18 with Dab1 links it to the Reelin signaling pathway, which controls neural positioning in the vertebrate brain.[52] Pcdh8/arcadlin is induced by neuronal activity,[30] whereas Pcdh19 and Pcdh20 are repressed by elevated activity in the rat cerebral cortex.[53] Interestingly, Pcdh8/arcadlin is involved in N-cadherin internalization from dendritic spine membranes, thereby directly affecting dendritic spine number in cultured hippocampal neurons.[10]

3.3. Molecular interaction partners

As the CDs of δ-Pcdhs contain at least three conserved motifs[6] besides PDZ-binding sites,[54] δ-Pcdhs likely share some core interaction partners. However, because the other sequences of their CDs differ, each δ-Pcdh might also interact with a different repertoire of intracellular proteins. This and their weak adhesive strength indicate that δ-Pcdhs are involved in various intracellular signaling pathways.

3.3.1 δ1-Pcdhs and PP1α

PP1α was isolated in a yeast two-hybrid screen for cytoplasmic interaction partners of PCDH7.[9] Later, the PP1α interaction motif RRVTF was shown to be present in all δ1 Pcdhs but not in δ2-Pcdhs and was called the CM3 motif.[6] In the interaction with PCDH7, the enzymatic activity of PP1α on glycogen phosphorylase is inhibited.[9] PP1α is strongly expressed in brain and enriched in medium-sized spiny neurons of the striatum (reviewed in Ref. 4). In this neuron type, several neurotransmitter pathways regulate the phosphorylation state of DARPP-32, which is a potent inhibitor of PP1α when phosphorylated. PP1α is thought to act as a negative regulator downstream of the neurotransmitter receptors, implicating it in the regulation of synaptic plasticity and both long-term potentiation and long-term depression, which are important in memory and learning. Since both DARPP-32 and PCDH7 inhibit the activity of PP1α, they might act synergistically (reviewed in Ref. 4) (Fig. 8.1E). PP1α is regulated in an isoform- and tissue-specific manner by a variety of interacting proteins. Thus, differentially expressed δ1-Pcdhs could influence PP1α function in different cell types of the nervous system.

3.3.2 δ-Pcdhs and RYK

Several C-PCDHs as well as δ-PCDHs were isolated in a screen for human receptor-like tyrosine kinase (RYK)-associated proteins[7] (Fig. 8.1F). RYK functions as a transmembrane receptor for ligands of the Wnt family and is required for β-catenin-dependent Wnt signaling. RYK binds Frizzled proteins but the latter are not always required for Wnt-triggered RYK signaling. The ectodomain or the transmembrane domain of RYK is required for interaction with δ-PCDHs, but the CD is not,[7] suggesting that the association occurs at the plasma membrane. However, whether PCDHs interact directly with RYK has not been reported. RYK has been shown to regulate cell movements during convergent extension and axon path finding *in vivo* and *in vitro* in multiple species (summarized in Ref. 7). As described above, these same phenomena require also δ-PCDHs.

3.3.3 *Pcdh7 and TAF1/Set*

TAF1/Set was identified in a GST pull-down experiment as a cytoplasmic interaction partner of Pcdh7/NFPC[8] (Fig. 8.1G). Ectopic expression of a dominant-negative form of Pcdh7 lacking the entire ectodomain in *Xenopus* embryos disrupts ectoderm integrity, leading to blister formation.[29] This phenotype can be rescued by expression of TAF1,[8] suggesting that the dominant-negative Pcdh7 sequesters endogenous TAF1. The interaction domain of Pcdh7 is likely the juxtamembrane domain.[8] Furthermore, Pcdh7 and TAF1 mediate cell adhesion within the neural ectoderm and are required for proper apical localization of F-actin in neural fold cells.[43] Pcdh7 and TAF1 are seemingly not involved in early convergent extension events but participate in later cell movements when the neural folds meet and fuse and the neural tube elongates along the anterior–posterior axis.[43] Moreover, Pcdh7 and TAF1 are both expressed in retinal ganglion cells (RGCs) of *Xenopus* and inhibiting Pcdh7 or TAF1 impairs RGC development.[50]

3.3.4 *Pcdh8 (Arcadlin) and TAO2β*

Stimulation of rat hippocampal neurons induces transcription of Pcdh8 (also called arcadlin in rats) and its transport to the synaptic membrane, where it forms a complex with N-cadherin[10,30] (Fig. 8.1H). Moreover, TAO2β, a splice variant of the MAPKKK TAO2β, forms a complex with the Pcdh8 intracellular domain.[10] Homophilic interaction of Pcdh8 on the cell surface activates p38 MAPK through activation of TAO2β. In turn, active p38 feeds back to TAO2β, phosphorylating its carboxy-terminal domain, which triggers coendocytosis of the N-cadherin–Pcdh8 complex. Thus, the activity-induced transport of Pcdh8 to the dendritic spine membrane and the consequent internalization of N-cadherin might disrupt synaptic adherens junctions. Interestingly, cultured hippocampal neurons from Pcdh8-knockout mice protrude more dendritic spines than wild-type neurons.[10] The Pcdh8-TAO2β-p38 pathway of activity-regulated endocytosis of N-cadherin may be important for the dynamics of neurotransmitter receptor exposure at the cell surface.

3.3.5 Pcdh10 and Nap1

Nakao *et al.*[18] screened for proteins interacting with mouse Pcdh10 and identified Nap1. Nap1 is conserved among various organisms and organizes a molecular complex containing itself, CYFIP2 (cytoplasmic interacting FMR1 protein), Abi (Abl interactor), HSPC300, and WAVE.[55] Further experiments showed that Pcdh10 also pulled down WAVE1 and Abi,[18] suggesting that Pcdh10-associated Nap1 can form a complex with its other partners (Fig. 8.1I). Nap1 regulates WAVE-mediated actin assembly and is important in morphogenetic events such as neurite extension, neural tube closure, and migration of endoderm and mesoderm.[36,56] Pcdh10 overexpression recruits Nap1 and WAVE1 to cell–cell contacts, which results in altered assembly of F-actin and N-cadherin at these locations.[18] Such ectopic expression of Pcdh10 did not alter the migratory behavior of individual cells but accelerated cell migration and uncoordinated movement of contacting cells.[18] Apparently, the redistributed N-cadherin molecules lose the ability to induce contact inhibition, resulting in uncooperative, accelerated cell movement.

Chicken Pcdh19 was also shown to interact with Nap1, and such interaction was as well suggested for Pcdh1 and Pcdh9.[22] Although all δ1- and δ2-Pcdhs contain the conserved CM1 and CM2 motifs, the Nap1 interaction domain of Pcdh10 is located at the C-terminus of Pcdh10, downstream of the CMs.[18] Nap1 probably binds to the CDs of different Pcdhs in different ways.

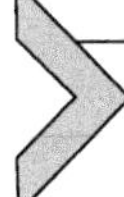

4. ABERRATIONS IN HUMAN DISEASES

4.1. Cancer

In recent years, direct and indirect evidence has accumulated that δ-PCDHs are important during human tumorigenesis, either as tumor suppressor genes (TSGs) or oncogenes. For instance, global genomic analyses of pancreatic cancers revealed cancer-associated missense mutations in *PCDH9*, *PCDH17*, and *PCDH18* (reviewed in Ref. 57), and promoter methylation of *PCDH1*, *PCDH10*, *PCDH11Y*, *PCDH17*, and *PCDH8*,[58] but the functional relevance of these aberrations remains unclear.

So far, there have been only limited reports on the relation of the δ1-PCDH subfamily to human cancer. In a large-scale microarray-based screening, *PCDH1* was identified as a gene whose expression correlates with favorable survival in 35 medulloblastoma patients (reviewed in Ref. 57), but functional studies remain to be done. It has been proposed that PCDH9 acts as a TSG protein in gliomas.[59,60] Loss of PCDH9 expression was associated with higher histological grades and shorter survival times in glioma patients and could be a useful prognostic biomarker.[59] In ovarian carcinomas, PCDH9 expression is weaker in metastases than in primary tumors,[61] but how this compares to normal tissue was not investigated.

PCDH8, a δ2-PCDH gene, has been proposed as a candidate TSG in breast cancer (reviewed in Ref. 57). Loss of PCDH8 functionality in tumor samples and cell lines was associated with increased proliferation, and this dysfunctionality was ascribed to somatic mutations, loss of heterozygosity, and promoter methylation. Somatic missense mutations affected the ectodomain or a conserved motif in the CD. The E146K mutation was suggested to affect the posttranslational processing of PCDH8 and to lead to reduced growth factor requirements of mammary epithelial cells. *PCDH8* has also been proposed to be a TSG in nasopharyngeal carcinoma, since frequent downregulation was found in nasopharyngeal cancer cell lines and primary tumors.[62] Ectopic overexpression of PCDH8 in such cell lines reduced clonogenicity and cell migration. Other tumor types with *PCDH8* promoter methylation and gene silencing are colorectal cancers,[63] renal cell carcinomas,[64] and mantle cell lymphomas.[65] Whether the putative tumor suppressor activity of PCDH8 is related to the reported triggering of N-cadherin endocytosis[10] is unknown.

There is strong evidence that *PCDH10* is silenced epigenetically by promoter methylation in various human cancers, but not in matched human tissues (reviewed in Ref. 57). Ectopic expression of PCDH10 in nasopharyngeal and esophageal cancer cell lines reduced migration and *in vitro* invasive potential and clonogenicity. Reexpression of PCDH10 in gastric cancer cells inhibited tumor growth in nude mice, induced cell apoptosis, and inhibited cell proliferation and invasion. The same study showed that *PCDH10* methylation was associated with shortened survival of gastric cancer patients (reviewed in Ref. 57). Moreover, silencing of *PCDH10* contributes to chemotherapy resistance of leukemia[66,67] and might therefore serve as an indicator of drug resistance. Cancer cells also use other epigenetic mechanisms to inactivate *PCDH10*. The expression of HOTAIR is increased in human primary breast tumors and metastases.[68] HOTAIR is a long, noncoding RNA transcribed from the *HOX* locus. It interacts with the Polycomb Repressive Complex 2 (PRC2) and is required for histone methylation of the *HOXD* locus. When HOTAIR is overexpressed in the breast carcinoma cell line MDA-MB-231, PRC2 is recruited to the *PCDH10* promoter, and this results in *PCDH10* downregulation and in increased cell invasion and tumor metastasis.[68]

Two more δ-PCDHs were shown to be deregulated by epigenetic mechanisms in human cancer. Silencing of *PCDH17* expression by promoter methylation or other mechanisms leads to loss of its tumor suppressive activities in esophageal squamous cell carcinoma,[69] laryngeal squamous cell carcinoma[70], and gastric cancer cell lines.[71] The *PCDH20* promoter is frequently silenced in non-small-cell lung cancers (NSCLC) with poor prognosis.[72] Restoration of *PCDH20* expression in NSCLC cells reduced clonogenicity and anchorage-independent growth. In prostate cancer xenografts, PCDH20 was upregulated in slow-growing, androgen-dependent tumors, confirming a putative tumor suppressor role.[73]

In contrast to the many reports on the roles of *δ-PCDHs* as TSGs, less has been reported on their roles as putative oncogenes. A human-specific cytoplasmic form of PCDH11Y, known as PCDH-PC, decreases the apoptotic sensitivity of prostate cancer cells (reviewed in Ref. 57). PCDH11Y can upregulate Wnt signaling by stabilizing β-catenin, which causes β-catenin to translocate to the nucleus and initiate TCF-mediated transcription. Although β-catenin coimmunoprecipitates with PCDH11Y, direct interaction between the two proteins has not been demonstrated. Activation of the Wnt signaling pathway by PCDH11Y drives neuroendocrine transdifferentiation and permits androgen-independent growth of prostate cancer cells (reviewed in Ref. 57). Recently, *PCDH11X* was picked up in a screen for protein-altering point mutations in highly aggressive lethal prostate cancers.[74]

In a screen for prometastatic genes in breast cancer, *PCDH7* was one of the strongest upregulated genes in brain metastases.[75] Strong upregulation of PCDH7 was seen also in ERBB2-overexpressing cells in medulloblastoma,[76] human breast cancer cells overexpressing RASSF1C,[77] and murine fibroblasts overexpressing S-adenosylmethionine decarboxylase.[78] In these cases, *in vitro* migration was increased. Inversely, when migration of HeLa cells is blocked by knockdown of ALR, a histone methyltransferase, the expression of PCDH7 is also decreased.[79] A putative role for PCDH7 in cell migration has also been observed in normal development, as *Pcdh7* was the strongest upregulated gene in migratory neuroblasts in the mouse.[80] *PCDH7* is also the most strongly upregulated gene in androgen-independent prostate cancer cells,[81] and it is more strongly expressed in melanomas with poor prognosis,[82] in primary ovarian tumors,[83] and in an *in vitro* model of ovarian carcinoma.[84] Single nucleotide polymorphisms (SNPs) in *PCDH7* are correlated with adverse prognosis in lung cancer,[85] but the implications of these SNPs are not clear. The promigratory effect of PCDH7 might be explained by its interaction with TAF1/Set. As PCDH7 can target TAF1 to the plasma - membrane in HeLa cells[8] and membrane targeting of TAF1 stimulates cell migration in an Rac1-dependent manner,[86] PCDH7 might promote cell migration by stimulating Rac.

In conclusion, there is emerging evidence that it is not only the dysregulation of classical cadherins that might be a useful target for anticancer therapy, but dysregulation of Pcdhs might also be. Unfortunately, the molecular mechanisms underlying tumor suppressor versus oncogenic functions of δ-Pcdhs are largely unknown, and research to unravel the probably complex regulatory pathways mediated by Pcdhs is still in its infancy.

4.2. Neurological disorders

Since Pcdhs are strongly expressed in the brain, it may not be surprising that many δ-PCDHs have been implicated in human neurological disorders like autism, mental retardation, Alzheimer's disease, and epilepsy.

A genome-microarray screen revealed that the protocadherin genes *PCDHB1* and *PCDH7* are repressed by methyl-CpG-binding protein 2,[87] which is mutated in Rett syndrome, a form of autism. Moreover, copy number variation in *PCDH9* introns was found in two autistic patients,[88] and *PCDH10* is located in a chromosomal region that suffers from large, inherited homozygous deletions in families with autism.[89] This study also indicated that homozygous deletions in autism patients may preferentially involve genes that are regulated by neuronal activity, including *PCDH10*, in line with previous findings.[49] Additionally, *PCDH10*, together with *PCDH9* and *PCDH17*, is regulated by the MEF2 transcription factor.[90] Activation of MEF2 results in rapid elimination of unused synapses, a process required for normal learning.[91] The mechanisms leading to this elimination are complex. It has been shown that the RNA-binding protein FMR1 (Fragile X mental retardation 1) functions downstream of MEF2.[92] Interestingly, PCDH10 interacts with CYFIP2,[18] an interactor of FMR1, suggesting an intricate interplay between all these molecules. How PCDH10 is involved in synaptic elimination was most recently resolved by Tsai *et al.*[93] These researchers found that nuclear MEF2 activation stimulates in some way the localization of the ubiquitin E3 ligase Mdm2 to the synaptic compartment where the postsynaptic scaffolding protein PSD-95 is polyubiquinated. Moreover, MEF2 activation promotes the association of the cytoplasmic domain of Pcdh10 with modified PSD-95, facilitating the delivery of the latter to the proteasome. Degradation of PSD-95 is expected to lead to synapse elimination. Recently, a single gene deletion in a boy with severe intellectual disability was mapped to the *PCDH18* gene,[94] which makes *PCDH18* a putative candidate gene for mental retardation.

The *PCDH11X/Y* genes have also been screened for involvement in schizophrenia and other psychiatric disorders. Human *PCDH11X* and *PCDH11Y* transcripts are predominantly expressed in the brain, but exhibit differences in expression regulation.[95] As *PCDH11Y* is found only in humans, it might be involved in the evolution of hominid-specific characteristics, including language capacity and cerebral asymmetry. Some *PCDH11X/Y* variations have been suggested to be relevant to psychosis. An intragenic deletion in both *PCDH11X* and *PCDH11Y* was associated with nonsyndromic language delay in a young boy.[96] This deletion on the X chromosome was inherited from the mother, who was unaffected. Because the protein expression levels of PCDH11X/Y in this boy could not be determined, it remains unclear if the deletions caused the disorder.[96] Human X-linked genes with functional paralogs on the Y chromosome tend to escape X inactivation, and it was demonstrated that *PCDH11X* transcripts are more abundant in women than in men, implicating this gene in sexually dimorphic traits in the human brain.[97] In line with these results, some PCDH11-positive brain structures, such as the globus pallidus, the amygdala, and the tuberal hypothalamus, exhibit a gender-specific development and differentiation.[6,98] Moreover, *PCDH11X* escapes X inactivation in Klinefelter syndrome,[99] a condition in which males have an extra X chromosome. This escape likely leads to overexpression of *PCDH11X*, which could account for some abnormalities in Klinefelter syndrome. A genome-wide association study to find novel late-onset Alzheimer's disease susceptibility genes identified an SNP in the *PCDH11X* gene.[100] However, several other studies (Ref. 101 and references therein) could not confirm these results. These discrepancies might be due to different study populations, but no conclusive explanations have been provided.

The strongest evidence for a role of a δ-PCDH in neurological pathologies comes from *PCDH19* studies. Several different *PCDH19* mutations have been identified in families with EFMR (epilepsy and mental retardation limited to females).[98] This disorder is characterized by an X-linked mode of inheritance and an unusual expression pattern, as it spares transmitting males and affects only carrier females. More than 40 distinct mutations of *PCDH19* have been reported in patients with EFMR, sporadic infantile-onset epileptic encephalopathy, and Dravet syndrome, including about 20 missense mutations in the PCDH19 ectodomain (summarized in Ref. 102). One mutation, a duplication of three residues (Ser135–Glu136–Ala137) in EC2, was shown to impair calcium-dependent adhesion of the Pcdh19–N-cadherin complex, although the association of Pcdh19 with N-cadherin was not affected.[20] Several other mutations tested in a Pcdh19–N-cadherin bead aggregation assay did not show adhesion defects, suggesting that various mechanisms underlie how *PCDH19* mutations might lead to EFMR development. Nonetheless, *PCDH19* has become the second most relevant gene in epilepsy after *SCN1A*, encoding the voltage-gated sodium channel type-1 alpha subunit. The fact that carrier males are not affected shows that PCDH19 function is dispensable for human development, even though it is required in zebrafish.[34] It has been hypothesized that the human male-specific *PCDH11Y*, which is related but not paralogous to *PCDH19*, may compensate for the loss of functional PCDH19 in these individuals.[98]

4.3. Other diseases

Bronchial hyperresponsiveness (BHR), a hallmark of asthma, is caused by multiple genetic and environmental factors. *PCDH1* gene variants were recently found to be linked to BHR.[103] Moreover, *PCDH1* mRNA was strongly expressed in primary epithelial cell cultures of asthma patients.[103] As PCDH1 mRNA and protein levels are strongly upregulated during mucociliary differentiation of primary bronchial epithelial cells,[104] the following model was proposed. Deregulation of PCDH1 expression in the lung would result in slower or incomplete differentiation of the epithelial layer, leading to weaker epithelial barrier function. It has been suggested that asthma and eczema share common genetic backgrounds. Therefore, *PCDH1* SNPs, previously shown to associate with asthma and BHR, were checked for association with eczema. Indeed, the *PCDH1* insertion deletion polymorphism IVS3-116 was significantly linked to increased risk of eczema.[105]

Another study looking for a link between asthma and gene methylation identified *PCDH20* as a gene aberrantly methylated in sputum DNA of smokers with asthma.[106] This finding could not be explained by environmental factors, but further studies are needed to determine whether gene methylation precedes airway disease or is a consequence of it.

5. CONCLUSIONS

From the work discussed here, it is evident that δ-Pcdhs play critical roles during vertebrate development. Although some δ-Pcdhs display weak homophilic adhesive properties, they clearly differ from classical cadherins, suggesting that δ-Pcdhs function as signaling and regulatory molecules, not as classical adhesion molecules. Moreover, it was demonstrated that several δ-Pcdhs can form heterophilic cis interactions with classical cadherins and thereby regulate the adhesive properties of these cadherins. Furthermore, the known intracellular interaction partners link δ-Pcdhs to several signaling pathways important in cell differentiation, tissue morphogenesis, and in particular, brain development. Although the involvement of δ-PCDHs in the pathogenesis of several human cancer types, and neural and other diseases is well established, our understanding of their molecular functions and their interplay with various signaling pathways is limited. More detailed functional analyses are required at the molecular, cellular, and organismal levels.

ACKNOWLEDGMENTS

We thank A. Bredan for critical reading and editing of the chapter. Research supported by the Research Foundation - Flanders (FWO), by the Geconcerteerde Onderzoeksacties (GOA) of Ghent University and by the Belgian Science Policy (Interuniversity attraction poles - IAP 7/07).

REFERENCES

1. Sano K, Tanihara H, Heimark RL, Obata S, Davidson M, St John T, et al. Protocadherins—a large family of cadherin-related molecules in central nervous system. *EMBO J* 1993;**12**:2249–56.
2. Hulpiau P, van Roy F. New insights into the evolution of metazoan cadherins. *Mol Biol Evol* 2011;**28**:647–57.
3. Hulpiau P, van Roy F. Molecular evolution of the cadherin superfamily. *Int J Biochem Cell Biol* 2009;**41**:349–69.
4. Redies C, Vanhalst K, van Roy F. Delta-protocadherins: unique structures and functions. *Cell Mol Life Sci* 2005;**62**:2840–52.
5. Morishita H, Yagi T. Protocadherin family: diversity, structure, and function. *Curr Opin Cell Biol* 2007;**19**:584–92.
6. Vanhalst K, Kools P, Staes K, van Roy F, Redies C. Delta-protocadherins: a gene family expressed differentially in the mouse brain. *Cell Mol Life Sci* 2005;**62**:1247–59.
7. Berndt JD, Aoyagi A, Yang P, Anastas JN, Tang L, Moon RT. Mindbomb 1, an E3 ubiquitin ligase, forms a complex with RYK to activate Wnt/beta-catenin signaling. *J Cell Biol* 2011;**194**:737–50.
8. Heggem MA, Bradley RS. The cytoplasmic domain of *Xenopus* NF-protocadherin interacts with TAF1/Set. *Dev Cell* 2003;**4**:419–29.
9. Yoshida K, Watanabe M, Kato H, Dutta A, Sugano S. BH-protocadherin-c, a member of the cadherin superfamily, interacts with protein phosphatase 1 alpha through its intracellular domain. *FEBS Lett* 1999;**460**:93–8.
10. Yasuda S, Tanaka H, Sugiura H, Okamura K, Sakaguchi T, Tran U, et al. Activity-induced protocadherin arcadlin regulates dendritic spine number by triggering N-cadherin endocytosis via TAO2 beta and p38 MAP kinases. *Neuron* 2007;**56**:456–71.
11. Yamamoto A, Kemp C, Bachiller D, Geissert C, DeRobertis EM. Mouse paraxial protocadherin is expressed in trunk mesoderm and is not essential for mouse development. *Genesis* 2000;**27**:49–57.
12. Chen X, Gumbiner BM. Paraxial protocadherin mediates cell sorting and tissue morphogenesis by regulating C-cadherin adhesion activity. *J Cell Biol* 2006;**174**:301–13.

13. Chen X, Koh E, Yoder M, Gumbiner BM. A protocadherin-cadherin-FLRT3 complex controls cell adhesion and morphogenesis. *PLoS One* 2009;**4**:e8411.
14. Medina A, Swain RK, Kuerner KM, Steinbeisser H. Xenopus paraxial protocadherin has signaling functions and is involved in tissue separation. *EMBO J* 2004;**23**:3249–58.
15. Chung HA, Yamamoto TS, Ueno N. ANR5, an FGF target gene product, regulates gastrulation in Xenopus. *Curr Biol* 2007;**17**:932–9.
16. Wang Y, Janicki P, Koster I, Berger CD, Wenzl C, Grosshans J, et al. Xenopus paraxial protocadherin regulates morphogenesis by antagonizing sprouty. *Genes Dev* 2008;**22**: 878–83.
17. Kietzmann A, Wang Y, Weber D, Steinbeisser H. Xenopus paraxial protocadherin inhibits Wnt/beta-catenin signalling via casein kinase 2beta. *EMBO Rep* 2012;**13**:129–34.
18. Nakao S, Platek A, Hirano S, Takeichi M. Contact-dependent promotion of cell migration by the OL-protocadherin-Nap1 interaction. *J Cell Biol* 2008;**182**:395–410.
19. Chen MW, Vacherot F, De La Taille A, Gil-Diez-De-Medina S, Shen R, Friedman RA, et al. The emergence of protocadherin-PC expression during the acquisition of apoptosis-resistance by prostate cancer cells. *Oncogene* 2002;**21**:7861–71.
20. Emond MR, Biswas S, Blevins CJ, Jontes JD. A complex of protocadherin-19 and N-cadherin mediates a novel mechanism of cell adhesion. *J Cell Biol* 2011;**195**:1115–21.
21. Homayouni R, Rice DS, Curran T. Disabled-1 interacts with a novel developmentally regulated protocadherin. *Biochem Biophys Res Commun* 2001;**289**:539–47.
22. Tai K, Kubota M, Shiono K, Tokutsu H, Suzuki ST. Adhesion properties and retinofugal expression of chicken protocadherin-19. *Brain Res* 2010;**1344**:13–24.
23. Biswas S, Emond MR, Jontes JD. Protocadherin-19 and N-cadherin interact to control cell movements during anterior neurulation. *J Cell Biol* 2010;**191**:1029–41.
24. Wolverton T, Lalande M. Identification and characterization of three members of a novel subclass of protocadherins. *Genomics* 2001;**76**:66–72.
25. Williams NA, Close JP, Giouzeli M, Crow TJ. Accelerated evolution of Protocadherin11X/Y: a candidate gene-pair for cerebral asymmetry and language. *Am J Med Genet B Neuropsychiatr Genet* 2006;**141b**:623–33.
26. Blanco-Arias P, Sargent CA, Affara NA. Protocadherin X (PCDHX) and Y (PCDHY) genes; multiple mRNA isoforms encoding variant signal peptides and cytoplasmic domains. *Mamm Genome* 2004;**15**:41–52.
27. Fukata Y, Fukata M. Protein palmitoylation in neuronal development and synaptic plasticity. *Nat Rev Neurosci* 2010;**11**:161–75.
28. Kuroda H, Inui M, Sugimoto K, Hayata T, Asashima M. Axial protocadherin is a mediator of prenotochord cell sorting in Xenopus. *Dev Biol* 2002;**244**:267–77.
29. Bradley RS, Espeseth A, Kintner C. NF-protocadherin, a novel member of the cadherin superfamily, is required for Xenopus ectodermal differentiation. *Curr Biol* 1998;**8**:325–34.
30. Yamagata K, Andreasson KI, Sugiura H, Maru E, Dominique M, Irie Y, et al. Arcadlin is a neural activity-regulated cadherin involved in long term potentiation. *J Biol Chem* 1999;**274**:19473–9.
31. Kim SH, Yamamoto A, Bouwmeester T, Agius E, De Robertis EM. The role of paraxial protocadherin in selective adhesion and cell movements of the mesoderm during Xenopus gastrulation. *Development* 1998;**125**:4681–90.
32. Hirano S, Yan Q, Suzuki ST. Expression of a novel protocadherin, OL-protocadherin, in a subset of functional systems of the developing mouse brain. *J Neurosci* 1999;**19**:995–1005.
33. Chen X, Molino C, Liu L, Gumbiner BM. Structural elements necessary for oligomerization, trafficking, and cell sorting function of paraxial protocadherin. *J Biol Chem* 2007;**282**:32128–37.

34. Emond MR, Biswas S, Jontes JD. Protocadherin-19 is essential for early steps in brain morphogenesis. *Dev Biol* 2009;**334**:72–83.
35. Hong E, Brewster R. N-cadherin is required for the polarized cell behaviors that drive neurulation in the zebrafish. *Development* 2006;**133**:3895–905.
36. Rakeman AS, Anderson KV. Axis specification and morphogenesis in the mouse embryo require Nap1, a regulator of WAVE-mediated actin branching. *Development* 2006;**133**:3075–83.
37. Yoder MD, Gumbiner BM. Axial protocadherin (AXPC) regulates cell fate during notochordal morphogenesis. *Dev Dyn* 2011;**240**:2495–504.
38. Bononi J, Cole A, Tewson P, Schumacher A, Bradley R. Chicken protocadherin-1 functions to localize neural crest cells to the dorsal root ganglia during PNS formation. *Mech Dev* 2008;**125**:1033–47.
39. Unterseher F, Hefele JA, Giehl K, DeRobertis EM, Wedlich D, Schambony A. Paraxial protocadherin coordinates cell polarity during convergent extension via Rho A and JNK. *EMBO J* 2004;**23**:3259–69.
40. Rhee J, Takahashi Y, Saga Y, Wilson-Rawls J, Rawls A. The protocadherin papc is involved in the organization of the epithelium along the segmental border during mouse somitogenesis. *Dev Biol* 2003;**254**:248–61.
41. Murakami T, Hijikata T, Matsukawa M, Ishikawa H, Yorifuji H. Zebrafish protocadherin 10 is involved in paraxial mesoderm development and somitogenesis. *Dev Dyn* 2006;**235**:506–14.
42. Schambony A, Wedlich D. Wnt-5A/Ror2 regulate expression of XPAPC through an alternative noncanonical signaling pathway. *Dev Cell* 2007;**12**:779–92.
43. Rashid D, Newell K, Shama L, Bradley R. A requirement for NF-protocadherin and TAF1/set in cell adhesion and neural tube formation. *Dev Biol* 2006;**291**:170–81.
44. Redies C, Neudert F, Lin J. Cadherins in cerebellar development: translation of embryonic patterning into mature functional compartmentalization. *Cerebellum* 2011;**10**:393–408.
45. Lin J, Wang C, Redies C. Expression of delta-protocadherins in the spinal cord of the chicken embryo. *J Comp Neurol* 2012;**520**:1509–31.
46. Hertel N, Redies C, Medina L. Cadherin expression delineates the divisions of the postnatal and adult mouse amygdala. *J Comp Neurol* 2012;**520**:3982–4012 Epub ahead of print.
47. Yagi T. Molecular codes for neuronal individuality and cell assembly in the brain. *Front Mol Neurosci* 2012;**5**:45.
48. Uemura M, Nakao S, Suzuki ST, Takeichi M, Hirano S. OL-protocadherin is essential for growth of striatal axons and thalamocortical projections. *Nat Neurosci* 2007;**10**:1151–9.
49. Williams E, Sickles H, Dooley A, Palumbos S, Bisogni A, Lin D. Delta-protocadherin 10 is regulated by activity in the mouse main olfactory system. *Front Neural Circuits* 2011;**5**:9.
50. Piper M, Dwivedy A, Leung L, Bradley RS, Holt CE. NF-Protocadherin and TAF1 regulate retinal axon initiation and elongation in vivo. *J Neurosci* 2008;**28**:100–5.
51. Mulkey RM, Endo S, Shenolikar S, Malenka RC. Involvement of a calcineurin/inhibitor-1 phosphatase cascade in hippocampal long-term depression. *Nature* 1994;**369**:486–8.
52. Bar I, Lambert de Rouvroit C, Goffinet AM. The evolution of cortical development. An hypothesis based on the role of the Reelin signaling pathway. *Trends Neurosci* 2000;**23**:633–8.
53. Kim SY, Mo JW, Han S, Choi SY, Han SB, Moon BH, et al. The expression of non-clustered protocadherins in adult rat hippocampal formation and the connecting brain regions. *Neuroscience* 2010;**170**:189–99.

54. Demontis F, Habermann B, Dahmann C. PDZ-domain-binding sites are common among cadherins. *Dev Genes Evol* 2006;**216**:737–41.
55. Ibarra N, Pollitt A, Insall RH. Regulation of actin assembly by SCAR/WAVE proteins. *Biochem Soc Trans* 2005;**33**:1243–6.
56. Yokota Y, Ring C, Cheung R, Pevny L, Anton ES. Nap1-regulated neuronal cytoskeletal dynamics is essential for the final differentiation of neurons in cerebral cortex. *Neuron* 2007;**54**:429–45.
57. Berx G, van Roy F. Involvement of members of the cadherin superfamily in cancer. *Cold Spring Harb Perspect Biol* 2009;**1**:a003129.
58. Vincent A, Omura N, Hong SM, Jaffe A, Eshleman J, Goggins M. Genome-wide analysis of promoter methylation associated with gene expression profile in pancreatic adenocarcinoma. *Clin Cancer Res* 2011;**17**:4341–54.
59. Wang C, Yu G, Liu J, Wang J, Zhang Y, Zhang X, et al. Downregulation of PCDH9 predicts prognosis for patients with glioma. *J Clin Neurosci* 2012;**19**:541–5.
60. de Tayrac M, Etcheverry A, Aubry M, Saikali S, Hamlat A, Quillien V, et al. Integrative genome-wide analysis reveals a robust genomic glioblastoma signature associated with copy number driving changes in gene expression. *Genes Chromosomes Cancer* 2009;**48**:55–68.
61. Lancaster JM, Dressman HK, Clarke JP, Sayer RA, Martino MA, Cragun JM, et al. Identification of genes associated with ovarian cancer metastasis using microarray expression analysis. *Int J Gynecol Cancer* 2006;**16**:1733–45.
62. He D, Zeng Q, Ren G, Xiang T, Qian Y, Hu Q, et al. Protocadherin8 is a functional tumor suppressor frequently inactivated by promoter methylation in nasopharyngeal carcinoma. *Eur J Cancer Prev* 2012;**21**:569–75 Epub ahead of print.
63. Suzuki H, Gabrielson E, Chen W, Anbazhagan R, Van Engeland M, Weijenberg MP, et al. A genomic screen for genes upregulated by demethylation and histone deacetylase inhibition in human colorectal cancer. *Nat Genet* 2002;**31**:141–9.
64. Morris MR, Ricketts CJ, Gentle D, McRonald F, Carli N, Khalili H, et al. Genome-wide methylation analysis identifies epigenetically inactivated candidate tumour suppressor genes in renal cell carcinoma. *Oncogene* 2011;**30**:1390–401.
65. Leshchenko VV, Kuo PY, Shaknovich R, Yang DT, Gellen T, Petrich A, et al. Genome-wide DNA methylation analysis reveals novel targets for drug development in mantle cell lymphoma. *Blood* 2010;**116**:1025–34.
66. Ding K, Su Y, Pang L, Lu Q, Wang Z, Zhang S, et al. Inhibition of apoptosis by downregulation of hBex1, a novel mechanism, contributes to the chemoresistance of Bcr/Abl+ leukemic cells. *Carcinogenesis* 2009;**30**:35–42.
67. Narayan G, Freddy AJ, Xie D, Liyanage H, Clark L, Kisselev S, et al. Promoter methylation-mediated inactivation of PCDH10 in acute lymphoblastic leukemia contributes to chemotherapy resistance. *Genes Chromosomes Cancer* 2011;**50**:1043–53.
68. Gupta RA, Shah N, Wang KC, Kim J, Horlings HM, Wong DJ, et al. Long non-coding RNA HOTAIR reprograms chromatin state to promote cancer metastasis. *Nature* 2010;**464**:1071–6.
69. Haruki S, Imoto I, Kozaki K, Matsui T, Kawachi H, Komatsu S, et al. Frequent silencing of protocadherin 17, a candidate tumour suppressor for esophageal squamous cell carcinoma. *Carcinogenesis* 2010;**31**:1027–36.
70. Giefing M, Zemke N, Brauze D, Kostrzewska-Poczekaj M, Luczak M, Szaumkessel M, et al. High resolution ArrayCGH and expression profiling identifies PTPRD and PCDH17/PCH68 as tumor suppressor gene candidates in laryngeal squamous cell carcinoma. *Genes Chromosomes Cancer* 2011;**50**:154–66.
71. Yang Y, Liu J, Li X, Li JC. PCDH17 gene promoter demethylation and cell cycle arrest by genistein in gastric cancer. *Histol Histopathol* 2012;**27**:217–24.
72. Imoto I, Izumi H, Yokoi S, Hosoda H, Shibata T, Hosoda F, et al. Frequent silencing of the candidate tumor suppressor PCDH20 by epigenetic mechanism in non-small-cell lung cancers. *Cancer Res* 2006;**66**:4617–26.

73. Jennbacken K, Gustavsson H, Tesan T, Horn M, Vallbo C, Welen K, et al. The prostatic environment suppresses growth of androgen-independent prostate cancer xenografts: an effect influenced by testosterone. *Prostate* 2009;**69**:1164–75.
74. Kumar A, White TA, MacKenzie AP, Clegg N, Lee C, Dumpit RF, et al. Exome sequencing identifies a spectrum of mutation frequencies in advanced and lethal prostate cancers. *Proc Natl Acad Sci USA* 2011;**108**:17087–92.
75. Bos PD, Zhang XH, Nadal C, Shu W, Gomis RR, Nguyen DX, et al. Genes that mediate breast cancer metastasis to the brain. *Nature* 2009;**459**:1005–9.
76. Hernan R, Fasheh R, Calabrese C, Frank AJ, Maclean KH, Allard D, et al. ERBB2 up-regulates S100A4 and several other prometastatic genes in medulloblastoma. *Cancer Res* 2003;**63**:140–8.
77. Reeves ME, Baldwin SW, Baldwin ML, Chen ST, Moretz JM, Aragon RJ, et al. Ras-association domain family 1C protein promotes breast cancer cell migration and attenuates apoptosis. *BMC Cancer* 2010;**10**:562.
78. Nummela P, Yin M, Kielosto M, Leaner V, Birrer MJ, Holtta E. Thymosin beta4 is a determinant of the transformed phenotype and invasiveness of S-adenosylmethionine decarboxylase-transfected fibroblasts. *Cancer Res* 2006;**66**:701–12.
79. Issaeva I, Zonis Y, Rozovskaia T, Orlovsky K, Croce CM, Nakamura T, et al. Knockdown of ALR (MLL2) reveals ALR target genes and leads to alterations in cell adhesion and growth. *Mol Cell Biol* 2007;**27**:1889–903.
80. Khodosevich K, Seeburg PH, Monyer H. Major signaling pathways in migrating neuroblasts. *Front Mol Neurosci* 2009;**2**:7.
81. Singh AP, Bafna S, Chaudhary K, Venkatraman G, Smith L, Eudy JD, et al. Genome-wide expression profiling reveals transcriptomic variation and perturbed gene networks in androgen-dependent and androgen-independent prostate cancer cells. *Cancer Lett* 2008;**259**:28–38.
82. Jaeger J, Koczan D, Thiesen HJ, Ibrahim SM, Gross G, Spang R, et al. Gene expression signatures for tumor progression, tumor subtype, and tumor thickness in laser-microdissected melanoma tissues. *Clin Cancer Res* 2007;**13**:806–15.
83. Tchagang AB, Tewfik AH, DeRycke MS, Skubitz KM, Skubitz AP. Early detection of ovarian cancer using group biomarkers. *Mol Cancer Ther* 2008;**7**:27–37.
84. Urzua U, Roby KF, Gangi LM, Cherry JM, Powell JI, Munroe DJ. Transcriptomic analysis of an in vitro murine model of ovarian carcinoma: functional similarity to the human disease and identification of prospective tumoral markers and targets. *J Cell Physiol* 2006;**206**:594–602.
85. Huang YT, Heist RS, Chirieac LR, Lin X, Skaug V, Zienolddiny S, et al. Genome-wide analysis of survival in early-stage non-small-cell lung cancer. *J Clin Oncol* 2009;**27**:2660–7.
86. ten Klooster JP, Leeuwen I, Scheres N, Anthony EC, Hordijk PL. Rac1-induced cell migration requires membrane recruitment of the nuclear oncogene SET. *EMBO J* 2007;**26**:336–45.
87. Miyake K, Hirasawa T, Soutome M, Itoh M, Goto Y, Endoh K, et al. The protocadherins, PCDHB1 and PCDH7, are regulated by MeCP2 in neuronal cells and brain tissues: implication for pathogenesis of Rett syndrome. *BMC Neurosci* 2011;**12**:81.
88. Marshall CR, Noor A, Vincent JB, Lionel AC, Feuk L, Skaug J, et al. Structural variation of chromosomes in autism spectrum disorder. *Am J Hum Genet* 2008;**82**:477–88.
89. Morrow EM, Yoo SY, Flavell SW, Kim TK, Lin Y, Hill RS, et al. Identifying autism loci and genes by tracing recent shared ancestry. *Science* 2008;**321**:218–23.
90. Flavell SW, Kim TK, Gray JM, Harmin DA, Hemberg M, Hong EJ, et al. Genome-wide analysis of MEF2 transcriptional program reveals synaptic target genes and neuronal activity-dependent polyadenylation site selection. *Neuron* 2008;**60**:1022–38.

91. Barbosa AC, Kim MS, Ertunc M, Adachi M, Nelson ED, McAnally J, et al. MEF2C, a transcription factor that facilitates learning and memory by negative regulation of synapse numbers and function. *Proc Natl Acad Sci USA* 2008;**105**:9391–6.
92. Pfeiffer BE, Zang T, Wilkerson JR, Taniguchi M, Maksimova MA, Smith LN, et al. Fragile X mental retardation protein is required for synapse elimination by the activity-dependent transcription factor MEF2. *Neuron* 2010;**66**:191–7.
93. Tsai NP, Wilkerson JR, Guo W, Maksimova MA, DeMartino GN, Cowan CW, Huber KM. Multiple autism-linked genes mediate synapse elimination via proteasomal degradation of a synaptic scaffold PSD-95. *Cell* 2012;**151**:1581–94.
94. Kasnauskiene J, Ciuladaite Z, Preiksaitiene E, Matuleviciene A, Alexandrou A, Koumbaris G, et al. A single gene deletion on 4q28.3: PCDH18—a new candidate gene for intellectual disability? *Eur J Med Genet* 2012;**55**:274–7.
95. Blanco P, Sargent CA, Boucher CA, Mitchell M, Affara NA. Conservation of PCDHX in mammals; expression of human X/Y genes predominantly in brain. *Mamm Genome* 2000;**11**:906–14.
96. Speevak MD, Farrell SA. Non-syndromic language delay in a child with disruption in the Protocadherin11X/Y gene pair. *Am J Med Genet B Neuropsychiatr Genet* 2011;**156B**:484–9.
97. Lopes AM, Ross N, Close J, Dagnall A, Amorim A, Crow TJ. Inactivation status of PCDH11X: sexual dimorphisms in gene expression levels in brain. *Hum Genet* 2006;**119**:267–75.
98. Dibbens LM, Tarpey PS, Hynes K, Bayly MA, Scheffer IE, Smith R, et al. X-linked protocadherin 19 mutations cause female-limited epilepsy and cognitive impairment. *Nat Genet* 2008;**40**:776–81.
99. Ross NLJ, Wadekar R, Lopes A, Dagnall A, Close J, Delisi LE, et al. Methylation of two *Homo sapiens*-specific X-Y homologous genes in Klinefelter's syndrome (XXY). *Am J Med Genet B Neuropsychiatr Genet* 2006;**141**:544–8.
100. Carrasquillo MM, Zou F, Pankratz VS, Wilcox SL, Ma L, Walker LP, et al. Genetic variation in PCDH11X is associated with susceptibility to late-onset Alzheimer's disease. *Nat Genet* 2009;**41**:192–8.
101. Miar A, Alvarez V, Corao AI, Alonso B, Diaz M, Menendez M, et al. Lack of association between protocadherin 11-X/Y (PCDH11X and PCDH11Y) polymorphisms and late onset Alzheimer's disease. *Brain Res* 2011;**1383**:252–6.
102. Depienne C, Leguern E. PCDH19-related infantile epileptic encephalopathy: an unusual X-linked inheritance disorder. *Hum Mutat* 2012;**33**:627–34.
103. Koppelman GH, Meyers DA, Howard TD, Zheng SL, Hawkins GA, Ampleford EJ, et al. Identification of PCDH1 as a novel susceptibility gene for bronchial hyperresponsiveness. *Am J Respir Crit Care Med* 2009;**180**:929–35.
104. Koning H, Sayers I, Stewart CE, de Jong D, Ten Hacken NH, Postma DS, et al. Characterization of protocadherin-1 expression in primary bronchial epithelial cells: association with epithelial cell differentiation. *FASEB J* 2011;**26**:439–48.
105. Koning H, Postma DS, Brunekreef B, Duiverman EJ, Smit HA, Thijs C, et al. Protocadherin-1 polymorphisms are associated with eczema in two Dutch birth cohorts. *Pediatr Allergy Immunol* 2011;**23**:270–7.
106. Sood A, Petersen H, Blanchette CM, Meek P, Picchi MA, Belinsky SA, et al. Methylated genes in sputum among older smokers with asthma. *Chest* 2012;**142**:425–31.

CHAPTER NINE

Atypical Cadherins Celsr1–3 and Planar Cell Polarity in Vertebrates

Fadel Tissir, André M. Goffinet
Developmental Neurobiology, University of Louvain, Institute of Neuroscience, Box B1.73.16, Brussels, Belgium

Contents

Abstract

Cadherin EGF LAG seven-pass G-type receptors 1, 2, and 3 (Celsr1–3) form a family of three atypical cadherins with multiple functions in epithelia and in the nervous system. During the past decade, evidence has accumulated for important and distinct roles of Celsr1–3 in planar cell polarity (PCP) during the development of the brain and some other organs. Although Celsr function in PCP is conserved from flies to mammals, other functions may be more distantly related, with Celsr working only with one or a subset of the classical core PCP partners. Here, we review the literature on Celsr, focusing on PCP and particularly on brain development.

1. INTRODUCTION TO Celsr PROTEINS AND PCP

Celsr genes were first identified in the mouse.[1–3] They form a highly conserved family with orthologs in ascidians, worms, flies, and vertebrates. In mammals, the *Celsr* family is composed of three members, whereas the chick apparently lacks *Celsr2*,[4] and the zebrafish has four members, including *Celsr1a* and *1b*.[5] In the mouse, *Celsr1–3* genes have similar genomic organization, with 35 exons (*Celsr1* & *Celsr3*) or 34 exons (*Celsr2*). Apart from

Progress in Molecular Biology and Translational Science, Volume 116
ISSN 1877-1173
http://dx.doi.org/10.1016/B978-0-12-394311-8.00009-1

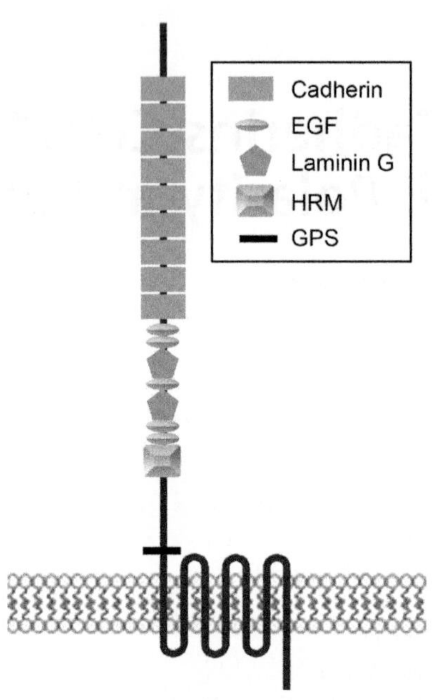

Figure 9.1 Schematic representation of the Celsr1-3 proteins. The extracellular domain includes nine cadherin repeats, six epidermal growth factor (EGF)-like domains, two laminin G repeats, one hormone receptor motif (HRM), and a G-protein-coupled receptor proteolytic site (GPS). In contrast to typical cadherins, which are single-pass proteins, Celsr1, 2, and 3 cadherins are anchored to the plasma membrane by seven transmembrane domains. The cytoplasmic tail varies in size and is poorly conserved among the three Celsrs. (See Color Insert.)

the studies on 3′ alternative exons in *Celsr2*, alternative splicing has not been investigated in detail. *Celsr1–3* genes encode atypical cadherins of more than 3000 amino acids (Fig. 9.1). The large ectodomain is composed of nine N-terminal cadherin repeats (typical cadherins have five repeats), six epidermal growth factor (EGF)-like domains, two laminin G repeats, one hormone receptor motif (HRM), and a G-protein-coupled receptor proteolytic site (GPS). This is followed by seven transmembrane domains (classic cadherins are single-pass type I proteins) and a cytoplasmic tail (CT) that varies in size and, in contrast to the ectodomain, is poorly conserved among the three members. Celsr1–3 belong to the "adhesion receptor" family of G-protein-coupled receptors (GPCRs), also referred to as long N-terminal group B (LNB) GPCRs. These receptors are thought to be natural chimeras of cell adhesion proteins and signaling receptors that can convert cell–cell communication cues into intracellular signals. Studies of the latrotoxin receptor latrophilin, a member of LNB-GPCR, showed that cleavage at the GPS proceeds intracellularly and, despite cleavage, the N-terminal cell adhesion domain and C-terminal (GPCR) fragments remain associated by noncovalent bonds at the plasma membrane.[6,7] There is clear evidence that Celsr3 is not processed at the GPS, but whether Celsr1 or Celsr2 undergo cleavage remains unclear. In addition to cadherin and EGF-like repeats, which confer adhesive properties, the extracellular domain of Celsr1–3 contains a HRM that may bind putative ligands, although none has been identified thus far.

The fruit fly Celsr ortholog is known as *Flamingo* (*fmi*) or *Starry night* (*stan*).[8,9] Genetic studies have established a critical role for *fmi*/*stan* in planar cell polarity (PCP), particularly in the stereotypic organization of wing hairs (trichomes) and sensory bristles,[8,9] the orientation of ommatidia in the eye,[10] and the asymmetric division of sensory organ precursors.[11,12] Fly PCP has been reviewed extensively.[11,13–15] Therefore, we summarize only some data that are likely to help a better understanding of PCP in vertebrates. In *Drosophila*, fmi/stan interacts genetically and functionally with other "core PCP proteins," which include serpentine receptors Frizzled (among four fz receptors in flies, fz and fz2 are partially redundant in PCP), the tetraspanin Van Gogh (vang, also named strabismus), the three cytoplasmic proteins Dishevelled (dsh), Prickle (pk), and Diego (dgo),[16–21] and probably other less clearly identified members.[22] In the wing, the distribution of core PCP proteins is tightly and dynamically regulated. Initially distributed in all subapical junctions, they adopt a transient polarized partition along the proximal–distal axis shortly before the growth of trichomes, and this partition is essential for the establishment of polarity and the proper orientation of hairs. Fmi/stan, fzd, dsh, and dgo are enriched at the distal border of wing cells, whereas fmi/stan, vang, and pk are enriched at the proximal border.[9,23,24] As fmi/stan localizes to both proximal and distal junctions, it was considered merely a permissive molecule: fmi/stan homodimers would act as a scaffold, promoting cell adhesion and bridging the distal (fz-expressing) side of a cell and the proximal (vang-expressing) side of an adjacent cell. However, there is evidence that fmi/stan plays an instructive role: its central portion containing the HRM and TM domains interacts physically with fz. Further, fmi/stan recruits fz and vang to opposite cell boundaries, thereby initiating bidirectional polarity signals.[13,25] In line with this, genetic analyses of flies demonstrated a mutual requirement for fmi/stan, fz, and dsh for achieving a correct partition of polarity complexes.[26] In the *Drosophila* eye, polarity results from the differentiation of the initially equipotent R3/R4 photoreceptor progenitors into an equatorial R3 and a polar R4. The key distinction is not between opposite sides of the same cell, but between adjacent sides of this pair of progenitor cells. Competition for Notch signaling activation between the R3/R4 pair becomes biased by the PCP signal, so the equatorial cell always expresses low Notch levels and becomes R3. PCP mutations lead either to incorrect R3/R4 fate decisions or, in some cases, to indistinctly differentiate pairs of R3/R4 cells.[13–15] On the *Drosophila* notum (dorsal thorax), PCP controls the orientation of an asymmetric division. pI cells differentiate within the epithelium and divide asymmetrically to produce an anterior pIIb daughter cell and a posterior pIIa daughter cell. The PCP pathway distinguishes anterior and posterior sides of the pI cell through asymmetric interactions with its anterior and posterior neighbors. The pI cell therefore seems to become polarized much like the surrounding epithelial cells, but it uses this polarity to position cell fate determinants and the mitotic spindle prior to an asymmetric division. In PCP mutants, this division is not correctly orientated.[11,27,28] Asymmetric localization of polarity complexes requires differential membrane addressing of at least one component. In wing cells, there is, indeed, a vectorial vesicular transport of Fz along microtubules to the distal membrane.[29] These data point to the importance of polarization of the apical microtubule network,[13,14] a process that still requires further investigation.

2. Celsr1–3 EXPRESSION PATTERNS

Celsr1–3 expression is finely regulated spatially and temporally, hinting at their importance during development. A striking feature is the complementary pattern of *Celsr1* and *Celsr3* expression in different systems.[2,30,31] In the mouse nervous system, *Celsr1* mRNA is strongly expressed in zones of neural stem cell (NSC) proliferation, namely all ventricular zones during embryonic and early postnatal development, and telencephalic ependymal zones as well as the subgranular layer of the dentate gyrus in the mature brain. In contrast, *Celsr3* mRNA is absent from NSC and is instead associated with most postmitotic neural cells, whereas *Celsr2* mRNA is found in both NSC and postmitotic cells. Expression of *Celsr1* abates during the early postnatal development, in parallel to decreasing numbers of NSC. Expression of *Celsr3* is sharply downregulated postnatally and persists in the cerebellar granular layer, the hilus of the dentate gyrus, the rostral migratory stream, and the central region of the olfactory bulb. By contrast, *Celsr2* expression remains quite stable throughout life. These expression patterns suggest that *Celsr1* functions in NSC regulation, *Celsr3* in neural cell maturation, and *Celsr2* in development and maintenance of the nervous system. *Celsr1–3* mRNAs are also variably expressed in nonneural tissue, such as the skin, lungs, kidney, and digestive and reproductive systems. In the rodent testis, the spatiotemporal pattern of *Celsr1–3* expression is somewhat reminiscent of that in the brain. *Celsr3* is expressed exclusively in postmeiotic germ cells, *Celsr1* and *Celsr2* in Sertoli cells, with postnatal downregulation of *Celsr1* and persistence of *Celsr2* in the adult.[32–34]

Data on protein localization remain scant. Polyclonal antibodies against the extra- and intracellular segments of Celsr1 allowed detection of two forms, full length (p400) and cleaved (p85 kDa), the latter generated by as yet unidentified proteolysis events not involving the GPS.[35] In the hindbrain and spinal cord, Celsr1 protein immunoreactivity was detected in the floor and roof plates, as well as in radial neuroepithelial progenitors. Intriguingly, Celsr1 is present not only in the apical junction belt as expected, but also in end-feet abutting the pial surface, which belongs to the basolateral domain.[35] In the embryonic skin, Celsr1 is asymmetrically expressed in hair germ cells and in basal layer epidermal cells, a pattern evocative of that of Fmi/stan in the *Drosophila* wing.[36] Unfortunately, thus far the cellular localization of Celsr2 and Celsr3 could not be investigated due to the lack of antibodies suitable for immunohistochemistry.

3. Celsr1: A MAJOR PLAYER IN VERTEBRATE PCP

The main phenotypic traits in *Celsr1–3* and some other PCP mutant mice are summarized in Table 9.1. Two *Celsr1* mutant alleles, *Crash* ($Celsr1^{Crsh}$) and *Spin Cycle* ($Celsr1^{Scy}$), were identified in an ENU screen.[37] In $Celsr1^{Crsh}$, a G-to-A mutation at nucleotide 3126 results in an aspartate to glycine substitution in codon 1040, within the eighth cadherin repeat. In $Celsr1^{Scy}$, a T-to-A point mutation at nucleotide 3337 results in an asparagine to lysine substitution in codon 1110, in a region connecting cadherin repeats 8 and 9. Heterozygous animals show abnormal head-shaking behavior. Both heterozygous and homozygous mice have defective organization of stereocilia bundles in inner ear hair cells. Normally, the apical surface of each cochlear hair cell is decorated with actin-filled stereocilia arranged into a "V" centered on one microtubular "kinocilium," with all "Vs" pointing to the external aspect of the cochlear canal. This organization is altered in $Celsr^{Crsh}$ and $Celsr1^{Scy}$ mutants, where stereociliary bundles are oriented nearly randomly, displaying up to 180° rotation.[37] The stereotyped orientation of ear hair bundles is a hallmark of PCP and parallels the polarized

Table 9.1 Summary of phenotypes of *Celsr1–3* and some other core PCP mutant mice

Gene	Allele	Phenotype[a]	References
Celsr1	*Celsr1 Crsh* and *Scy*	Heterozygotes have inner ear PCP abnormalities, head-shaking, looped tail, abnormal FBM neuron migration, and lung alveologenesis defects. Homozygotes are embryonic lethal with open neural tube	37,38
	Celsr1 KO	Heterozygotes are normal and fertile. Homozygotes viable but mostly sterile. About 20% die *in utero* with neural tube closure defects. Adults have abnormal behavioral traits, a looping tail and abnormal skin hair patterning. The direction of FBM neuron migration is affected	36,39,40
Celsr2	*Celsr2gt*	Heterozygotes are normal and fertile. Homozygotes fertile, except for some females with vaginal atresia. There is abnormal trajectory of FBM neuron migration and hydrocephalus due to defective ependymal ciliogenesis	40,41
	Celsr2 KO and *Celsr2 F*	Heterozygotes are normal and fertile. Homozygotes are fertile and have abnormal FBM neuron migration	40
Celsr3	*Celsr3 KO* and *Celsr3 F*	Heterozygotes are normal and fertile. Homozygotes die in a few hours after birth and have severe axonal defects	42,43
Fzd3	*Fzd3 KO*	Heterozygotes are normal and fertile. Homozygotes die in a few hours after birth and have axonal defects similar to those in *Celsr3* mutants. Some have looped tail or open neural tube and PCP phenotype in inner ear. The trajectory of FBM neuron migration is affected and mimics *Celsr2* and *3* double KO	40,44,45
Fzd6	*Fzd6 KO*	Homozygous mutants survive and breed, and have abnormal skin hair patterning	46
Vangl2	*Vangl2 Lp (Looptail)*	Some heterozygotes have kinked or looped tails. Homozygotes do not survive due to open neural tube, loop-tail, inner-ear PCP defects, FBM neuron migration defects and other defects in cardiovascular, skeletal, and respiratory systems	47–49
	Vangl2 KO and *Vangl2 F*	Heterozygotes are normal and fertile. Homozygote nulls have open neural tube and other PCP defects in inner ear. Regulates nodal cilia beats and laterality	50
Vangl1	*Vangl1 gt*	Homozygotes normal and fertile	50

[a]FBM, facial branchiomotor; KO, knockout; and PCP, planar cell polarity.

distribution of core PCP proteins such as Fzd3, Fzd6, Vangl2, and pk2.[44,47,51,52] PCP proteins localize asymmetrically to one edge of the apical cortex of the cells by mechanisms involving selective targeting and protein stabilization or degradation. Intriguingly, studies of the distribution of PCP proteins were carried out in late embryogenesis, starting from embryonic day 18 when the kinocilium had already migrated from the center to the lateral edge of the cell. This raises the question of whether the asymmetric partition of PCP proteins is the initial event that sets up the polarity and organizes the epithelium, or simply a molecular readout of polarity signals induced by as yet unidentified cues.

Celsr1^{Crsh} and *Celsr1*^{Scy} homozygotes as well as compound *Celsr1*$^{Crsh/Scy}$ heterozygotes exhibit craniorachischisis, a severe neural tube defect due to a failure to initiate neural tube closure in the cervical region.[37] The role of *Celsr1* in neural tube closure was recently confirmed by identification of six *Celsr1* mutations in human fetuses with craniorachischisis. In *in vitro* assays, all these mutations impaired trafficking of Celsr1 protein, reducing its membrane localization.[53] In the bending neural plates, Celsr1 is concentrated in adherens junctions oriented toward the mediolateral axes of the plates, and this was proposed to explain its role during neural tube closure.[54] Neural tube closure defects are observed in *Vangl2*, double *Dvl1* and *-2*, and *Fzd3* and *-6* mutant mice.[44,55,56] Similarly, in zebrafish, inactivation of Vangl2 also leads to a neural tube closure defect,[57] further confirming the importance of classical PCP signaling in this developmental event.

To palliate embryonic lethality of *Celsr1*^{Crsh} and *Celsr1*^{Scy} mutants, a conditional allele (*Celsr1*^{f}) was generated,[39] from which a null allele *Celsr1*^{ko} was derived by crosses with PGK-Cre mice. Western blot with an antibody against the N-terminal region of Celsr1 confirmed the absence of Celsr1 protein in embryonic brain extracts. Unlike heterozygous *Celsr1*^{Crsh} and *Celsr1*^{Scy} mice, heterozygous *Celsr1*$^{ko/+}$ mice have no perceptible phenotypic abnormality. In contrast, homozygous *Celsr1*$^{ko/ko}$ mutants display abnormal behavior such as turning/circling, hyperactivity, and abnormal squeaking. About 20% die *in utero* with various degrees of neural tube closure defects, and many have a looping tail.

Some Celsr1 null mutants have striking skin hair patterning defects, with whorls and crests instead of regular caudal and distal hair orientation on the body and limbs.[39] This phenotype is identical to that reported in Fzd6 null mice.[46] When *Celsr1* is inactivated in crosses with *Emx1-Cre* mice, absence of the protein in the apical ectodermal ridge induces a whorl in distal hind limbs, showing that the action of Celsr1 is autonomous of ectodermal cells.

The mechanism of Celsr1 action in the hair bulb has been studied in detail in the Fuchs lab.[36] Prior to hair growth, Celsr1 becomes asymmetrically localized along the anterior/posterior (A/P) axis in basal epidermal cells in hair follicles and in interfollicular epithelium. This asymmetric localization is essential for A/P orientation of skin hairs. Hair follicles fail to adopt the A/P orientation in skin explants isolated from E13.5 embryos, before the polarization of Celsr1. In contrast, when explants are isolated from E14.5 embryos, when Celsr1 is fully polarized, epidermis keeps the Celsr1 polarization established *in vivo* and the A/P polarity. Consistent with this, hair follicles are misaligned in E18.5 embryos with homozygous *Crsh* mutation. In mutant embryos, Celsr1 protein is produced as in the wild type but is no longer asymmetrically distributed. Moreover, the membrane recruitment of Fzd6 and the asymmetric localization of Vangl2 along the A/P axis are compromised, with some Vangl2 immunoreactivity forming intracellular puncta. Taken together, these results demonstrate that Celsr1 plays a critical role for PCP establishment in the developing skin and hair follicles.[36] During hair development, Vangl1 and Vangl2 act redundantly: whorls on hind limbs are invariably seen following double inactivation of Vangl1 and Vangl2, but inconsistently when one allele remains intact (Yibo Qu *et al.*, unpublished observation).

These data provide strong evidence that Celsr1 is a key partner in classical PCP, where it works with Fzd3 and -6, Dvl1 and -2, Vangl1 and -2, and probably with other PCP proteins that have not yet been studied in that context.

4. Celsr2 AND -3 IN CILIOGENESIS

Based on their structure and motility, cilia are classified into three types: motile, primary, and nodal cilia.[58] In mice, motile cilia garnish the apical surface of epithelial cells lining airways, reproductive tracts, and cerebral ventricles; in *Xenopus*, some skin cells bear multicilia, providing an accessible model system.[59–61] Motile cilia have a central pair of microtubules surrounded by 9 doublets, an organization known as 9+2, and are anchored to a basal body in the apical cortex. They form tufts and their concerted beats generate directional flow. In the airways, they are crucial for clearing mucus and debris, and in genital tracts they assist in the transit of sperm and eggs.[62,63] In the mouse brain at late embryonic stages and during the first postnatal week, neuroepithelial cells that line the cerebral ventricles differentiate into a monolayer of ependymal cells. At birth, ependymal cells are not multiciliated yet. Motile cilia on these cells develop progressively and reach their mature shape around P10.

The basal body of each cilium has a lateral extension called "basal foot" that can be seen by transmission electron microscopy. Basal feet point in the effective beat direction and are used as a hallmark of cilia polarity.[64–66] To generate an efficient directional flow, cilia coordinate their beats within the same cell (each cell has dozens of motile cilia) and in all cells in the epithelial sheet. Thus, basal feet are aligned in the same orientation with respect to the tissue polarity axis, a process referred to as "rotational polarity".[67] Elegant studies on explants of *Xenopus* ciliated epidermis, isolated at different developmental stages, showed that motile cilia acquire their rotational polarity in two steps. Early in development, genetic cues specify a rough planar axis that allows cells to produce a directional flow, thereby inducing a positive feedback loop that tunes basal feet polarity.[68] In mice, *Celsr1*, *2*, and *3* are expressed in the developing ependymal layer. Inactivation of *Celsr2* alone or in combination with *Celsr3* ablation impairs ependymal ciliogenesis.[41] Although differentiation of ependymal cells occurs normally, their motile cilia never develop in normal numbers, and those that develop display abnormalities in position and planar organization. Basal feet are misoriented, a rotational polarity defect,[67] and some basal bodies assemble deep in the cytoplasm. At the tissue level, ciliary tufts from neighboring cells point aberrantly in divergent directions, generating a "translational polarity" defect.[67] Residual mutant cilia display the typical "9 + 2" structure and are still able to beat with the same frequency as the controls. Yet, because of their abnormal orientation, they fail to generate a robust and directed flow. The membrane localization of Vangl2 and Fzd3 is disrupted in mutant ependymal cells, providing strong evidence that Celsr2 and Celsr3 regulate ciliogenesis via PCP signaling. In accord with this, downregulation of core PCP genes *Disheveled1–3*, and PCP effectors *Inturned* and *Fuzzy* affects the orientation of multicilia in *Xenopus*.[60,61] Further, mouse ependymal cells with the *loop-tail* Vangl2 mutation fail to align their motile cilia in response to hydrodynamic forces *in vitro*.[69]

5. Celsr1–3 IN NEURONAL MIGRATION

The migration of facial branchiomotor (FBM) neurons in the developing rhombencephalon is an intriguing case that combines tangential and radial migration modes. FBM neurons, which innervate muscles responsible for facial expression,[70–72] are generated in medial rhombomere 4 (r4) at

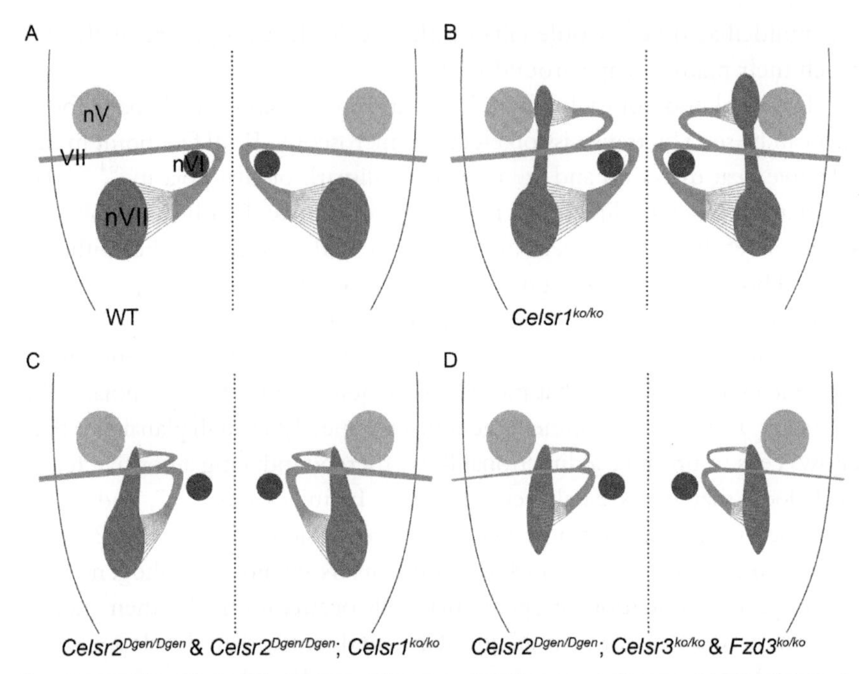

Figure 9.2 Migration of facial branchiomotor (FBM) neurons in normal and *Celsr1*, *2*, and *3* knockout mice. Drawings summarize the phenotypes at P0. In wild-type animals (A), FBM neurons form a single nucleus (nVII, blue) in rhombomere (r) r6. Their axons loop around the abducens nucleus (nVI, red) before exiting the rhombencephalon caudally to the trigeminal nucleus (nV, green). In *Celsr1*$^{ko/ko}$ mutants (B), in addition to the normal nVII, FBM neurons form another ectopic nucleus at the level of nV, but their axons leave the rhombencephalon at the right position. This is due to aberrant migration of FBM neurons in r2 during embryogenesis. In *Celsr2*$^{Dgen/Dgen}$ and *Celsr2*$^{Dgen/Dgen}$; *Celsr1*$^{ko/ko}$ (C), the facial nerve *genu* is abnormal and axons do not loop around nVI, because FBM neurons migrate prematurely in lateral r4–r5, forming lateral ectopias. In *Celsr2*$^{Dgen/Dgen}$; *Celsr3*$^{ko/ko}$ and in *Fzd3*$^{ko/ko}$, in addition to the absence of the *genu* of the facial nerve, the size of nVII is reduced because of cell death (D). "r," rhombomere; nV, motor trigeminal nucleus; nVI, abducens nucleus; nVII, facial nerve nucleus. (See Color Insert.)

E9.5–10.5 and immediately extend their axons laterally toward muscle targets. At E11.5, their cell bodies initiate a tangential caudal migration from r4 to r6. They migrate in the subventricular region, pass medial to the nucleus abducens (nVI) in r5,[73] and then move laterally and dorsally in r6. Finally, they undergo a radial migration in r6 to reach their subpial location, where they form the motor nucleus of the facial nerve (nVII) (Fig. 9.2A).[70] The caudal soma translocation of FBM neurons, with looping of their axons (so-called *genu* of facial nerve), is conserved from fish to mammals, with important species differences; for example, it is blunted in chick.[74] The first indication that *Celsr* genes are involved in FBM neuron migration came from an ENU mutagenesis screen in zebrafish, which identified four point mutations in the *Celsr2/off road* locus.[5] In these mutants, FBM neurons fail to migrate caudally to r6, moving instead prematurely into lateral r4–r5. Morpholino knockdown experiments showed that *Celsr1a* and *1b* have adjunct functions in FBM neuron migration. Whereas combined down-regulation of *Celsr1a* and *Celsr1b* has little effect on its own, it worsens the *off road* mutant phenotype, with more cells stacked in r4 in *Celsr1a; Celsr1b; Celsr2* triple mutants than in *Celsr2* mutants. In normal mice, migrating FBM neurons form streams from medial r4 to lateral r6, with a sharp rostral edge. In *Celsr1*$^{ko/ko}$ mice, FBM neurons are still able to move out of r4, but a subset migrates rostrally into r3 and r2 rather than caudally, a phenotype never seen in any other vertebrate (Fig. 9.2B). This rostral migration phenotype is fully penetrant, although with variable expressivity. Caudally directed *Celsr1*$^{ko/ko}$ neurons move through r5, medially to the abducens nucleus (nVI), before moving laterally in r6 like their wild-type counterparts. Hence, a facial (nVII) nucleus forms in its normal location in lateral r6 by E13.5. In addition, rostrally migrating FBM neurons form an ectopic nucleus adjacent or even rostral to the trigeminal (nV) nucleus, but send their axons normally in the facial nerve.[40] In the hindbrain, *Celsr1* is expressed in progenitors and in the floor plate, but not in postmitotic FBM neurons. Consistent with this, conditional inactivation of *Celsr1* in FBM neuron progenitors upon *Nk6.2-Cre*-mediated recombination induces abnormal rostral migration, whereas its deletion in FBM neurons using *Isl1-Cre* does not.

In *Celsr2* knockout mice, as in the *off road* mutant fish,[5] the caudal FBM neuron migration stream is severely truncated.[40] Cells turn laterally, rostral to the abducens nucleus, so that axons do not loop around nVI, and an elongated facial nucleus forms in r4 and r5 instead of r6 (Fig. 9.2C). A similar phenotype is generated when *Celsr2* is specifically inactivated in FBM neurons using *Isl1-Cre* mice, suggesting that *Celsr2* is required cell autonomously in FBM neurons for their caudal migration. This does not rule out the possibility that *Celsr2* may also be necessary in neuroepithelial cells along the migration path, as suggested for zebrafish.[5] The aberrant migration of FBM neurons in *Celsr2* mutant mice is reminiscent of the situation in birds, which have no facial nerve *genu* and, intriguingly, lack the *Celsr2* gene.[4] Whether there is a causal link between the absence of *Celsr2* and blunted FBM neuron migration in chick remains an open question. In zebrafish,

Celsr2 and *Celsr1a* and *-1b* regulate FBM neuron migration redundantly. In mouse, double inactivation of *Celsr1* and *Celsr2* phenocopies single *Celsr2* mutation, with no obvious rostral migration of FBM neurons, suggesting that *Celsr2* is epistatic to *Celsr1*. It looks as if *Celsr2* deficiency hampers or slows down neuronal migration, thus masking the *Celsr1* mutant phenotype.

Although *Celsr3* seems to be required for tangential migration of calretinin-positive interneurons and radial migration of calbindin-positive interneurons in the forebrain,[75] its deficiency does not affect the migration of FBM neurons. It does, however, exacerbate the *Celsr2* phenotype: the facial nucleus is greatly reduced in size in the double knockouts $Celsr2^{ko/ko}$; $Celsr3^{ko/ko}$ and $Celsr2^{ko/ko}$; $Celsr3^{Isl1cko}$ as compared to $Celsr2^{ko/ko}$ mutants, and Celsr3 deficiency increases FBM neuron apoptosis (Fig. 9.2D).[40] The phenotype of $Celsr2^{ko/ko}$; $Celsr3^{ko/ko}$ double mutants is similar to that of *Fzd3* mutants, and other PCP-related genes such as *Vangl2* and *Wnt5a* have been implicated in FBM neuron migration, suggesting that *Celsr1–3* regulate neuronal migration along the rostrocaudal axis by PCP-dependent mechanisms.[40,48] In line with this, normal functionality of PCP genes such as *van gogh-like 2* (*vangl2*), *fzd3a*, *celsr2*, *prickle1a*, and *prickle1b* is required for caudal migration of FBM neurons in fish.[5,76–80]

6. Celsr2 AND Celsr3 IN BRAIN WIRING

Functional neuronal networks are crucial for brain function. Network formation is finely orchestrated at the cellular and molecular levels by genetic programs and interactions with the environment. After reaching their location, postmitotic neurons extend axons that are guided to their targets by intrinsic programs, guidepost cells, and attractive or repulsive molecular cues. They ramify receptive dendritic fields according to tiling and self-avoidance rules. In addition to their role in neuronal migration, *Celsrs* are implicated in dendrite development and axon guidance, from *Caenorhabditis elegans* and *Drosophila* to mammals.[81–83]

In flies, sensory neurons extend dendrites dorsally toward the midline. Dendrites from homologous neurons in two opposite hemisegments avoid each other, leading to a dendrite-free zone near the dorsal midline. This reciprocal inhibition of dendrite growth is defective in *Fmi/stan* mutants, where dendrites do grow across the midline to occupy fields overlapping those of homologous neurons.[82] A role for Fmi/stan in the tiling of dendrites is further supported by the identification of Fmi/stan mutants in genetic screens for aberrant dendritic extension of mushroom body neurons.[84]

The function of Fmi/stan in dendrite growth does not rely solely on adhesion mediated by the extracellular domain, because overexpressing a N-terminally truncated Fmi that lacks cadherin, EGF-like, and LamininG motifs rescues the dendritic phenotype partially.[85] Intriguingly, neither loss nor gain of function of *frizzled*, nor overexpression of *dsh* phenocopy the *Fmi/stan* dendritic phenotype. Moreover, to our knowledge, no dendrite phenotype has been described in other fly PCP mutants. In the mammalian nervous system, *Celsr2* is required for the maintenance of dendritic arbors: RNAi-induced downregulation of *Celsr2* in rat brain slices reduces the length of dendrites in cortical pyramidal neurons and the complexity of dendritic trees of Purkinje cells.[86] By contrast, silencing of *Celsr3* leads to dendritic overgrowth. The opposite effects of Celsr2 and Celsr3 are related to a single amino acid change in the first intracellular loop: like Fmi/stan, the mammalian Celsr3 has a histidine at position 2413, and both repress growth and/or induce retraction of dendrites. That histidine residue is replaced with arginine in Celsr2, which promotes dendrite growth and maintenance.[87]

Early observations showed that *Fmi/stan* is essential for the development of axonal tracts. Null mutations fmi^{E45} and fmi^{E59} are embryonic lethal, and, in addition to displaying PCP defects, mutant embryos display abnormalities in longitudinal axonal tracts. Rescue of axonal, but not PCP defects, by brain-specific expression of *fmi* cDNA restores viability.[9] Subsequent elegant studies in the visual system showed that fmi/stan mediates axon–axon and axon–target interactions required for guidance of photoreceptor axons.[88,89] It looks as if growth cones "compare" their fmi/stan levels with that of their neighbors to "decide" whether to establish contact or not.[90]

In mice, *Celsr3* is crucial for axon guidance. *Celsr3* mutant mice display marked defects in major tracts such as the anterior commissure, internal capsule, medial lemniscus, and corticospinal tract. *Celsr3* is also essential for the AP organization of monoaminergic axons in the brainstem,[91] and for the rostral turning of commissural axons after midline crossing in the spinal cord.[92,93] Celsr3 deficiency does not affect axonal growth *per se*, but rather guidance, resulting in stalling at intermediate targets, or misrouting of axons.[42] For instance, during the development of reciprocal corticothalamic projections, mutant corticofugal axons travel normally in the subcortical layer but stall at the pallial–subpallial boundary. Reciprocally, thalamic axons never reach their cortical targets in layer 4; instead, they course ventrally along the hypothalamus and then turn externally toward the cortical marginal zone. Conditional removal of *Celsr3* in a stream that extends in the basal forebrain and ventral diencephalon precludes the entry of corticosubcortical and thalamocortical fibers in the basal telencephalic "corridor" and results in a defective internal capsule. This result demonstrates that *Celsr3* is required in intermediate targets to connect the cortex with subcortical structures. Targets cells are molecularly defined by expression of Dlx5/6 and qualify as "guidepost cells".[43] To test whether guidance of cortifugal axons is mediated by homotypic interactions of *Celsr3* in, respectively, navigating growth cones and guidepost cells, *Celsr3* was specifically deleted in corticofugal axons by crossing the $Celsr3^{fl}$ allele with *Emx1-Cre*. Celsr3-defective corticothalamic axons develop normally, strongly suggesting that those fibers use guidance mechanisms independently of Celsr3–Celsr3 homophilic interactions.

The axonal Celsr3 phenotype is very similar to that generated by constitutive inactivation of Fzd3,[45,94] indicating that a PCP-like mechanism might be involved. However, the role of other PCP genes such *Vangl1* and *-2* remains to be investigated. Consistent with the role of Celsr3 in axon guidance, a phenotype of axons stalled at intermediate target cells was reported for fmi-defective sensory neurons in flies.[95] This phenotype can be rescued by fmi constructs that lack most of the extracellular domains, indicating that the advance of sensory axons in flies does not depend on fmi–fmi homophilic interactions.[95] Contrary to corticothalamic axons, corticospinal axons do not develop in $Celsr3^{Emx1cKO}$ mice, suggesting that *Celsr3* mediates homophilic interactions between corticospinal growth cones and guidepost cells. In support of this, in the fly visual system, fmi mediates interactions between the growth cones of photoreceptor axons and their targets in the medulla.[96] Like in flies and mammals, FMI-1, the sole *C. elegans* fmi/stan ortholog, enables navigation of both pioneer and follower axons in the worm ventral nerve cord. Rescue of the mutant phenotype with different portions of FMI-1 revealed that the C- and N-termini are required for guidance of, respectively, pioneer and follower axons.[97] Taken together, these data show that Fmi and its mammalian ortholog *Celsr3* are major players in axon guidance, and that their mechanisms of action are context-dependent and involve both homophilic and heterophilic interactions.

7. STRUCTURE–FUNCTION DATA ON Celsr

As mentioned in the introduction, Celsr proteins contain a large extracellular N-terminus with nine cadherin repeats, several EGF-like and laminin G-like motifs, a HRM, and a G-proteolysis site, followed by seven transmembrane domains, and a variable intracellular C-terminal tail (Fig. 9.1). Although the functional relevance of these motifs remains poorly understood, some studies reveal a few hints. In *Drosophila* S2 cells, which have no self-aggregation properties, expression of full-length fmi induces cell aggregation, whereas expression of a form lacking most of the ectodomain does not, showing that the extracellular domain, indeed, promotes homophilic cell adhesion.[9] Expression of mutant forms of mammalian Celsr2 in S2 cells demonstrated cadherin domain-mediated homophilic interactions compatible with adhesive properties of the cadherin repeats.[86]

Dendritic overgrowth and tiling defects of *fmi* mutants, as well as fly viability, can be rescued by expression of full-length fmi, but the rescue can be ascribed to two different functions of the protein. In transgenic rescue experiments, ΔNfmi—lacking all extracellular domains except the HRM—can rescue the dendrite overgrowth but not the tiling phenotype. Unlike full-length fmi, ΔNfmi does not promote S2 cell aggregation *in vitro*. By contrast, C-terminally deleted fmi (ΔCfmi) retains the ability to mediate homotypic binding *in vitro* and rescues the tiling phenotype but neither dendritic overgrowth nor fly viability, pointing to a key role of the C-terminus. Thus, a dual molecular function of fmi plays pivotal roles in dendrite morphogenesis. In the initial growth phase, fmi might function as a receptor for an unidentified ligand, and this hypothetical heterophilic interaction would be responsible for limiting branch elongation. At a later stage, homophilic fmi binding at dendro–dendritic interfaces would elicit avoidance between dendritic terminals from opposing neurons.[85] Likewise, studies on worms, flies and mice show that the function of fmi/Celsr in axon guidance does not rely solely on its ability to mediate homophilic binding via cadherin repeats, but requires interactions of the C-terminus with unknown molecules.[43,95,97]

The concept of distinct roles for the extra- and intracellular domains of fmi/Celsr is supported by mutation analysis in vertebrates. In rodents, Celsr2 and Celsr3 have opposing roles on dendrites in the brain slice in culture, with Celsr2 promoting and Celsr3 restricting dendritic growth.[86,87] Studies of chimeric constructs in which the ectodomains of Celsr2 and Celsr3 are swapped show that, whereas homophilic interactions (Celsr2–Celsr2 and Celsr3–Celsr3) are important, the transmembrane domains and C-terminus determine the dendrite enhancing or suppressing action. Celsr2 and Celsr3 have different effects on calcium release and activate, respectively, CamKII and Calcineurin signaling, and an amino acid change (R2413H) in the first intracellular loop is crucial for these distinct functions.[87]

Other evidence for the importance of the C-terminus comes from a study on zebrafish.[98] A serine acidic amino acid-rich domain (SE/D) in the CT of Celsr is required for membrane localization of the Frizzled–Dishevelled complex, PCP signaling, and convergent extension.[98] Injection of a C-terminally truncated form of Celsr (ΔC-Celsr; laking 6 TM and tail) in wild-type embryos generates epiboly defects. This is accompanied by sequestration of ΔC-Celsr in the Golgi, where the protein behaves as a dominant-negative molecule, dimerizing with and precluding trafficking of the wild-type protein. Dimerization is thought to involve a conserved arginine-rich sequence N-terminal to the first cadherin repeat. Injection of the CT of Celsr2 fused to the membrane localization signal from the Lyn tyrosine kinase (Lyn-Celsr) generates convergent extension defects without affecting epiboly. Lyn-Celsr perturbs the Frizzled-induced membrane localization of Disheveled and PCP signaling. Interestingly, when the conserved SE/D domain is deleted from Lyn-Celsr, the membrane localization of Disheveled is restored.[98] Whereas the extracellular and TM domains are important for the distribution of Frizzled–Disheveled complexes at the membrane, this process depends critically on the C-terminal SE/D domain. The ability of Celsr to regulate epiboly is closely associated with its ability to modulate cell cohesive properties, whereas its ability to interact with the PCP pathway to regulate convergent extension may not require cell adhesion mediated by the cadherin repeats.

In the mouse skin, Celsr1 is internalized during epidermal basal progenitor division, a process that is crucial for balanced distribution of PCP proteins Fzd6 and Vangl2 in daughter cells and requires a cytoplasmic dileucine motif.[99] Whereas E-Cadherin is normally not internalized, its fusion with the Celsr1 cytoplasmic domain induces internalization during mitosis and enables the chimeric protein to recruit Fzd6 and Vangl2. When two consecutive leucines (2748–2749) are mutated to alanines, chimeric E-Cad-Celsr1, like mutated Celsr1, no longer translocates to endosomes.[99]

The similarity between the CTs of Celsr1, 2, and 3 is much less than the similarity between their extracellular and TM regions. As Celsr1 is directly implicated in PCP and Celsr2–3 is implicated in more distantly related processes, could differences in the C-terminus account for functional differences? In *Drosophila*, fmi, fz, and dsh depend on each other for their membrane localization.[26] Fmi interacts physically with fz via a region encompassing its HRM and TM domains and selectively recruits fz and vang (strabismus) to opposing cell boundaries.[25] In mice, the membrane localization of Fzd3 and Vangl2 in ependymal cells, and localization of Fzd6 and

Vangl2 in skin epithelial cells, depends on Celsr2 and Celsr1, respectively.[36,41] In *Celsr1$^{Crsh/Crsh}$* mutant mice, a single amino acid substitution results in failure of the Celsr1 protein to reach the apical membrane.[35,36,39] Further, all six Celsr1 mutations recently identified in human fetuses with craniorachischisis impair membrane trafficking of Celsr1 in *in vitro* assays.[53] Although not directly useful for characterization of structure–function relationships, this exquisite sensitivity to minor sequence changes underscores the importance of the fmi/Celsr conformations for their membrane localization as well as for the correct targeting of partner proteins. This raises the question of the role of putative proteins assisting in folding, as receptor-activating protein does for the lipoprotein receptor family,[100] or the beta-2 microglobulin for major histocompatibility complexes.[101] Do Celsr, Fzd, and Vangl help each other's folding and complex formation in the endoplasmic reticulum and traffic as a complex to the Golgi and to the membrane? Or are they assisted by other unidentified partners? In the latter case, what molecules would act as chaperonins?

Clearly, we are still a long way from integrating mechanistically the action of Celsr proteins, particularly their interactions with other core PCP components and downstream signaling pathways and terminal cell effectors. Elucidating those questions will further our understanding of PCP during normal development, in pathological conditions such as ciliopathies, and in stem cell biology and regeneration.

REFERENCES

1. Formstone CJ, Barclay J, Rees M, Little PF. Chromosomal localization of Celsr2 and Celsr3 in the mouse; Celsr3 is a candidate for the tippy (tip) lethal mutant on chromosome 9. *Mamm Genome* 2000;**11**:392–4.
2. Formstone CJ, Little PF. The flamingo-related mouse Celsr family (Celsr1-3) genes exhibit distinct patterns of expression during embryonic development. *Mech Dev* 2001;**109**:91–4.
3. Hadjantonakis AK, Sheward WJ, Harmar AJ, de Galan L, Hoovers JM, Little PF. Celsr1, a neural-specific gene encoding an unusual seven-pass transmembrane receptor, maps to mouse chromosome 15 and human chromosome 22qter. *Genomics* 1997;**45**:97–104.
4. Formstone CJ. 7TM-Cadherins: developmental roles and future challenges. *Adv Exp Med Biol* 2010;**706**:14–36.
5. Wada H, Tanaka H, Nakayama S, Iwasaki M, Okamoto H. Frizzled3a and Celsr2 function in the neuroepithelium to regulate migration of facial motor neurons in the developing zebrafish hindbrain. *Development* 2006;**133**:4749–59.
6. Volynski KE, Silva JP, Lelianova VG, Atiqur Rahman M, Hopkins C, Ushkaryov YA. Latrophilin fragments behave as independent proteins that associate and signal on binding of LTX(N4C). *EMBO J* 2004;**23**:4423–33.
7. Krasnoperov V, Lu Y, Buryanovsky L, Neubert TA, Ichtchenko K, Petrenko AG. Post-translational proteolytic processing of the calciumin-dependent receptor of

alpha-latrotoxin (CIRL), a natural chimera of the cell adhesion protein and the G protein-coupled recent—role of the G protein-coupled receptor proteolysis site (GPS) motif. *J Biol Chem* 2002;**277**:46518–26.
8. Chae J, Kim MJ, Goo JH, Collier S, Gubb D, Charlton J, et al. The Drosophila tissue polarity gene starry night encodes a member of the protocadherin family. *Development* 1999;**126**:5421–9.
9. Usui T, Shima Y, Shimada Y, Hirano S, Burgess RW, Schwarz TL, et al. Flamingo, a seven-pass transmembrane cadherin, regulates planar cell polarity under the control of Frizzled. *Cell* 1999;**98**:585–95.
10. Mlodzik M. Planar polarity in the Drosophila eye: a multifaceted view of signaling specificity and cross-talk. *EMBO J* 1999;**18**:6873–9.
11. Segalen M, Bellaiche Y. Cell division orientation and planar cell polarity pathways. *Semin Cell Dev Biol* 2009;**20**:972–7.
12. Bellaiche Y, Gho M, Kaltschmidt JA, Brand AH, Schweisguth F. Frizzled regulates localization of cell-fate determinants and mitotic spindle rotation during asymmetric cell division. *Nat Cell Biol* 2001;**3**:50–7.
13. Axelrod JD. Progress and challenges in understanding planar cell polarity signaling. *Semin Cell Dev Biol* 2009;**20**:964–71.
14. Simons M, Mlodzik M. Planar cell polarity signaling: from fly development to human disease. *Annu Rev Genet* 2008;**42**:517–40.
15. Strutt H, Strutt D. Asymmetric localisation of planar polarity proteins: mechanisms and consequences. *Semin Cell Dev Biol* 2009;**20**:957–63.
16. Adler PN. Planar signaling and morphogenesis in Drosophila. *Dev Cell* 2002;**2**:525–35.
17. Strutt D, Johnson R, Cooper K, Bray S. Asymmetric localization of frizzled and the determination of notch-dependent cell fate in the Drosophila eye. *Curr Biol* 2002;**12**:813–24.
18. Taylor J, Abramova N, Charlton J, Adler PN. Van Gogh: a new Drosophila tissue polarity gene. *Genetics* 1998;**150**:199–210.
19. Wolff T, Rubin GM. Strabismus, a novel gene that regulates tissue polarity and cell fate decisions in Drosophila. *Development* 1998;**125**:1149–59.
20. Feiguin F, Hannus M, Mlodzik M, Eaton S. The ankyrin repeat protein Diego mediates Frizzled-dependent planar polarization. *Dev Cell* 2001;**1**:93–101
21. Gubb D, Green C, Huen D, Coulson D, Johnson G, Tree D, et al. The balance between isoforms of the prickle LIM domain protein is critical for planar polarity in Drosophila imaginal discs. *Genes Dev* 1999;**13**:2315–27.
22. Weber U, Gault WJ, Olguin P, Serysheva E, Mlodzik M. Novel regulators of planar cell polarity: a genetic analysis in Drosophila. *Genetics* 2012;**191**:145–62.
23. Bastock R, Strutt H, Strutt D. Strabismus is asymmetrically localised and binds to Prickle and Dishevelled during Drosophila planar polarity patterning. *Development* 2003;**130**:3007–14.
24. Jenny A, Darken RS, Wilson PA, Mlodzik M. Prickle and Strabismus form a functional complex to generate a correct axis during planar cell polarity signaling. *EMBO J* 2003;**22**:4409–20.
25. Chen WS, Antic D, Matis M, Logan CY, Povelones M, Anderson GA, et al. Asymmetric homotypic interactions of the atypical cadherin flamingo mediate intercellular polarity signaling. *Cell* 2008;**133**:1093–105.
26. Das G, Reynolds-Kenneally J, Mlodzik M. The atypical cadherin Flamingo links Frizzled and Notch signaling in planar polarity establishment in the Drosophila eye. *Dev Cell* 2002;**2**:655–66.
27. Bellaiche Y, Beaudoin-Massiani O, Stuttem I, Schweisguth F. The planar cell polarity protein Strabismus promotes Pins anterior localization during asymmetric division of sensory organ precursor cells in Drosophila. *Development* 2004;**131**:469–78.

28. Segalen M, Johnston CA, Martin CA, Dumortier JG, Prehoda KE, David NB, et al. The Fz-Dsh planar cell polarity pathway induces oriented cell division via Mud/NuMA in Drosophila and zebrafish. *Dev Cell* 2010;**19**:740–52.
29. Shimada Y, Yonemura S, Ohkura H, Strutt D, Uemura T. Polarized transport of Frizzled along the planar microtubule arrays in Drosophila wing epithelium. *Dev Cell* 2006;**10**:209–22.
30. Shima Y, Copeland NG, Gilbert DJ, Jenkins NA, Chisaka O, Takeichi M, et al. Differential expression of the seven-pass transmembrane cadherin genes Celsr1-3 and distribution of the Celsr2 protein during mouse development. *Dev Dyn* 2002;**223**:321–32.
31. Tissir F, De-Backer O, Goffinet AM, Lambert de Rouvroit C. Developmental expression profiles of Celsr (Flamingo) genes in the mouse. *Mech Dev* 2002;**112**:157–60.
32. Hadjantonakis AK, Formstone CJ, Little PF. mCelsr1 is an evolutionarily conserved seven-pass transmembrane receptor and is expressed during mouse embryonic development. *Mech Dev* 1998;**78**:91–5.
33. Beall SA, Boekelheide K, Johnson KJ. Hybrid GPCR/cadherin (Celsr) proteins in rat testis are expressed with cell type specificity and exhibit differential Sertoli cell-germ cell adhesion activity. *J Androl* 2005;**26**:529–38.
34. Johnson KJ, Patel SR, Boekelheide K. Multiple cadherin superfamily members with unique expression profiles are produced in rat testis. *Endocrinology* 2000;**141**:675–83.
35. Formstone CJ, Moxon C, Murdoch J, Little P, Mason I. Basal enrichment within neuroepithelia suggests novel function(s) for Celsr1 protein. *Mol Cell Neurosci* 2010; **44**:210–22.
36. Devenport D, Fuchs E. Planar polarization in embryonic epidermis orchestrates global asymmetric morphogenesis of hair follicles. *Nat Cell Biol* 2008;**10**:1257–68.
37. Curtin JA, Quint E, Tsipouri V, Arkell RM, Cattanach B, Copp AJ, et al. Mutation of Celsr1 disrupts planar polarity of inner ear hair cells and causes severe neural tube defects in the mouse. *Curr Biol* 2003;**13**:1129–33.
38. Yates LL, Papakrivopoulou J, Long DA, Goggolidou P, Connolly JO, Woolf AS, et al. The planar cell polarity gene Vangl2 is required for mammalian kidney-branching morphogenesis and glomerular maturation. *Hum Mol Genet* 2010;**19**:4663–76.
39. Ravni A, Qu Y, Goffinet AM, Tissir F. Planar cell polarity cadherin Celsr1 regulates skin hair patterning in the mouse. *J Invest Dermatol* 2009;**129**:2507–9.
40. Qu Y, Glasco DM, Zhou L, Sawant A, Ravni A, Fritzsch B, et al. Atypical cadherins Celsr1-3 differentially regulate migration of facial branchiomotor neurons in mice. *J Neurosci* 2010;**30**:9392–401.
41. Tissir F, Qu Y, Montcouquiol M, Zhou L, Komatsu K, Shi D, et al. Lack of cadherins Celsr2 and Celsr3 impairs ependymal ciliogenesis, leading to fatal hydrocephalus. *Nat Neurosci* 2010;**13**:700–7.
42. Tissir F, Bar I, Jossin Y, De Backer O, Goffinet AM. Protocadherin Celsr3 is crucial in axonal tract development. *Nat Neurosci* 2005;**8**:451–7.
43. Zhou L, Bar I, Achouri Y, Campbell K, De Backer O, Hebert JM, et al. Early forebrain wiring: genetic dissection using conditional Celsr3 mutant mice. *Science* 2008;**320**:946–9.
44. Wang Y, Guo N, Nathans J. The role of Frizzled3 and Frizzled6 in neural tube closure and in the planar polarity of inner-ear sensory hair cells. *J Neurosci* 2006;**26**:2147–56.
45. Wang Y, Thekdi N, Smallwood PM, Macke JP, Nathans J. Frizzled-3 is required for the development of major fiber tracts in the rostral CNS. *J Neurosci* 2002;**22**:8563–73.
46. Guo N, Hawkins C, Nathans J. Frizzled6 controls hair patterning in mice. *Proc Natl Acad Sci USA* 2004;**101**:9277–81.
47. Montcouquiol M, Rachel RA, Lanford PJ, Copeland NG, Jenkins NA, Kelley MW. Identification of Vangl2 and Scrb1 as planar polarity genes in mammals. *Nature* 2003;**423**:173–7.

48. Vivancos V, Chen P, Spassky N, Qian D, Dabdoub A, Kelley M, et al. Wnt activity guides facial branchiomotor neuron migration, and involves the PCP pathway and JNK and ROCK kinases. *Neural Dev* 2009;**4**:7–15.
49. Torban E, Wang HJ, Groulx N, Gros P. Independent mutations in mouse Vangl2 that cause neural tube defects in looptail mice impair interaction with members of the Dishevelled family. *J Biol Chem* 2004;**279**:52703–13.
50. Song H, Hu J, Chen W, Elliott G, Andre P, Gao B, et al. Planar cell polarity breaks bilateral symmetry by controlling ciliary positioning. *Nature* 2010;**466**:378–82.
51. Montcouquiol M, Sans N, Huss D, Kach J, Dickman JD, Forge A, et al. Asymmetric localization of Vangl2 and Fz3 indicate novel mechanisms for planar cell polarity in mammals. *J Neurosci* 2006;**26**:5265–75.
52. Deans MR, Antic D, Suyama K, Scott MP, Axelrod JD, Goodrich LV. Asymmetric distribution of prickle-like 2 reveals an early underlying polarization of vestibular sensory epithelia in the inner ear. *J Neurosci* 2007;**27**:3139–47.
53. Robinson A, Escuin S, Doudney K, Vekemans M, Stevenson RE, Greene ND, et al. Mutations in the planar cell polarity genes CELSR1 and SCRIB are associated with the severe neural tube defect craniorachischisis. *Hum Mutat* 2012;**33**:440–7.
54. Nishimura T, Honda H, Takeichi M. Planar cell polarity links axes of spatial dynamics in neural-tube closure. *Cell* 2012;**149**:1084–97.
55. Kibar Z, Vogan KJ, Groulx N, Justice MJ, Underhill DA, Gros P. Ltap, a mammalian homolog of Drosophila Strabismus/Van Gogh, is altered in the mouse neural tube mutant Loop-tail. *Nat Genet* 2001;**28**:251–5.
56. Ybot-Gonzalez P, Savery D, Gerrelli D, Signore M, Mitchell CE, Faux CH, et al. Convergent extension, planar-cell-polarity signalling and initiation of mouse neural tube closure. *Development* 2007;**134**:789–99.
57. Ciruna B, Jenny A, Lee D, Mlodzik M, Schier AF. Planar cell polarity signalling couples cell division and morphogenesis during neurulation. *Nature* 2006;**439**:220–4.
58. Davenport JR, Yoder BK. An incredible decade for the primary cilium: a look at a once-forgotten organelle. *Am J Physiol Renal Physiol* 2005;**289**:F1159–F1169.
59. Marshall WF, Kintner C. Cilia orientation and the fluid mechanics of development. *Curr Opin Cell Biol* 2008;**20**:48–52.
60. Park TJ, Haigo SL, Wallingford JB. Ciliogenesis defects in embryos lacking inturned or fuzzy function are associated with failure of planar cell polarity and Hedgehog signaling. *Nat Genet* 2006;**38**:303–11.
61. Park TJ, Mitchell BJ, Abitua PB, Kintner C, Wallingford JB. Dishevelled controls apical docking and planar polarization of basal bodies in ciliated epithelial cells. *Nat Genet* 2008;**40**:871–9.
62. Salathe M. Regulation of mammalian ciliary beating. *Annu Rev Physiol* 2007;**69**: 401–22.
63. Voronina VA, Takemaru KI, Treuting P, Love D, Grubb BR, Hajjar AM, et al. Inactivation of Chibby affects function of motile airway cilia. *J Cell Biol* 2009; **185**:225–33.
64. Hagiwara H, Ohwada N, Aoki T, Suzuki T, Takata K. The primary cilia of secretory cells in the human oviduct mucosa. *Med Mol Morphol* 2008;**41**:193–8.
65. Gibbons IR. The relationship between the fine structure and direction of beat in gill cilia of a lamellibranch mollusc. *J Biophys Biochem Cytol* 1961;**11**:179–205.
66. Wallingford JB. Planar cell polarity signaling, cilia and polarized ciliary beating. *Curr Opin Cell Biol* 2010;**22**:597–604.
67. Mirzadeh Z, Han YG, Soriano-Navarro M, Garcia-Verdugo JM, Alvarez-Buylla A. Cilia organize ependymal planar polarity. *J Neurosci* 2010;**30**:2600–10.
68. Mitchell B, Jacobs R, Li J, Chien S, Kintner C. A positive feedback mechanism governs the polarity and motion of motile cilia. *Nature* 2007;**447**:97–101.

69. Guirao B, Meunier A, Mortaud S, Aguilar A, Corsi JM, Strehl L, et al. Coupling between hydrodynamic forces and planar cell polarity orients mammalian motile cilia. *Nat Cell Biol* 2010;**12**:341–50.
70. Chandrasekhar A. Turning heads: development of vertebrate branchiomotor neurons. *Dev Dyn* 2004;**229**:143–61.
71. Garel S, Garcia-Dominguez M, Charnay P. Control of the migratory pathway of facial branchiomotor neurones. *Development* 2000;**127**:5297–307.
72. Guthrie S. Patterning and axon guidance of cranial motor neurons. *Nat Rev Neurosci* 2007;**8**:859–71.
73. Song MR, Shirasaki R, Cai CL, Ruiz EC, Evans SM, Lee SK, et al. T-Box transcription factor Tbx20 regulates a genetic program for cranial motor neuron cell body migration. *Development* 2006;**133**:4945–55.
74. Gilland E, Baker R. Evolutionary patterns of cranial nerve efferent nuclei in vertebrates. *Brain Behav Evol* 2005;**66**:234–54.
75. Ying G, Wu S, Hou R, Huang W, Capecchi MR, Wu Q. The protocadherin gene Celsr3 is required for interneuron migration in the mouse forebrain. *Mol Cell Biol* 2009;**29**:3045–61.
76. Bingham S, Higashijima S, Okamoto H, Chandrasekhar A. The Zebrafish trilobite gene is essential for tangential migration of branchiomotor neurons. *Dev Biol* 2002;**242**:149–60.
77. Carreira-Barbosa F, Concha ML, Takeuchi M, Ueno N, Wilson SW, Tada M. Prickle 1 regulates cell movements during gastrulation and neuronal migration in zebrafish. *Development* 2003;**130**:4037–46.
78. Jessen JR, Topczewski J, Bingham S, Sepich DS, Marlow F, Chandrasekhar A, et al. Zebrafish trilobite identifies new roles for Strabismus in gastrulation and neuronal movements. *Nat Cell Biol* 2002;**4**:610–5.
79. Rohrschneider MR, Elsen GE, Prince VE. Zebrafish Hoxb1a regulates multiple downstream genes including prickle1b. *Dev Biol* 2007;**309**:358–72.
80. Mapp OM, Wanner SJ, Rohrschneider MR, Prince VE. Prickle1b mediates interpretation of migratory cues during zebrafish facial branchiomotor neuron migration. *Dev Dyn* 2010;**239**:1596–608.
81. Matsubara D, Horiuchi SY, Shimono K, Usui T, Uemura T. The seven-pass transmembrane cadherin Flamingo controls dendritic self-avoidance via its binding to a LIM domain protein, Espinas, in Drosophila sensory neurons. *Genes Dev* 2011; **25**:1982–96.
82. Gao FB, Kohwi M, Brenman JE, Jan LY, Jan YN. Control of dendritic field formation in Drosophila: the roles of flamingo and competition between homologous neurons. *Neuron* 2000;**28**:91–101.
83. Berger-Muller S, Suzuki T. Seven-pass transmembrane cadherins: roles and emerging mechanisms in axonal and dendritic patterning. *Mol Neurobiol* 2011;**44**:313–20.
84. Reuter JE, Nardine TM, Penton A, Billuart P, Scott EK, Usui T, et al. A mosaic genetic screen for genes necessary for Drosophila mushroom body neuronal morphogenesis. *Development* 2003;**130**:1203–13.
85. Kimura H, Usui T, Tsubouchi A, Uemura T. Potential dual molecular interaction of the Drosophila 7-pass transmembrane cadherin Flamingo in dendritic morphogenesis. *J Cell Sci* 2006;**119**:1118–29.
86. Shima Y, Kengaku M, Hirano T, Takeichi M, Uemura T. Regulation of dendritic maintenance and growth by a mammalian 7-pass transmembrane cadherin. *Dev Cell* 2004;**7**:205–16.
87. Shima Y, Kawaguchi SY, Kosaka K, Nakayama M, Hoshino M, Nabeshima Y, et al. Opposing roles in neurite growth control by two seven-pass transmembrane cadherins. *Nat Neurosci* 2007;**10**:963–9.

88. Senti KA, Usui T, Boucke K, Greber U, Uemura T, Dickson BJ. Flamingo regulates R8 axon-axon and axon-target interactions in the Drosophila visual system. *Curr Biol* 2003;**13**:828–32.
89. Lee RC, Clandinin TR, Lee CH, Chen PL, Meinertzhagen IA, Zipursky SL. The protocadherin Flamingo is required for axon target selection in the Drosophila visual system. *Nat Neurosci* 2003;**6**:557–63.
90. Chen PL, Clandinin TR. The cadherin Flamingo mediates level-dependent interactions that guide photoreceptor target choice in Drosophila. *Neuron* 2008;**58**:26–33.
91. Fenstermaker AG, Prasad AA, Bechara A, Adolfs Y, Tissir F, Goffinet A, et al. Wnt/planar cell polarity signaling controls the anterior-posterior organization of monoaminergic axons in the brainstem. *J Neurosci* 2010;**30**:16053–64.
92. Price DJ, Kennedy H, Dehay C, Zhou LB, Mercier M, Jossin Y, et al. The development of cortical connections. *Eur J Neurosci* 2006;**23**:910–20.
93. Wang Y, Nathans J. Tissue/planar cell polarity in vertebrates: new insights and new questions. *Development* 2007;**134**:647–58.
94. Lyuksyutova AI, Lu CC, Milanesio N, King LA, Guo N, Wang Y, et al. Anterior-posterior guidance of commissural axons by Wnt-frizzled signaling. *Science* 2003;**302**:1984–8.
95. Steinel MC, Whitington PM. The atypical cadherin Flamingo is required for sensory axon advance beyond intermediate target cells. *Dev Biol* 2009;**327**:447–57.
96. Hakeda-Suzuki S, Berger-Muller S, Tomasi T, Usui T, Horiuchi SY, Uemura T, et al. Golden Goal collaborates with Flamingo in conferring synaptic-layer specificity in the visual system. *Nat Neurosci* 2011;**14**:314–23.
97. Steimel A, Wong L, Najarro EH, Ackley BD, Garriga G, Hutter H. The Flamingo ortholog FMI-1 controls pioneer-dependent navigation of follower axons in C. elegans. *Development* 2010;**137**:3663–73.
98. Carreira-Barbosa F, Kajita M, Morel V, Wada H, Okamoto H, Martinez Arias A, et al. Flamingo regulates epiboly and convergence/extension movements through cell cohesive and signalling functions during zebrafish gastrulation. *Development* 2009;**136**:383–92.
99. Devenport D, Oristian D, Heller E, Fuchs E. Mitotic internalization of planar cell polarity proteins preserves tissue polarity. *Nat Cell Biol* 2011;**13**:893–902.
100. Herz J. The switch on the RAPper's necklace. *Mol Cell* 2006;**23**:451–5.
101. Hansen TH, Connolly JM, Gould KG, Fremont DH. Basic and translational applications of engineered MHC class I proteins. *Trends Immunol* 2010;**31**:363–9.

CHAPTER TEN

Fat and Dachsous Cadherins

Praveer Sharma[*,†], **Helen McNeill**[*,†]
[*]Samuel Lunenfeld Research Institute, Toronto, Canada
[†]Department of Molecular Genetics, University of Toronto, Toronto, Canada

Contents

Abstract

Fat and Dachsous (Ds) are very large cell adhesion molecules. They bind each other and have important, highly conserved roles in planar cell polarity (PCP) and growth control. PCP is defined as the directionally coordinated development of cellular structures or behavior. Cellular and tissue growth needs to be modulated in terms of rate and final size, and the Hippo pathway regulates growth in a variety of developmental contexts. Fat and Ds are important upstream regulators of these pathways. There are two Fat proteins in *Drosophila*, Fat and Fat2, and four in vertebrates, Fat1–4. There is one Ds protein in *Drosophila* and two in vertebrates, Dachsous1–2. In this chapter, we discuss the roles of Fat and Ds family members, focusing on *Drosophila* and mouse development.

Progress in Molecular Biology and Translational Science, Volume 116
ISSN 1877-1173
http://dx.doi.org/10.1016/B978-0-12-394311-8.00010-8

1. INTRODUCTION

Drosophila Fat (Ft) is a very large (560 kDa) type-I transmembrane protein with 34 cadherin domains, 4 EGF-like domains, and 2 laminin G domains in the extracellular region, followed by the transmembrane and intracellular regions.[1–3] The presence of 34 cadherin domains is characteristic of all members of the Fat family (see also Chapter 4), although an alternate method of sequence alignment predicts 35 cadherin domains.[4] Mature Ft exists in a stable form that is cleaved N-terminal to the transmembrane region.[5,6] The intracellular region of Fat does not contain any identifiable domains. However, several regions have been functionally defined as being involved in PCP and growth regulation (Fig. 10.1).

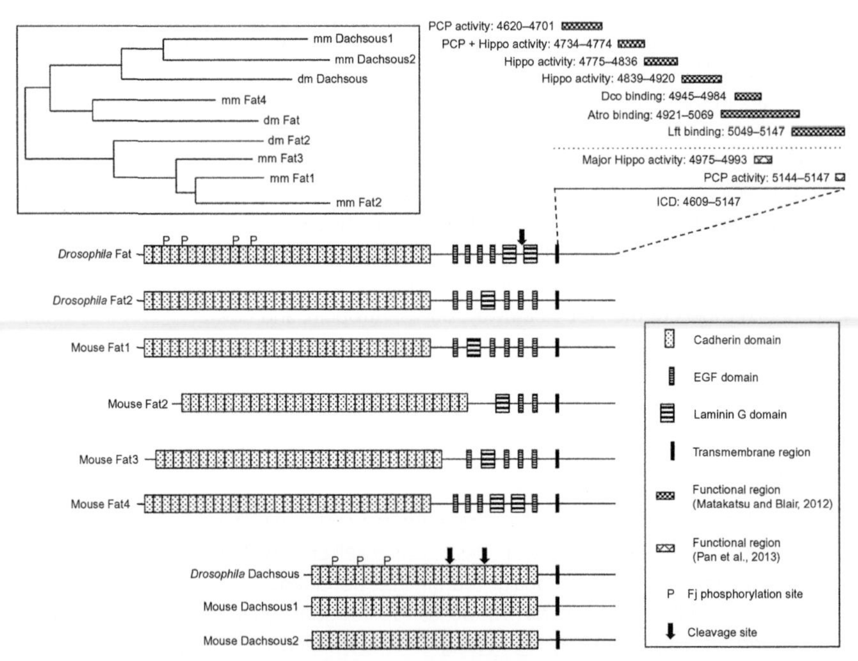

Figure 10.1 Structural representations of Fat and Dachsous homologs. Domain numbers and positions are based on Ref. 2. Functional regions of Fat ICD are based on biochemical and structure–function studies, as described in the text. Though mFat1 and mFat4 are known to be cleaved in their extracellular domains, their cleavage sites are not precisely known, and so they are not indicated. Inset shows phylogenetic relationships of Fat and Dachsous homologs of mouse (mm, *Mus musculus*) and fruit fly (dm, *Drosophila melanogaster*). Full-length protein sequences were aligned using Clustal Omega, and a Maximum Likelihood tree was constructed using MEGA 5.

Drosophila Dachsous (Ds) is another large (379 kDa) type-I transmembrane protein with 27 cadherin domains in the extracellular region, followed by the transmembrane and intracellular regions (ICDs).[7] Ds can also be cleaved between cadherin repeats at two sites in its extracellular region[8] (Fig. 10.1).

Drosophila Fat2 (also known as Fat-like) is a homolog of Ft.[9] Its intracellular region is, however, substantially different from that of Fat, and it has one laminin G domain in the extracellular region. There are four vertebrate Fat proteins. Fat4 is orthologous to Ft and has two laminin G domains. Fat1, Fat2, and Fat3 are closer to *Drosophila* Fat2 and each has one laminin G domain.[1] Like Ft, Fat1 and Fat4 are also cleaved N-terminal to the transmembrane region.[6,10]

There are two vertebrate Dachsous homologs, Dachsous1 and Dachsous2,[1,11,12] each with 27 cadherin domains (see also Chapter 4). The ICDs of Ds and its vertebrate homologs contain a putative β-catenin binding site.[1]

fat (ft) and *dachsous (ds)* were, along with *four-jointed (fj)*, which encodes a Golgi kinase, and *dachs (d)*, which encodes an atypical myosin, part of a group of genes described in the early days of *Drosophila* genetics as regulators of growth, limb development, and tissue patterning.[13,14] *ft* is a hyperplastic tumor suppressor gene: in *ft* mutants, epithelial structure and cell differentiation are maintained even though the epithelium is greatly overgrown.[15] *ds* mutants have patterning defects similar to *ft* mutants.[7] *ft* and *ds* are important in at least two major developmental processes: regulation of tissue growth through the Hippo pathway and directional control of cellular morphology and behavior, called planar cell polarity (PCP) (reviewed in Ref. 16–26).

This chapter focuses on the roles of Ft and Ds in regulating PCP and Hippo signaling. However, Ft–Ds signaling is also involved in other processes, including neural differentiation[27–29] and degeneration,[30] stem cell regulation,[31] cell–cell interactions,[32,33] cancer,[34–37] and neurological disorders.[38]

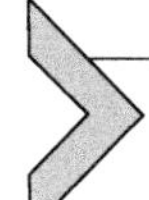

2. FAT–DACHSOUS AND PCP

2.1. Introduction to PCP

Many tissues display directionally organized cell structures or behaviors, which are called PCP.[18,39] Ft–Ds signaling and Frizzled/PCP (Fz/PCP) signaling are the main PCP pathways regulating PCP in contexts as disparate as ommatidial orientation in the *Drosophila* eye and orientation of stereocilia in the mouse cochlea.[40,41] PCP is also evident in dynamic processes such as oriented cell division (OCD) in *Drosophila* imaginal discs[42] and cell intercalation during zebrafish neurulation.[43] Ft and Ds cadherins act together with Fj, D, and the transcriptional corepressor Atrophin (Atro) to regulate PCP.[18] Vertebrate orthologs of *Drosophila ft* and *ds*, such as *fat1*, *fat4*, and *dachsous1* also regulate PCP.[40,44,45]

A key technique for analyzing planar polarity is analysis of clonal phenotypes. Clones are patches of cells mutant for a particular gene, which are surrounded by wild-type tissue. Mutant clones of PCP regulators can affect polarity in their genetically wild-type neighboring tissue.[41,46,47] These are known as nonautonomous phenotypes. They suggest that the PCP regulator in question is involved not only in interpretation of PCP signals but also in long-range transmission of such signals, as the change in signal caused by a mutant clone is propagated across multiple cells.

In *Drosophila*, Ft and Ds localize to the subapical region of epithelial cells and positively regulate each other's membrane localization.[48–50] Data from cell culture studies and from overexpression and mutant clones suggest that Ft and Ds form transheterodimers.[48,50,51] Ft and Ds can also localize on opposite sides of a cell, along the polarity axis.[8,52] Unlike classical cadherins, there is no evidence of homodimers of either Ft or Ds forming across cells.[51,53] Fj phosphorylates Ft and Ds in specific N-terminal cadherin domains[54] (Fig. 10.1). This phosphorylation changes Ft and Ds binding affinities: binding of cadherin-domain phosphorylated Ft to Ds increases, while binding of cadherin-domain phosphorylated Ds to Ft decreases. *ds* and *fj* are expressed in complementary gradients in epithelial tissue,[7,41,55–57] which is hypothesized to set up a gradient of Ft–Ds interaction across the tissue.[58,59] Thus, different cells, and different sides of the same cell, would have different levels of Ft–Ds signaling. This could then lead to differential activity of Ft–Ds downstream effectors. For example, Atro activity is proposed to be higher in a cell with greater Ft–Ds activity compared to its neighbor,[60] and subcellular localization of D is regulated by Ft–Ds signaling.[8,52,61–63]

2.2. Ft and Ds in the *Drosophila* eye

ds was first shown to regulate PCP in the Drosophila wing[64], and the roles of *ft*, *ds*, and *fj* were further elucidated in the Drosophila eye.[41,47,50,57] This insect eye consists of ~800 clusters of photoreceptors called ommatidia. Ommatidia have a trapezoidal shape and manifest PCP by consistently pointing upward in the dorsal half of the eye, and downward in the ventral half of the eye (Fig. 10.2A). The precise midline between the dorsal and ventral halves where ommatidial orientation abruptly switches is called the

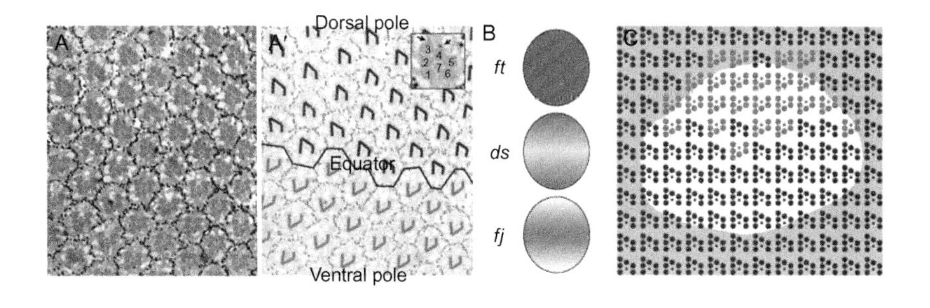

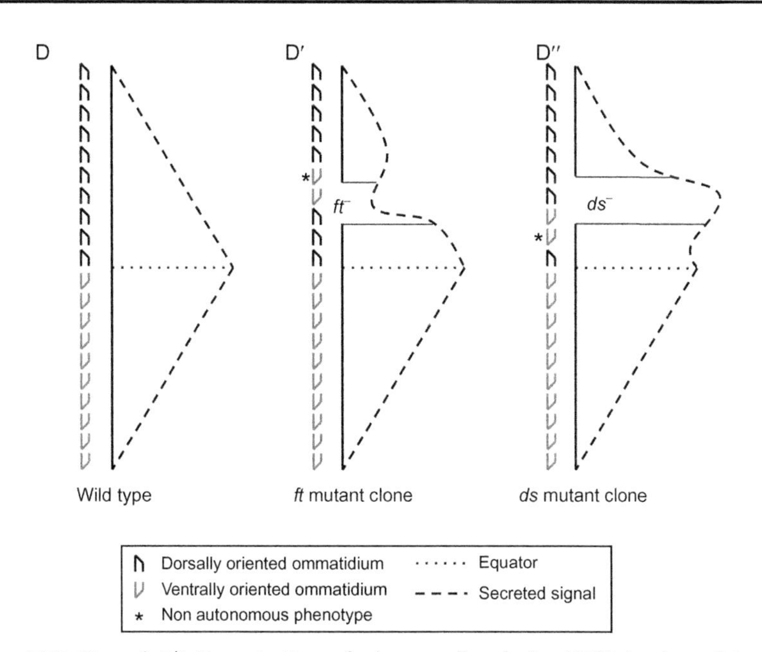

Figure 10.2 (A and A′) Organization of planar cell polarity (PCP) in the wild-type *Drosophila* eye. The outer photoreceptors in the ommatidia (numbers 1–6 in inset) are outlined in A′, and a black line marks the equator of the eye. In both the dorsal and ventral halves of the eye, the direction toward the equator is "equatorial," and away from the equator it is "polar." (B) Diagrams of *ft, ds,* and *fj* expression patterns in the imaginal disc of the eye. *ft* is expressed evenly throughout the disc. *ds* is expressed at high levels at the poles and less around the equator. *fj* has the reverse gradient, with high levels around the equator and lower levels around the poles. (C) A diagram of an idealized *fat* mutant clone illustrating nonautonomous polarity disruption (gray instead of black photoreceptors) on the polar side of the clone. (D) In the wild-type eye, the hypothesized polarity signal is at high levels at the equator, and low levels at the pole, reflecting Ft activity. Ommatidia orient so as to appear to point "down" the second signal gradient. (D′) Loss of *ft* in a mutant clone causes a reduction in second signal levels. As the signal is diffusible, its levels in neighboring wild-type cells are also reduced.

equator. By analogy, the dorsal and ventral extremes of the eye are called the poles. In either half, ommatidia normally point in the polar direction. *ft* is expressed evenly in the developing eye, while *ds* and *fj* are expressed in complementary gradients: Ds levels are high near the poles and Fj levels are high near the equator (Fig. 10.2B) (see also Section 2.8). Importantly, PCP is disrupted both inside and outside clones with *ft* or *ds* mutations (Fig. 10.2C).

Ommatidial polarity is initially established through a cell-fate specification event, wherein one of a pair of photoreceptor precursors closer to the equator is specified as photoreceptor R3, while the polar precursor is specified as photoreceptor R4. Loss of *ft* activity in one precursor specifies it as R4, while loss of *ds* specifies it as R3, regardless of position.[41] *ds ft* double mutant precursors are specified as R4, suggesting that *ft* is epistatic to *ds* in this process and that *ds* negatively regulates *ft*. If the polar precursor is misspecified as R3, or the equatorial precursor as R4, ommatidial polarity is reversed. If *ft* or *ds* is lost in both precursors, ommatidial polarity is randomized.

Phenotypes outside *ft or ds* mutant clones (nonautonomous phenotypes) are directionally specific: *ft* mutant clones disrupt wild-type polarity on the polar side, while *ds* mutant clones disrupt polarity on the equatorial side (Fig. 10.2D–D″). This occurs at a longer distance than can be explained by redistribution of Ft–Ds effectors (see Section 2.3). Instead, such nonautonomous polarity phenotypes of *ft* and ds mutant clones may be better explained by changes in a hypothetical secreted effector that results from Ft–Ds activity (referred to in the literature as the second signal or Factor X).[60] That second signal can travel a distance of 10 or more cells and affect R3/4 fate specification. Positive regulation of this effector by Ft and negative regulation by Ds (perhaps through its action on Ft) can explain the opposite nonautonomous phenotypes of *ft* and *ds* mutant clones (see Fig. 10.2D–D″).

2.3. Ft and Ds in the *Drosophila* wing

Wing cells have actin-rich protrusions on the apical surface called hairs or trichomes. All wing hairs point in the distal direction. Loss of PCP signaling causes hair to misorient and form "swirling" patterns, reflecting loss of

---

As ommatidia point "down" the signal gradient, this can cause nonautonomous reversal of polarity on the polar side of the clone. (D″) Loss of *ds* in a mutant clone causes an increase in second signal levels (possibly through derepression of Ft activity). This can result in nonautonomous reversal of polarity on the equatorial side of the clone.

long-range polarity, though short-range organization survives.[46,64] As in the eye, *ft* and *ds* mutant clones have both autonomous and nonautonomous polarity phenotypes. In particular, *ds* clones cause hairs outside the clone to point in toward it (most noticeable on the distal side of the clone), and *ft* clones cause hair outside the clone to point away from it (most noticeable on the proximal side of the clone).[50] Though wing PCP appears simple, its regulation is complex, with many different regional and temporal aspects. *ft* and *ds* clones show size dependence: small clones of *ds* and very small clones of *ft* have no polarity phenotype.[48,50] Ft–Ds signaling appears to regulate polarity through the whole wing in the early stage of its establishment, but only in the proximal wing in the later stages of development.[65,70,107,108]

Wing and eye polarity phenotypes of *ft* mutants can be partially rescued by loss of *d*.[62] Asymmetric localization of D to the distal side serves as a marker of Ft–Ds polarity signaling. In the wing, clones that change Ft–Ds signaling, for example, *ds* overexpression clones, repolarize the asymmetric distribution of D 4–5 cells away from the clone.[8,52] This distance corresponds well with the degree of nonautonomy seen around *ft* and *ds* mutant clones. Ft and Ds are distributed asymmetrically in wing cells, with Ft localized to the proximal side of the cell, and Ds localized to the distal side, where it colocalizes with D.[8,52] These data are consistent with the reduced localization of D on the side of greater Ft activity and with the propagation of this differential localization away from clone borders.

2.4. Ft and Ds in the *Drosophila* abdomen and embryonic and larval epidermis

The abdomen consists of repeated segments of anterior (A) and posterior (P) compartments, which maintain consistent hair and bristle polarity. *ds* and *ft* clones have directional nonautonomous polarity phenotypes that are the reverse of each other, though the direction depends on whether the clone is in the A or P compartment.[66] Interestingly, *ds ft* double mutants have no clear nonautonomous polarity phenotype.[58] Although *ft* is epistatic to *ds* in R3/4 fate specification in the eye,[41] this seems not to be the case in nonautonomous polarity regulation in the abdomen, as *ds ft* double mutant clones do not affect surrounding wild-type tissue.

The ventral epidermis of the *Drosophila* embryo and larva contains actin-rich cuticular projections called denticles, arrayed in rows in each abdominal segment. The denticles of each row point either proximally or distally, thus manifesting PCP. The Ft–Ds pathway regulates this orientation, and changing pathway activity by mutating or overexpressing *ft* or *ds* can randomize or reverse denticle orientation.[67,68,109] The Fz/PCP pathway has a minor, redundant role in controlling denticle polarity.

2.5. Ft and Ds in OCD

OCD refers to cell division in which the resultant daughter cells are aligned along a particular body axis. In combination with cell migration, OCD helps establish the shape of tissues and organs, for example, in *Drosophila* eyes and wings[42] and in vertebrate kidneys.[69] The Ft–Ds pathway regulates OCD, as the orientation of spindle axes is randomized in *Drosophila* mutants of the Ft–Ds pathway.[70,71] In mice, Ft–Ds signaling also regulates OCD in kidney development, and its loss leads to cystic kidneys.[40,44,45] Recent results suggest that polarity signaling might not control spindle orientation directly but instead influence cell shape.[72] The spindle axis is proposed to align with the long axis of an elongated cell, thus leading to OCD. Ft–Ds signaling is proposed to regulate cell shape by directionally modulating D activity.[72]

2.6. Fat2 in the *Drosophila* ovary

The shape of the *Drosophila* oocyte is influenced by follicle cells of the ovary, which align their actin cytoskeletons perpendicular to the long axis of the ovary, thus forming a "molecular corset," forcing the growing oocyte into a cylindrical shape. The cytoskeleton of follicle cells is planar polarized. Although mutations of *ft*, *ds*, or Fz/PCP components do not affect this process, *fat2* mutants lose polarization of follicle cytoskeletons and the resulting eggs are spherical. *fat2* mutant cells can also nonautonomously alter cytoskeletal polarity in genetically wild-type neighbors.[73] Interestingly, as long as less than 50% of cells are *fat2* mutants, the entire tissue can polarize correctly.[74]

2.7. Structure–function analysis of Ft–Ds and polarity regulation

Ft and Ds both have large extracellular domains (ECDs) with multiple cadherin repeats and unique ICDs (Fig. 10.1). In the *Drosophila* wing, membrane-bound versions of the Ft ICD and the Ds ECD are sufficient to rescue most aspects of growth control and PCP in *ft* or *ds* mutant backgrounds, respectively.[75] Overexpressing Ft ECD in a wild-type background results in a dominant-negative phenotype, resulting in overgrowth like *ft* mutants.[75,76] This indicates that the Ft ECD might interfere with normal *ft* function, perhaps forming unproductive dimers with Ft or Ds. In the context of abdominal PCP, the Ft and Ds ECDs are more important for regulating polarity than the

ICDs, as fusion proteins of the Ft ECD with the Ds ICD, or the Ds ECD with the Ft ICD behave like Ft ECD alone or Ds ECD alone, respectively.[58]

Recent results suggest that the Ds ICD has signaling functions. In the eye, overexpression of Ds in a clone leads to increased activity of a *fj–lacZ* reporter around the clone boundary.[63] Expression of Ds ECD also has this effect. However, in a *ds* mutant background, overexpression of Ds in a clone leads to increased *fj–lacZ* reporter only inside the clone. The discontinuity at the clone border is thus detected only inside the clone, where *ds* is present. If the Ds ECD is expressed in a clone in the *ds* mutant background, no *fj–lacZ* reporter activity is observed. This suggests that while the Ds ECD is sufficient to create the boundary signal, the Ds ICD must be present to receive the signal.[63] The Ds ICD also has dominant-negative effects.[77]

Recent studies[77,110] have conducted extensive structure–function analyses of the Ft ICD to identify regions important for Ft activity in different pathways (Fig. 10.1). The authors of one study[77] searched for the minimal regions required to rescue abdominal polarity and wing disc growth phenotypes of *ft*. They found that an 82-aa region of the Ft ICD just C-terminal to the transmembrane region was sufficient to mostly rescue polarity. They termed this region the PCP domain of the Ft ICD. They further found that two regions near the middle of the Ft ICD, Hippo N (62 aa) and Hippo C (82 aa), were sufficient to mostly rescue growth. A 41-aa region called the PCP-Hippo (PH) domain, located roughly between the PCP and Hippo N domains, had a minor role in regulating both polarity and growth.

The other study[110] examined growth and polarity changes in the adult wing. The authors found that a region of just 4 amino acids at the extreme C-terminus of Ft was important in regulating polarity. The region of Ft they found primarily responsible for growth regulation was also different and more C-terminal from those found by Matakatsu and Blair. It is thus likely that different parts of Ft regulate polarity and growth in different developmental contexts.

The binding sites of many Ft interactors such as Atrophin, Dco, and Lowfat (Lft) are located near the C-terminus of Ft.[5,6,60,78] Surprisingly, a Ft construct (FtΔECDΔ6-C) lacking the binding sites for all these factors as well as the ECD was sufficient to rescue most abdominal polarity and wing disc growth phenotypes of *ft* mutant flies.[77] This suggests that at least when Ft constructs are overexpressed in these contexts, most of the known interactors of Ft are not needed for normal function, and that other effectors must be interacting with Ft.

In summation, both ECDs and ICDs of Ft and Ds have context-dependent roles in regulating polarity and growth control.

2.8. Interaction between Ft and Ds and polarity regulation

Cells in culture that are singly transfected with Ft or Ds are unable to adhere to each other, but cells that are cotransfected with both, or that are singly transfected and then mixed, do adhere, showing that Ft and Ds can mediate heterophilic interactions between cells.[51] In imaginal discs, Ft and Ds promote each other's membrane localization; loss of one leads to diffuse localization of the other.[79] Wild-type cells immediately surrounding a *ft* mutant clone show depletion of Ds, as its distribution changes outward to the side on which Ft is available for binding; the reverse is also true.[48] Ft molecules can also bind each other in *cis* and Ds promotes this clustering.[6] In the wing, Ft localizes to the proximal sides of cells, while Ds is on the distal sides, suggesting that binding across cells promotes their asymmetric localization.[8,52] In the mouse brain, loss of *fat4* reduces membrane localization of Ds1, and vice versa, consistent with the data from flies.[33] In contrast, Fat4 and Ds1 negatively regulate each other's levels in the lung and kidney.[45] Thus, Ft and Ds interact with each other across cell boundaries and in most contexts promote each other's membrane localization.

Fj phosphorylates specific cadherin repeats in the Ft and Ds ECDs (Fig. 10.1).[54,80] Biochemical and cell culture experiments show that phosphorylation by Fj increases the binding affinity of Ft for Ds, while decreasing the binding affinity of Ds for Ft.[59,81] The change in distribution of Ft and Ds around *fj* mutant or overexpression clones is consistent with these data. For example, loss of *fj* in a clone causes the Ds inside the clone to bind Ft with higher affinity, while the Ft inside the clone binds Ds with lower affinity; the Ds immediately inside the clone preferentially localizes to the clone border, where it can bind "high-affinity" Ft in the wild-type cells.[48,50]

While *ft* is expressed ubiquitously in imaginal discs, *ds* and *fj* are expressed in complementary gradients. For example, *ds* is expressed at a high level at the poles and at a low level near the equator, while *fj* is expressed in the reverse pattern. In the eye (Fig. 10.2B), these gradients are established by Wingless, JAK/STAT, and Notch signaling,[41,57] in the wing by Wingless and Dpp signaling,[82] and in the abdomen by Wingless and Hedgehog signaling.[66,83] As Ds binds Ft, and Fj modulates Ds and Ft binding, the expression gradients of *ds* and *fj* combine to create a gradient of Ft–Ds binding across the tissue. This modulates downstream signaling through Ft and Ds, leading to asymmetric distribution of Ft, Ds, and D in the wing.[8,52,62] Indeed, the gradients act redundantly as either one is sufficient to establish polarity. This is evident in the lack of polarity phenotype in *fj* mutant flies,[57] or when the expression gradients of either *fj* or *ds* are flattened.[84] Only when both gradients are abolished, or when a sharp discontinuity is created through clones, is polarity disrupted. Interestingly, loss of *fj* strengthens nonautonomous phenotypes of overexpressing *ft* or *ds*,[58] suggesting that the presence of the *fj* gradient helps to suppress the nonautonomous effects of clones.

Long-range transmission of polarity information downstream of Ft–Ds signaling has been proposed to occur both by a secreted signaling molecule and by changes in distribution of Ft, Ds, and D. There is evidence to support both models in different tissues, but more experiments are needed to clarify the situation.

Gradients of *fj* and *ds* and the changes in signaling at clone borders are also important for regulation of the Hippo pathway by Ft–Ds signaling (see also Section 3). Expression of Hippo pathway reporters is increased at boundaries of *ds* or *fj* clones, *d* and *ft* are required for these effects, and abolition of the gradient by uniform overexpression of *fj* or *ds* decreases growth.[63,85]

2.9. Interaction of Ft–Ds signaling with Fz/PCP

The Fz/PCP pathway consists of the transmembrane proteins Fz, Van Gogh (Vang, also known as Strabismus), and Flamingo (Fmi, also known as Starry night), as well as the cytoplasmic proteins Disheveled (Dsh), Diego (Dgo), and Prickle (Pk). Current models[20,39,111] for Fz/PCP function involves the formation of one complex by Fz, Dsh, and Dgo and another complex by Vang and Pk. Fmi is part of both complexes. Each complex is localized to the apical cell membrane but on opposite sides of the cell. For example, in the wing Fz/Dsh/Dgo is at distal boundaries whereas Vang/Pk is at the proximal borders. Each complex inhibits the other in its own cell but recruits the other one in the neighboring cell. This leads to establishment of polarity at the cellular level, and also enables polarity to propagate between cells, as loss of one factor changes localization of its binding partner in the neighboring cell. The nonautonomous phenotypes of Fz/PCP mutants result from propagation of this effect across several rows of cells.[86–88]

How do the Ft–Ds and Fz/PCP pathways interact with each other? Fz/PCP signaling leads to asymmetric localization of its components. In the *Drosophila* eye disc, asymmetric localization of Fmi is maintained in *ft* mutant tissue, though its directional orientation is disrupted.[41] Furthermore, mosaic analysis suggests that biasing of R3/4 fate specification by *ft* does not occur in tissue mutant for *fz*. These data suggest that in the eye, Ft–Ds signaling acts by biasing Fz/PCP asymmetry. However, other data suggest that Ds and Fz act in parallel during the early stages of polarity establishment in the eye. In particular, *ds fz* double mutant clones can cause nonautonomous polarity changes on both the polar side of the clone (as with *fz* mutants) and the equatorial side (as with *ds* mutants),[50] which would not be the case if *fz* was strictly downstream of *ds*.

The interaction between the two pathways also varies by tissue. In the *Drosophila* wing, Ft-Ds signaling is mostly upstream of Fz/PCP activity[48,51,64,65,70,107,108]. In the abdomen Ft-Ds and Fz/PCP act in parallel[58], with each pathway able to regulate polarity while the other is disrupted. In mice, loss of one copy of *vangl2* enhances *fat4* mutant phenotypes, suggesting that the two pathways have partially redundant functions in vertebrates.[40,44]

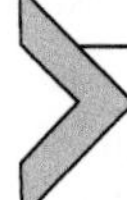

3. FAT–DACHSOUS, GROWTH CONTROL, AND THE HIPPO PATHWAY

3.1. Regulation of Hippo signaling by the Fat–Dachsous pathway

Proper development relies on tissues growing at an appropriate rate and ceasing growth when they have reached the required size. A well-conserved kinase cascade signaling pathway called the Hippo pathway regulates this process.[17,22,25,89] In *Drosophila*, the core of the Hippo pathway consists of the Ser/Thr kinases Warts (Wts) and Hippo (Hpo), and the adaptor proteins Salvador (Sav) and Mob as tumor suppressor (Mats). Signaling by the Hippo pathway results in phosphorylation of the transcriptional coactivator Yorkie (Yki), which promotes its binding to the 14-3-3 protein. 14-3-3 is localized in the cytoplasm, and so Hippo activity restricts Yki to the cytoplasm. In the absence of Hippo signaling, Yki is free to move to the nucleus, where it acts with transcription factors to control expression of a variety of proliferation and antiapoptosis effectors.[25] Experiments in *Drosophila* showed that *ft* is an upstream activator of Hippo signaling,[49,76,90,91] and that *ds*, *fj* and *d* also regulate Hippo signaling.[62,63,82,85] *ft* mutants phenocopy Hippo pathway mutants in hypertrophy, and in upregulating growth promoting factors such as Cyclin E and antiapoptotic factors such as Diap1. Ft has complex interactions with another upstream regulator of Hippo signaling, the protein Expanded (Ex), which is localized at the subapical membrane. Some reports suggest, based on changes in Ex localization in *ft* mutants and on the apparent lack of an additive phenotype in *ex ft* double mutants, that *ft* and *ex* act in the same pathway.[49,76,90] However, other reports suggest that *ft* and *ex* act in parallel, as the double mutant has an additive growth phenotype.[92]

Discs overgrown (Dco) is *Drosophila* casein kinase I δ/ε.[93] Dco binds the C-terminus of Ft and phosphorylates it.[5,6] This interaction is positively regulated by Ds, possibly through promoting formation of Ft homodimer clusters. Null mutants of *dco* have impaired growth of imaginal discs, reflecting its importance in multiple pathways. However, a hypomorphic allele of *dco*, *dco*^{3}, which cannot phosphorylate Ft, has growth phenotypes similar to *ft* mutants and upregulates the same Hippo target genes.[91,93] These effects can be rescued by overexpressing *wts*,[5] suggesting that Dco promotes Ft activity in the Hippo pathway.

Gradients of *ds* and *fj* can modulate Hippo activity through Ft and D.[63,85] *d* negatively regulates *wts* levels genetically downstream of *ft* and *dco*, suggesting that Ft primarily regulates Hippo signaling through repressing negative regulation of Wts by D[5,62,91] (see also Section 4).

3.2. Neurodegeration in *ft* and Hippo mutants

Loss of *ft* or overexpression of Atro in *Drosophila* eyes leads to degeneration of ommatidia.[30] This neurodegeneration can be reduced by loss of *d* or *yki*, while loss of *wts* or *sav* causes degeneration. *atro* and Hippo signaling both regulate autophagy,[94,95] and degenerating eyes in these mutants show an increased number of autophagic vesicles, suggesting that neurodegeneration in *ft* mutant or Atro-overexpressing eyes occurs through aberrant Hippo-mediated autophagy.[30]

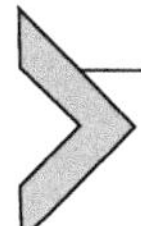

4. OTHER MEMBERS OF THE Ft–Ds PATHWAY: Dachs, ATROPHIN, AND Lowfat

The unconventional myosin *dachs (d)* suppresses *ft* phenotypes in polarity and growth control.[62,82,91] *ft* positively regulates *wts* levels while *d* negatively regulates *wts* levels, and *d* physically interacts with *wts*.[91] These data suggest that *ft* negatively regulates *d*, which in turn negatively regulates *wts*. This is supported by genetic epistasis studies placing *d* downstream of *ft* and *dco* but upstream of *wts*.[5,62,91] Thus, *d* is proposed to repress *wts* activity and promote growth, unless this repression is suppressed by *ft* activity.

Membrane localization of D is affected by changes in Ft–Ds signaling,[62] and asymmetric localization of D serves as a readout of Ft–Ds gradient activity.[8,52,61,110] D membrane localization is also influenced by *approximated*, a palmitoyltransferase that genetically interacts with *ft*.[51,96] Loss of *d* partially rescues polarity inversions in *ft* and *ds* mutant abdomen and eyes.[52,62]

d is also involved in cross talk between the regulations of PCP and Hippo signaling. Overexpression of *wts* in *ft* mutant eyes and wings can partially rescue *ft* mutant polarity defects,[52,92] and overexpression of *yki* can reverse polarity rescue in *ft d* double mutant eyes.[52] This suggests that regulation of *yki* activity by *ft* through *d* and *wts* influences planar polarity. How might this occur? Hippo signaling affects *fj* expression,[76,91] so a feedback loop might exist wherein changes in Hippo signaling alter Fj levels, which in turn modulates Ft and Ds binding to effect polarity. As the rescue of *ft* and *ds* polarity phenotypes by *d* is incomplete, and as *ds* overexpression can repolarize ommatidia in the absence of *d*[52], Ft–Ds signaling likely also regulates polarity independently of *d*.

Another Ft–Ds PCP effector is the transcriptional corepressor *atrophin (atro)*. *atro* mutants have defects in multiple developmental pathways, including eye and wing polarity.[97,98] *atro* phenocopies *ft* in regulating eye polarity, including nonautonomous polarity disruption on the polar side of mutant clones,[60] and in orienting cell division in response to cell death.[71] Atro physically interacts with the C-terminus of Ft and, like *ft*, *atro* biases R3/4 fate specification toward the R3 fate. *atro* and *ft* mutants both upregulate expression of the *fj–lacZ* reporter. Atro protein is known to localize to the nucleus.[97,98] Thus, *atro* could be a *d*-independent Ft–Ds effector, transducing Ft–Ds signals to the nucleus.

lowfat (lft) is a conserved gene that shows only very mild polarity and growth phenotypes on its own, but genetically interacts with *ft* and *ds* in these processes.[78] It also physically binds Ft and Ds and posttranscriptionally promotes their protein levels.

5. Ft–Ds SIGNALING AND VERTEBRATE DEVELOPMENT

Vertebrate Fat homologs are involved in a variety of developmental processes, some not obviously linked to PCP. Fat1 binds actin cytoskeleton regulators Mena and Vasp and influences cell–cell contacts and cytoskeletal polarity in cell culture.[99,100] *fat1* mutant mice die soon after birth and do not form proper glomerular slit junctions in kidneys.[101] In the developing retina, interneurons called amacrine cells lose their normal unipolar morphology in *fat3* mutant animals and instead acquire multipolar morphology.[102] *fat3* genetically interacts with the vertebrate *fj* ortholog *fjx1* in regulating this process.

fat4 mutant mice have extensive developmental defects and usually die at birth.[40,44,45] They have curly tails, small kidneys, lungs, and intestines, and broader neural tubes and cochleae. Parts of their skeleton are wider and shorter. The growth and orientation of hair cells in the cochleae are disrupted. Their kidneys are cystic and show reduced ureteric epithelium branching and defects in oriented cell division. Their hearts have atrial septation defects. Mouse *fat4* genetically interacts with other members of the Ft–Ds pathway in these processes.[40,44] Partial loss of *fat1* aggravates kidney cysts, cochlear widening, and hair cell polarity changes of *fat4* mutants. *fat1 fat4* double mutant mice display exencephaly, a cranial neural tube morphogenesis defect. Partial loss of *atn2l*, the vertebrate ortholog of *atro*, aggravates kidney cysts and hair cell phenotypes of *fat4* mutants. Loss of *fat4* increases *fjx1* expression, as is the case for *Drosophila ft* and *fj*.

fat4 also interacts with other pathways. Partial loss of *vangl2*, a member of the Fz/PCP pathway, aggravates kidney cysts, cochlear widening, and hair cell polarity changes of *fat4* mutant mice.[40,44] In the developing mouse cerebral cortex, Fat4 binds MUPP1 and Pals1, vertebrate homologs of *Drosophila* Patj and Stardust, respectively. MUPP1 and Pals1 regulate apical cell membrane organization, and knocking down Fat4 indeed disrupts apical architecture.[33]

Loss of mouse Dachsous1 (*dchs1*) phenocopies loss of *fat4*.[45] *dchs1* mutant mice mostly die at birth with curly tails, small kidneys, lungs, and intestines, broader neural tubes and cochleae, defects in their skeleton, abnormalities in some cochlear hair cells, and cystic kidneys. These phenotypes are similar to or slightly weaker than those of *fat4* mutants. Notably, they are not stronger in *fat4 dchs1* double mutants, suggesting that they act in the same pathway and that they interact as *Drosophila ft* and *ds* do.

Although Hippo signaling regulates organ size in vertebrates,[103,104] there is so far no clear link between Ft–Ds and Hippo mammalian orthologs. However, in zebrafish, a cystic pronephros phenotype caused by knockdown of *fat1* is rescued by knockdown of *yap1*, indicating that Ft and Hippo signaling can also interact in vertebrates.[105]

6. SUMMARY

Fat and Dachsous are critical for a variety of developmental processes in vertebrates and invertebrates. Ft–Ds signaling provides directional information to establish PCP and modulates Hippo signaling for appropriate growth control. However, important aspects of this pathway are still not understood. How do Ft and Ds affect localization or activity of effectors such as D and Atro? Is protein redistribution enough to transmit the Ft–Ds signal across multiple cells or is a secreted signal required? Is graded Ft–Ds activity important for Hippo signaling, or does it only play a permissive role[106]? What exact roles do Ft and Ds vertebrate homologs play, and to what degree is Ft–Ds signaling and its effect on polarity and growth conserved in vertebrates? Is the enormous size of Ft and Ds ECDs functionally significant? Answers to these questions will help illuminate the roles of these highly conserved adhesion molecules.

REFERENCES

1. Rock R, Schrauth S, Gessler M. Expression of mouse dchs1, fjx1, and fat-j suggests conservation of the planar cell polarity pathway identified in Drosophila. *Dev Dyn* 2005;**234**:747–55.
2. Tanoue T, Takeichi M. New insights into fat cadherins. *J Cell Sci* 2005;**118**:2347–53.
3. Mahoney PA, Weber U, Onofrechuk P, Biessmann H, Bryant PJ, Goodman CS. The fat tumor suppressor gene in Drosophila encodes a novel member of the cadherin gene superfamily. *Cell* 1991;**67**:853–68.
4. Jin X, Walker MA, Felsovalyi K, Vendome J, Bahna F, Mannepalli S, et al. Crystal structures of Drosophila N-cadherin ectodomain regions reveal a widely used class of Ca(2)+−free interdomain linkers. *Proc Natl Acad Sci USA* 2012;**109**:E127–E134.
5. Feng Y, Irvine KD. Processing and phosphorylation of the Fat receptor. *Proc Natl Acad Sci U S A* 2009;**106**:11989–94.
6. Sopko R, Silva E, Clayton L, Gardano L, Barrios-Rodiles M, Wrana J, et al. Phosphorylation of the tumor suppressor fat is regulated by its ligand Dachsous and the kinase discs overgrown. *Curr Biol* 2009;**19**:1112–7.
7. Clark HF, Brentrup D, Schneitz K, Bieber A, Goodman C, Noll M. Dachsous encodes a member of the cadherin superfamily that controls imaginal disc morphogenesis in Drosophila. *Genes Dev* 1995;**9**:1530–42.
8. Ambegaonkar AA, Pan G, Mani M, Feng Y, Irvine KD. Propagation of dachsous-fat planar cell polarity. *Curr Biol* 2012;**22**:1302–8.
9. Castillejo-Lopez C, Arias WM, Baumgartner S. The fat-like gene of Drosophila is the true orthologue of vertebrate fat cadherins and is involved in the formation of tubular organs. *J Biol Chem* 2004;**279**:24034–43.
10. Sadeqzadeh E, de Bock CE, Zhang XD, Shipman KL, Scott NM, Song C, et al. Dual processing of FAT1 cadherin protein by human melanoma cells generates distinct protein products. *J Biol Chem* 2011;**286**:28181–91.
11. Nakajima D, Nakayama M, Kikuno R, Hirosawa M, Nagase T, Ohara O. Identification of three novel non-classical cadherin genes through comprehensive analysis of large cDNAs. *Mol Brain Res* 2001;**94**:85–95.
12. Hong JC, Ivanov NV, Hodor P, Xia M, Wei N, Blevins R, et al. Identification of new human cadherin genes using a combination of protein motif search and gene finding methods. *J Mol Biol* 2004;**337**:307–17.
13. Mohr OL. Exaggeration and inhibition phenomena encountered in the analysis of an autosomal dominant. *Z Indukt Abstamm Vererblehre* 1929;**50**:113–200.
14. Stern C, Bridges CB. The mutants of the extreme left end of the second chromosome of Drosophila melanogaster. *Genetics* 1926;**11**:503–30.
15. Bryant PJ, Huettner B, Held Jr LI, Ryerse J, Szidonya J. Mutations at the fat locus interfere with cell proliferation control and epithelial morphogenesis in Drosophila. *Dev Biol* 1988;**129**:541–54.
16. Goodrich LV, Strutt D. Principles of planar polarity in animal development. *Development* 2011;**138**:1877–92.
17. Grusche FA, Richardson HE, Harvey KF. Upstream regulation of the hippo size control pathway. *Curr Biol* 2010;**20**:R574–R582.
18. Thomas C, Strutt D. The roles of the cadherins Fat and Dachsous in planar polarity specification in Drosophila. *Dev Dyn* 2012;**241**:27–39.

19. Saburi S, McNeill H. Organising cells into tissues: new roles for cell adhesion molecules in planar cell polarity. *Curr Opin Cell Biol* 2005;**17**:482–8.
20. Maung SM, Jenny A. Planar cell polarity in Drosophila. *Organogenesis* 2011;**7**:165–79.
21. Gray RS, Roszko I, Solnica-Krezel L. Planar cell polarity: coordinating morphogenetic cell behaviors with embryonic polarity. *Dev Cell* 2011;**21**:120–33.
22. Staley BK, Irvine KD. Hippo signaling in Drosophila: recent advances and insights. *Dev Dyn* 2012;**241**:3–15.
23. Hergovich A. Mammalian Hippo signalling: a kinase network regulated by protein-protein interactions. *Biochem Soc Trans* 2012;**40**:124–8.
24. Zhao B, Tumaneng K, Guan KL. The Hippo pathway in organ size control, tissue regeneration and stem cell self-renewal. *Nat Cell Biol* 2011;**13**:877–83.
25. Halder G, Johnson RL. Hippo signaling: growth control and beyond. *Development* 2011;**138**:9–22.
26. Sopko R, McNeill H. The skinny on Fat: an enormous cadherin that regulates cell adhesion, tissue growth, and planar cell polarity. *Curr Opin Cell Biol* 2009;**21**:717–23.
27. Dearborn Jr R, Kunes S. An axon scaffold induced by retinal axons directs glia to destinations in the Drosophila optic lobe. *Development* 2004;**131**:2291–303.
28. Kawamori H, Tai M, Sato M, Yasugi T, Tabata T. Fat/Hippo pathway regulates the progress of neural differentiation signaling in the Drosophila optic lobe. *Dev Growth Differ* 2011;**53**:653–67.
29. Reddy BV, Rauskolb C, Irvine KD. Influence of fat-hippo and notch signaling on the proliferation and differentiation of Drosophila optic neuroepithelia. *Development* 2010;**137**:2397–408.
30. Napoletano F, Occhi S, Calamita P, Volpi V, Blanc E, Charroux B, et al. Polyglutamine Atrophin provokes neurodegeneration in Drosophila by repressing fat. *EMBO J* 2011;**30**:945–58.
31. Karpowicz P, Perez J, Perrimon N. The Hippo tumor suppressor pathway regulates intestinal stem cell regeneration. *Development* 2010;**137**:4135–45.
32. Hamaratoglu F, Gajewski K, Sansores-Garcia L, Morrison C, Tao C, Halder G. The Hippo tumor-suppressor pathway regulates apical-domain size in parallel to tissue growth. *J Cell Sci* 2009;**122**:2351–9.
33. Ishiuchi T, Misaki K, Yonemura S, Takeichi M, Tanoue T. Mammalian Fat and Dachsous cadherins regulate apical membrane organization in the embryonic cerebral cortex. *J Cell Biol* 2009;**185**:959–67.
34. Chosdol K, Misra A, Puri S, Srivastava T, Chattopadhyay P, Sarkar C, et al. Frequent loss of heterozygosity and altered expression of the candidate tumor suppressor gene 'FAT' in human astrocytic tumors. *BMC Cancer* 2009;**9**:5.
35. Nakaya K, Yamagata HD, Arita N, Nakashiro KI, Nose M, Miki T, et al. Identification of homozygous deletions of tumor suppressor gene FAT in oral cancer using CGH-array. *Oncogene* 2007;**26**:5300–8.
36. Qi C, Zhu YT, Hu L, Zhu YJ. Identification of Fat4 as a candidate tumor suppressor gene in breast cancers. *Int J Cancer* 2009;**124**:793–8.
37. Settakorn J, Kaewpila N, Burns GF, Leong AS. FAT, E-cadherin, beta catenin, HER 2/neu, Ki67 immuno-expression, and histological grade in intrahepatic cholangiocarcinoma. *J Clin Pathol* 2005;**58**:1249–54.
38. Blair IP, Chetcuti AF, Badenhop RF, Scimone A, Moses MJ, Adams LJ, et al. Positional cloning, association analysis and expression studies provide convergent evidence that the cadherin gene FAT contains a bipolar disorder susceptibility allele. *Mol Psychiatry* 2006;**11**:372–83.
39. Bayly R, Axelrod JD. Pointing in the right direction: new developments in the field of planar cell polarity. *Nat Rev Genet* 2011;**12**:385–91.

40. Saburi S, Hester I, Fischer E, Pontoglio M, Eremina V, Gessler M, et al. Loss of Fat4 disrupts PCP signaling and oriented cell division and leads to cystic kidney disease. *Nat Genet* 2008;**40**:1010–5.
41. Yang CH, Axelrod JD, Simon MA. Regulation of Frizzled by fat-like cadherins during planar polarity signaling in the Drosophila compound eye. *Cell* 2002;**108**:675–88.
42. Baena-Lopez LA, Baonza A, Garcia-Bellido A. The orientation of cell divisions determines the shape of Drosophila organs. *Curr Biol* 2005;**15**:1640–4.
43. Ciruna B, Jenny A, Lee D, Mlodzik M, Schier AF. Planar cell polarity signalling couples cell division and morphogenesis during neurulation. *Nature* 2006;**439**:220–4.
44. Saburi S, Hester I, Goodrich L, McNeill H. Functional interactions between Fat family cadherins in tissue morphogenesis and planar polarity. *Development* 2012;**139**:1806–20.
45. Mao Y, Mulvaney J, Zakaria S, Yu T, Morgan KM, Allen S, et al. Characterization of a Dchs1 mutant mouse reveals requirements for Dchs1-Fat4 signaling during mammalian development. *Development* 2011;**138**:947–57.
46. Gubb D, Garcia-Bellido A. A genetic analysis of the determination of cuticular polarity during development in Drosophila melanogaster. *J Embryol Exp Morphol* 1982;**68**:37–57.
47. Rawls AS, Guinto JB, Wolff T. The cadherins fat and dachsous regulate dorsal/ventral signaling in the Drosophila eye. *Curr Biol* 2002;**12**:1021–6.
48. Ma D, Yang CH, McNeill H, Simon MA, Axelrod JD. Fidelity in planar cell polarity signalling. *Nature* 2003;**421**:543–7.
49. Silva E, Tsatskis Y, Gardano L, Tapon N, McNeill H. The tumor-suppressor gene fat controls tissue growth upstream of expanded in the hippo signaling pathway. *Curr Biol* 2006;**16**:2081–9.
50. Strutt H, Strutt D. Nonautonomous planar polarity patterning in Drosophila: dishevelled-independent functions of frizzled. *Dev Cell* 2002;**3**:851–63.
51. Matakatsu H, Blair SS. Interactions between Fat and Dachsous and the regulation of planar cell polarity in the Drosophila wing. *Development* 2004;**131**:3785–94.
52. Brittle A, Thomas C, Strutt D. Planar polarity specification through asymmetric subcellular localization of Fat and Dachsous. *Curr Biol* 2012;**22**:907–14.
53. Brasch J, Harrison OJ, Honig B, Shapiro L. Thinking outside the cell: how cadherins drive adhesion. *Trends Cell Biol* 2012;**22**:299–310.
54. Ishikawa HO, Takeuchi H, Haltiwanger RS, Irvine KD. Four-jointed is a golgi kinase that phosphorylates a subset of cadherin domains. *Science* 2008;**321**:401–4.
55. Brodsky MH, Steller H. Positional information along the dorsal-ventral axis of the Drosophila eye: graded expression of the four-jointed gene. *Dev Biol* 1996;**173**:428–46.
56. Villano JL, Katz FN. Four-jointed is required for intermediate growth in the proximal-distal axis in Drosophila. *Development* 1995;**121**:2767–77.
57. Zeidler MP, Perrimon N, Strutt DI. The four-jointed gene is required in the Drosophila eye for ommatidial polarity specification. *Curr Biol* 1999;**9**:1363–72.
58. Casal J, Lawrence PA, Struhl G. Two separate molecular systems, Dachsous/Fat and Starry night/Frizzled, act independently to confer planar cell polarity. *Development* 2006;**133**:4561–72.
59. Simon MA, Xu A, Ishikawa HO, Irvine KD. Modulation of fat:dachsous binding by the cadherin domain kinase four-jointed. *Curr Biol* 2010;**20**:811–7.
60. Fanto M, Clayton L, Meredith J, Hardiman K, Charroux B, Kerridge S, et al. The tumor-suppressor and cell adhesion molecule Fat controls planar polarity via physical interactions with Atrophin, a transcriptional co-repressor. *Development* 2003;**130**:763–74.
61. Bosveld F, Bonnet I, Guirao B, Tlili S, Wang Z, Petitalot A, et al. Mechanical control of morphogenesis by Fat/Dachsous/Four-jointed planar cell polarity pathway. *Science* 2012;**336**:724–7.

62. Mao Y, Rauskolb C, Cho E, Hu WL, Hayter H, Minihan G, et al. Dachs: an unconventional myosin that functions downstream of Fat to regulate growth, affinity and gene expression in Drosophila. *Development* 2006;**133**:2539–51.
63. Willecke M, Hamaratoglu F, Sansores-Garcia L, Tao C, Halder G. Boundaries of Dachsous Cadherin activity modulate the Hippo signaling pathway to induce cell proliferation. *Proc Natl Acad Sci USA* 2008;**105**:14897–902.
64. Adler PN, Charlton J, Liu JC. Mutations in the cadherin superfamily member gene dachsous cause a tissue polarity phenotype by altering frizzled signaling. *Development* 1998;**125**:959–68.
65. Hogan J, Valentine M, Cox C, Doyle K, Collier S. Two frizzled planar cell polarity signals in the Drosophila wing are differentially organized by the Fat/Dachsous pathway. *PLoS Genet* 2011;**7**:e1001305.
66. Casal J, Struhl G, Lawrence PA. Developmental compartments and planar polarity in Drosophila. *Curr Biol* 2002;**12**:1189–98.
67. Donoughe S, DiNardo S. Dachsous and frizzled contribute separately to planar polarity in the Drosophila ventral epidermis. *Development* 2011;**138**:2751–9.
68. Repiso A, Saavedra P, Casal J, Lawrence PA. Planar cell polarity: the orientation of larval denticles in Drosophila appears to depend on gradients of Dachsous and Fat. *Development* 2010;**137**:3411–5.
69. Fischer E, Legue E, Doyen A, Nato F, Nicolas JF, Torres V, et al. Defective planar cell polarity in polycystic kidney disease. *Nat Genet* 2006;**38**:21–3.
70. Aigouy B, Farhadifar R, Staple DB, Sagner A, Roper JC, Julicher F, et al. Cell flow reorients the axis of planar polarity in the wing epithelium of Drosophila. *Cell* 2010;**142**:773–86.
71. Li W, Kale A, Baker NE. Oriented cell division as a response to cell death and cell competition. *Curr Biol* 2009;**19**:1821–6.
72. Mao Y, Tournier AL, Bates PA, Gale JE, Tapon N, Thompson BJ. Planar polarization of the atypical myosin Dachs orients cell divisions in Drosophila. *Genes Dev* 2011;**25**:131–6.
73. Viktorinova I, Konig T, Schlichting K, Dahmann C. The cadherin Fat2 is required for planar cell polarity in the Drosophila ovary. *Development* 2009;**136**:4123–32.
74. Viktorinova I, Pismen LM, Aigouy B, Dahmann C. Modelling planar polarity of epithelia: the role of signal relay in collective cell polarization. *J R Soc Interface* 2011;**8**:1059–63.
75. Matakatsu H, Blair SS. Separating the adhesive and signaling functions of the Fat and Dachsous protocadherins. *Development* 2006;**133**:2315–24.
76. Willecke M, Hamaratoglu F, Kango-Singh M, Udan R, Chen CL, Tao C, et al. The fat cadherin acts through the hippo tumor-suppressor pathway to regulate tissue size. *Curr Biol* 2006;**16**:2090–100.
77. Matakatsu H, Blair SS. Separating planar cell polarity and Hippo pathway activities of the protocadherins Fat and Dachsous. *Development* 2012;**139**:1498–508.
78. Mao Y, Kucuk B, Irvine KD. Drosophila lowfat, a novel modulator of Fat signaling. *Development* 2009;**136**:3223–33.
79. Strutt D, Johnson R, Cooper K, Bray S. Asymmetric localization of frizzled and the determination of notch-dependent cell fate in the Drosophila eye. *Curr Biol* 2002;**12**:813–24.
80. Strutt H, Mundy J, Hofstra K, Strutt D. Cleavage and secretion is not required for Four-jointed function in Drosophila patterning. *Development* 2004;**131**:881–90.
81. Brittle AL, Repiso A, Casal J, Lawrence PA, Strutt D. Four-jointed modulates growth and planar polarity by reducing the affinity of dachsous for fat. *Curr Biol* 2010;**20**:803–10.
82. Cho E, Irvine KD. Action of fat, four-jointed, dachsous and dachs in distal-to-proximal wing signaling. *Development* 2004;**131**:4489–500.

83. Lawrence PA, Casal J, Struhl G. Towards a model of the organisation of planar polarity and pattern in the Drosophila abdomen. *Development* 2002;**129**:2749–60.
84. Simon MA. Planar cell polarity in the Drosophila eye is directed by graded four-jointed and Dachsous expression. *Development* 2004;**131**:6175–84.
85. Rogulja D, Rauskolb C, Irvine KD. Morphogen control of wing growth through the Fat signaling pathway. *Dev Cell* 2008;**15**:309–21.
86. Chen WS, Antic D, Matis M, Logan CY, Povelones M, Anderson GA, et al. Asymmetric homotypic interactions of the atypical cadherin flamingo mediate intercellular polarity signaling. *Cell* 2008;**133**:1093–105.
87. Wu J, Mlodzik M. The frizzled extracellular domain is a ligand for Van Gogh/Stbm during nonautonomous planar cell polarity signaling. *Dev Cell* 2008;**15**:462–9.
88. Strutt H, Warrington SJ, Strutt D. Dynamics of core planar polarity protein turnover and stable assembly into discrete membrane subdomains. *Dev Cell* 2011;**20**:511–25.
89. Bao Y, Hata Y, Ikeda M, Withanage K. Mammalian Hippo pathway: from development to cancer and beyond. *J Biochem* 2011;**149**:361–79.
90. Bennett FC, Harvey KF. Fat cadherin modulates organ size in Drosophila via the Salvador/Warts/Hippo signaling pathway. *Curr Biol* 2006;**16**:2101–10.
91. Cho E, Feng Y, Rauskolb C, Maitra S, Fehon R, Irvine KD. Delineation of a Fat tumor suppressor pathway. *Nat Genet* 2006;**38**:1142–50.
92. Feng Y, Irvine KD. Fat and expanded act in parallel to regulate growth through warts. *Proc Natl Acad Sci USA* 2007;**104**:20362–7.
93. Zilian O, Frei E, Burke R, Brentrup D, Gutjahr T, Bryant PJ, et al. Double-time is identical to discs overgrown, which is required for cell survival, proliferation and growth arrest in Drosophila imaginal discs. *Development* 1999;**126**:5409–20.
94. Nisoli I, Chauvin JP, Napoletano F, Calamita P, Zanin V, Fanto M, et al. Neurodegeneration by polyglutamine Atrophin is not rescued by induction of autophagy. *Cell Death Differ* 2010;**17**:1577–87.
95. Dutta S, Baehrecke EH. Warts is required for PI3K-regulated growth arrest, autophagy, and autophagic cell death in Drosophila. *Curr Biol* 2008;**18**:1466–75.
96. Matakatsu H, Blair SS. The DHHC palmitoyltransferase approximated regulates Fat signaling and Dachs localization and activity. *Curr Biol* 2008;**18**:1390–5.
97. Erkner A, Roure A, Charroux B, Delaage M, Holway N, Core N, et al. Grunge, related to human Atrophin-like proteins, has multiple functions in Drosophila development. *Development* 2002;**129**:1119–29.
98. Zhang S, Xu L, Lee J, Xu T. Drosophila atrophin homolog functions as a transcriptional corepressor in multiple developmental processes. *Cell* 2002;**108**:45–56.
99. Moeller MJ, Soofi A, Braun GS, Li X, Watzl C, Kriz W, et al. Protocadherin FAT1 binds Ena/VASP proteins and is necessary for actin dynamics and cell polarization. *EMBO J* 2004;**23**:3769–79.
100. Tanoue T, Takeichi M. Mammalian Fat1 cadherin regulates actin dynamics and cell-cell contact. *J Cell Biol* 2004;**165**:517–28.
101. Ciani L, Patel A, Allen ND, ffrench-Constant C. Mice lacking the giant protocadherin mFAT1 exhibit renal slit junction abnormalities and a partially penetrant cyclopia and anophthalmia phenotype. *Mol Cell Biol* 2003;**23**:3575–82.
102. Deans MR, Krol A, Abraira VE, Copley CO, Tucker AF, Goodrich LV. Control of neuronal morphology by the atypical cadherin Fat3. *Neuron* 2011;**71**:820–32.
103. Camargo FD, Gokhale S, Johnnidis JB, Fu D, Bell GW, Jaenisch R, et al. YAP1 increases organ size and expands undifferentiated progenitor cells. *Curr Biol* 2007;**17**:2054–60.
104. Dong J, Feldmann G, Huang J, Wu S, Zhang N, Comerford SA, et al. Elucidation of a universal size-control mechanism in Drosophila and mammals. *Cell* 2007;**130**:1120–33.

105. Skouloudaki K, Puetz M, Simons M, Courbard JR, Boehlke C, Hartleben B, et al. Scribble participates in Hippo signaling and is required for normal zebrafish pronephros development. *Proc Natl Acad Sci USA* 2009;**106**:8579–84.
106. Lawrence PA, Struhl G, Casal J. Do the protocadherins Fat and Dachsous link up to determine both planar cell polarity and the dimensions of organs? *Nat Cell Biol* 2008;**10**:1379–82.
107. Harumoto T, Ito M, Shimada Y, Kobayashi TJ, Ueda HR, Lu B, Uemura T. Atypical cadherins Dachsous and Fat control dynamics of noncentrosomal microtubules in planar cell polarity. *Developmental cell.* 2010;**19**:389–401.
108. Sagner A, Merkel M, Aigouy B, Gaebel J, Brankatschk M, Jülicher F, Eaton S. Establishment of global patterns of planar polarity during growth of the Drosophila wing epithelium. *Current biology: CB.* 2012;**22**:1296–1301.
109. Marcinkevicius E, Zallen JA. Regulation of cytoskeletal organization and junctional remodeling by the atypical cadherin Fat. *Development.* 2013;**140**:433–43.
110. Pan G, Feng Y, Ambegaonkar AA, Sun G, Huff M, Rauskolb C, Irvine KD. Signal transduction by the Fat cytoplasmic domain. *Development.* 2013;**140**:831–42.
111. Struhl G, Casal J, Lawrence PA. Dissecting the molecular bridges that mediate the function of Frizzled in planar cell polarity. *Development.* 2012;**139**:3665–74.

Functions of Cadherins in Development, Health, and Disease

CHAPTER ELEVEN

Cadherins and Their Partners in the Nematode Worm *Caenorhabditis elegans*

Jeff Hardin[*,†], Allison Lynch[*], Timothy Loveless[†], Jonathan Pettitt[‡]
[*]Department of Zoology, University of Wisconsin, Madison, Wisconsin, USA
[†]Program in Cellular and Molecular Biology, University of Wisconsin, Madison, Wisconsin, USA
[‡]Department of Molecular and Cell Biology, University of Aberdeen Institute of Medical Sciences, Aberdeen, United Kingdom

Contents

Progress in Molecular Biology and Translational Science, Volume 116
ISSN 1877-1173
http://dx.doi.org/10.1016/B978-0-12-394311-8.00011-X

Abstract

The extreme simplicity of *Caenorhabditis elegans* makes it an ideal system to study the basic principles of cadherin function at the level of single cells within the physiologically relevant context of a developing animal. The genetic tractability of *C. elegans* also means that components of cadherin complexes can be identified through genetic modifier screens, allowing a comprehensive *in vivo* characterization of the macromolecular assemblies involved in cadherin function during tissue formation and maintenance in *C. elegans*. This work shows that a single cadherin system, the classical cadherin–catenin complex, is essential for diverse morphogenetic events during embryogenesis through its interactions with a range of mostly conserved proteins that act to modulate its function. The role of other members of the cadherin family in *C. elegans*, including members of the Fat-like, Flamingo/CELSR and calsyntenin families is less well characterized, but they have clear roles in neuronal development and function.

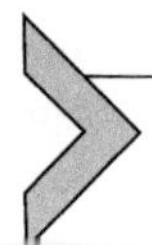

1. *CAENORHABDITIS ELEGANS*: A MODEL FOR STUDYING CADHERIN FUNCTION *IN VIVO*

The relatively small number of cadherin superfamily proteins encoded in the *C. elegans* genome, combined with its well-known transparency, simplicity of organization, and robust forward and reverse genetics makes *C. elegans* an appealing model organism for understanding how cadherins and their binding partners function in defined cellular contexts. *C. elegans* has 12 genes encoding 13 cadherins.[1–3] *C. elegans* has representatives of most of the main cadherin families that are conserved throughout the metazoa, including members of the classical cadherin, Fat-like cadherin, Dachsous, Flamingo/CELSR, and Calsytenin families (Fig. 11.1). *C. elegans* lacks protocadherins,[4] a feature that it shares with *Drosophila*; however, the presence of this family in other major invertebrate groups suggests that they may have been lost in the Ecdysozoa.[4] As is the case for other invertebrates and nonvertebrate chordates, *C. elegans* lacks desmosomal cadherins.[5] While overall, single loss-of-function experiments suggest that most cadherins have surprisingly limited functions (J. Pettitt, unpublished[6]), the classical cadherin/catenin complex of *C. elegans* is crucial for morphogenesis. We discuss it and other cadherins with known functions in the following sections.

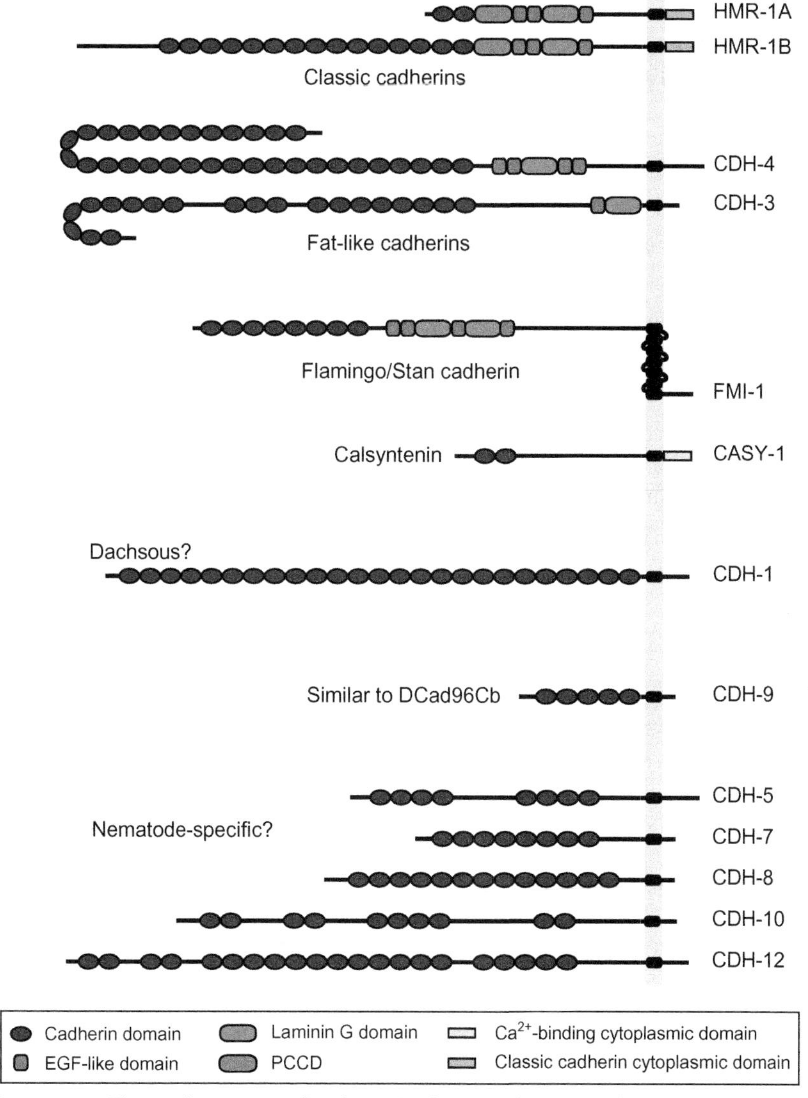

Figure 11.1 The cadherin superfamily in *C. elegans*. The 13 *C. elegans* cadherins are grouped according to their structural similarity with cadherins from other organisms. Each cadherin is positioned with its N-terminus to the left. PCCD, primitive classic cadherin domain. *Adapted from Ref. 3 with permission.* (See Color Insert.)

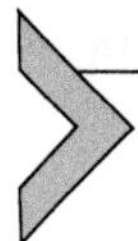

2. THE CLASSICAL CADHERIN/CATENIN COMPLEX IN *C. ELEGANS*

Like all nonchordates characterized to date, *C. elegans* has single representatives of the α- and p120-catenin families, and these are encoded by the *hmp-1* and *jac-1* genes, respectively.[7,8] Unusually, it possesses four β-catenin homologs, but only one, HMP-2, is known to participate in CCC junction formation (Fig. 11.2A). Its single, type III classical cadherin gene, *hmr-1*, encodes two proteins, HMR-1A and HMR-1B, which differ in their ectodomains (Fig. 11.1), and are expressed by alternative splicing and alternative promoter use.[9] HMR-1A is expressed in all epithelia (Fig. 11.2B) plus some neurons, while HMR-1B appears to be confined to neurons. Thus, *C. elegans* has both epithelial and neuronal classical cadherins, but their origin from a single locus appears to be unique to Caenorhabditids. Based on colocalization and epistasis,[7] as well as molecular interactions,[8,10,11]

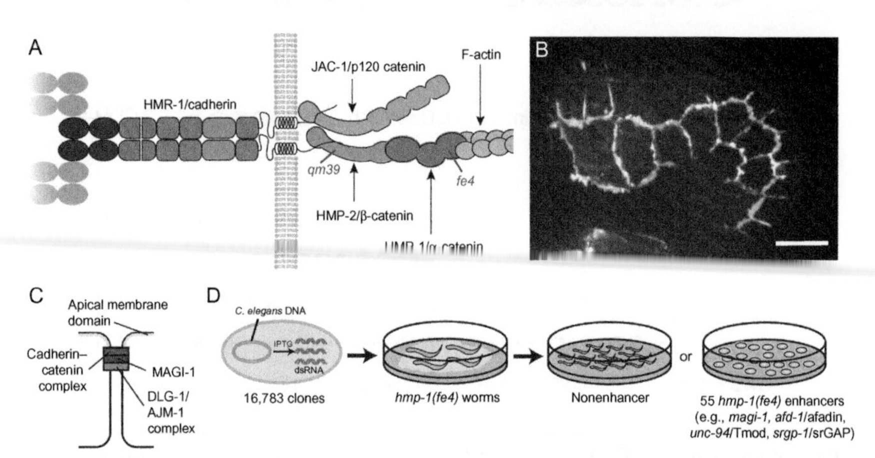

Figure 11.2 The core components of the *C. elegans* classical cadherin/catenin complex (CCC). (A) HMR-1A is the epithelial cadherin, HMP-2 the junctional β-catenin, HMP-1 the *C. elegans* α-catenin, and JAC-1 the p120-catenin homolog. (B) A comma stage embryo expressing *hmr-1::gfp* (image courtesy of B. Lucas). Bar = 10 μm. (C) Schematic of the *C. elegans* apical junction. MAGI-1 localizes to a domain between the CCC and the DLG-1/AJM-1 complex. (D) Genome-wide screening in *hmp-1(fe4)* mutants identifies CCC partners during morphogenesis. A library of 16,783 RNAi feeding clones was screened to identify enhancers of a hypomorphic allele of α-catenin, *hmp-1(fe4)*. Each bacterial clone was fed to wild-type and *hmp-1(fe4)* worms and phenotypes were assessed in the next generation. Enhancers increased lethality in *hmp-1(fe4)* embryos to >83% and caused <10% lethality in wild-type embryos. Fifty-five enhancers were identified, including *magi-1*, *afd-1*/afadin/Canoe, *srgp-1*/srGAP, and *unc-94*/Tropomodulin. *Panel A: Adapted from Ref. 3; C-D adapted from Ref. 71, with permission.* (See Color Insert.)

HMR-1A, HMP-1, HMP-2, and JAC-1 form a CCC that is a component of junctions in all *C. elegans* epithelia.

While *C. elegans* has a core CCC, many proteins associated with the vertebrate CCC are missing, including vinculin and α-actinin.[12,13] Other effectors, such as *C. elegans* Enabled, have surprisingly minor roles (see below). Such surprises suggest that *C. elegans* may be extremely useful as a "minimalist" system for examining the core features of the CCC. Morphologically, epithelial junctions in *C. elegans* are minimalist as well. Unlike in *Drosophila* and vertebrates, they do not possess structural specializations that lead to different electron-dense structures. Instead, the *C. elegans* apical junction possesses a single electron-dense structure, with a stratified set of molecular complexes, including the CCC (Fig. 11.2C).[14]

2.1. HMR-1A: A classical cadherin required for cell–cell adhesion

HMR-1 proteins possess all of the canonical domains of other invertebrate classical cadherins. HMR-1A has two extracellular cadherin (EC) repeats as well as a primitive classical cadherin domain (PCCD; see Fig. 11.1), which, together with EGF-like and laminin G repeats, lies between the EC repeats and the transmembrane helix. The PCCD is proteolytically cleaved during the maturation of *Drosophila* E-cadherin[15]; this has not been shown for HMR-1.

HMR-1 behaves like a *bona fide* classical cadherin. Based on yeast two-hybrid[8,10] and biochemical assays,[11] the HMR-1 C-terminus can bind HMP-2. Significantly, this association is only efficient when HMR-1 is prephosphorylated by casein kinase II.[11] Mammalian E-cadherin's association with β-catenin increases in a similar, dramatic fashion when it is phosphorylated at a key serine (S840 in human E-cadherin[16]; see also Ref. 17). A candidate serine in the same location in HMR-1 (S1212) could potentially serve the same function in *C. elegans*.

HMR-1 also contains a well-conserved juxtamembrane domain that binds JAC-1/p120-catenin.[8] p120-catenin is an important regulator of trafficking of the CCC in vertebrates.[18] Current evidence does not favor a major role for JAC-1 in similar events in *C. elegans* (see below); it remains to be determined whether HMR-1 undergoes regulation via endocytic trafficking in ways that are similar to mammalian classical cadherins.[19]

2.2. HMR-1A regulates cell ingression during gastrulation

C. elegans gastrulation begins with the ingression of the endodermal precursors, Ea and Ep (Fig. 11.3A). This process is partly driven by actomyosin-mediated

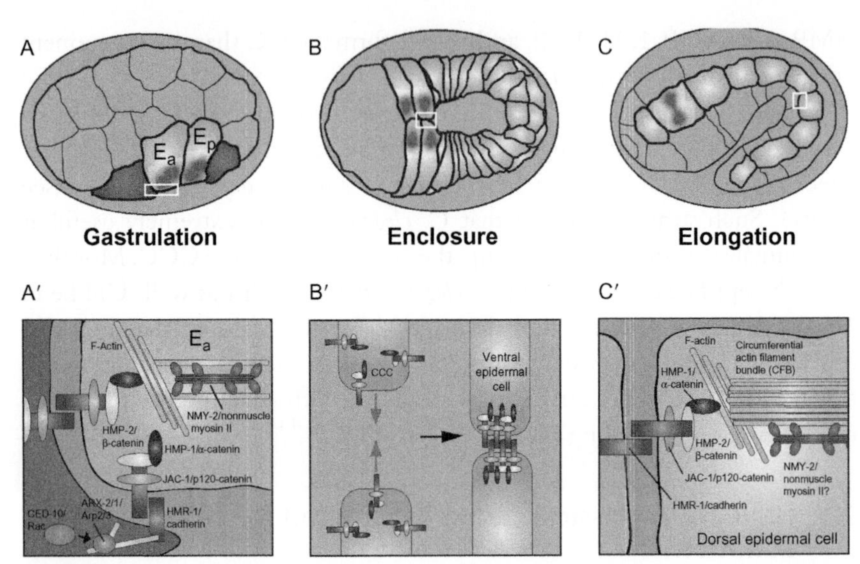

Figure 11.3 Roles of the classical CCC in epidermal morphogenesis. Developmental stages are depicted in A–C, and postulated subcellular events in the boxed regions are depicted in A′–C′. A, A′. Gastrulation involves internalization of the endodermal precursors, Ea and Ep, via apical constriction (A, arrows), which internalizes Ea and Ep and aids covering of the space they vacate by pulling on neighboring cells. Normal apical constriction requires the CCC, as well as SAX-7/L1CAM, and components of Rac signaling. (B, B′) Ventral enclosure involves migration of epithelial cells toward the ventral midline (B, arrows). The first cells to meet at the midline (leading cells) rapidly make cadherin-dependent contacts. (C, C′) Elongation of the embryo from a bean to a worm shape requires coordinated, actomyosin-mediated changes in the shape of epidermal cells (C, arrows). The CCC is required for attachment of actin, including circumferential filament bundles (CFBs) to the cell surface, and hence for transmission of forces to the cell surface. HMP-1/α-catenin is particularly crucial in this process. Anterior is to the left in each panel. *Adapted from Ref. 20 with permission.* (See Color Insert.)

apical constriction, which helps to pull adjacent cells over the apical surface of Ea/Ep and displaces them internally.[21] Cadherin-based adhesion appears to be involved in this process because knockdown of maternal and zygotic *hmr-1* mRNA by RNAi reduces the rate of Ea/Ep apical constriction by approximately 50%.[22] In such embryos with reduced HMR-1 function, endodermal precursors show diminished membrane movements concomitant with myosin translocation.[22] Given the role of the CCC in stabilizing F-actin linkages to adherens junctions in other tension-driven processes,[23,24] HMR-1 may stabilize the cytoskeletal machinery required for apical constriction (Fig. 11.3A′). Consistent with this possibility, the CCC interacts genetically with the Rac pathway in endodermal precursors. Knockdown of CCC components along with Rac pathway components abrogates Ea/Ep apical constriction without loss of apical nonmuscle myosin.[6,22] Rac may function as part of a molecular "clutch," linking actomyosin contractions to the CCC within Ea/Ep. Alternatively, the CCC may synergize with Rac by providing adhesive traction for Arp2/3-dependent lamellipodia extended by neighboring mesodermal precursors.[25,26]

HMR-1 is also involved in ingression of primordial germ cells (PGCs). Following Ea/Ep ingression, PGCs maintain contact with endodermal daughters, as they are drawn into the interior by the dorsal movement of the endoderm.[27] PGC internalization requires surface enrichment of HMR-1 in the PGCs; such enrichment appears to be mediated post-transcriptionally via a specific element in the 3′-UTR of the *hmr-1* mRNA. Surprisingly, however, HMR-1 expression in PGCs alone is both necessary and sufficient for their internalization.[27] This suggests that HMR-1 participates in heterotypic interactions in this context. Confirming this interesting possibility and identifying the heterotypic binding partner involved await additional experiments.

2.3. HMR-1A and the CCC are crucial during ventral enclosure and embryonic elongation

Postgastrula morphogenesis in *C. elegans* involves three major events in the epidermis: dorsal intercalation, ventral enclosure, and elongation.[28] To date, no function for the CCC during dorsal intercalation has been identified. In contrast, the CCC is crucial for ventral enclosure and elongation.[7] During enclosure, the free, ventral edges of the epidermis migrate around the embryo toward the ventral midline, where they meet and form nascent junctions[29–31] (Fig. 11.3B and B′). These events require the CCC: zygotic loss of HMR-1 or maternal + zygotic removal of HMP-2 and HMP-1 leads to failure of ventral enclosure and the Hammerhead (Hmr) phenotype, as anterior cells spill out of the embryo[7,29] (Fig. 11.4A and B). Zygotic *hmp-1* and *hmp-2* mutants enclose, but as the embryo elongates, it develops characteristic dorsal folds, known as the Humpback (Hmp) phenotype[7] (Fig. 11.4C), which results from the detachment of circumferential filament bundles (CFBs; Fig. 11.3C and C′) from the junction-associated actin network (Fig. 11.4D–F). CFBs presumably transmit forces generated by actomyosin-mediated contraction in the epidermis (reviewed in Ref. 28).

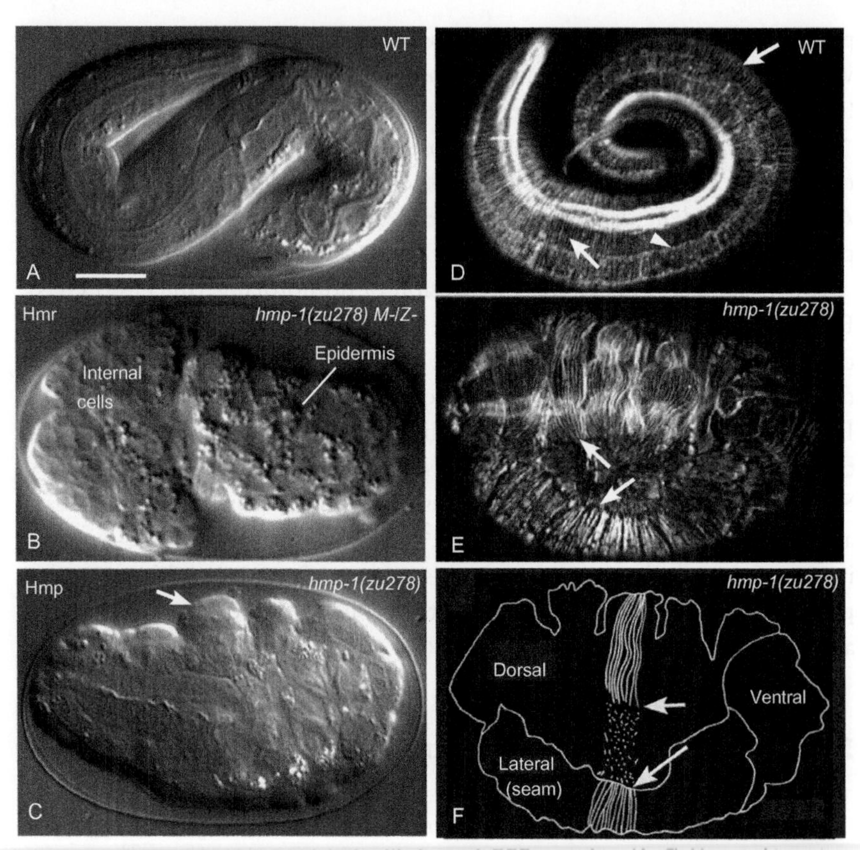

Figure 11.4 Phenotypes associated with loss of CCC proteins. (A–C) Nomarski optics; (D–E) phalloidin staining of F-actin imaged using confocal microscopy. (A) Wild-type embryo at approximately the threefold stage. (B) An embryo from a *hmp-1(zu278)* germline mosaic mother, which therefore lacks both maternal and zygotic HMP-1, displaying the Hmr (*Hammer*head) phenotype. Enclosure failure in the anterior results in rupture of the embryo and ejection of internal cells through the opening in the epidermis. (C) A *hmp-1(zu278)* homozygote displaying the Hmp (*Hump*back) phenotype. Prominent dorsal bulges in the epidermis are visible (arrow). (D) Phalloidin staining of a wild-type embryo at approximately the threefold stage of elongation. Circumferential filament bundles (CFBs) (arrows) and junctional proximal actin (arrowhead) are clearly visible; the bright staining is one of the four muscle quadrants. (E) A *hmp-1(zu278)* homozygote stained with phalloidin. CFBs have detached from cell–cell junctions (arrows). (F) A schematic of (E). Scale bar = 5 μm. *Images adapted from Ref. 7 with permission.*

2.4. HMP-2: A β-catenin specialized for adhesion

Whereas β-catenins in most animals have dual functions, in the CCC and in Wnt signaling[32] (see also Chapter 17), in *C. elegans* these activities are distributed into separate proteins.[10] HMP-2 has clear functions in cell–cell adhesion as part of the classical CCC. However, it is one of four β-catenin homologs in *C. elegans*, which have undergone subfunctionalization during their evolution. HMP-2 interacts strongly with HMR-1, but not with the single Tcf/Lef homolog in *C. elegans*, POP-1.[10,33] Instead, β-catenin's role in canonical Wnt signaling is fulfilled by BAR-1.[34–36] The remaining two β-catenin homologs, WRM-1 and SYS-1, participate in what appears to be a highly diverted Wnt signaling pathway, the "Wnt/β-catenin asymmetry" pathway,[36] which is involved in cell differentiation during anterior–posterior cell divisions. The subfunctionalization of β-catenins extends to other members of the nematode phylum, but importantly does not appear to be the case for the most distant relatives of *C. elegans*: these nematodes possess dual function β-catenins similar to those found in other animal groups, indicating subfunctionalization events occurred after the diversification of the nematode phylum (D. Sarkar and J. Pettitt, unpublished).

Structural comparison of HMP-2 with "dual function" β-catenins from vertebrates and *Drosophila* confirms that HMP-2 retains key features required for interactions with α-catenin and cadherin. The interaction sites with cadherin and α-catenin have been mapped for β-catenin of both vertebrates and *Drosophila* (reviewed in Ref. 32). β-catenin's binding to the CCC can be regulated by phosphorylation of residues in these regions. When Y654 of human β-catenin is phosphorylated, β-catenin's affinity for cadherin is significantly reduced. Phosphorylation of Y142 likewise decreases β-catenin's affinity for α-catenin.[37–41] HMP-2 possesses similar subdomains (H.-J. Choi and W. Weis, personal communication; T. Loveless and J. Hardin, unpublished); so, it will be interesting to see whether mutating these sites has important functional consequences *in vivo*.

Vertebrate and *Drosophila* β-catenins are well-known transcriptional coactivators acting downstream of canonical Wnt signaling. Residues in the C-terminus of β-catenin just terminal to the 12th armadillo repeat, termed "Helix C",[42] are conserved in all β-catenins known to be involved in transcriptional coactivation.[32] This domain is conspicuously lacking in HMP-2,[32,43] suggesting that it should have reduced transcriptional coactivator activities. Indeed, HMP-2 binds POP-1/Tcf weakly, if at all in yeast two-hybrid assays.[10] HMP-2 nevertheless has non-adhesive roles under some circumstances. When FRK-1/Fer kinase is knocked down, HMP-2 can function in the Wnt/β-catenin asymmetry pathway.[44] Immunostaining also indicates that HMP-2 is present in the nucleus in the early embryo (Ref. 45; T. Loveless and J. Hardin, unpublished). Moreover, when overexpressed, HMP-2 can rescue *bar-1* mutants.[33] Taken together, the data suggest that HMP-2 normally acts as an "adhesion-only" β-catenin, but that it can function weakly in the nucleus. The physiological significance of such weak activity is unknown.

2.5. HMP-1/α-catenin: Connecting to the actin cytoskeleton

The traditional view of the cadherin/catenin complex (CCC) assumes that its core components directly tether F-actin to the CCC via a quaternary complex of cadherin/β-catenin/α-catenin/F-actin. In favor of this classical model are the findings that direct molecular fusion of α-catenin to cadherin rescues some adhesive functions of adhesion-defective cells, both in vertebrate tissue culture cells[24] and *in vivo* in *Drosophila*.[46] However, the traditional view of the situation in vertebrates has been questioned (see also Chapter 1), leading to a model in which E-cadherin and β-catenin recruit α-catenin to junctions via transient interactions.[24,47] The subsequent dissociation of α-catenin from the complex would then result in the formation of α-catenin homodimers in the vicinity of junctional actin, where they can influence filament dynamics, possibly by antagonizing the Arp2/3 complex.[24] Beyond these questions about whether the core CCC functions as a stoichiometric complex, recent experiments indicate that the vertebrate CCC can undergo mechanosensitive adhesive strengthening involving the actin-binding protein vinculin.[48] αE-catenin may undergo conformational changes as part of this process that allows it to bind and recruit vinculin.[49]

Given these complexities, *C. elegans* may be an outstanding system in which to address how α-catenin functions. Based on sequence comparisons, several domains of vertebrate α-catenin, including its three vinculin homology domains, an N-terminal β-catenin binding domain, and an F-actin binding site at the C-terminus, are highly conserved.[11] Moreover, the tension-bearing mechanical requirements on HMP-1 due to the stresses of elongation are stringent, making it a good system for studying essential features of α-catenin.

Recent experiments provide unambiguous evidence that both linkage to the CCC via HMP-2 and F-actin binding are essential for HMP-1 function. Mutations that truncate HMP-1 before the C-terminal F-actin binding domain result in complete failure of elongation.[11] Deletion of the N-terminal β-catenin binding domain also results in complete morphogenetic failure, with loss of recruitment of HMP-1 to junctions.[11] Interestingly, HMP-1 appears conformationally autoinhibited *in vitro*: full-length HMP-1 is very compact and only binds F-actin weakly *in vitro*, whereas C-terminal fragments bind actin avidly.[11] Although the details remain to be worked out, HMP-1 shows that a common theme among α-catenins is that their conformational state regulates the CCC.

2.6. JAC-1: A modulator of the other components of the core CCC

Like other p120-catenins, JAC-1 (*Juxtamembrane domain-associated catenin*) has 10 armadillo repeats in its central region, but unlike the N-termini of mammalian and *Drosophila* p120-catenins, the N-terminus of JAC-1 contains four fibronectin type III domains.[8] The functional significance of these domains is unclear, but they are found in all characterized nematode p120-catenins. JAC-1 is not essential for cadherin-mediated events in the *C. elegans* epidermis.[8] However, it does positively contribute to CCC function, since reducing its function enhances the phenotype of a weak *hmp-1* loss-of-function mutant.[8] Indeed, loss of *jac-1* function in this genetic background leads to exacerbated clumping of CFBs and phenocopies the strong *hmp-1* loss-of-function phenotype. Interestingly, *Drosophila* p120-catenin plays a similar modulatory role.[50] In contrast, vertebrate p120-catenins appear to play a more essential function[51] (see also Chapter 18).

2.7. HMR-1B: A classical neuronal cadherin involved in axon guidance

Functional analysis of the HMR-1B isoform indicates that a classical cadherin also acts during neuronal development in *C. elegans*. Animals with reduced or absent HMR-1B function are viable, but they display incompletely penetrant defects in axon guidance of a subset of motor neurons.[9] Strong *hmp-1* mutants that express HMP-1 only in the epithelia typically have mild uncoordinated (Unc) phenotypes.[11] However, since the penetrance of axonal guidance defects caused by loss of HMR-1B function is relatively low, it is likely that HMR-1B-based adhesion only augments other more important guidance cues.

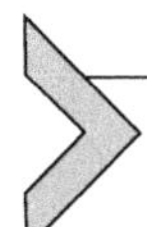

3. REGULATING THE CLASSICAL CCC IN *C. ELEGANS*

3.1. The PAR/aPKC complex and LET-413/Scribble: Regulators of apical junctional organization

As in other metazoans, PAR-3, PAR-6, and PKC-3/atypical Protein Kinase C3 constitute an apical polarity complex in *C. elegans* epithelia.[52] PAR protein constructs possessing a C-terminal sequence that allows PAR proteins to persist in early embryos, but leads to their destruction before the birth of epithelial cells, have been used to study PAR proteins in embryonic epithelia.[53] Loss of PAR-3 leads to ectopic separations between lateral surfaces of cells and to abnormal HMP-1 localization; it colocalizes with DLG-1/Discs Large, rather than apical to it. In the absence of PAR-3, intestinal HMR-1 is initially dispersed and mislocalized, whereas HMP-1 is still recruited into foci. Basolateral (as opposed to apical) foci of HMP-1 and DLG-1 accumulate, despite unperturbed localization of LET-413/Scribble,[54] which normally excludes AJ components from basolateral surfaces.[55,56] Similar localization defects are seen in the pharynx in *par-3* mutants.[54] Loss of PAR-6 prior to gastrulation results in elongation failure despite successful ventral enclosure.[57] The distribution of DLG-1, AJM-1, HMP-1, and HMR-1 is fragmentary, with the CCC being affected less severely.

While PAR-3 is critical for CCC establishment in endodermal and mesodermal epithelia, it is relatively less important in the epidermis; *par-3* embryos are still able to correctly localize CCC proteins and form mature epidermal junctions.[54] This difference may be due to an inherent difference in tissue organization: the epidermis is in contact with the developing cuticle. Eventually, however, localization of HMR-1 and DLG-1 is lost in *par-3* embryos, resulting in ripping of the epidermis.[57] This may be due to loss of apical PAR-6 in the *par-3* mutant epidermis,[57] suggesting a role in junctional maintenance rather than formation in this tissue.

The LAP (leucine-rich repeat and PDZ) protein LET-413/Scribble is also involved in apicobasal positioning of junctional proteins. LET-413 localizes to basolateral membranes of *C. elegans* epithelia.[58] *let-413* loss of function leads to a lack of apical-basal "focusing" of junctional proteins in the epidermis and gut (reviewed in Ref. 14). *let-413* mutants display occasional ventral enclosure defects,[56] suggesting effects on the CCC as well.

3.2. FRK-1/Fer kinase: A regulator of HMP-2/β-catenin

FRK-1 is the ortholog of the mammalian nonreceptor tyrosine kinase, Fer. In the early embryo, abrogation of *frk-1* function results in relocalization of HMP-2 to the nucleus in endodermal precursors, and allows it to functionally substitute for WRM-1 during endoderm development.[44] Later, during morphogenesis, FRK-1 localizes to epithelial junctions and is reported to be necessary for embryonic enclosure and morphogenesis.[59] FRK-1 mislocalizes in *hmp-2* mutants and *jac-1(RNAi)* embryos, and coimmunoprecipitates with HMP-2.[59] Surprisingly, some functions of FRK-1 appear to be independent of its function as a *bona fide* kinase,[59] and it is not known if FRK-1 kinase activity plays a role in regulating the phosphorylation of HMP-2 or other β-catenins.

4. NEW FUNCTIONAL PARTNERS WITH THE CLASSICAL CCC IN *C. ELEGANS*

4.1. SAX-7/L1CAM: A redundant partner of HMR-1 in blastomere adhesion

Because the *C. elegans* CCC is not absolutely essential for general cell–cell adhesion in embryos, it is a useful system for identifying other proteins that act redundantly with the CCC during development. One promising class of adhesion molecules that may function alongside the cadherin complex is the L1CAM family of proteins, which are members of the immunoglobulin superfamily of cell surface proteins.[60] In *Drosophila*, hypomorphic mutations in the L1CAM homolog neuroglian result in disrupted septate junctions, suggesting a role in cell adhesion.[61] In *C. elegans*, SAX-7/L1CAM localizes to cell–cell contacts in the early embryo and in epithelia at all stages,[62] and transgenic embryos expressing a dominant-negative SAX-7 display cells with altered shapes and locations.[62]

Recent work implicates HMR-1 as a functionally redundant partner with SAX-7/L1CAM in the early embryo. Knockdown of *hmr-1*, *hmp-1*, or *hmp-2* in *sax-7* mutants results in failure of apical constriction and defective ingression of Ea and Ep.[63] Moreover, removal of HMR-1 and SAX-7 together leads to loss of apical NMY-2/nonmuscle myosin II in endodermal precursors. Double loss-of-function embryos show reduced adhesion between isolated blastomeres.[63] How the combined effects of loss of HMR-1 and SAX-7 leads to widespread failure of NMY-2 recruitment is unclear, but these results highlight the utility of *C. elegans* for uncovering functionally interacting networks that affect cell–cell adhesion during embryogenesis.

4.2. VAB-9: A divergent claudin family member regulated by the cadherin complex

Positional cloning of a mutant with variable abnormal (Vab) defects identified a divergent member of the claudin superfamily, *vab-9*.[64] *vab-9* mutants show a variety of body shape defects, including mild disruption of CFBs similar to hypomorphic *hmp-1* mutants, suggesting *vab-9* functions within the same pathway. Consistent with this possibility, VAB-9 is expressed in all *C. elegans* epithelia and colocalizes with the CCC. HMR-1 is required for VAB-9 membrane localization and HMP-1 is required to maintain uniform VAB-9 distribution.[64] Mutations in *vab-9* enhance morphological defects in weak *hmp-1* mutants and enhance cell adhesion defects following loss of AJM-1 or DLG-1. Thus, *vab-9* participates in the organization of F-actin at epithelial junctions and, alongside the AJM-1/DLG-1 complex, maintains proper epithelial adhesion.

4.3. ZOO-1/ZO-1: An associate of VAB-9 and the CCC that promotes strong adhesion

Further insight into the function of *vab-9* came from an analysis of the ZO-1 ortholog, ZOO-1.[65] Similar to *vab-9* mutants, loss of *zoo-1* function enhances the morphogenetic phenotypes of a weak *hmp-1* loss-of-function allele, *hmp-1(fe4)*. Several data indicate that *zoo-1* closely interacts with *vab-9*. ZOO-1 localizes to cell junctions in the epidermis and, like VAB-9, requires HMR-1, but not HMP-1 or HMP-2, for junctional localization. Moreover, ZOO-1 localization requires VAB-9, and loss of either affects the organization of F-actin in the epidermis of elongating embryos. However, loss of *zoo-1* function does not enhance the phenotypes of *vab-9* mutants, suggesting that they act in a common pathway. Both *vab-9* and *zoo-1* interact with mutations affecting the actomyosin contractile machinery. Loss of *zoo-1* activity enhances *mel-11* (myosin light chain phosphatase) mutants and suppresses *let-502* (ROCK) mutants,[65] whereas *vab-9* mutations suppress *mel-11* alleles (J. Simske, personal communication). Together these results suggest a regulatory pathway in which HMR-1 regulates VAB-9, and VAB-9, in turn, regulates epidermal morphogenetic processes, some of which are mediated by *zoo-1* and *mel-11*.

4.4. UNC-34/Enabled: A potential integrator of the cadherin and DLG-1/AJM-1 complexes

UNC-34 is the *C. elegans* homolog of *Drosophila* Enabled (Ena), a member of the Ena/VASP family of proteins.[66] In tissue culture, Ena/VASP proteins have been shown to play important roles in epithelial sealing events in culture,[67] and during neurulation.[68] Surprisingly, UNC-34 plays only a minor role in modulating motility in epithelial cells in *C. elegans*, apparently because *unc-34* is genetically redundant with the N-WASP homolog, *wsp-1*, during ventral enclosure.[30,66] During enclosure, UNC-34::GFP localizes to the leading edges of migrating epidermal cells, but it is also redistributed to junctions when contralateral cells meet.[30] In primary mouse keratinocytes, junctional association of VASP depends on CCC function.[67] However, this does not appear to be the case in *C. elegans* because disrupting CCC function has little effect on UNC-34 localization.[30] Surprisingly,

UNC-34 junctional localization requires the activity of the DLG-1/Discs large binding partner AJM-1, which contains a putative consensus Ena/VASP binding motif.[30]

Interestingly, *unc-34* synergizes with *hmp-1* during morphogenesis. Depleting UNC-34 in *hmp-1(fe4)* embryos results in frequent failure of ventral enclosure, suggesting that UNC-34 might partially compensate for reduced HMP-1 activity.[30] The reason for this is unclear; however, if α-catenin acts to regulate actin dynamics partly by antagonizing actin filament branching and promoting filament bundling,[69,70] it is possible that UNC-34 could compensate for compromised α-catenin function by promoting actin bundles via a parallel pathway. Further experiments will be necessary to test this or alternative models.

4.5. MAGI-1: A SAX-7/L1CAM-dependent regulator of AFD-1/afadin that stabilizes junctions

The functional redundancy between HMR-1 and SAX-7 suggests that other proteins act synergistically or in parallel with the CCC during morphogenesis. A genome-wide RNAi screen identified 55 such enhancers[71] (Fig. 11.2D). Among them was *magi-1*, a highly conserved inverted MAGUK protein.[72] Loss of *magi-1* function in wild-type embryos results in disorganized epithelial migration and occasional morphogenetic failure during ventral enclosure, with loss of leading edge organization and excessive protrusive activity at the ventral midline. Knocking down *magi-1* in *hmp-1(fe4)* mutants leads to complete morphogenetic failure during elongation and actin disorganization in the epidermis.[71]

MAGI-1 localizes to cell junctions in all epithelial tissues in *C. elegans*[71–73] at a position basal to the CCC and apical to the DLG-1/AJM-1 complex (see Fig. 11.2C). Although MAGI-1 is capable of binding HMP-2 in cellular extracts,[74] this interaction is apparently unimportant for localization of MAGI-1 in epithelia, since MAGI-1::GFP is initially targeted to junctions following depletion of HMR-1 or DLG-1.[71,73] However, as embryos age, MAGI-1::GFP begins to lose its tight association with junctions under these conditions,[71] indicating that the CCC and DLG-1/AJM-1 complexes are not crucial for MAGI-1's initial recruitment, but contribute to its maintenance at junctions.

MAGI-1 physically interacts with AFD-1/afadin/Canoe, which was also identified in the same screen, and which colocalizes with MAGI-1.[71] Normal accumulation of MAGI-1 at junctions requires SAX-7, which can bind MAGI-1 via its C-terminus.[71] This MAGI-1/AFD-1/SAX-7 middle layer may help to spatially organize the other components of the apical junction. Depletion of MAGI-1 leads to loss of spatial segregation and expansion of the CCC and DLG-1/AJM-1 domains and greater mobility of junctional proteins in fluorescence recovery after photobleaching assays.[71] These newly discovered molecular interactions between MAGI-1, AFD-1/afadin, and SAX-7/L1CAM appear to be part of a larger functional interactome that acts in concert with the core components of the CCC. Two other members of this interactome, SRGP-1/srGAP and UNC-94/tropomodulin, are described in the next two sections.

4.6. SRGP-1/srGAP: Bending adhesions into shape

SRGP-1, the *C. elegans* Slit/Robo GAP homolog, is another lethal enhancer of *hmp-1(fe4)*, and also enhances lethality in *hmp-2(qm39)* mutants.[75] Like its vertebrate counterparts, SRGP-1 contains an N-terminal F-BAR (Bin1, Amphiphysin, RVS167) domain, and a central GTPase activating (GAP) domain, which are thought to allow such proteins to regulate membrane curvature and modulate Rho family GTPase activity, respectively.[75] SRGP-1 colocalizes with the CCC at junctions; interestingly, expression of only the SRGP-1 F-BAR domain plus 200 downstream amino acids is sufficient to target SRGP-1 to junctions. This downstream sequence may be responsible for SRGP-1 interaction with the CCC as the F-BAR domain alone is not targeted to junctions.[75] SRGP-1 transgenes lacking GAP activity can rescue synergistic defects in *hmp-2(qm39)* mutants, suggesting that some of its functions are independent of its GAP activity. Whether SRGP-1's GAP activity is required in the more stringent *hmp-1(fe4)* background is currently unknown.

Overexpression of SRGP-1 leads to tubulations in the junctional membrane, which correlate with the level of SRGP-1 overexpression. Significantly, these tubulations contain HMR-1 and HMP-1, but not DLG-1,[75] suggesting that SRGP-1 aids adhesion by increasing the surface area of contact between adjacent cells at the level of the CCC. F-BAR proteins can also modulate actin dynamics more directly through their C-terminal domains[76]; it remains unclear whether SRGP-1 has this capacity.

4.7. UNC-94: A regulator of HMP-1-dependent actin networks

UNC-94 is a tropomodulin (Tmod) ortholog[77] that is structurally very similar to vertebrate Tmods,[78] which cap the minus ends of actin filaments in a tropomyosin-dependent manner.[79] In body wall muscle sarcomeres,

UNC-94 is found at the minus (pointed) ends of thin filaments.[77] Biochemically, UNC-94 behaves like other Tmods, protecting pointed ends of actin filaments from depolymerization by UNC-60B/ADF/cofilin *in vitro*[80] and against minus-end subunit loss when actin monomer concentration is limiting.[81]

Like several other *hmp-1(fe4)* enhancers, *unc-94(RNAi);hmp-1(fe4)* embryos fail to elongate.[81] Remarkably, UNC-94 localizes to a subset of cell–cell junctions in the epidermis in a *hmp-1*-dependent manner, at sites where lateral (seam) cells connect to dorsal and ventral cells; these junctions are under considerable mechanical stress.[81] High-speed filming of GFP-tagged proteins at cell–cell junctions confirms the instability of junctions in double loss-of-function embryos: prior to complete mechanical failure, junctions in *unc-94 (RNAi);hmp-1(fe4)* embryos exhibit "excursions" of JAC-1::GFP that are pulled away from the junction. Phalloidin staining of *unc-94(RNAi);hmp-1 (fe4)* embryos reveals separation of the CCC and its associated junctional actin, and CFB detachment. *In vitro*, UNC-94 promotes formation of F-actin bundles that are longer than in the presence of HMP-1 alone. Taken together, these results suggest that UNC-94 stabilizes actin filaments against minus-end subunit loss at junctions, leading to a more mechanically rigid junction that is better able to withstand stress during elongation.

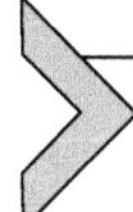

5. BEYOND THE CLASSICAL CCC: OTHER CADHERIN FAMILY MEMBERS IN *C. ELEGANS*

The nonclassical cadherins in *C. elegans* are much less well understood, in part because single loss-of-function mutations do not result in strong, easily interpretable phenotypes, and in some cases, have no observable phenotype at all. In addition, many of the *C. elegans* cadherins are highly divergent. Coupled with the fact that many of the pathways that their putative homologs impinge upon in other animals are significantly altered in *C. elegans*, this has made functional comparisons difficult. We have thus restricted our discussion to those cadherins that have clear mutant phenotypes, and/or have clear homologs in other organisms.

5.1. CDH-3 and CDH-4: Fat-like cadherins that do not modulate planar cell polarity

Two *C. elegans* cadherins, CDH-3 and CDH-4, structurally resemble the large *Drosophila* cadherin, Fat, with regard to the organization of their ectodomains.[82] However, their cytoplasmic domains are highly divergent from Fat or Fat-like cadherins found in other animals and lack sequence motifs conserved across many metazoan phyla[83] (see also Chapters 4 and 10). Thus, unambiguous assignment of orthology is not possible. In addition, while *Drosophila* Fat is a key component of the planar cell polarity (PCP) and Hippo signaling pathways,[84] there is no functional evidence implicating either CDH-3 or CDH-4 in the regulation of cell polarity through these pathways. Moreover, several members of the Hippo pathway are missing in *C. elegans*,[85] and despite concerted efforts, there is currently no clear evidence for a true PCP pathway in *C. elegans*.[34,35,86]

Loss of *cdh-3* function causes mild, variably penetrant defects in the morphogenesis of a single epidermal cell, together with weakly penetrant defects in the morphology of the excretory system[3] (L. Hodgson and J. Pettitt, unpublished). In addition, overexpression of CDH-3 enhances the penetrance of defects in the morphogenesis of the buccal epithelium in animals lacking the miR-51 family of microRNAs.[87] *cdh-4* loss-of-function mutants show defects in axon fasciculation, synapse formation, and the migration of a pair of neuroblasts, as well as incompletely penetrant defects involving epithelial cells.[88] The precise cellular bases of the *cdh-3*/*-4* loss-of-function phenotypes are not known, but defects in cell polarity cannot be ruled out.

5.2. FMI-1: A Flamingo/Stan protein that modulates fasciculation in follower axons

fmi-1 encodes the sole *C. elegans* homolog of Flamingo/Starry night, a seven-pass transmembrane protein that is a core PCP pathway component in *Drosophila* and mice.[89] In addition, *Drosophila* Flamingo can mediate neuronal morphogenesis through both PCP-dependent and -independent mechanisms.[90,91] One of the Flamingo orthologs in mice is likewise involved in forebrain axon guidance[92] (see also Chapter 9). In *C. elegans*, it seems unlikely that FMI-1 functions in a conventional PCP pathway (see above), but recent experiments show that it is involved in axon guidance.

FMI-1 contributes to navigation of both pioneer and follower axons during embryogenesis.[93] Guidance of PVQ and HSN axons is dependent on FMI-1's extracellular domain.[93] Intriguingly, genetic evidence from VD motor neurons suggests that FMI-1 may interact with CDH-4.[94] The possibility of such heterophilic interactions has some precedent in other systems.[95] The intracellular domain of FMI-1 is cell-autonomously necessary for PVP pathfinding but dispensable for PVQ and HSN fasciculation,[93] suggesting that FMI-1 may be involved in signaling required for pioneer axon guidance.

5.3. CASY-1: A calsyntenin involved in associative learning

Comparison of CASY-1 to its mammalian and *Drosophila* homologs shows that these proteins share significant sequence similarity along their entire lengths.[2] The mammalian homologs are termed calsyntenins because their cytoplasmic domains can bind synaptic calcium.[96] Calsyntenins are associated with the postsynaptic membranes of excitatory CNS synapses. The extracellular domains of calsyntenins are proteolytically cleaved close to the membrane, and the transmembrane-intracellular portion is then internalized. This has led to a model whereby calsyntenins modulate postsynaptic calcium levels.[96] CASY-1 may play a similar role in *C. elegans*. Based on transgene assays, CASY-1 is widely expressed in the nervous system,[97] and loss-of-function studies show that it plays a role during associative learning in *C. elegans*.[97,98] Increasing the gene dosage of the glutamate receptor subunit GLR-1 can compensate for behavioral defects of *casy-1* mutants, suggesting that CASY-1 positively regulates GLR-1 signaling during olfactory associative learning.[98] Like its vertebrate homologs, CASY-1 is cleaved and released from neurons; indeed, the ectodomain of CASY-1 is sufficient to rescue defects in salt chemotaxis learning.[97]

6. CONCLUSIONS

C. elegans has emerged as an outstanding model system for the *in vivo* functional analysis of the cadherin/catenin complex during morphogenesis. By analyzing this basal metazoan, it should be possible to identify core functionalities of the CCC that allow cells to specifically adhere to each other, to modulate those adhesions, and to withstand the inherent stresses during embryonic morphogenesis. Recent genome-wide functional screens have been especially fruitful for identifying new molecular complexes that partner with the core components of the CCC during morphogenesis. Understanding in detail how these complexes function in *C. elegans* and in vertebrates is an exciting challenge for the future.

ACKNOWLEDGMENTS

This work was supported by NIH grants R01GM58038 and R21HD072769 awarded to J. H. T. L. was supported by Molecular Biosciences predoctoral training grant T32GM007215.

REFERENCES

1. Cox EA, Tuskey C, Hardin J. Cell adhesion receptors in *C. elegans*. *J Cell Sci* 2004;**117**:1867–70.

2. Hill E, Broadbent ID, Chothia C, Pettitt J. Cadherin superfamily proteins in *Caenorhabditis elegans* and *Drosophila melanogaster*. *J Mol Biol* 2001;**305**:1011–24.
3. Pettitt J. The cadherin superfamily. *WormBook* 2005;1–9.
4. (a) Morishita H, Yagi T. Protocadherin family: diversity, structure, and function. *Curr Opin Cell Biol* 2007;**19**:584–92. (b) Hulpiau P, van Roy F. New insights into the evolution of metazoan cadherins. *Mol Biol Evol* 2011;**28**:647–57.
5. Garrod DR, Merritt AJ, Nie Z. Desmosomal cadherins. *Curr Opin Cell Biol* 2002;**14**:537–45.
6. Sawyer JM, Glass S, Li T, Shemer G, White ND, Starostina NG, et al. Overcoming redundancy: an RNAi enhancer screen for morphogenesis genes in *Caenorhabditis elegans*. *Genetics* 2011;**188**:549–64.
7. Costa M, Raich W, Agbunag C, Leung B, Hardin J, Priess JR. A putative catenin-cadherin system mediates morphogenesis of the *Caenorhabditis elegans* embryo. *J Cell Biol* 1998;**141**:297–308.
8. Pettitt J, Cox EA, Broadbent ID, Flett A, Hardin J. The *Caenorhabditis elegans* p120 catenin homologue, JAC-1, modulates cadherin-catenin function during epidermal morphogenesis. *J Cell Biol* 2003;**62**:15–22.
9. Broadbent ID, Pettitt J. The *C. elegans hmr-1* gene can encode a neuronal classic cadherin involved in the regulation of axon fasciculation. *Curr Biol* 2002;**12**:59–63.
10. Korswagen H, Herman M, Clevers H. Distinct β-catenins mediate adhesion and signalling functions in *C. elegans*. *Nature* 2000;**406**:527–32.
11. Kwiatkowski AV, Maiden SL, Pokutta S, Choi HJ, Benjamin JM, Lynch AM, et al. In vitro and in vivo reconstitution of the cadherin-catenin-actin complex from *Caenorhabditis elegans*. *Proc Natl Acad Sci USA* 2010;**107**:14591–6.
12. Barstead RJ, Kleiman L, Waterston RH. Cloning, sequencing, and mapping of an alpha-actinin gene from the nematode *Caenorhabditis elegans*. *Cell Motil Cytoskeleton* 1991;**20**:69–78.
13. Barstead RJ, Waterston RH. The basal component of the nematode dense-body is vinculin. *J Biol Chem* 1989;**264**:10177–85.
14. Lynch AM, Hardin J. The assembly and maintenance of epithelial junctions in *C. elegans*. *Front Biosci* 2009;**14**:1414–32.
15. Oda H, Tsukita S. Dynamic features of adherens junctions during Drosophila embryonic epithelial morphogenesis revealed by a Dα-catenin-GFP fusion protein. *Dev Genes Evol* 1999;**209**:218–25.
16. Lickert H, Bauer A, Kemler R, Stappert J. Casein kinase II phosphorylation of E-cadherin increases E-cadherin/β-catenin interaction and strengthens cell-cell adhesion. *J Biol Chem* 2000;**275**:5090–5.
17. Huber AH, Weis WI. The structure of the beta-catenin/E-cadherin complex and the molecular basis of diverse ligand recognition by β-catenin. *Cell* 2001;**105**:391–402.
18. Xiao K, Oas RG, Chiasson CM, Kowalczyk AP. Role of p120-catenin in cadherin trafficking. *Biochim Biophys Acta* 2007;**1773**:8–16.
19. Delva E, Kowalczyk AP. Regulation of cadherin trafficking. *Traffic* 2009;**10**:259–67.
20. Loveless T, Hardin J. Cadherin complexity: recent insights into cadherin superfamily function in *C. elegans*. *Curr Opin Cell Biol* 2012;**24**:695–701.
21. Sawyer JM, Harrell JR, Shemer G, Sullivan-Brown J, Roh-Johnson M, Goldstein B. Apical constriction: a cell shape change that can drive morphogenesis. *Dev Biol* 2010;**341**:5–19.
22. Roh-Johnson M, Shemer G, Higgins CD, McClellan JH, Werts AD, Tulu US, et al. Triggering a cell shape change by exploiting preexisting actomyosin contractions. *Science* 2012;**335**:1232–5.
23. Gomez GA, McLachlan RW, Yap AS. Productive tension: force-sensing and homeostasis of cell-cell junctions. *Trends Cell Biol* 2011;**21**:499–505.

24. Maiden SL, Hardin J. The secret life of α-catenin: moonlighting in morphogenesis. *J Cell Biol* 2011;**195**:543–52.
25. Nance J, Priess JR. Cell polarity and gastrulation in *C. elegans*. *Development* 2002;**129**:387–97.
26. Roh-Johnson M, Goldstein B. In vivo roles for Arp2/3 in cortical actin organization during *C. elegans* gastrulation. *J Cell Sci* 2009;**122**:3983–93.
27. Chihara D, Nance J. An E-cadherin-mediated hitchhiking mechanism for *C. elegans* germ cell internalization during gastrulation. *Development* 2012;**139**:2547–56.
28. Chisholm AD, Hardin J. Epidermal morphogenesis. *WormBook* 2005;1–22.
29. Raich WB, Agbunag C, Hardin J. Rapid epithelial-sheet sealing in the *Caenorhabditis elegans* embryo requires cadherin-dependent filopodial priming. *Curr Biol* 1999;**9**: 1139–46.
30. Sheffield M, Loveless T, Hardin J, Pettitt J. C. elegans Enabled exhibits novel interactions with N-WASP, Abl, and cell-cell junctions. *Curr Biol* 2007;**17**:1791–6.
31. Williams-Masson EM, Malik AN, Hardin J. An actin-mediated two-step mechanism is required for ventral enclosure of the *C. elegans* hypodermis. *Development* 1997;**124**: 2889–901.
32. Gottardi CJ, Peifer M. Terminal regions of β-catenin come into view. *Structure* 2008;**16**:336–8.
33. Natarajan L, Witwer NE, Eisenmann DM. The divergent *Caenorhabditis elegans* β-catenin proteins BAR-1, WRM-1 and HMP-2 make distinct protein interactions but retain functional redundancy in vivo. *Genetics* 2001;**159**:159–72.
34. Herman M, Wu M. Noncanonical Wnt signaling pathways in *C. elegans* converge on POP-1/TCF and control cell polarity. *Front Biosci* 2004;**9**:1530–9.
35. Korswagen H. Canonical and non-canonical Wnt signaling pathways in *Caenorhabditis elegans*: variations on a common signaling theme. *Bioessays* 2002;**24**:801–10.
36. Mizumoto K, Sawa H. Two betas or not two betas: regulation of asymmetric division by β-catenin. *Trends Cell Biol* 2007;**17**:465–73.
37. Piedra J, Miravet S, Castano J, Palmer HG, Heisterkamp N, Garcia de Herreros A, et al. p120catenin-associated Fer and Fyn tyrosine kinases regulate β-catenin Tyr-142 phosphorylation and β-catenin-α-catenin interaction. *Mol Cell Biol* 2003;**23**:2287–97.
38. Rosato R, Veltmaat JM, Groffen J, Heisterkamp N. Involvement of the tyrosine kinase fer in cell adhesion. *Mol Cell Biol* 1998;**18**:5762–70.
39. Roura S, Miravet S, Piedra J, Garcia de Herreros A, Dunach M. Regulation of E-cadherin/catenin association by tyrosine phosphorylation. *J Biol Chem* 1999;**274**: 36734–40.
40. Tominaga J, Fukunaga Y, Abelardo E, Nagafuchi A. Defining the function of b-catenin tyrosine phosphorylation in cadherin-mediated cell-cell adhesion. *Genes Cells* 2008;**13**: 67–77.
41. Xu G, Craig AW, Greer P, Miller M, Anastasiadis PZ, Lilien J, et al. Continuous association of cadherin with beta-catenin requires the non-receptor tyrosine-kinase Fer. *J Cell Sci* 2004;**117**:3207–19.
42. Xing Y, Takemaru K, Liu J, Berndt JD, Zheng JJ, Moon RT, et al. Crystal structure of a full-length beta-catenin. *Structure* 2008;**16**:478–87.
43. Schneider SQ, Finnerty JR, Martindale MQ. Protein evolution: structure-function relationships of the oncogene β-catenin in the evolution of multicellular animals. *J Exp Zool B Mol Dev Evol* 2003;**295**:25–44.
44. Putzke AP, Rothman JH. Repression of Wnt signaling by a Fer-type nonreceptor tyrosine kinase. *Proc Natl Acad Sci USA* 2010;**107**:16154–9.
45. Sumiyoshi E, Takahashi S, Obata H, Sugimoto A, Kohara Y. The β-catenin HMP-2 functions downstream of Src in parallel with the Wnt pathway in early embryogenesis of *C. elegans*. *Dev Biol* 2011;**355**:302–12.

46. Sarpal R, Pellikka M, Patel RR, Hui FY, Godt D, Tepass U. Mutational analysis supports a core role for *Drosophila* α-catenin in adherens junction function. *J Cell Sci* 2012;**125**:233–45.
47. Niessen CM, Leckband D, Yap AS. Tissue organization by cadherin adhesion molecules: dynamic molecular and cellular mechanisms of morphogenetic regulation. *Physiol Rev* 2011;**91**:691–731.
48. le Duc Q, Shi Q, Blonk I, Sonnenberg A, Wang N, Leckband D, et al. Vinculin potentiates E-cadherin mechanosensing and is recruited to actin-anchored sites within adherens junctions in a myosin II-dependent manner. *J Cell Biol* 2010;**189**:1107–15.
49. Yonemura S, Wada Y, Watanabe T, Nagafuchi A, Shibata M. α-Catenin as a tension transducer that induces adherens junction development. *Nat Cell Biol* 2010;**12**:533–42.
50. Myster SH, Cavallo R, Anderson CT, Fox DT, Peifer M. *Drosophila* p120catenin plays a supporting role in cell adhesion but is not an essential adherens junction component. *J Cell Biol* 2003;**160**:433–49.
51. Fang X, Ji H, Kim SW, Park JI, Vaught TG, Anastasiadis PZ, et al. Vertebrate development requires ARVCF and p120 catenins and their interplay with RhoA and Rac. *J Cell Biol* 2004;**165**:87–98.
52. Nance J, Zallen JA. Elaborating polarity: PAR proteins and the cytoskeleton. *Development* 2011;**138**:799–809.
53. Nance J, Munro EM, Priess JR. *C. elegans* PAR-3 and PAR-6 are required for apicobasal asymmetries associated with cell adhesion and gastrulation. *Development* 2003;**130**: 5339–50.
54. Achilleos A, Wehman AM, Nance J. PAR-3 mediates the initial clustering and apical localization of junction and polarity proteins during *C. elegans* intestinal epithelial cell polarization. *Development* 2010;**137**:1833–42.
55. Koppen M, Simske JS, Sims PA, Firestein BL, Hall DH, Radice AD, et al. Cooperative regulation of AJM-1 controls junctional integrity in *Caenorhabditis elegans* epithelia. *Nat Cell Biol* 2001;**3**:983–91.
56. Legouis R, Gansmuller A, Sookhareea S, Bosher JM, Baillie DL, Labouesse M. LET-413 is a basolateral protein required for the assembly of adherens junctions in *Caenorhabditis elegans*. *Nat Cell Biol* 2000;**2**:415–22.
57. Totong R, Achilleos A, Nance J. PAR-6 is required for junction formation but not apicobasal polarization in *C. elegans* embryonic epithelial cells. *Development* 2007;**134**: 1259–68.
58. Legouis R, Jaulin-Bastard F, Schott S, Navarro C, Borg J, Labouesse M. Basolateral targeting by leucine-rich repeat domains in epithelial cells. *EMBO Rep* 2003;**4**:1096–102.
59. Putzke AP, Hikita ST, Clegg DO, Rothman JH. Essential kinase-independent role of a Fer-like non-receptor tyrosine kinase in *Caenorhabditis elegans* morphogenesis. *Development* 2005;**132**:3185–95.
60. Chen L, Zhou S. "CRASH"ing with the worm: insights into L1CAM functions and mechanisms. *Dev Dyn* 2010;**239**:1490–501.
61. Genova J, Fehon R. Neuroglian, gliotactin, and the Na+/K+ ATPase are essential for septate junction function in *Drosophila*. *J Cell Biol* 2003;**161**:979–89.
62. Chen L, Ong B, Bennett V. LAD-1, the *Caenorhabditis elegans* L1CAM homologue, participates in embryonic and gonadal morphogenesis and is a substrate for fibroblast growth factor receptor pathway-dependent phosphotyrosine-based signaling. *J Cell Biol* 2001;**154**:841–55.
63. Grana TM, Cox EA, Lynch AM, Hardin J. SAX-7/L1CAM and HMR-1/cadherin function redundantly in blastomere compaction and non-muscle myosin accumulation during *Caenorhabditis elegans* gastrulation. *Dev Biol* 2010;**344**:731–44.

64. Simske JS, Koppen M, Sims P, Hodgkin J, Yonkof A, Hardin J. The cell junction protein VAB-9 regulates adhesion and epidermal morphology in *C. elegans*. *Nat Cell Biol* 2003;**5**:619–25.
65. Lockwood C, Zaidel-Bar R, Hardin J. The *C. elegans* zonula occludens ortholog cooperates with the cadherin complex to recruit actin during morphogenesis. *Curr Biol* 2008;**18**:1333–7.
66. Withee J, Galligan B, Hawkins N, Garriga G. *Caenorhabditis elegans* WASP and Ena/VASP proteins play compensatory roles in morphogenesis and neuronal cell migration. *Genetics* 2004;**167**:1165–76.
67. Vasioukhin V, Bauer C, Yin M, Fuchs E. Directed actin polymerization is the driving force for epithelial cell-cell adhesion. *Cell* 2000;**100**:209–19.
68. Vanderzalm P, Garriga G. Losing their minds: Mena/VASP/EVL triple knockout mice. *Dev Cell* 2007;**13**:757–8.
69. Drees F, Pokutta S, Yamada S, Nelson WJ, Weis WI. α-Catenin is a molecular switch that binds E-cadherin-β-catenin and regulates actin-filament assembly. *Cell* 2005;**123**: 903–15.
70. Yamada S, Pokutta S, Drees F, Weis WI, Nelson WJ. Deconstructing the cadherin-catenin-actin complex. *Cell* 2005;**123**:889–901.
71. Lynch AM, Grana T, Cox-Paulson E, Couthier A, Cameron M, Chin-Sang I, et al. A genome-wide functional screen identifies MAGI-1 as an L1CAM-dependent stabilizer of apical junctions in C. elegans. *Curr Biol* 2012;**22**:1891–9.
72. Emtage L, Chang H, Tiver R, Rongo C. MAGI-1 modulates AMPA receptor synaptic localization and behavioral plasticity in response to prior experience. *PLoS One* 2009;**4**: e4613.
73. Stetak A, Hajnal A. The *C. elegans* MAGI-1 protein is a novel component of cell junctions that is required for junctional compartmentalization. *Dev Biol* 2011;**350**:24–31.
74. Stetak A, Horndli F, Maricq AV, van den Heuvel S, Hajnal A. Neuron-specific regulation of associative learning and memory by MAGI-1 in *C. elegans*. *PLoS One* 2009;**4**: e6019.
75. Zaidel-Bar R, Joyce MJ, Lynch AM, Witte K, Audhya A, Hardin J. The F-BAR domain of SRGP-1 facilitates cell-cell adhesion during *C. elegans* morphogenesis. *J Cell Biol* 2010;**191**:761–9.
76. Aspenstrom P. Roles of F-BAR/PCH proteins in the regulation of membrane dynamics and actin reorganization. *Int Rev Cell Mol Biol* 2009;**272**:1–31.
77. Stevenson TO, Mercer KB, Cox EA, Szewczyk NJ, Conley CA, Hardin JD, et al. *unc-94* encodes a tropomodulin in *Caenorhabditis elegans*. *J Mol Biol* 2007;**374**:936–50.
78. Lu S, Symersky J, Li S, Carson M, Chen L, Meehan E, et al. Structural genomics of *Caenorhabditis elegans*: crystal structure of the tropomodulin C-terminal domain. *Proteins* 2004;**56**:384–6.
79. Fischer RS, Fowler VM. Tropomodulins: life at the slow end. *Trends Cell Biol* 2003;**13**:593–601.
80. Yamashiro S, Cox EA, Baillie DL, Hardin JD, Ono S. Sarcomeric actin organization is synergistically promoted by tropomodulin, ADF/cofilin, AIP1 and profilin in *C. elegans*. *J Cell Sci* 2008;**121**:3867–77.
81. Cox-Paulson EA, Walck-Shannon E, Lynch AM, Yamashiro S, Zaidel-Bar R, Eno CC, et al. Tropomodulin protects α-catenin-dependent junctional-actin networks under stress during epithelial morphogenesis. *Curr Biol* 2012;**22**:1500–5.
82. Mahoney PA, Weber U, Onofrechuk P, Biessmann H, Bryant PJ, Goodman CS. The fat tumor suppressor gene in *Drosophila* encodes a novel member of the cadherin gene superfamily. *Cell* 1991;**67**:853–68.

83. Castillejo-Lopez C, Arias WM, Baumgartner S. The fat-like gene of *Drosophila* is the true orthologue of vertebrate fat cadherins and is involved in the formation of tubular organs. *J Biol Chem* 2004;**279**:24034–43.
84. Matakatsu H, Blair SS. Separating planar cell polarity and Hippo pathway activities of the protocadherins Fat and Dachsous. *Development* 2012;**139**:1498–508.
85. Hilman D, Gat U. The evolutionary history of YAP and the hippo/YAP pathway. *Mol Biol Evol* 2011;**28**:2403–17.
86. King RS, Maiden SL, Hawkins NC, Kidd 3rd AR, Kimble J, Hardin J, et al. The N- or C-terminal domains of DSH-2 can activate the *C. elegans* Wnt/beta-catenin asymmetry pathway. *Dev Biol* 2009;**328**:234–44.
87. Shaw WR, Armisen J, Lehrbach NJ, Miska EA. The conserved miR-51 microRNA family is redundantly required for embryonic development and pharynx attachment in *Caenorhabditis elegans*. *Genetics* 2010;**185**:897–905.
88. Schmitz C, Wacker I, Hutter H. The Fat-like cadherin CDH-4 controls axon fasciculation, cell migration and hypodermis and pharynx development in *Caenorhabditis elegans*. *Dev Biol* 2008;**316**:249–59.
89. Wallingford JB. Planar cell polarity and the developmental control of cell behavior in vertebrate embryos. *Annu Rev Cell Dev Biol* 2012;**28**:627–53.
90. Matsubara D, Horiuchi SY, Shimono K, Usui T, Uemura T. The seven-pass transmembrane cadherin Flamingo controls dendritic self-avoidance via its binding to a LIM domain protein, Espinas, in *Drosophila* sensory neurons. *Genes Dev* 2011;**25**:1982–96.
91. Mrkusich EM, Flanagan DJ, Whitington PM. The core planar cell polarity gene prickle interacts with flamingo to promote sensory axon advance in the *Drosophila* embryo. *Dev Biol* 2011;**358**:224–30.
92. Zhou L, Bar I, Achouri Y, Campbell K, De Backer O, Hebert JM, et al. Early forebrain wiring: genetic dissection using conditional Celsr3 mutant mice. *Science* 2008;**320**: 946–9.
93. Steimel A, Wong L, Najarro EH, Ackley BD, Garriga G, Hutter H. The Flamingo ortholog FMI-1 controls pioneer-dependent navigation of follower axons in *C. elegans*. *Development* 2010;**137**:3663–73.
94. Najarro EH, Wong L, Zhen M, Carpio EP, Goncharov A, Garriga G, et al. *Caenorhabditis elegans* Flamingo cadherin *fmi-1* Regulates GABAergic neuronal development. *J Neurosci* 2012;**32**:4196–211.
95. Ishiuchi T, Misaki K, Yonemura S, Takeichi M, Tanoue T. Mammalian Fat and Dachsous cadherins regulate apical membrane organization in the embryonic cerebral cortex. *J Cell Biol* 2009;**185**:959–67.
96. Vogt L, Schrimpf SP, Meskenaite V, Frischknecht R, Kinter J, Leone DP, et al. Calsyntenin-1, a proteolytically processed postsynaptic membrane protein with a cytoplasmic calcium-binding domain. *Mol Cell Neurosci* 2001;**17**:151–66.
97. Ikeda DD, Duan Y, Matsuki M, Kunitomo H, Hutter H, Hedgecock EM, et al. CASY-1, an ortholog of calsyntenins/alcadeins, is essential for learning in *Caenorhabditis elegans*. *Proc Natl Acad Sci USA* 2008;**105**:5260–5.
98. Hoerndli FJ, Walser M, Frohli Hoier E, de Quervain D, Papassotiropoulos A, Hajnal A. A conserved function of *C. elegans* CASY-1 calsyntenin in associative learning. *PLoS One* 2009;**4**:e4880.

N-Cadherin-Mediated Adhesion and Signaling from Development to Disease: Lessons from Mice

Glenn L. Radice
Department of Medicine, Center for Translational Medicine, Jefferson Medical College, Philadelphia, Pennsylvania, USA

Contents

Abstract

Of the 20 classical cadherin subtypes identified in mammals, the functions of the two initially identified family members E- (epithelial) and N- (neural) cadherin have been most extensively studied. E- and N-Cadherin have mostly mutually exclusive expression patterns, with E-cadherin expressed primarily in epithelial cells, whereas N-cadherin is found in a variety of cells, including neural, muscle, and mesenchymal cells. N-Cadherin function, in particular, appears to be cell context-dependent, as it can mediate strong

Progress in Molecular Biology and Translational Science, Volume 116
ISSN 1877-1173
http://dx.doi.org/10.1016/B978-0-12-394311-8.00012-1

cell–cell adhesion in the heart but induces changes in cell behavior in favor of a migratory phenotype in the context of epithelial–mesenchymal transition (EMT). The ability of tumor cells to alter their cadherin expression profile, for example, E- to N-cadherin, is critical for malignant progression. Recent advances in mouse molecular genetics, and specifically tissue-specific knockout and knockin alleles of N-cadherin, have provided some unexpected results. This chapter highlights some of the genetic studies that explored the complex role of N-cadherin in embryonic development and disease.

1. INTRODUCTION

In the early 1980s, several groups identified a 135-kDa cell-surface protein responsible for calcium-dependent cell–cell adhesion in neural retina,[1] heart muscle,[2] and brain.[3] The subsequent cloning of the gene encoding this cell adhesion molecule and the realization that it belongs to a family of related molecules eventually led to the name N (neural)-cadherin.[4] The function of N-cadherin depends to a great extent on the cellular context. N-Cadherin plays a critical role in maintaining the structural integrity of the heart by mediating strong adhesion between cardiomyocytes. On the other hand, N-cadherin is expressed in migratory fibroblastic or mesenchymal cells. Tumor cells take advantage of this latter characteristic of N-cadherin to promote their invasive and metastatic behavior (see also Chapter 14). These different adhesive characteristics are likely controlled by modification of protein interactions with its cytoplasmic domain, thus affecting the linkage with the actin cytoskeleton and the overall adhesion strength of the cadherin/catenin complex. Thirty years after the identification of N-cadherin, researchers continue to actively investigate the complex relationship between N-cadherin-mediated adhesion and signaling in development and disease.

In most cell types, N-cadherin is concentrated at cell–cell contact sites called adherens junctions. Cadherins form *cis*-dimers in the plasma membrane that interact in an antiparallel fashion with like cadherins on a neighboring cell, creating an adhesion zipper between the cells.[5] Intracellularly, cadherins interact with the armadillo proteins β-catenin or γ-catenin (plakoglobin), and with p120 catenin. Structurally unrelated to these direct binding partners, α-catenin indirectly links the cadherin/catenin complex to the actin cytoskeleton. Besides mediating intercellular adhesion, N-cadherin might influence signaling networks involved in various cellular processes, including cell proliferation, differentiation, and apoptosis (Fig. 12.1).

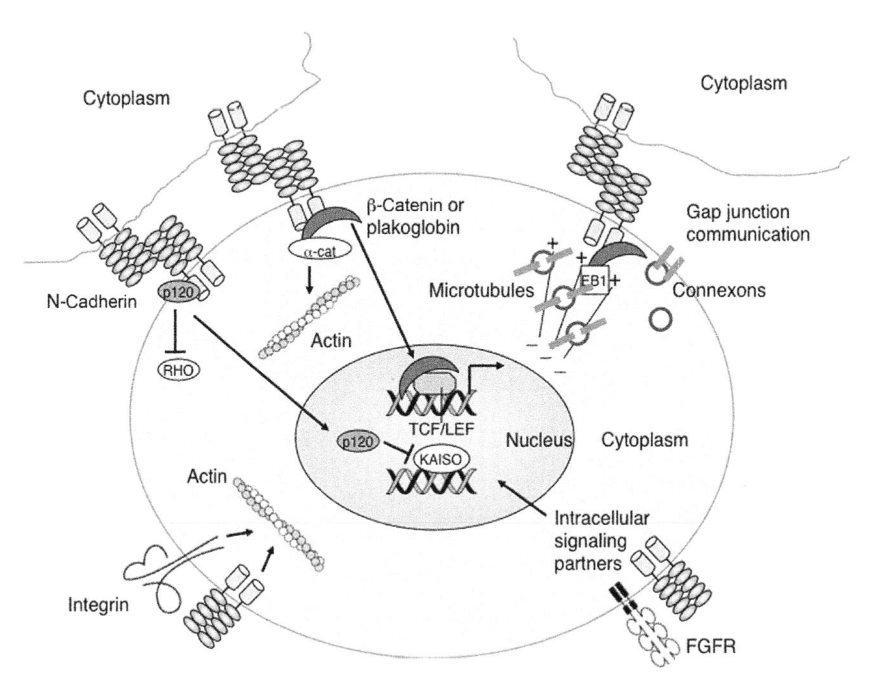

Figure 12.1 N-Cadherin affects a range of biological activities. Schematic representation of the multiple functions of N-cadherin. Besides mediating cell–cell interactions between neighboring cells, N-cadherin may influence cell signaling by binding β-catenin, plakoglobin, and p120 catenin, thus regulating their ability to enter the nucleus and modulate gene expression. Additionally, unbound p120 can regulate cell motility by inhibiting the activity of Rho GTPases. The forward trafficking of connexon-containing vesicles to the membrane is dependent on N-cadherin/β-catenin interactions with microtubule plus-end-tracking proteins, including EB1. N-Cadherin stabilizes fibroblast growth factor receptor (FGFR) at the membrane thus enhancing its downstream signaling pathways. Moreover, N-cadherin and integrins cooperate to regulate actin dynamics, what is important for cell motility. See text for details and references.

In addition to its cell adhesive role, β-catenin is also involved in signal transduction and developmental patterning as part of the canonical Wnt signaling pathway (see also Chapter 17). By sequestering β-catenin at the plasma membrane, N-cadherin can regulate the cytosolic pool of β-catenin and thereby modulate the Wnt pathway. As discussed later in this chapter, N-cadherin can stabilize the fibroblast growth factor receptor (FGFR) on the cell surface, thus facilitating intracellular signaling pathways.

In mammals, the classical cadherin family is composed of 20 members (see also Chapter 4), which exhibit specific spatiotemporal expression patterns associated with embryonic morphogenesis. Interestingly, only three cadherin subtypes are essential for embryonic development of, for instance, mouse embryos: E (epithelial), N, and VE (vascular endothelial) cadherin. In contrast, germline deletion of other classical cadherins such as P(placental)-cadherin, R(retinal)-cadherin, and M(muscle)-cadherin is compatible with embryonic development and these mutant mice are viable, with only subtle phenotypic changes.[6–8] On the other hand, N-cadherin's importance is illustrated by the various tissue-specific knockout experiments that affect embryonic morphogenesis and tissue homeostasis (Table 12.1).

Many studies over the years have employed function-blocking antibodies or dominant-negative cadherin molecules to perturb N-cadherin function both *in vitro* and *in vivo*. Such antibody-mediated blocking

Table 12.1 N-Cadherin mutant phenotypes in mice

Mutation	Tissue	Phenotypes	References
N-cad$^{-/-}$	Global	Embryonic lethal, multiple abnormalities, severe cardiovascular defect	9,10
N-cad$^{-/-}$; Cad11$^{-/-}$	Global	Embryonic lethal, more severe somite defects compared to N-cad$^{-/-}$	11
N-cad$^{+/-}$	Global	Altered gap junction organization in the heart, susceptibility to arrhythmia; bone loss in ovarectomized females	12,13
N-cad$^{+/-}$; Cx43$^{+/-}$	Global	Severe gap junction remodeling in the heart, increase susceptibility to arrhythmias compared to N-cad$^{+/-}$	12
N-cad$^{+/-}$; Cad11$^{-/-}$	Global	Viable, small bones, osteopenia, and reduced bone strength	14
N-cad$^{fl/fl}$;α MHC-MerCre	Adult myocardium	Cardiomyopathy, sudden arrhythmic death, gap junction, and ion channel remodeling	15–17
N-cad$^{fl/fl}$;α MHC-Cre	Embryonic myocardium	Embryonic lethal, cardiomyocyte cell adhesion defect	18
N-cad$^{fl/fl}$; Tie2-Cre	Endothelium	Embryonic lethal, disruption of EC–EC interactions, vascular defects	19
N-cad$^{fl/fl}$; Wnt1-Cre	Neural crest cells/ epicardium	Embryonic lethal, malformed outflow tract, hypoplastic myocardium	20

Table 12.1 N-Cadherin mutant phenotypes in mice—cont'd

Mutation	Tissue	Phenotypes	References
N-cad$^{fl/fl}$; D6-Cre	Cerebral cortex	Disorganized cerebral cortical structures, loss of laminar organization	21
N-cad$^{fl/fl}$; Col2.3-Cre	Osteoblast	Viable, modest trabecular osteopenia	14
N-cad$^{fl/fl}$; Pdx1-Cre	Pancreas	Viable, pancreas develops normally, defect in beta-cell granule turnover	22
N-cad$^{fl/fl}$; Ht-PA-Cre	Enteric neural crest cells	Directionality of migration perturbed, delay in colonization of the developing gut	23
N-cad$^{fl/fl}$; β1-integrin$^{fl/fl}$; Ht-PA-Cre	Enteric neural crest cells	Failure to colonize the distal part of the gut, severe aganglionosis in the proximal hindgut compared to single mutants	23
N-cad$^{fl/fl}$; Lens-Cre	Embryonic lens	Microphthalmia, defects Microphthalmia, defects in lens epithelia and fiber cell integrity	24
N-cad$^{fl/fl}$; Ecad$^{fl/fl}$; Lens-Cre	Embryonic lens	Failure in lens vesicle separation	24
N-cad$^{fl/fl}$; POMC-Cre	Pituitary	Viable, aberrant corticotrope cell clustering	25

fl, floxed (loxP-flanked) allele; EC, endothelial cell.

experiments have been reported to interfere with cellular processes, whereas, in some cases, the genetic loss-of-function (LOF) approach had no apparent effect on the same processes.[26] All studies have their limitations and it is important not to misinterpret LOF studies. When a gene is eliminated and a cellular process continues in its absence, it proves that the gene is not essential for that process. However, it remains possible that the gene (and its product) is normally involved in that process, but that in its absence, its function is taken over by other genes, which is called biological redundancy. One explanation for this phenomenon is that two (or more) genes have overlapping functions. Both genes might be simultaneously expressed and their products active in the process. However, in the absence of one of the genes (G1), the other one (G2) is sufficient for the process to occur normally. Alternatively, in the case of compensation, gene G2 might not be normally involved in a particular process, but it is upregulated in response to elimination of G1, thus substituting for the function of G1. In either scenario, overlapping or compensatory function, the generation of double (or triple) mutants might be necessary to define the genes involved in a particular biological process.

This chapter focuses on the role of N-cadherin in regulating cell adhesion, cellular communication, and signaling in different cellular contexts. The most dramatic cell-adhesion defects observed *in vivo* are associated with tissues undergoing active cellular rearrangements and/or increased mechanical stress (e.g., heart muscle). Genetic manipulation of N-cadherin expression in various murine models has provided new insight into the complex role of N-cadherin in disease pathogenesis. In this regard, knockin (KI) of the N-cadherin gene into the E-cadherin locus has generated some unexpected phenotypes that may help us understand the significance of ectopic expression of this mesenchymal cadherin with respect to different signaling pathways.

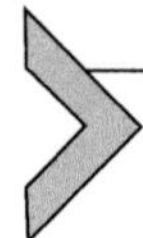

2. N-CADHERIN FUNCTION IN CARDIOVASCULAR DEVELOPMENT AND DISEASE

2.1. N-Cadherin function in embryonic myocardium

Most cells express multiple cadherin subtypes. For example, skeletal muscle expresses R-cadherin, M-cadherin, and N-cadherin. In contrast, heart muscle depends on one classical cadherin, N-cadherin. The heart is the first organ to form in the embryo with the establishment of the primitive circulatory system. N-Cadherin has been implicated in various aspects of cardiac development, including sorting out of the precardiac mesoderm,[27] establishment of left–right asymmetry,[28] cardiac looping morphogenesis,[29] and trabeculation of the myocardial wall.[30] N-Cadherin is strongly expressed in precardiac mesoderm and its expression continues throughout cardiomyocyte development and in the adult differentiated myocardium. Despite this expression in precardiac mesoderm, cardiomyocytes develop in N-cadherin-null embryos and the primitive heart tube forms normally, indicating that N-cadherin is not required for specification of the cardiac lineage.[9] However, N-cadherin-deficient myocytes are unable to maintain their cell–cell contacts when the heart begins to pump, resulting in dis-associated rounded myocytes surrounding the endocardial cell layer.

Obstruction of the cardiac outflow tract is likely the primary cause of the dramatic pericardial swelling seen in the N-cadherin-null embryos.[10] The cell adhesion defect in N-cadherin-null embryos is rescued by cardiac-specific expression of either N-cadherin or E-cadherin, demonstrating that these two classical cadherins are functionally interchangeable in the primitive heart.[10] In another study, chimeric mouse embryos derived from N-cadherin-null embryonic stem (ES) cells showed that N-cadherin-deficient cardiomyocytes initially interact with N-cadherin-positive cells, but as the primitive heart tube develops, they are excluded from the myocardial wall. The timing of the segregation and exclusion of the N-cadherin-null cells from the wild-type cells corresponds to the transition from simple cuboidal cells to a flattened, tightly associated myocardial cell layer.[31] Myofibrillogenesis occurs in the absence of N-cadherin, but alignment of the myofibrils through regions of cell–cell contact is lost, resulting in their random orientation.[32] As earlier studies demonstrated the importance of N-cadherin for myofibril formation and maintenance, this result was not expected. These former studies employed function-blocking antibodies[33,34] or dominant-negative cadherin molecules[35] to perturb N-cadherin function in cultured cardiomyocytes. The discrepancy in these results may be explained by the different experimental approaches. In the case of cultured N-cadherin-null cardiomyocytes, integrins might have compensated for the loss of N-cadherin by providing sufficient cytoskeletal integrity to support myofibril formation.[32]

2.2. N-Cadherin is required in multiple cell lineages during cardiovascular development

In addition to mediating adhesion between cardiomyocytes, N-cadherin plays an essential role in other cell lineages required for mouse cardiovascular development. Using a Wnt1-Cre transgene, it was shown that N-cadherin-deficient cardiac neural crest cells are unable to undergo the normal morphological changes associated with outflow tract remodeling, and this resulted in persistent truncus arteriosus in the majority of mutant embryos.[20] Some mutant embryos did undergo septation, but rotation of the outflow tract was incomplete, suggesting that alignment of the aortic and pulmonary channels is dependent on N-cadherin-generated cytoskeletal forces. Furthermore, N-cadherin is required for heterotypic epicardial–myocardial cell–cell interactions in the developing heart.[20] Disruption of these cellular interactions leads to myocardial cell hypoplasia that is likely caused by the loss of proliferative signaling from the epicardium.

In endothelial cells (ECs), N-cadherin exhibits an unusual nonjunctional distribution thought to be important for interactions of the EC with mural cells (i.e., pericyte and smooth muscle cells).[36] However, more recently it was shown that N-cadherin localizes to EC–EC junctions in addition to its well-known diffuse membrane expression.[19] Unexpectedly, deletion of N-cadherin in EC using Tie2-Cre caused embryonic death at midgestation due to severe vascular defects.[19] This phenotype manifested before mural cell investment, a finding that supports a previously unappreciated role for N-cadherin in EC–EC interactions. The Tie2-Cre;N-cad phenotype is remarkably similar to the VE-cadherin-null phenotype.[37,38] Interestingly, loss of N-cadherin caused a significant decrease in VE-cadherin and p120ctn levels. These studies demonstrate that N-cadherin function is required for angiogenesis and redefine the hierarchical relationship of the structural components of the interendothelial cell junction.

2.3. N-Cadherin is essential for the structural integrity of the adult myocardium

Efficient cardiac contractile function is highly dependent on the coordinated mechanical and electrical activation of myocardial tissue. In this regard, it is essential that the structural integrity of the myocardium is tightly maintained. This is largely achieved by the end-to-end connections between myocytes, called intercalated discs (ICD). The ICD is classically defined as containing three distinct junctional complexes: adherens junctions and desmosomes provide strong cell–cell adhesion, and gap junctions electrically couple myocytes. Recently, a novel, exclusive type of "hybrid adhering junction" or "area composita," was identified in the mammalian heart.[39,40] In this mixed-type junctional complex, classical adherens junction components, including N-cadherin, β-catenin, and αE-catenin, are colocalized with desmosomal proteins. Interestingly, the area composita is not found in lower vertebrates,[41] which suggests that it might have evolved to support the increased mechanical load of the mammalian four-chamber heart by anchoring both actin and intermediate filaments over an extended junctional area of the ICD.

A cardiac-specific, inducible Cre transgene (i.e., tamoxifen administration required for gene deletion) was used to ablate N-cadherin in the working myocardium to generate what is known as the N-cadherin conditional knockout (N-cad CKO) model.[15] Loss of N-cadherin in the adult heart led to disassembly of the ICD, including adherens junctions and desmosomes (Fig. 12.2). N-cad CKO mice exhibit a modestly dilated cardiomyopathy

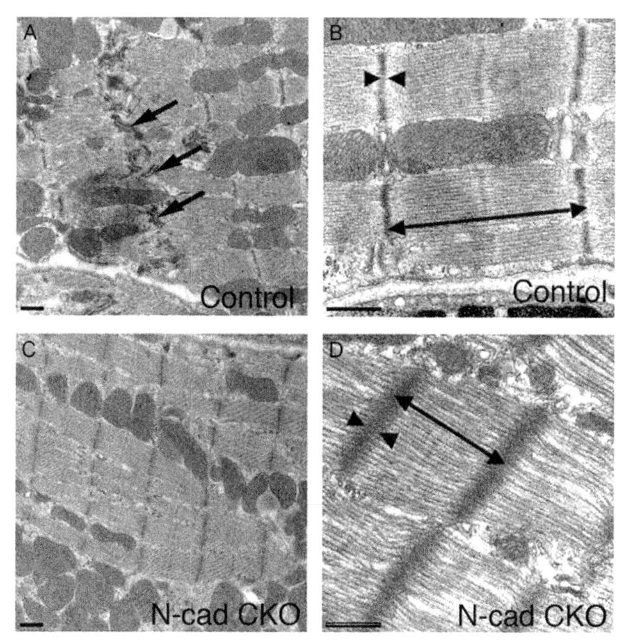

Figure 12.2 Transmission electron microscopy of N-cadherin CKO hearts. Electron micrographs of ventricular myocardium from control (N-cad$^{fl/-}$, αMHC/MerCreMer mouse minus tamoxifen) and N-cad CKO (N-cad$^{fl/-}$, αMHC/MerCreMer mouse plus tamoxifen) mice. Intercalated discs (arrows) were readily visible in the control (A). In contrast, these structures were absent in the N-cad CKO heart (C). The myofibrils appear distorted in the N-cad CKO (D) compared to control (B), with increased sarcomere length (double-headed arrow) and wider, less dense Z-lines (arrowheads). Bars, 500 nm. *Reprinted with permission from Ref. 15.*

and impaired cardiac function, with most animals dying suddenly within 2 months of deletion of the N-cadherin gene. Telemetric monitoring of conscious free moving N-cad CKO animals captured the abrupt onset of ventricular tachyarrhythmia coincident with sudden death (Fig. 12.3). It is quite remarkable that the heart can function at all without the ICD structure and its adhesive junctions connecting the individual cardiomyocytes. In this regard, it is worth noting that β1-integrin is upregulated in the N-cadherin-depleted myocardium, which suggests that the structural integrity of the heart might be partially maintained by enhanced cell–extracellular matrix (ECM) interactions. This animal model provided the first demonstration of the hierarchical relationship of the structural components of the ICD in the working myocardium, thus establishing N-cadherin's paramount importance in maintaining the structural integrity of the heart.

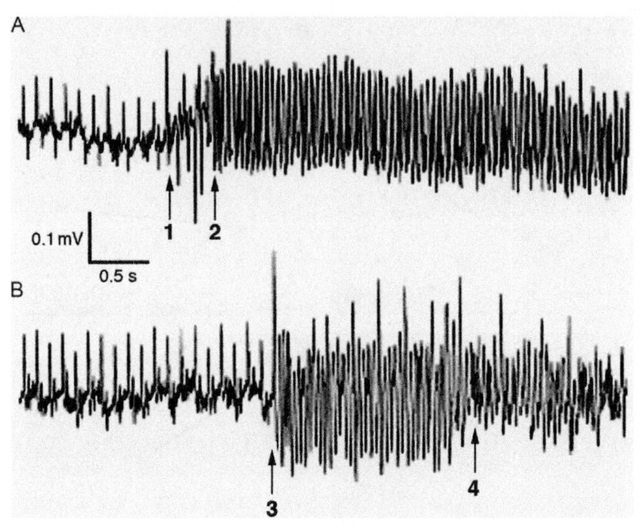

Figure 12.3 Spontaneous ventricular arrhythmias recorded from N-cadherin CKO mice that died suddenly. Continuous electrocardiographic recordings from a miniaturized transmitter implanted in conscious, freely mobile animals. (A) The left-hand side of the panel shows normal sinus activity followed by the onset of several premature ventricular beats (arrow #1), which then suddenly develops into ventricular tachycardia (arrow #2). (B) Again, the left-hand side of the panel shows normal sinus activity followed by the sudden onset of ventricular tachycardia (arrow #3), which then quickly degenerates into ventricular fibrillation (arrow #4). *Reprinted with permission from Ref. 15.*

2.4. Relationship between mechanical and electrical junctions in the heart

The loss of large gap junction plaques from the termini of cardiomyocytes, referred to as gap junction remodeling, is commonly observed in many forms of heart disease, including hypertrophy, ischemia, and dilated cardiomyopathies.[42] A major role of gap junctions in the myocardium is to enable rapid and coordinated electrical excitation, a prerequisite for normal rhythmic cardiac function. There is compelling evidence from human disease and animal models linking gap junction remodeling with a highly proarrhythmic substrate.[43,44] Gap junctions are plaques of multiple intercellular channels that connect the cytoplasm of adjacent cells. An individual channel is created by stable, noncovalent interactions of two hemichannels, referred to as connexons. Each connexon is composed of six transmembrane connexin proteins. Many connexins, including those expressed in the heart, have been found to turn over quite rapidly. The principal connexin expressed in the working ventricular myocardium, Cx43, is surprisingly short-lived in the intact adult heart (half-life = 1.5 h).[45] Cx43 oligomerizes in the post-ER/Golgi compartment and vesicles containing connexons are transported to the plasma membrane. The forward trafficking of these Cx43-containing vesicles to the junctional membrane is dependent on N-cadherin/β-catenin interactions with microtubule plus-end-tracking proteins, including EB1, p150 (Glued), and the dynein/dynactin complex.[46]

The carboxy terminus of Cx43 contains numerous sites that are subject to posttranslational phosphorylation, and these modifications can affect gap junction activity.[47] Cx43 is phosphorylated when it forms an active gap junctional complex at the cell surface. In contrast, it is generally dephosphorylated during trafficking/endocytosis in the cytoplasm. The time course of N-cadherin depletion directly correlates with a reduction in Cx43-containing gap junctions in the N-cad CKO heart (Fig. 12.4).[16] Moreover, a significant increase in dephosphorylated Cx43 is observed immediately after deleting N-cadherin, consistent with an increased turnover of Cx43 coincident with disassembly of the N-cadherin/catenin

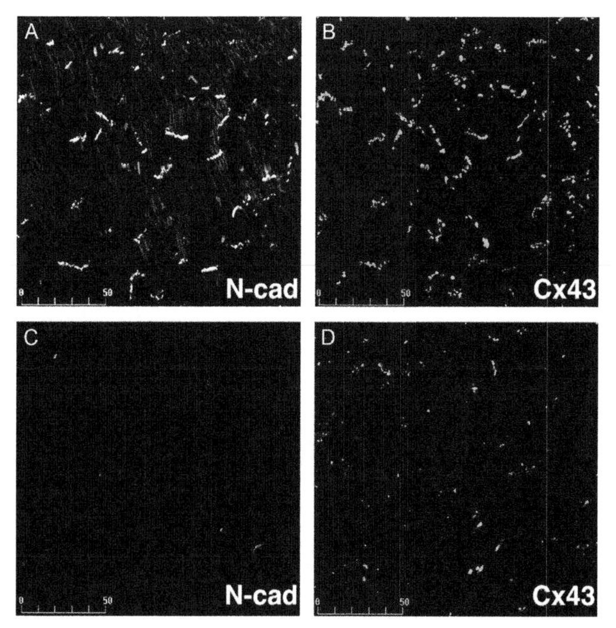

Figure 12.4 Expression of Cx43 in N-cadherin CKO hearts. Heart ventricles from wild-type (A, B) and N-cadherin CKO (C, D) animals 7 weeks posttamoxifen treatment were coimmunostained for N-cadherin (A, C) and Cx43 (B, D). N-Cadherin was lost from the intercalated discs in the CKO heart, while Cx43 (D) was significantly decreased in the ventricular myocardium. Bar, 50 μm. *Reprinted with permission from Ref. 16.*

complex. Like the Cx43 CKO mice,[48] N-cad CKO mice have decreased epicardial conduction velocity associated with sudden arrhythmic death. Loss of N-cadherin contributes to the incidence of arrhythmias not only by slowing conduction but also through significant remodeling of the major outward potassium currents.[17] The actin-binding scaffold protein, cortactin, is required for N-cadherin regulation of Kv1.5 channel function.

The severity of gap junction remodeling in N-cad CKO hearts suggested that animals expressing reduced levels of N-cadherin might exhibit an intermediate phenotype. Indeed, even in the absence of structural heart disease, mice heterozygous for the N-cadherin-null allele exhibit a reduced number of large Cx43-containing plaques and increased arrhythmic susceptibility.[12] Gap junctions were further decreased in number and size in N-cad$^{+/-}$; Cx43$^{+/-}$ compound heterozygous mice, which exhibit increased arrhythmogenicity compared to the single mutants. These double-mutant mice also show an increase in dephosphorylated Cx43. The cytoskeletal scaffold protein zona occludens-1 (ZO-1) binds to the carboxy tail of Cx43[49] and it was shown to regulate plaque size.[50] The smaller gap junctions in the N-cad$^{+/-}$ cardiomyocytes might be explained by the reduced ZO-1 at the ICD in these myocytes. In future studies, it will be important to understand how myocardial stress leads to uncoupling of the N-cadherin/Cx43 macromolecular complex and subsequent gap junction remodeling in diseased myocardium. In this regard, a recent study suggests that oxidative stress disrupts the forward trafficking of Cx43 by interfering with N-cadherin–microtubule interactions at the ICD.[51] Further investigation of the cytoskeletal changes associated with heart disease may provide clues into the underlying mechanism(s) responsible for gap junction remodeling and arrhythmogenesis.

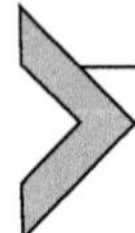

3. COOPERATION BETWEEN THE MESENCHYMAL CADHERINS, N-CADHERIN AND CADHERIN-11

N-Cadherin is first expressed in the nascent mesoderm migrating from the primitive streak in the gastrulating mouse embryo.[9] A day later (neurula stage), there is widespread expression of N-cadherin in the neural tube, notochord, somites, and precardiac mesoderm. Despite the widespread expression at this stage, embryos lacking N-cadherin undergo gastrulation, neurulation, and somitogenesis, but the neural tube and somites are often malformed.[9] N-Cadherin plays an important role in the coalescence of smaller presomitic clusters of cells into a well-organized epithelial somite structure.[11,52] The expression of other cell adhesion molecules, including cadherin-11, might partially substitute for the lack of N-cadherin in the paraxial mesoderm.[53] Cadherin-11 KO embryos have normal epithelial somite structure.[11] However, in embryos lacking both N-cadherin and cadherin-11, somites are further fragmented into smaller clusters than in the N-cad-null mice, suggesting that these two cadherins cooperate in the maintenance of epithelial somites.[11]

Osteoblasts are the skeletal cells responsible for the synthesis, deposition, and mineralization of the bone matrix. Cells of the osteoblast lineage express both N-cadherin and cadherin-11. Germline deletion of the cadherin-11 gene (*Cdh11*) does not affect skeletal development, but there is a modest reduction in bone density and cranial sutures show minor calcification defects.[54] Interestingly, cadherin-11 is required for the maintenance of the synovial lining of joints and has been implicated in the destructive tissue response in inflammatory arthritis.[55] Expression of a dominant-negative N-cadherin protein in differentiated osteoblasts results in delayed peak bone mass acquisition, impaired osteogenic differentiation, and an osteogenic-to-adipogenic shift in bone marrow stromal cell precursors.[56] The more severe skeletal phenotype in mice expressing dominant-negative N-cadherin suggested that another cadherin might compensate for the lack of cadherin-11. N-Cadherin haploinsufficiency by itself does not alter postnatal skeletal growth, but it accentuates ovariectomy-induced bone loss, the result of attenuated activation of bone formation following estrogen deprivation.[13] Conditional deletion of the N-cadherin gene in differentiated osteoblasts results in modest trabecular osteopenia with age.[14] However, double-mutant mice (N-cad$^{+/-}$;Cad11$^{-/-}$) exhibit a more severe skeletal phenotype of smaller bones, osteopenia, and reduced bone strength relative to Cad11$^{-/-}$ mice, demonstrating that N-cadherin and cadherin-11 cooperate to maintain bone homeostasis (Fig. 12.5). Functional overlap between N-cadherin and cadherin-11 likely explains the ability of N-cadherin-null limb buds to undergo mesenchymal cell condensation and chondrogenesis, resulting in skeletal structures.[57]

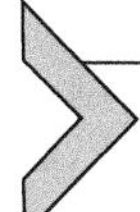

4. N-CADHERIN-MEDIATED SIGNALING IN CANCER

4.1. Cadherin switching and EMT

Changes in the expression of cell adhesion receptors accompany the transition from benign tumors to invasive, malignant cancer and the subsequent metastatic dissemination of tumor cells. E-Cadherin is expressed in epithelial

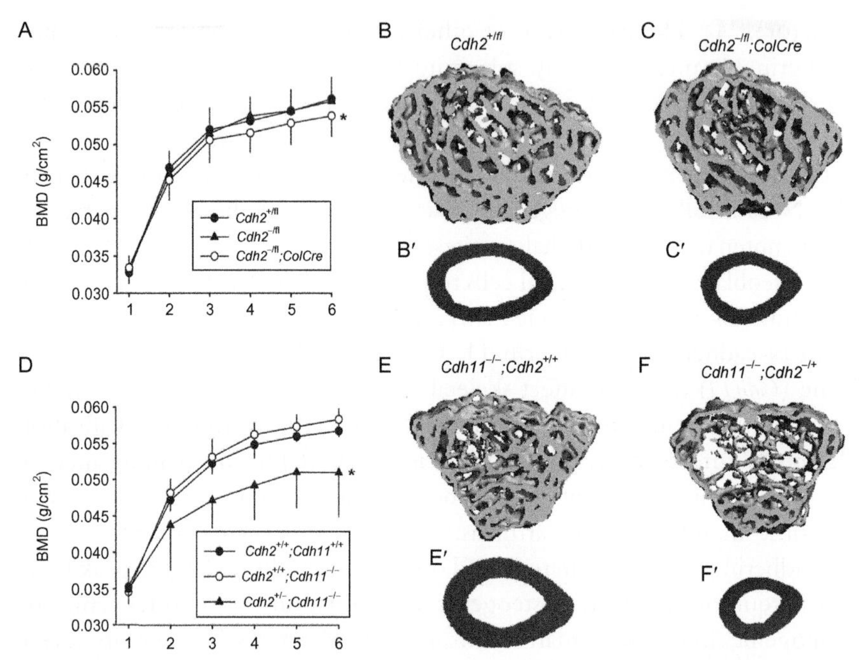

Figure 12.5 Osteopenia in cadherin-deficient mice. (A) Whole-bone and mineral density (BMD), monitored by DXA bone scans at monthly intervals, was lower in *Cdh2*$^{-/fl}$;*ColCre* than in *Cdh2*$^{+/fl}$ and wild-type littermates from the age of 4 months on (*$p<0.05$ for genotype, 2-way ANOVA; $n=4$–5). (B) MicroCT reconstruction of the tibial epiphysis and (B′) cross section of the middiaphysis of control *Cdh2*$^{+/fl}$ mice, contrasting with more rarefied trabecular pattern (C) and smaller cortical cross-sectional area (C′) of *Cdh2*$^{-/fl}$;*ColCre* mice. (D) BMD was significantly lower in *Cdh2*$^{+/-}$;*Cdh11*$^{-/-}$ mice than in the other two genotype groups from the age of 2 months on (*$p<0.05$ for genotypes, 2-way ANOVA; $n=4$–7). Relative to *Cdh11*$^{-/-}$ mice (E, E′), *Cdh2*$^{+/-}$;*Cdh11*$^{-/-}$ mice exhibit severely disrupted trabecular microarchitecture (F) and severely diminished cortical cross section (F′). *Reprinted with permission from Ref. 14.*

tissues, where it maintains cell–cell adhesion and tissue architecture, and it is often downregulated during malignant progression. In contrast, *de novo* expression of N-cadherin is often observed in tumors of epithelial origin. This so-called cadherin switch from E-cadherin to N-cadherin is associated with the epithelial–mesenchymal transition (EMT) observed during cancer invasion[58,59] (see also Chapter 14). As a result of EMT, epithelial cells lose their normal cell polarity and become spindle-shaped and more motile. Cadherin switching is a common occurrence in most cancers, including breast, prostate, and pancreatic cancer, as well as melanoma. N-Cadherin is not required for the morphological changes associated with TGF-β1-induced EMT in mammary epithelial cells, but rather it is important for the change in cell behavior (i.e., increased motility).[60] However, recent genetic data (N-cad KI model discussed later in this chapter) indicate that ectopic expression of N-cadherin in mammary epithelial cells is sufficient to promote EMT *in vivo*.[61]

The mechanisms by which N-cadherin promotes cell motility are not completely understood. However, cross talk with growth factor receptors likely contributes to this cellular behavior. FGFRs comprise a subfamily of receptor tyrosine kinases that are master regulators of a broad spectrum of cellular and developmental processes, including apoptosis, proliferation, migration, and angiogenesis.[62] Upregulation of N-cadherin in breast cancer cells was shown to enhance cell motility, invasion, and metastasis.[63–66] Importantly, N-cadherin attenuates the MAPK-ERK pathway by stabilizing FGFR-1 on the cell surface, thus enhancing receptor signaling.[66] Interestingly, the matrix metalloprotease, MMP-9, was upregulated in N-cadherin-expressing mammary tumor cells in association with increased invasion.[63,66] In a PyMT mouse model, N-cadherin was shown to mediate tumor cell migration by suppressing Akt3 function, whereas Akt1 and Akt2 were unaffected.[67] Hence, N-cadherin may influence multiple signaling pathways that cooperate to mediate invasion and metastasis.

Pancreatic ductal adenocarcinoma (PDA) is an aggressive and deadly disease that is characterized by invasive, metastatic progression and profound resistance to conventional therapeutics. The ability of the KPC (LSL-K-ras^{G12D};LSL-p53^{R17H};Pdx-1/Cre) murine model to recapitulate—histologically and genetically—the cognate features of human pancreatic cancer[68] has allowed investigators to directly test the requirement for N-cadherin in this highly lethal disease. Remarkably, KPC mice with reduced levels of N-cadherin (i.e., N-cad$^{+/-}$) survive 25% longer than animals expressing two wild-type N-cadherin (*Cdh2*) alleles.[69] The survival benefit is likely due to a cumulative effect of N-cadherin's role in different aspects of tumorigenesis, including tumor cell survival, growth, migration, and invasion. The matrix metalloproteinase MMP-7, the expression of which is associated with poor survival in PDA patients,[70] is decreased in the KPC;N-cad$^{+/-}$ tumors. Aberrant FGF signaling is associated with poor prognosis of pancreatic cancer patients,[71,72] and activation of FGF signaling induces cell migration and invasion of pancreatic cancer cells *in vitro*.[73] KPC; N-cad$^{+/-}$ tumor cells exhibit a decrease in FGF-stimulated ERK1/2 activation and a decrease in migration and invasion, consistent with N-cadherin's ability to promote FGFR signaling in PDA.[69] The survival benefit observed in the KPC model is consistent with previous results showing that knockdown of N-cadherin leads to smaller tumors in an orthotopic model of pancreatic cancer.[74]

Paradoxically, KPC N-cad$^{-/-}$ tumors devoid of N-cadherin did not result in a survival advantage, whereas partly reducing N-cadherin levels prolonged survival of the animals.[69] As the N-cadherin gene is presumably deleted before tumor initiation, the KPC N-cad$^{-/-}$ tumors must rely on an alternative mechanism for the development and progression of PDA. The Ig superfamily member NCAM is a potential candidate to mediate FGFR signaling in the absence of N-cadherin. Indeed, NCAM was upregulated in KPC;N-cad$^{-/-}$ tumor cells and these cells had a similar response to FGF stimulation as N-cad$^{+/+}$ tumor cells. Additional studies will be necessary to determine how KPC N-cad$^{-/-}$ tumor cells overcome the requirement for N-cadherin in tumorigenesis.

N-Cadherin has also been implicated in tumor cell–EC interactions critical for the transit of tumor cells into and out of blood vessels. Transendothelial migration is a dynamic process that involves the constant breaking and remaking of intercellular contacts and is accompanied by drastic changes in cell shape and cytoskeletal reorganization in both the tumor cell and its neighboring ECs. Two groups demonstrated that N-cadherin function is important for transendothelial migration of melanoma cells[75] and prostate cancer cells.[76] In melanoma cells, tyrosine phosphorylation of the N-cadherin cytoplasmic domain by Src family kinases resulted in β-catenin dissociation leading to translocation of β-catenin to the nucleus and activation of gene transcription.[77] Collectively, these *in vitro* and *in vivo* studies provide compelling evidence that N-cadherin plays a critical role at one or more steps in cancer progression.

4.2. N-Cadherin as a therapeutic target for cancer

The ability of N-cadherin antagonists to act as anticancer drugs has gained considerable attention in recent years. Synthetic peptides containing the sequence His-Ala-Val (HAV), which is found in the first extracellular (EC1) domains of classical cadherins, were initially shown to inhibit compaction of mouse embryos at the eight-cell stage.[78] Cadherin antagonists have since been refined, including a cyclic pentapeptide containing HAV (known as ADH-1) that inhibits N-cadherin but not E-cadherin function.[79] ADH-1 recently received orphan drug designation from the United States Food and Drug Administration for its use in conjunction with melphalan for the treatment of Stage IIB/C, III, and IV malignant melanoma. ADH-1 has shown promise in controlling metastatic disease in preclinical and initial clinical trials. In animal studies, ADH-1 reduced tumor growth in an orthotopic mouse model of pancreatic cancer using N-cadherin-overexpressing BxPC-3 cells.[80] In a melanoma xenograft model, the combination therapy of ADH-1 and melphalan significantly slowed growth of N-cadherin-positive tumor cells.[81] Additionally, monoclonal antibodies against the ectodomain of N-cadherin slowed growth, invasion, and metastasis of prostate cancer cells in a xenograft model.[82] Whether an N-cadherin antagonist can block metastatic dissemination of tumor cells in the aggressive pancreatic cancer model described above (i.e., KPC mice) remains to be determined.

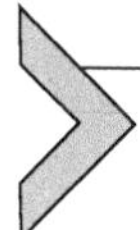

5. FUNCTIONAL OVERLAP AND DISTINCT FUNCTIONS FOR N- AND E-CADHERIN

5.1. Are N- and E-cadherin functionally interchangeable?

Changes in cadherin expression are often associated with changes in cellular morphology and tissue architecture. E- and N-Cadherin exhibit unique and mostly mutually exclusive spatiotemporal expression patterns during embryonic morphogenesis.[83] E-Cadherin is present in the unfertilized egg as maternal mRNA and protein and is essential for preimplantation development.[84] Specifically, E-cadherin is required for the compaction of the blastomeres at the morula stage and the formation of the trophectoderm, the first polarized epithelial cell layer in the mouse embryo. N-Cadherin is initially expressed at the gastrulation stage when epiblast cells downregulate E-cadherin and undergo an EMT. *De novo* expression of N-cadherin occurs at this time in the nascent mesoderm. N- and E-Cadherin share several structural and functional features related to their adhesive activity. The extracellular domains of N- and E-cadherin both contain five cadherin domains, but these are less conserved (46%) than their respective cytoplasmic domains (62%), which interact with the catenins.[85] The differences in the extracellular domains of the cadherins may provide cell-context-dependent functions via specific interactions with distinct growth factor receptors besides homophilic interactions with like cadherins.

One of the central questions in mammalian developmental biology is why the organism uses N- and E-cadherin differentially during specific developmental processes. Studies manipulating cadherin expression *in vivo* are beginning to provide insight into the functional similarities and differences between these two classical cadherins. The ability of a cardiac-specific expression of the E-cadherin transgene to rescue early cardiac morphogenesis (i.e., looping stage) in the N-cadherin-null embryo was the first example of functional replacement of a cadherin subtype during organogenesis.[10] A limitation of this study was the restricted period of expression of the transgene; so, it was not possible to assess E-cadherin function later in cardiac development. In contrast to the above rescue experiment, it was discovered that misexpression of E-cadherin together with N-cadherin in the adult heart is not compatible with normal cardiac function, as such animals develop dilated cardiomyopathy and succumb to sudden cardiac death.[86] Although the cellular mechanism responsible for this phenotype is unknown, it was proposed that the E-cadherin/catenin complex at the ICD might interfere with the normal transmission of the contractile force across the sarcolemma, leading to cardiomyopathy. Interestingly, cyclin D1 expression and DNA synthesis were increased in E-cadherin transgenic hearts compared to N-cadherin-overexpressing animals, suggesting a specific proliferative signal from the epithelial cadherin. Of potential relevance to these transgenic studies, E-cadherin was reported to be upregulated in the heart of hypertensive rats, suggesting that inappropriate E-cadherin expression might indeed play a role in natural heart disease.[87]

5.2. KI of N-cadherin into the E-cadherin locus

Differences in the ectodomain structure of the different cadherins might affect their strength of adhesion. Using a quantitative dual pipette assay, the adhesive strength of E-cadherin was determined to be 3–4 × higher than that of N-cadherin.[88,89] In the above-mentioned transgenic experiments, E-cadherin was expressed in heart muscle, which normally expresses only N-cadherin.[86] In a reverse experiment, Kemler and colleagues generated a gene replacement of N-cadherin into the endogenous E-cadherin locus in mice, referred to as N-cad KI.[90] In this approach, N-cadherin is subjected to the same transcriptional regulation as E-cadherin. Somewhat surprisingly, heterozygous mice coexpressing E- and N-cadherin in all epithelial tissues of the body are viable, with no gross phenotypic abnormality. The lack of an obvious aberration in these mice is in contrast to the transgenic model discussed above, where coexpression of E- and N-cadherin in the working myocardium leads to cardiomyopathy.[86] This suggests that heart muscle, in particular, is sensitive to changes in the cadherin expression profile. However, homozygous N-cad KI mice have a similar lethal phenotype as the E-cad null embryos, demonstrating that E-cadherin provides a specific function(s) that cannot be replaced by N-cadherin in the preimplantation embryo.[90] N-cad KI embryos genetically depleted of maternal E-cadherin undergo compaction at the morula stage, but they are unable to maintain a polarized trophectoderm layer and die at the blastocyst stage. The adhesive force generated by N-cadherin might not be sufficient to maintain the trophectoderm, which must withstand the pressure generated by the blastocoel fluid. In contrast, when ES cells derived from the N-cad KI blastocysts are injected subcutaneously, they can form teratomas containing various epithelial cell lineages, which demonstrates that N-cadherin can support epithelial cell differentiation. In regard to cell-context-dependent adhesion, the mechanical stress on the tumor epithelium is likely much less than on the highly specialized trophectoderm, consistent with the idea that the adhesive force generated by the N-cadherin/catenin complex is less than in the case of E-cadherin.

5.3. N-Cadherin function in intestinal epithelium

The consequences of replacing E-cadherin with N-cadherin in specific epithelial lineages later in development and in the adult have provided some unexpected phenotypes. This experiment can be performed by combining the N-cad KI allele, a floxed E-cad allele and a tissue-specific Cre transgene to delete the wild-type E-cadherin allele, leaving only N-cadherin expressed from the E-cadherin locus. In one study, it was shown that N-cadherin structurally substitutes for E-cadherin in the intestinal epithelial cell layer, but mutant pups are malnourished and die around weaning age.[91] The presence of N-cadherin replacing E-cadherin was sufficient to induce hyperplasia of the intestinal epithelium, leading to polyps that presumably interfered with the animals' ability to feed normally. The self-renewing intestinal epithelium is divided into two functional compartments: the crypts, and the villi that protrude into the lumen and absorb nutrients. Previous studies demonstrated the requirement for the canonical Wnt signaling pathway for homeostasis of the intestinal epithelium.[92] For reasons that are not clear, N-cad KI mice exhibit increased nuclear localization of β-catenin in the crypt cells, providing a possible explanation for the hyperplastic phenotype of the small intestine. Additionally, BMP signaling might play a role in the pathophysiology of the mutant intestine. Although the molecular explanation for the mutant phenotype is unknown, this experiment clearly demonstrates that N-cadherin elicits signals in the stem cell compartment of the intestine that are distinct from those elicited by E-cadherin.

5.4. N-Cadherin function in the mammary epithelium

In another intriguing study using the N-cad KI allele, expression of N-cadherin in the mammary epithelium of pregnant dams interfered with lactation due to precocious involution.[61] As in the gut, this phenotype appears to be due to aberrant signaling and not to a structural defect. However, in the mammary gland, N-cadherin enhances FGF signaling, not Wnt/β-catenin signaling. Of interest, phospho-Stat3, a key player in the involution process, is strongly activated in the N-cad KI alveolar epithelial cells. Furthermore, a functional relationship between N-cadherin and p53 may exist, as heterozygous deletion of the p53 gene rescues the lactation defect in N-cad KI mice. Multiple rounds of lactation led to accumulation of fibrotic tissue and cysts in N-cad KI mice and to invasive tumors when in addition one p53 allele was deleted in the mammary gland. Most surprisingly, using the Rosa26 reporter allele, N-cad KI cells were visualized as they disseminated from the epithelial duct and invaded the surrounding stroma, a process likely driven by activation of FGF signaling. Collectively, these genetic studies show that N-cadherin is functionally interchangeable with E-cadherin as a cell adhesion molecule in epithelial cells but that it stimulates different signaling pathways, which can lead to hyperplasia (intestine) or invasive growth (mammary gland), depending on the cellular context.

6. COOPERATION BETWEEN N-CADHERIN AND INTEGRINS

Cell–cell and cell–ECM adhesions represent highly integrated networks of protein interactions that are crucial for tissue homeostasis and the responses of individual cells to their microenvironments.[93] Through the actin cytoskeleton, cadherins and integrins cooperate to form an interdependent functional network that translates mechanical inputs into intracellular signals. It is generally accepted that N-cadherin promotes cell migration; however, it has also been reported that interfering with N-cadherin expression can increase cell motility in neural crest cells,[94] ECs,[19] and tumor cells.[95] It is possible that increased β1-integrin function may positively influence cell motility in N-cadherin-deficient ECs.[19] Depending on the type of migration assay used, subtle differences in cell behavior may not be appreciated. For example, although N-cadherin-null neural crest cells were shown to migrate faster in culture, their directionality was abnormal so that the total distance they traveled was less than in case of wild-type cells.[94] Moreover, using time-lapse imaging of *ex vivo* whole guts, directionality of migration was perturbed in N-cadherin-deficient enteric neural crest cells, although the speed of locomotion of the mutant cells was not affected.[23] In the same study, it was shown that depletion of both N-cadherin and β1-integrin in enteric neural crest cells led to a failure to colonize the distal part of the gut and that there was more severe aganglionosis in the double-mutant mice compared to the single mutants, thus demonstrating that these cell–cell and cell–ECM adhesion molecules cooperate to mediate cell migration.

7. CONCLUSIONS/FUTURE DIRECTIONS

The goal of this chapter was to highlight recent advances in our understanding of N-cadherin biology, particularly in relation to human disease. To overcome the requirement for N-cadherin in the early mouse embryo, it is necessary to utilize a conditional mutagenesis approach (i.e., Cre/loxP technology) to investigate N-cadherin function in specific cell lineages later in development as well as in the adult organism. The ability to ablate N-cadherin in a tissue-specific and inducible manner is beginning to shed new light on the complex role of this cell-adhesion receptor in both normal physiological and pathological conditions. N-Cadherin function, in particular, appears to be cell context-dependent, as it can mediate strong cell–cell adhesion in the heart, but induces changes in cell behavior toward a more migratory phenotype in other contexts. The importance of N-cadherin in maintaining mechanoelectrical coupling in the heart is evident. However, how the N-cadherin/catenin complex and other ICD proteins are affected by heart disease is poorly understood. Recent studies suggest that oxidative stress can lead to uncoupling of N-cadherin from Cx43 though there is no obvious change in N-cadherin expression at the ICD.[51] The underlying cytoskeleton may be affected in cardiomyocytes experiencing stress[96] so that the normal protein interactions mediated by the N-cadherin/catenin complex are altered causing impaired gap junction trafficking to the ICD. The challenge is to understand the molecular changes that occur in the failing heart that might modify protein interactions with the cytoplasmic domain of N-cadherin, thus affecting linkage with the actin cytoskeleton and overall adhesion strength of the N-cadherin/catenin complex.

The KPC;N-cad$^{+/-}$ mice are an ideal model system to further investigate the role of N-cadherin in tumor development and progression. In contrast to its structural role in the heart, N-cadherin haploinsufficiency affects tumor cell survival, growth, migration, and invasion in the KPC model. It will be important to determine which signaling pathways are required for the survival benefit in the KPC;N-cad$^{+/-}$ animals. This could lead to a new combination therapy targeting different points in the same pathway or two different pathways (e.g., N-cadherin and FGF signaling pathways) to achieve maximum benefit. According to typical Boyden chamber assays, KPC;N-cad$^{+/-}$ tumor cells exhibit decreased migration and invasion.[69] Whether this *in vitro* result translates to a less invasive phenotype *in vivo* awaits further investigation using methods of cell lineage tracing in the KPC mouse model. Considering the wide distribution of N-cadherin expression, it is somewhat surprising that the N-cadherin antagonist ADH-1 is well tolerated by patients. This lack of toxicity might mean that the N-cadherin levels and/or molecular accessibility to the drug may be different on the surface of tumor cells in comparison to normal cells.

In conclusion, loss- and gain-of-function genetic analyses in the mouse have provided new insight into how N-cadherin can influence cell adhesion, cellular communication, and signaling in different cellular contexts. The identification of N-cadherin haploinsufficiency phenotypes further substantiates the importance of this cell adhesion receptor in mammalian physiology. The mouse models discussed in this chapter represent only the beginning as N-cadherin is implicated in various biological processes and in many organ systems. As most studies to date have examined N-cadherin function under normal physiological conditions, it will be particularly informative to manipulate N-cadherin levels in mouse models of human disease.

ACKNOWLEDGMENTS

The research in the author's laboratory is supported by National Institutes of Health grant (HL111788) and American Heart Association grant (GRNT12030343).

REFERENCES

1. Grunwald GB, Pratt RS, Lilien J. Enzymic dissection of embryonic cell adhesive mechanisms. III. Immunological identification of a component of the calcium-dependent adhesive system of embryonic chick neural retina cells. *J Cell Sci* 1982;**55**:69–83.
2. Volk T, Geiger B. A 135-kd membrane protein of intercellular adherens junctions. *EMBO J* 1984;**3**:2249–60.
3. Hatta K, Okada TS, Takeichi M. A monoclonal antibody disrupting calcium-dependent cell-cell adhesion of brain tissues: possible role of its target antigen in animal pattern formation. *Proc Natl Acad Sci USA* 1985;**82**:2789–93.
4. Hatta K, Nose A, Nagafuchi A, Takeichi M. Cloning and expression of cDNA encoding a neural calcium-dependent cell adhesion molecule: its identity in the cadherin gene family. *J Cell Biol* 1988;**106**:873–81.

5. Shapiro L, Fannon AM, Kwong PD, Thompson A, Lehmann MS, Grubel G, et al. Structural basis of cell-cell adhesion by cadherins. *Nature* 1995;**374**:327–37.
6. Dahl U, Sjodin A, Larue L, Radice GL, Cajander S, Takeichi M, et al. Genetic dissection of cadherin function during nephrogenesis. *Mol Cell Biol* 2002;**22**:1474–87.
7. Hollnagel A, Grund C, Franke WW, Arnold HH. The cell adhesion molecule M-cadherin is not essential for muscle development and regeneration. *Mol Cell Biol* 2002;**22**:4760–70.
8. Radice GL, Ferreira-Cornwell MC, Robinson SD, Rayburn H, Chodosh LA, Takeichi M, et al. Precocious mammary gland development in P-cadherin-deficient mice. *J Cell Biol* 1997;**139**:1025–32.
9. Radice GL, Rayburn H, Matsunami H, Knudsen KA, Takeichi M, Hynes RO. Developmental defects in mouse embryos lacking N-cadherin. *Dev Biol* 1997;**181**:64–78.
10. Luo Y, Ferreira-Cornwell M, Baldwin H, Kostetskii I, Lenox J, Lieberman M, et al. Rescuing the N-cadherin knockout by cardiac-specific expression of N- or E-cadherin. *Development* 2001;**128**:459–69.
11. Horikawa K, Radice G, Takeichi M, Chisaka O. Adhesive subdivisions intrinsic to the epithelial somites. *Dev Biol* 1999;**215**:182–9.
12. Li J, Levin MD, Xiong Y, Petrenko N, Patel VV, Radice GL. N-Cadherin haploinsufficiency affects cardiac gap junctions and arrhythmic susceptibility. *J Mol Cell Cardiol* 2008;**44**:597–606.
13. Lai CF, Cheng SL, Mbalaviele G, Donsante C, Watkins M, Radice GL, et al. Accentuated ovariectomy-induced bone loss and altered osteogenesis in heterozygous N-cadherin null mice. *J Bone Miner Res* 2006;**21**:1897–906.
14. Di Benedetto A, Watkins M, Grimston S, Salazar V, Donsante C, Mbalaviele G, et al. N-Cadherin and cadherin 11 modulate postnatal bone growth and osteoblast differentiation by distinct mechanisms. *J Cell Sci* 2010;**123**:2640–8.
15. Kostetskii I, Li J, Xiong Y, Zhou R, Ferrari VA, Patel VV, et al. Induced deletion of the N-cadherin gene in the heart leads to dissolution of the intercalated disc structure. *Circ Res* 2005;**96**:346–54.
16. Li J, Patel VV, Kostetskii I, Xiong Y, Chu AF, Jacobson JT, et al. Cardiac-specific loss of N-cadherin leads to alteration in connexins with conduction slowing and arrhythmogenesis. *Circ Res* 2005;**97**:474–81.
17. Cheng L, Yung A, Covarrubias M, Radice GL. Cortactin is required for N-cadherin regulation of Kv1.5 channel function. *J Biol Chem* 2011;**286**:20478–89.
18. Piven OO, Kostetskii IE, Macewicz LL, Kolomiets YM, Radice GL, Lukash LL. Requirement for N-cadherin-catenin complex in heart development. *Exp Biol Med (Maywood)* 2011;**236**:816–22.
19. Luo Y, Radice GL. N-Cadherin acts upstream of VE-cadherin in controlling vascular morphogenesis. *J Cell Biol* 2005;**169**:29–34.
20. Luo Y, High FA, Epstein JA, Radice GL. N-Cadherin is required for neural crest remodeling of the cardiac outflow tract. *Dev Biol* 2006;**299**:517–28.
21. Kadowaki M, Nakamura S, Machon O, Krauss S, Radice GL, Takeichi M. N-Cadherin mediates cortical organization in the mouse brain. *Dev Biol* 2007;**304**:22–33.
22. Johansson JK, Voss U, Kesavan G, Kostetskii I, Wierup N, Radice GL, et al. N-Cadherin is dispensable for pancreas development but required for beta-cell granule turnover. *Genesis* 2010;**48**:374–81.
23. Broders-Bondon F, Paul-Gilloteaux P, Carlier C, Radice GL, Dufour S. N-Cadherin and beta1-integrins cooperate during the development of the enteric nervous system. *Dev Biol* 2012;**364**:178–91.
24. Pontoriero GF, Smith AN, Miller LA, Radice GL, West-Mays JA, Lang RA. Co-operative roles for E-cadherin and N-cadherin during lens vesicle separation and lens epithelial cell survival. *Dev Biol* 2009;**326**:403–17.

25. Himes AD, Fiddler RM, Raetzman LT. N-Cadherin loss in POMC-expressing cells leads to pituitary disorganization. *Mol Endocrinol* 2011;**25**:482–91.
26. Hynes RO. Targeted mutations in cell adhesion genes: what have we learned from them? *Dev Biol* 1996;**180**:402–12.
27. Linask KK, Knudsen KA, Gui YH. N-Cadherin-catenin interaction: necessary component of cardiac cell compartmentalization during early vertebrate heart development. *Dev Biol* 1997;**185**:148–64.
28. Garcia-Castro MI, Vielmetter E, Bronner-Fraser M. N-Cadherin, a cell adhesion molecule involved in establishment of embryonic left-right asymmetry. *Science* 2000;**288**:1047–51.
29. Shiraishi I, Takamatsu T, Fujita S. 3-D observation of N-cadherin expression during cardiac myofibrillogenesis of the chick embryo using a confocal laser scanning microscope. *Anat Embryol (Berl)* 1993;**187**:115–20.
30. Ong LL, Kim N, Mima T, Cohen-Gould L, Mikawa T. Trabecular myocytes of the embryonic heart require N-cadherin for migratory unit identity. *Dev Biol* 1998;**193**:1–9.
31. Kostetskii I, Moore R, Kemler R, Radice GL. Differential adhesion leads to segregation and exclusion of N-cadherin-deficient cells in chimeric embryos. *Dev Biol* 2001;**234**:72–9.
32. Luo Y, Radice GL. Cadherin-mediated adhesion is essential for myofibril continuity across the plasma membrane but not for assembly of the contractile apparatus. *J Cell Sci* 2003;**116**:1471–9.
33. Goncharova EJ, Kam Z, Geiger B. The involvement of adherens junction components in myofibrillogenesis in cultured cardiac myocytes. *Development* 1992;**114**:173–83.
34. Soler AP, Knudsen KA. N-Cadherin involvement in cardiac myocyte interaction and myofibrillogenesis. *Dev Biol* 1994;**162**:9–17.
35. Hertig CM, Eppenberger-Eberhardt M, Koch S, Eppenberger HM. N-Cadherin in adult rat cardiomyocytes in culture. I. Functional role of N-cadherin and impairment of cell-cell contact by a truncated N-cadherin mutant. *J Cell Sci* 1996;**109**(Pt. 1):1–10.
36. Navarro P, Ruco L, Dejana E. Differential localization of VE- and N-cadherins in human endothelial cells: VE-cadherin competes with N-cadherin for junctional localization. *J Cell Biol* 1998;**140**:1475–84.
37. Carmeliet P, Lampugnani MG, Moons L, Breviario F, Compernolle V, Bono F, et al. Targeted deficiency or cytosolic truncation of the VE-cadherin gene in mice impairs VEGF-mediated endothelial survival and angiogenesis. *Cell* 1999;**98**:147–57.
38. Gory-Faure S, Prandini MH, Pointu H, Roullot V, Pignot-Paintrand I, Vernet M, et al. Role of vascular endothelial-cadherin in vascular morphogenesis. *Development* 1999;**126**:2093–102.
39. Borrmann CM, Grund C, Kuhn C, Hofmann I, Pieperhoff S, Franke WW. The area composita of adhering junctions connecting heart muscle cells of vertebrates. II. Colocalizations of desmosomal and fascia adhaerens molecules in the intercalated disk. *Eur J Cell Biol* 2006;**85**:469–85.
40. Franke WW, Borrmann CM, Grund C, Pieperhoff S. The area composita of adhering junctions connecting heart muscle cells of vertebrates. I. Molecular definition in intercalated disks of cardiomyocytes by immunoelectron microscopy of desmosomal proteins. *Eur J Cell Biol* 2006;**85**:69–82.
41. Pieperhoff S, Franke WW. The area composita of adhering junctions connecting heart muscle cells of vertebrates. VI. Different precursor structures in non-mammalian species. *Eur J Cell Biol* 2008;**87**:413–30.
42. Saffitz JE, Hames KY, Kanno S. Remodeling of gap junctions in ischemic and nonischemic forms of heart disease. *J Membr Biol* 2007;**218**:65–71.
43. Saffitz JE. Arrhythmogenic cardiomyopathy and abnormalities of cell-to-cell coupling. *Heart Rhythm* 2009;**6**:S62–S65.

44. Severs NJ, Bruce AF, Dupont E, Rothery S. Remodelling of gap junctions and connexin expression in diseased myocardium. *Cardiovasc Res* 2008;**80**:9–19.
45. Beardslee MA, Laing JG, Beyer EC, Saffitz JE. Rapid turnover of connexin43 in the adult rat heart. *Circ Res* 1998;**83**:629–35.
46. Shaw RM, Fay AJ, Puthenveedu MA, von Zastrow M, Jan YN, Jan LY. Microtubule plus-end-tracking proteins target gap junctions directly from the cell interior to adherens junctions. *Cell* 2007;**128**:547–60.
47. Solan JL, Lampe PD. Connexin43 phosphorylation: structural changes and biological effects. *Biochem J* 2009;**419**:261–72.
48. Gutstein DE, Morley GE, Tamaddon H, Vaidya D, Schneider MD, Chen J, et al. Conduction slowing and sudden arrhythmic death in mice with cardiac-restricted inactivation of connexin43. *Circ Res* 2001;**88**:333–9.
49. Toyofuku T, Yabuki M, Otsu K, Kuzuya T, Hori M, Tada M. Direct association of the gap junction protein connexin-43 with ZO-1 in cardiac myocytes. *J Biol Chem* 1998;**273**:12725–31.
50. Hunter AW, Barker RJ, Zhu C, Gourdie RG. Zonula occludens-1 alters connexin43 gap junction size and organization by influencing channel accretion. *Mol Biol Cell* 2005;**16**:5686–98.
51. Smyth JW, Hong TT, Gao D, Vogan JM, Jensen BC, Fong TS, et al. Limited forward trafficking of connexin 43 reduces cell-cell coupling in stressed human and mouse myocardium. *J Clin Invest* 2010;**120**:266–79.
52. Linask KK, Ludwig C, Han MD, Liu X, Radice GL, Knudsen KA. N-Cadherin/catenin-mediated morphoregulation of somite formation. *Dev Biol* 1998;**202**:85–102.
53. Kimura Y, Matsunami H, Inoue T, Shimamura K, Uchida N, Ueno T, et al. Cadherin-11 expressed in association with mesenchymal morphogenesis in the head, somite, and limb bud of early mouse embryos. *Dev Biol* 1995;**169**:347–58.
54. Kawaguchi J, Azuma Y, Hoshi K, Kii I, Takeshita S, Ohta T, et al. Targeted disruption of cadherin-11 leads to a reduction in bone density in calvaria and long bone metaphyses. *J Bone Miner Res* 2001;**16**:1265–71.
55. Lee DM, Kiener HP, Agarwal SK, Noss EH, Watts GF, Chisaka O, et al. Cadherin-11 in synovial lining formation and pathology in arthritis. *Science* 2007;**315**:1006–10.
56. Castro CH, Shin CS, Stains JP, Cheng SL, Sheikh S, Mbalaviele G, et al. Targeted expression of a dominant-negative N-cadherin in vivo delays peak bone mass and increases adipogenesis. *J Cell Sci* 2004;**117**:2853–64.
57. Luo Y, Kostetskii I, Radice GL. N-Cadherin is not essential for limb mesenchymal chondrogenesis. *Dev Dyn* 2005;**232**:336–44.
58. Huber MA, Kraut N, Beug H. Molecular requirements for epithelial-mesenchymal transition during tumor progression. *Curr Opin Cell Biol* 2005;**17**:548–58.
59. Wheelock MJ, Shintani Y, Maeda M, Fukumoto Y, Johnson KR. Cadherin switching. *J Cell Sci* 2008;**121**:727–35.
60. Maeda M, Johnson KR, Wheelock MJ. Cadherin switching: essential for behavioral but not morphological changes during an epithelium-to-mesenchyme transition. *J Cell Sci* 2005;**118**:873–87.
61. Kotb AM, Hierholzer A, Kemler R. Replacement of E-cadherin by N-cadherin in the mammary gland leads to fibrocystic changes and tumor formation. *Breast Cancer Res* 2011;**13**:R104.
62. Eswarakumar VP, Lax I, Schlessinger J. Cellular signaling by fibroblast growth factor receptors. *Cytokine Growth Factor Rev* 2005;**16**:139–49.
63. Hazan RB, Phillips GR, Qiao RF, Norton L, Aaronson SA. Exogenous expression of N-cadherin in breast cancer cells induces cell migration, invasion, and metastasis. *J Cell Biol* 2000;**148**:779–90.

64. Hulit J, Suyama K, Chung S, Keren R, Agiostratidou G, Shan W, et al. N-Cadherin signaling potentiates mammary tumor metastasis via enhanced extracellular signal-regulated kinase activation. *Cancer Res* 2007;**67**:3106–16.
65. Nieman MT, Prudoff RS, Johnson KR, Wheelock MJ. N-Cadherin promotes motility in human breast cancer cells regardless of their E-cadherin expression. *J Cell Biol* 1999;**147**:631–44.
66. Suyama K, Shapiro I, Guttman M, Hazan RB. A signaling pathway leading to metastasis is controlled by N-cadherin and the FGF receptor. *Cancer Cell* 2002;**2**:301–14.
67. Chung S, Yao J, Suyama K, Bajaj S, Qian X, Loudig OD, et al. N-Cadherin regulates mammary tumor cell migration through Akt3 suppression. *Oncogene* 2013;**32**:422–30.
68. Hingorani SR, Wang L, Multani AS, Combs C, Deramaudt TB, Hruban RH, et al. Trp53R172H and KrasG12D cooperate to promote chromosomal instability and widely metastatic pancreatic ductal adenocarcinoma in mice. *Cancer Cell* 2005;**7**:469–83.
69. Su Y, Li J, Witkiewicz AK, Brennan D, Neill T, Talarico J, et al. N-Cadherin haploinsufficiency increases survival in a mouse model of pancreatic cancer. *Oncogene* 2012;**31**:4484–9.
70. Fukuda A, Wang SC, Morris JPT, Folias AE, Liou A, Kim GE, et al. Stat3 and MMP7 contribute to pancreatic ductal adenocarcinoma initiation and progression. *Cancer Cell* 2011;**19**:441–55.
71. Ishiwata T, Friess H, Buchler MW, Lopez ME, Korc M. Characterization of keratinocyte growth factor and receptor expression in human pancreatic cancer. *Am J Pathol* 1998;**153**:213–22.
72. Wagner M, Lopez ME, Cahn M, Korc M. Suppression of fibroblast growth factor receptor signaling inhibits pancreatic cancer growth in vitro and in vivo. *Gastroenterology* 1998;**114**:798–807.
73. Nomura S, Yoshitomi H, Takano S, Shida T, Kobayashi S, Ohtsuka M, et al. FGF10/FGFR2 signal induces cell migration and invasion in pancreatic cancer. *Br J Cancer* 2008;**99**:305–13.
74. Shintani Y, Hollingsworth MA, Wheelock MJ, Johnson KR. Collagen I promotes metastasis in pancreatic cancer by activating c-Jun NH(2)-terminal kinase 1 and up-regulating N-cadherin expression. *Cancer Res* 2006;**66**:11745–53.
75. Qi J, Chen N, Wang J, Siu CH. Transendothelial migration of melanoma cells involves N-cadherin-mediated adhesion and activation of the beta-catenin signaling pathway. *Mol Biol Cell* 2005;**16**:4386–97.
76. Drake JM, Strohbehn G, Bair TB, Moreland JG, Henry MD. ZEB1 enhances transendothelial migration and represses the epithelial phenotype of prostate cancer cells. *Mol Biol Cell* 2009;**20**:2207–17.
77. Qi J, Wang J, Romanyuk O, Siu CH. Involvement of Src family kinases in N-cadherin phosphorylation and beta-catenin dissociation during transendothelial migration of melanoma cells. *Mol Biol Cell* 2006;**17**:1261–72.
78. Blaschuk OW, Sullivan R, David S, Pouliot Y. Identification of a cadherin cell adhesion recognition sequence. *Dev Biol* 1990;**139**:227–9.
79. Williams E, Williams G, Gour BJ, Blaschuk OW, Doherty P. A novel family of cyclic peptide antagonists suggests that N-cadherin specificity is determined by amino acids that flank the HAV motif. *J Biol Chem* 2000;**275**:4007–12.
80. Shintani Y, Fukumoto Y, Chaika N, Grandgenett PM, Hollingsworth MA, Wheelock MJ, et al. ADH-1 suppresses N-cadherin-dependent pancreatic cancer progression. *Int J Cancer* 2008;**122**:71–7.
81. Augustine CK, Yoshimoto Y, Gupta M, Zipfel PA, Selim MA, Febbo P, et al. Targeting N-cadherin enhances antitumor activity of cytotoxic therapies in melanoma treatment. *Cancer Res* 2008;**68**:3777–84.

82. Tanaka H, Kono E, Tran CP, Miyazaki H, Yamashiro J, Shimomura T, et al. Monoclonal antibody targeting of N-cadherin inhibits prostate cancer growth, metastasis and castration resistance. *Nat Med* 2010;**16**:1414–20.
83. Takeichi M. The cadherins: cell-cell adhesion molecules controlling animal morphogenesis. *Development* 1988;**102**:639–55.
84. Larue L, Ohsugi M, Hirchenhain J, Kemler R. E-Cadherin null mutant embryos fail to form a trophectoderm epithelium. *Proc Natl Acad Sci USA* 1994;**91**:8263–7.
85. Miyatani S, Shimamura K, Hatta M, Nagafuchi A, Nose A, Matsunaga M, et al. Neural cadherin: role in selective cell-cell adhesion. *Science* 1989;**245**:631–5.
86. Ferreira-Cornwell MC, Luo Y, Narula N, Lenox JM, Lieberman M, Radice GL. Remodeling the intercalated disc leads to cardiomyopathy in mice misexpressing cadherins in the heart. *J Cell Sci* 2002;**115**:1623–34.
87. Craig MA, McBride MW, Smith G, George SJ, Baker A. Dysregulation of cadherins in the intercalated disc of the spontaneously hypertensive stroke-prone rat. *J Mol Cell Cardiol* 2010;**48**:1121–8.
88. Chu YS, Thomas WA, Eder O, Pincet F, Perez E, Thiery JP, et al. Force measurements in E-cadherin-mediated cell doublets reveal rapid adhesion strengthened by actin cytoskeleton remodeling through Rac and Cdc42. *J Cell Biol* 2004;**167**:1183–94.
89. Chu YS, Eder O, Thomas WA, Simcha I, Pincet F, Ben-Ze'ev A, et al. Prototypical type I E-cadherin and type II cadherin-7 mediate very distinct adhesiveness through their extracellular domains. *J Biol Chem* 2006;**281**:2901–10.
90. Kan NG, Stemmler MP, Junghans D, Kanzler B, de Vries WN, Dominis M, et al. Gene replacement reveals a specific role for E-cadherin in the formation of a functional trophectoderm. *Development* 2007;**134**:31–41.
91. Libusova L, Stemmler MP, Hierholzer A, Schwarz H, Kemler R. N-Cadherin can structurally substitute for E-cadherin during intestinal development but leads to polyp formation. *Development* 2010;**137**:2297–305.
92. Pinto D, Gregorieff A, Begthel H, Clevers H. Canonical Wnt signals are essential for homeostasis of the intestinal epithelium. *Genes Dev* 2003;**17**:1709–13.
93. Weber GF, Bjerke MA, DeSimone DW. Integrins and cadherins join forces to form adhesive networks. *J Cell Sci* 2011;**124**:1183–93.
94. Xu X, Li WE, Huang GY, Meyer R, Chen T, Luo Y, et al. Modulation of mouse neural crest cell motility by N-cadherin and connexin 43 gap junctions. *J Cell Biol* 2001;**154**:217–30.
95. Deramaudt TB, Takaoka M, Upadhyay R, Bowser MJ, Porter J, Lee A, et al. N-Cadherin and keratinocyte growth factor receptor mediate the functional interplay between Ki-RASG12V and p53V143A in promoting pancreatic cell migration, invasion, and tissue architecture disruption. *Mol Cell Biol* 2006;**26**:4185–200.
96. Smyth JW, Vogan JM, Buch PJ, Zhang SS, Fong TS, Hong TT, et al. Actin cytoskeleton rest stops regulate anterograde traffic of connexin 43 vesicles to the plasma membrane. *Circ Res* 2012;**110**:978–89.

CHAPTER THIRTEEN

Cadherin Dynamics During Neural Crest Cell Ontogeny

Lisa A. Taneyhill, Andrew T. Schiffmacher
Department of Animal and Avian Sciences, University of Maryland, 1405 Animal Sciences Center, College Park, Maryland, USA

Contents

Abstract

Cell membrane-associated junctional complexes mediate cell–cell adhesion, intercellular interactions, and other fundamental processes required for proper embryo morphogenesis. Cadherins are calcium-dependent transmembrane proteins at the core of adherens junctions and are expressed in distinct spatiotemporal patterns throughout the development of an important vertebrate cell type, the neural crest. Multipotent neural crest cells arise from the ectoderm as epithelial cells under the influence of inductive cues, undergo an epithelial-to-mesenchymal transition, migrate throughout the embryonic body, and then differentiate into multiple derivatives at predetermined destinations. Neural crest cells change their expressed cadherin repertoires as they undergo each new morphogenetic transition, providing insight into distinct functions of expressed cadherins that are essential for proper completion of each specific stage. Cadherins modulate neural crest cell morphology, segregation, migration, and tissue

Progress in Molecular Biology and Translational Science, Volume 116
ISSN 1877-1173
http://dx.doi.org/10.1016/B978-0-12-394311-8.00013-3

formation. This chapter reviews the knowledge base of cadherin regulation, expression, and function during the ontogeny of the neural crest.

1. AN INTRODUCTION TO THE NEURAL CREST

Neural crest cells are an evolutionary marvel and are essential for vertebrate body formation.[1] As multipotent cells, neural crest cells differentiate into a wide array of tissues, yet their duration as a distinct embryonic lineage is short. The biogenesis of the neural crest begins very early in development following the establishment of the embryonic axes. Neural crest specification occurs within early epiblast cells at the neural plate border[2,3] that undergo changes in gene expression and become loosely defined as premigratory neural crest cells (Fig. 13.1). In all species studied except mice, premigratory neural crest cells arise within the dorsal region of the embryonic neural tube.[4] Premigratory neural crest cells initiate an epithelial-to-mesenchymal transition (EMT) that transforms these stationary cells into motile cells. During EMT, premigratory neural crest cells lose apicobasal polarity, downregulate junctional complexes, and reorganize their cytoskeleton to facilitate emigration.[5–7] Migratory neural crest cells communicate with each other in transit as they home in on their target sites. Upon reaching these destinations, they differentiate into specialized cell types, including neurons and glia of peripheral and sensory ganglia, odontoblasts, craniofacial tissues, adrenal cells, and melanocytes. At nearly every step of neural crest ontogeny, cadherins play a critical role (Fig. 13.1 and Table 13.1), whether they function in well-characterized mechanisms such as regulation of cell–cell adhesion, cell movements or cell polarity, or in newly discovered areas, including gene expression feedback and cell–cell signaling.

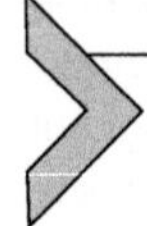

2. CADHERINS IN PREMIGRATORY NEURAL CREST CELLS AND THEIR EMT

2.1. E-cadherin and N-cadherin in early neural crest cell formation

Prior to neurulation, ectodermal cells, including future neural crest cells, express E-cadherin (previously identified as uvomorulin or L-CAM in the chick), the major type-I cadherin.[37–41] Downregulation of E-cadherin and gradual induction of N-cadherin beginning in the neural groove of the developing neural plate define the neural plate border between the neural and the nonneural ectoderm (Fig. 13.1A). This border designates the future

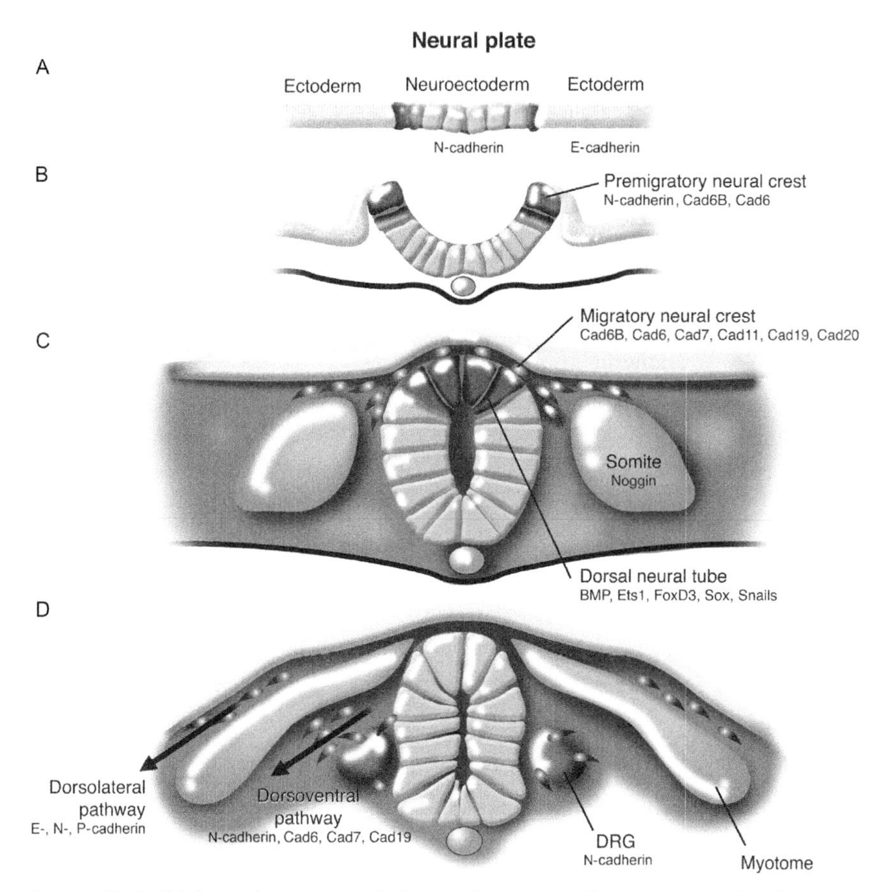

Figure 13.1 Chick trunk anatomy during major stages of neural crest development. (A) Following neurulation, the neural plate border (green) separates neuroectoderm (pink) and nonneural ectoderm (tan) and contains epithelial cells specified to become neural crest cells. (B) As invagination occurs, the premigratory neural crest domain (green) is formed within the elevating neural folds. (C) Later, neural folds fuse to form the dorsal neural tube where premigratory neural crest cells reside. Neural crest cells (blue) delaminate from the dorsal neural tube, undergo an epithelial-to-mesenchymal transition and emigrate away from the dorsal midline. (D) The neuroepithelium fully fuses to form the roof plate of the neural tube. Migratory neural crest cells travel in waves along distinct trajectories, first along the dorsoventral pathway followed by the dorsolateral pathway. Expression of cadherins and relevant signaling molecules and transcription factors are indicated in (A–D). *Illustration courtesy of Paul W. Schiffmacher.* (See Color Insert.)

dorsal neural tube in which premigratory neural crest cells will eventually reside. Elevation of the neural folds occurs, and at this time in the chick head (four-somite stage, 4ss), N-cadherin diminishes in the neural folds.[8,9] N-cadherin expression in the remainder of the neural tube is maintained in a rostrocaudal direction through the trunk.[10] In all vertebrates, N-cadherin

Table 13.1 Chapter-reviewed cadherins expressed in neural crest cells

Species	Cadherin	Neural crest cell type (NCC)	Location	References
Chicken (*Gallus gallus*)	N-cadherin	Premigratory	Head/trunk	8–13
	Cadherin-6B	Premigratory	Head	11,14,15
		Premigratory/early migratory	Trunk	11,14,16
	Cadherin-7	Migratory	Head/trunk	11,14,17
	Cadherin-11	Migratory	Trunk	18
Mouse (*Mus musculus*)	N-cadherin	Premigratory	Head/trunk	19
		NCC-derived enteric ganglia	Gut	20,21
		NCC-derived sympathetic ganglia	Trunk	8–10,22
		NCC-derived cardiac ganglia	Heart	23
	Cadherin-6	Premigratory/ migratory	Head/trunk	24
	Cadherin-20	Migratory	Trunk	25
	Cadherin-11	Migratory	Head/trunk	26,27
Rat (*Rattus norvegicus*)	Cadherin-7	Migratory	Head	28
	Cadherin-19	Migratory	Head/trunk	29
	Cadherin-11	Migratory	Head	30
Frog (*Xenopus laevis*)	N-cadherin	Premigratory	Head/trunk	31
	Cadherin-11	Migratory	Head/trunk	32–35
Zebrafish (*Danio rerio*)	N-Cadherin	Premigratory	Head/trunk	36

is downregulated in the premigratory neural crest domain prior to EMT (Fig. 13.1B). In the chick and frog, premigratory neural crest cells initiate EMT around the time of neural fold fusion to form the neural tube.[8,10–12,31] In mice, however, neural crest EMT is initiated well before neural tube formation, yet N-cadherin is still downregulated in the neural folds.[19] *In vivo* N-cadherin overexpression studies in the chick neural tube/crest provide solid evidence for both direct and indirect functions of N-cadherin in regulating neural crest development. Mesenchymal injection of an N-cadherin function-blocking antibody (raised against the N-cadherin ectodomain) disrupts neural tube integrity and causes ectopic neural crest emigration.[42] N-cadherin-mediated adhesion also prevents premature emigration of more ventrally located neural tube cells that are later specified to become neural crest cells after initial loss of dorsal cells via EMT.[43] Conversely, N-cadherin overexpression in the chick blocks neural crest emigration[11,13] but does not affect differentiation potential as continued expression of Mitf, a marker for early melanocytes, is observed in neural crest cells that fail to migrate beyond the dorsal neural tube midline. Interestingly, overexpression of N-cadherin mutants that contain deletions in the N-terminal, juxtamembrane, or C-terminal domains disrupt neural tube integrity but do not affect neural crest emigration, indicating that all domains are collectively required to elicit the N-cadherin overexpression phenotype.[13]

2.2. N-cadherin and neural crest cell EMT/delamination

Data from *in vivo* studies involving manipulation of cadherin expression in premigratory neural crest cells support the theory that cadherin downregulation is important for the emigration process.[13,44] Recent evidence obtained from live-cell imaging studies, however, demonstrates that disassembly of adherens junctions is not the only event essential for emigration.[45] In addition to downregulating specific cadherins (dependent upon the species), neural crest cells employ other methods to delaminate. By using confocal time-lapse imaging, Ahlstrom and colleagues observed that chick trunk neural crest cells actively retracted their trailing edges upon exit from the neural tube and had the ability to either rupture this edge from the cell body or to maintain it if adherens junction disassembly had not occurred.[46] Although adherens junctions were indirectly lost in the former case as the result of a mechanical event, the presence of adherens junction components in the trailing edge of some emigrating neural crest cells indicates that trailing edge retraction occurred with presumptive adherens junctions still intact. Similar results have been suggested by electron microscopy and observed by live-cell imaging of mouse and zebrafish cells undergoing EMT at other developmental stages.[47–49] These insights highlight a level of plasticity in the individual molecular mechanisms that together comprise EMT.

Delamination, the process by which premigratory neural crest cells segregate and emigrate from the dorsal neural tube, is often temporally associated with the processes comprising EMT.[7,50] The timing of delamination and its overlap with completion of EMT are species-dependent. Cranial neural crest cells in all vertebrates delaminate *en masse*. For instance, *Xenopus* cranial neural crest cells undergo delamination before EMT and maintain cell–cell adhesions as a group segregated from the neural and nonneural ectoderm. EMT and migration begin just before neural tube closure, and EMT is completed only after migration has commenced.[51–53] In the case of mouse cranial neural crest cells, EMT begins in the open neural plate, followed soon afterward by *en masse* delamination of neural crest cells.[7,54,55] Chick cranial neural crest cells undergo EMT and delamination simultaneously after the onset of neural fold fusion.[56] While the synchronicity of delamination and neural tube closure varies across teleost species,[57] zebrafish cranial neural crest cells undergo EMT and exit the dorsal neural tube at the same time.[58]

In contrast to cranial neural crest cells, trunk neural crest cells of all the above-mentioned organisms progressively delaminate as individual cells. In the chick trunk, neural crest cells simultaneously delaminate and undergo EMT and then emigrate, beginning at the rostral end adjacent to dissociating somites.[6,31,46,50,59–61] This timing is controlled by the downregulation of Noggin, which normally serves to antagonize the delamination activator BMP4[62–64] (Fig. 13.1C). BMP4 upregulation inversely associates with N-cadherin protein turnover, yet *N-cadherin* transcript abundance remains unchanged.[13] To achieve this, BMP4 negatively regulates N-cadherin posttranslationally through the activation of the A Disintegrin And Metalloproteinase (ADAM) family member ADAM10, which proteolytically cleaves membrane-bound N-cadherin in premigratory neural crest cells.[13] These results tie together previous evidence of a role for ADAM10 in regulating avian neural crest migration[65] as well as biochemical data demonstrating ADAM10-mediated N-cadherin cleavage in neurons.[66,67] A recent study has also shown that secreted matrix metalloproteinase 9 regulates chick neural crest cell delamination and migration through processing of N-cadherin and laminin.[68]

Other posttranslational mechanisms may exist to modulate N-cadherin protein levels on emigrating neural crest cell membranes and reduce cell–cell adhesion. *In vitro* immunostaining of N-cadherin in quail trunk neural tube explants reveals N-cadherin at the leading tip of processes of migrating neural crest cells but not at the sites of cell–cell contact.[69] Following addition of inhibitors of serine/threonine kinases, protein kinase C and tyrosine kinases to the media, N-cadherin was trafficked to cell–cell contacts, presumably to restore adherens junctions. Interestingly, kinase or phosphatase inhibitor supplementation had no effect on overall N-cadherin phosphorylation,[69] suggesting that N-cadherin intracellular trafficking is controlled by the phosphorylation states of other proteins. While the presence of N-cadherin protein in cultured migratory neural crest cells does not recapitulate the observed N-cadherin downregulation within emigrating chick neural crest cells *in vivo*, it is likely that emigrating neural crest cells utilize similar mechanisms to modulate N-cadherin adherens junction stability and disassembly.

Work conducted in zebrafish further demonstrates this in that the Wnt target transcription factor Ovo1 represses secretory pathway effector genes, including *Rab3c*, *Rab12*, *Rab11fip2*, and *Sec6*, all encoding Rab GTPases.[36] Knockdown of Ovo1 leads to derepression of these *Rab GTPases* and abnormal N-cadherin trafficking. Eventually, hyperactive N-cadherin endocytosis depletes membrane pools and disrupts neural crest migration. A similar mechanism may be employed to regulate N-cadherin trafficking in mouse neural crest cells, as cranial neural crest emigration is also disrupted in *Ovo-like2* null mutants.[70]

2.3. Type-II cadherins in premigratory and delaminating neural crest cells

Loss of N-cadherin protein from the presumptive neural crest domain is concomitant with the induction of type-II cadherins. As N-cadherin is downregulated, mouse premigratory neural crest cells upregulate the type-II cadherin Cadherin-6 (Cad6, K-Cad) and maintain its expression throughout EMT and the early migratory phase (Fig. 13.1). Mouse Cad6 protein is 97% identical to human Cadherin-6 (CDH6), 95% identical to rat Cad6, and 86% identical to chick Cadherin-6B (Cad6B).[24] Differences in spatiotemporal expression profiles between mouse Cad6 and chick Cad6B called into question whether they were true orthologs. Unlike mouse Cad6 expression, chick Cad6B expression is restricted to premigratory neural crest cells and is not retained in the migratory population. Because of this difference, the "B" designation was added to Cad6 by Nakagawa and colleagues upon their discovery of it.[14]

Mouse midbrain and anterior hindbrain *Cad6* expression is restricted to the neural fold ridge containing the premigratory neural crest domain.[24] On embryonic day 8 (E8), actively migrating midbrain neural crest cells express *Cad6*. These cells move through the mandibular arches by E8.5 and differentiate into neurons of developing cranial ganglia by E9.5. They also aggregate alongside nerves as Schwann cell precursors. Beginning in the hindbrain at E8.5 and extending caudally, *Cad6*-positive neural crest cells are observed migrating laterally in streams.[24] Cad6B expression during chick neural crest formation is similar but not entirely conserved with mouse Cad6 patterns. *Cad6B* is expressed along the entire rostrocaudal axis of the emerging neural folds/premigratory neural crest in HH stage 6 embryos, and its expression in the cranial neural folds peaks around the time of neural fold fusion. As cranial neural crest cells initiate EMT, *Cad6B* is downregulated to undetectable levels as soon as the cells actively emigrate from the dorsal midline.[14] *Cad6B* remains depleted in the dorsal neural tube, even as midbrain neural crest cells continue to undergo EMT, until emigration ceases at HH stage 11.[71–74] Intriguingly, Nakagawa and Takeichi hypothesized that restricted Cad6B expression in premigratory neural crest cells may help segregate this special subpopulation of neural tube cells from the rest of the neuroepithelium.[14]

Cad6B immunohistochemical analysis confirms previous *in situ* data demonstrating *Cad6B*-positive expression in premigratory neural crest cells,[11,14] yet transcript and protein spatiotemporal profiles do not entirely overlap.[14–16] In the cranial neural crest, membrane-localized Cad6B protein turnover does not coincide with *Cad6B* repression at the 6ss. Instead, Cad6B protein loss is detected only within neural crest cells that have completed EMT and have emigrated. A subpopulation of cells, including presumably premigratory neural crest cells at the initial phase of EMT as well as dorsal neural tube cells, continues to exhibit membrane-localized Cad6B protein. Only a few hours later, at the 7–8ss when a significant number of cranial neural crest cells are emigrating, are Cad6B protein levels decreased in cells residing in the midbrain dorsal neuroepithelium.[15] From the 8ss onward, Cad6B expression becomes restricted to apical regions of cells interfacing the lumen and contributing to the roof plate (unpublished data[6]).

In the chick trunk, Cad6B expression also demarcates the premigratory neural crest domain. Trunk neural crest cells actively transitioning to the migratory phase, however, exhibit a different temporal pattern of downregulation than that observed in midbrain neural crest cells, with Cad6B present in newly migratory trunk neural crest cells in 10–13ss embryos.[16] This pattern is reminiscent of Cad6 expression in mouse migratory neural crest cells,[24] although Cad6 is maintained even longer in advanced migratory cells proceeding dorsolaterally at comparable trunk axial levels.[24] As such, Cad6 expression recapitulates the cumulative expression profiles of Cad6B and another type-II cadherin, Cadherin-7 (Cad7), which is observed in the chick migratory neural crest cell population.[14] At trunk axial levels possessing both premigratory and migratory neural crest cell populations, migratory neural crest cells express Cad7 (Fig. 13.1C). Interestingly, Cad7 is expressed in newly emigrating midbrain neural crest cells that are negative for Cad6B, yet Cad7 expression in trunk migratory neural crest is present only in a subpopulation migrating between the dermomyotome and adjacent neural tube.[14] These reciprocal expression patterns of Cad6B and Cad7 suggest that Cad6B turnover and Cad7 induction in migratory neural crest cells could be coupled mechanisms.

The difference in the developmental timing at which Cad6B protein levels are lost in cranial and trunk neural crest cells suggests alternate roles of Cad6B during EMT at these different axial levels. The rapid loss of Cad6B protein from midbrain neural crest cells undergoing EMT suggests that Cad6B turnover is required for completion of EMT and successful transitioning to a migratory state.[44] Manipulation of endogenous Cad6B levels in the midbrain prior to EMT supports this hypothesized function. By electroporating Cad6B morpholinos into the premigratory neural crest domain, Coles and colleagues demonstrated that a reduction in Cad6B expanded the migratory neural crest cell domain later in development. Thus, it was deduced that morpholino-induced premature reduction of Cad6B-containing adherens junctions, prior to commencement of EMT, quickens the processing duration of one EMT component (disassembly of Cad6B interactions between cells), and results in overall premature emigration and an increase in the migratory neural crest cell domain.[44] This hypothesis was further supported by *in vitro* observation of a significant increase, compared to controls, in the number of neural crest cells actively migrating away from Cad6B morpholino-treated neural fold explants.

Because trunk neural crest delamination progresses with a continuous migration of individual cells in a drip-like fashion,[6] the need for Cad6B to maintain domain cohesiveness would only be biologically relevant to sustain the premigratory neural crest cell population. Therefore, additional functions for Cad6B may be utilized by emigrating trunk neural crest cells. The persistent expression of Cad6B in fully delaminated chick rostral trunk neural crest cells supports the notion that Cad6B downregulation and membrane turnover are not so critical for trunk neural crest cell EMT. Instead, the data support a more facilitative pro-EMT function, as downregulation occurs only after EMT and delamination take place. Not surprisingly, the results obtained following Cad6B knockdown in trunk neural crest cells (inhibition of neural crest emigration) are different from what is observed upon Cad6B depletion in the head.[16] Though only individual cells were counted and the size of the migratory neural crest domain itself was not determined, these findings suggest that sustained Cad6B levels may regulate emigration.

Cad6B overexpression in midbrain premigratory neural crest cells disrupts neural crest migration, causing emigrating neural crest cells to cease migration and clump adjacent to the neural tube, while overexpression of Cad6B in the trunk enhances migration.[16,44] Disruption of migration in neural crest cells overexpressing other cadherins, including N-cadherin[11,13] and Cad7,[11] indicates that this cessation of migration and neural crest cell aggregation phenotype is a result of exaggerated cadherin-mediated cell–cell adhesion and not due to other cadherin functions. Interestingly, neural crest cell aggregation via ectopic cadherin expression may affect the migratory capacity of these cells but not their differentiation potential. Trunk neural crest cells overexpressing N-cadherin or Cad7 are at times observed in the neural tube lumen, but these cells still express Mitf.[11] Similar results were obtained with Cad6B overexpression in cranial neural crest cells.[44]

In contrast to the effects of BMP4 on N-cadherin, inhibition of BMP signaling by ectopic Noggin downregulates *Cad6B* transcripts *in vivo*.[62] Further, Cad6B and BMP signaling interact positively with each other via feedback, and thus provide insight into the mechanism behind the pro-EMT activities of Cad6B in the trunk. Coelectroporation of a Cad6B expression construct with a BMP signaling-sensitive β-galactosidase reporter leads to increased Cad6B-induced ectopic BMP signaling in neural tube cells and disrupts epithelial integrity.[16] This BMP signaling induction occurs in the presence of Noggin, indicating that Cad6B may act by inhibiting Noggin and/or by functioning downstream of ligand-receptor binding. BMP and Wnt also induce *Cad6B* in cultured explants of intermediate neural plate tissue.[75,76] While the mechanism remains to be elucidated, it is not known whether this function of BMP is specific to Cad6B or is a universal function with respect to other type-II cadherins that modulate EMT. Thus, Cad6B functions differently in cranial and trunk neural crest cell EMT, possibly due to axial level molecular and physical differences.

2.4. Cadherin transcriptional regulation

Differences in the timing of cadherin expression between head and trunk neural crest cells reflect the usage of separate regulatory mechanisms within these populations. These differences are not at all surprising, given that cranial and trunk neural crest cells are governed by similar yet distinct gene regulatory networks that direct delamination and EMT.[7,15,77] Prior work indicates that N-cadherin is negatively regulated by FoxD3,[74,78,79] with FoxD3 overexpression resulting in appreciable reduction of membrane-bound N-cadherin and upregulation of β1-integrins and Cad7, but without promoting neural crest emigration.[78] It is likely that FoxD3 acts as a co-regulator of cadherin switching and functions together with other transcription factors, along with other posttranscriptional mechanisms, to facilitate emigration.

Snail1 and Snail2 act together with other transcription factors to promote EMT (see also Chapter 14), and in the chick, Snail2 represses the transcription of *Cad6B* and *αN-catenin*, genes whose protein products form adherens junctions.[15,80] *Cad6B* repression by Snail2 exemplifies a conserved role during normal development for Snail family transcription factors in negatively regulating cadherins during EMT.[81–84] The spatiotemporal expression patterns of Snail2 and *Cad6B* are reciprocal within cranial neural crest cells, and this finding led to the hypothesis that Snail2 negatively regulates *Cad6B* in premigratory neural crest cells *in vivo*.[15] Indeed, Snail2 knockdown within cranial (and trunk) premigratory neural crest cells results in *Cad6B* upregulation within 30 minutes. Neural fold explants in which Snail2 has been depleted also show a significant reduction in *Snail2* morpholino-positive cells emigrating from the explants. Importantly, this reduction can be partially rescued by concomitant electroporation of *Cad6B* morpholino. Chromatin immunoprecipitation and quantitative PCR analysis for Snail2 binding within the *Cad6B* regulatory region reveal that Snail2 binds to three distinct sets of E boxes to repress *Cad6B*. Within the trunk, it is unclear if Snail2 expression is sufficient to completely reduce *Cad6B*, as *Cad6B* transcripts and protein show some spatiotemporal overlap with Snail2.[15,16] It remains to be elucidated whether this persistent Cad6B protein is due to incomplete repression of *Cad6B* expression via Snail2 (and below the detection threshold of *in situ* hybridizations) or to a long protein half-life, thereby accounting for the lag observed between mRNA down-regulation and protein turnover at the membrane. Failure to detect *Cad6B* transcripts in newly migrating neural crest cells positive for Cad6B protein indicates that Cad6B protein may be stabilized.[14,15,75] It is unknown whether Cad6B protein levels in emigrating cranial neural crest cells are regulated by a proteolytic processing mechanism similar to N-cadherin in the trunk.[13] Alternatively, a common mechanism of Cad6B posttranslational regulation could be universally utilized in all emigrating neural crest cells, yet only trunk neural crest cells may possess an extra level of proteolytic inhibition to delay Cad6B turnover.

It is not known whether Snail proteins in other species regulate *Cad6B* orthologs or other cadherins. While *Xenopus* Snails play an equally vital and conserved role as transcriptional repressors in neural crest cell specification and migration, it remains to be determined whether Snails regulate other unknown cadherins in this species.[85] A reduction in *Xenopus* Snail2 decreases neural crest migration,[86] yet the *Xenopus* CDH6 ortholog Xcad6 is not expressed in neural crest cells but rather in placode cells.[87] In mice, *Snail1* is expressed in developing neural folds in premigratory neural crest cells likely expressing *Cad6*.[88] *Snail1* and/or *Snail2* knock-out mice, however, do not exhibit any defects in neural crest cell formation or delamination.[89,90] In contrast, a neural crest cell-specific knockout of *Snail1* on a *Snail2* null background results in later craniofacial defects.[91]

Snail2 overexpression in the chick, alone or with Sox9, does not repress *N-cadherin* transcription,[78] and chick trunk *N-cadherin* transcripts are not significantly elevated within 30 minutes after electroporation of a *Snail2* morpholino.[15] Despite these findings, Snail2 may regulate chick trunk *N-cadherin* levels through an interaction with Ets1, which also controls cadherin expression during cranial neural crest development. Overexpression of both Snail2 and Ets1 depletes N-cadherin in ectopically delaminating trunk neural tube cells and also downregulates *Cad6B*.[77] Biochemical studies detailing the potential interaction of Snail2 and Ets1 in regulating cadherins thus warrant further attention.

Inoue and colleagues recently dissected out potential *cis*-regulatory elements within the mouse *Cad6* upstream sequence.[92] Transgenic mouse lines containing LacZ or GFP reporters driven by various lengths of the genomic *Cad6* sequence were created to determine which sequences could recapitulate endogenous *Cad6* spatiotemporal patterns.[92,93] Separate modules of the genomic sequence elicited restricted *LacZ* induction in either midbrain/hindbrain neural ridges, in trunk neural crest cells, or in both. It remains to be elucidated, however, whether the same set of transcription factors regulate *Cad6* in both cranial and trunk neural crest cells yet bind different enhancers, or if neural crest-specific *Cad6* expression is controlled by different transcription factors. Putative transcription factor binding sites for Pax6 and Sox9 are located within the segment that induced cranial neural crest-specific *LacZ* expression, while a potential Sox5 binding site controls *LacZ* induction in the trunk. This Sox5 site is conserved within the genomic sequence spanning chick *Cad6B*, implying a potential role for Sox5 regulation of *Cad6* orthologs in trunk neural crest cells.[92]

3. CADHERINS IN NEURAL CREST MIGRATION

3.1. Cadherin-7, Cadherin-19, and Cadherin-20

Cadherin-7 (Cad7) expression is induced in a subset of chick migratory neural crest cells and early migratory cephalic neural crest cells in the rat.[11,14,28] Cad7 or Cad7-like cadherin expression, however, is conserved in other neural and nonneural crest tissues among frogs (F-cadherin or Cad20 in *Xenopus*),[94,95] mice,[25] and zebrafish.[96,97] Although rat *Cad7* transcripts were detected in a subset of migratory neural crest cells,[28] no comparable expression pattern was detected in mouse neural crest cells, as mice express *Cad7* in the brain only.[25]

Both chick cranial and trunk migratory neural crest cells express Cad7 (Fig. 13.1). Midbrain neural crest cells express detectable levels of *Cad7* transcripts soon after emigrating from the dorsal neural tube at HH stage 10.[11,14] Following delamination, trunk neural crest cells within the first wave emigrating along the ventral trajectory express Cad7, and expression is maintained as the trajectory changes to the more lateral route. Although transcripts are undetectable by *in situ* hybridization, migratory neural crest-derived melanocyte precursors following the dorsolateral trajectory in the last wave of trunk migration remain positive for Cad7 protein.[11,14] Immunostaining with Cad7 and HNK-1 antibodies shows considerable colocalization of these two proteins, which was not as evident in earlier experiments combining colabeling of *Cad7* transcripts with HNK-1 immunostaining. The incomplete overlap of *Cad7* transcripts and protein localization has been verified in other studies where Cad7 immunostaining was positive in the entire migratory domain, yet *Cad7* transcripts were diminished in the leading trunk ventral migratory population.[17] These data suggest a discrepancy in detection thresholds between *Cad7 in situ* hybridization and immunohistochemistry techniques or imply a long half-life for Cad7 protein. Cad7 overexpression in the trunk mimics the N-cadherin overexpression phenotype, whereby late emigrating neural crest cells fated to become melanocytes fail to travel between the dorsal dermomyotome and the somite and therefore do not emigrate beyond the dorsal neural tube.

Cad7 induction may promote the emigration process in conjunction with downregulation of N-cadherin and Cad6B in delaminating neural crest cells. Cells expressing Cad7 or N-cadherin tend to not intermix and aggregate separately, demonstrating that homophilic interactions are the dominant forces in promoting cellular segregation.[14] Cad7-expressing cells reaggregated at a slower rate, and they exhibited enhanced dispersion rates when plated on fibronectin,[98] which suggested that Cad7 imparts a lower homotypic affinity than N-cadherin to neural crest cells *in vitro* and does not significantly impact forward migration of neural crest cells *in vivo*.

Multiple studies provide insight into the regulatory mechanisms underlying *Cad7* expression in chick migratory neural crest cells. Sox9 overexpression induces ectopic *Cad7* and *Cad6B* within 6 hours of electroporation, suggesting that Sox9 may be a transcriptional activator for both cadherins.[99] *Cad6B* expression, however, was only transient, while *Cad7* was not maximally expressed until 24 hours postelectroporation. These data suggest that induction of *Cad7* and *Cad6B* requires other potential cadherin-specific coregulators to activate expression. In support of this, FoxD3 and Sox10 overexpression also induces ectopic *Cad7* in the neural tube.[78,100] As Sox9 is expressed early during neural crest specification, and *Cad6B* and Cad7 are expressed in premigratory and migratory neural crest cells, respectively, the timing of ectopic cadherin induction and continual expression *in vitro* correlates with induction *in vivo*.

Other differences exist between mouse and rat cadherin profiles in migratory neural crest cells. In the rat, *Cadherin-19* (*Cad19*) is induced in neural crest cells migrating throughout the facial region and down into the first pharyngeal arch between E10.5 (12ss) and E10.75 (16–18ss), and a subset of these cells coexpress *Sox10*.[29] These data suggest that *Cad7* expression in newly migratory cephalic neural crest cells precedes *Cad19* expression,[28,29] but further studies are needed to determine if there is any spatiotemporal overlap. Another observed species difference involves *Cadherin-20* (*Cad20*) expression, which is observed in mouse trunk neural crest cells migrating along the dorsolateral trajectory. Rat neural crest cells and mouse cephalic migratory neural crest cells, however, do not express *Cad20*.[25,28] In the mouse, cephalic *Cad20* expression is limited to the anterior neural plate and newly forming cephalic folds, and later in the cephalic neuroectoderm.[25] The partially overlapping expression patterns among chick *Cad7*,[14] rat *Cad7* and *Cad19*,[28,29] and mouse *Cad20*[25] in various subpopulations of migratory neural crest cells imply that these cadherins may be regulated, in part, by common, conserved transcription factors. Indeed, the *Cad7*, *Cad19*, and *Cad20* genes are universally clustered at genetic loci in chick, rat, mouse, and human, although synteny is not conserved in chick as it is in human and rodents.[28,101,102] The fact that there is not more overlap in the expression of this *Cad7*/*Cad19*/*Cad20* gene cluster among species indicates that these genes are more likely to be regulated by species-specific transcriptional regulators and/or species-specific *cis*-regulatory elements within their promoters.[28]

3.2. Cadherin-11

In addition to the *Cad7/Cad19/Cad20* cluster, *Cadherin-11* (*Cad11*) is another conserved type-II cadherin expressed in the migratory neural crest populations of mice, rats, chick, and frogs (Fig. 13.1C). Chick Cad11 is limited to a subset of HNK1-positive migratory trunk neural crest cells as well as the surrounding mesoderm.[18] It is not known if Cad11 is expressed in chick cranial neural crest cells. In the mouse, *Cad11* is observed in both migratory neural crest cells and mesoderm of the cranial mesenchyme by E7.5–E8, and its expression is maintained as neural crest cells migrate through the mandibular and maxillary arches.[26,27] *Cad11* is then downregulated in neural crest cells as they differentiate into cranial ganglia. Further, *Cad11* expression is conserved in rat and mouse,[30] and together these early *Cad11* profiling studies support a role for Cad11 as a major mesenchyme-specific cadherin.[26,27,30] These spatiotemporal expression patterns are also conserved across species because *Xenopus Cad11* expression parallels rodent *Cad11* patterns.[32,33] In addition to mesoderm expression, *Xenopus Cad11* transcripts are detected in cranial neural crest populations migrating within the mandibular, branchial, and hyoid arches, and also within a subset of trunk neural crest cells migrating into the dorsal fin at later stages.[32,33]

Xenopus Cad11 knockdown disrupts completion of cranial neural crest migration into the hyoid and branchial arches but not migration into the mandibular arch.[34] Failure to enter the pharyngeal pouches results in increased melanocyte differentiation at the expense of facial cartilage formation and can be attributed to loss of Cad11 localization to the leading cell membrane edge. Phenotypic rescue is achieved by coinjection of human or *Xenopus* Cad11 or various *Xenopus* Cad11 deletion mutants (partial rescue only), but not *Xenopus* Cad6. In agreement with Cad11 knockdown observations, neural crest cells with elevated Cad11 levels also fail to reach their destination and differentiate appropriately.[35] This block in migration can be rescued by coinjection of *β-catenin* mRNA, implying that ectopic Cad11 expression not only affects proper cell–cell adhesion, but also results in abnormal sequestration of endogenous β-catenin and a reduction in Wnt signaling.[34] Interestingly, *Cad11* is also under positive control of Wnt signaling.[32] Injection of a mutant *Cad11* lacking a β-catenin-binding domain elicited a similar phenotype to wild-type overexpression but to a lesser degree, further demonstrating that an equilibrium of Cad11 binding to β-catenin is important for overall cadherin function.[35] The Cad11 overexpression phenotype is also rescued by coinjection of *ADAM13* mRNA, which proteolytically processes Cad11 in *Xenopus* cranial neural crest cells *in vivo*.[103] Because McCusker and colleagues found that Cad11 proteolytic processing does not solely decrease homophilic interactions and destabilize cell–cell adhesion, they postulated that the cleaved Cad11 ectodomain possesses a specific novel function.[103] Cad11 ectodomain overexpression rescues the disrupted migration phenotype caused by elevated Cad11 levels (due to Cad11 overexpression or ADAM13 knockdown) in a non-cell-autonomous fashion that is independent of intercellular cadherin–protein interactions. Although the mechanism underlying the promigration function of the Cad11 ectodomain remains unresolved, McCusker and colleagues posit that the ectodomain acts as both a competitor of wild-type Cad11 and a cofactor with other membrane receptors to induce signaling cascades, as observed for type-I cadherins in other systems.[104]

Recent data also demonstrate an equally critical role for the Cad11 cytoplasmic domain in promoting proper *Xenopus* cranial neural crest migration. Embryos expressing a mutant Cad11 protein containing the membrane-bound cytoplasmic domain, but not the soluble cytoplasmic domain, enhanced rescue of Cad11 knockdown *Xenopus* morphants better than other Cad11 mutants possessing an intact ectodomain but lacking the cytoplasmic domain.[34] The Cad11 cytoplasmic domain facilitates membrane protrusions by interacting with multiple proteins, including β-catenin, Trio, and the GTPases RhoA, Rac1, and Cdc42.[34] These findings are further supported by data showing that ADAM13-mediated proteolysis of Cad11 produces a membrane-bound cytoplasmic domain during cranial neural crest cell migration.[103]

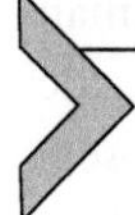

4. CADHERINS IN NEURAL CREST DIFFERENTIATION

4.1. Ganglia formation

Many neural crest-derived ganglia depend on N-cadherin for their appropriate formation. N-cadherin induction generally correlates with a loss of type-II cadherin expression that is essential during migration. N-cadherin is expressed in the dorsal root and sympathetic ganglia,[8–10,22,105–107] in chick sensory nerve fibers,[107] and in mouse cranial nerves.[106] The precursory origin of N-cadherin-expressing cells within zebrafish cranial ganglia, however, remains to be elucidated.[108,109]

Enteric neural crest cells originating from the hindbrain region and derived from the vagal neural crest migrate long distances to invade the gut and progressively differentiate, along with neural crest cells from the sacral region, into cells of the enteric nervous system. Proper formation of the enteric nervous system requires both appropriate enteric neural crest cell migration and proliferation.[110,111] Aberrant enteric neural crest migration leads to insufficient ganglia formation in the terminal region of the gut, resulting in Hirschsprung's disease, a pathology characterized by distal bowel distension and frequent intestinal obstruction.[112] Enteric neural crest cells expressing N-cadherin migrate through the gut in waves, progressively differentiating into neurons as they exit the wave front.[20,113] They extend axonal growth cones to associate with other neurons and participate in ganglia formation. From the time when enteric neural crest cells are within a migratory wave front to later when they aggregate to form ganglia, N-cadherin expression increases and remains localized to intercellular membrane contacts.[20,114] In contrast, Cad11 and Cad6 levels are decreased, with Cad6 becoming localized primarily to the cytoplasm as opposed to the membrane.[20]

In wild-type embryos, enteric neural crest cells located at the leading front form a scaffold along which lagging, more rostral, enteric neural crest cells migrate and extend axonal processes.[115,116] To investigate the role of N-cadherin in these processes, conditional knockout mice were created that lack *N-cadherin* in cephalic neural crest cells beginning at E8.[21] These mice only transiently exhibit abnormal gut migration, manifested as a temporary delay in colon colonization. Migrating neural crest cells lacked directionality and did not form chains. Interestingly, *N-cadherin*-deficient enteric neural crest cells did not exhibit increased β-catenin nuclear localization and still differentiated normally into neurons.[21] Migrating neural crest cells from *N-cadherin–β1-integrin* double-conditional knockout mice exhibited a complete lack of proximal hindgut colonization and severe agangliogenesis, which is not observed in either of the single conditional knockouts.[21] These observations highlight a cooperative interaction between N-cadherin-based adherens junctions and β1-integrin-containing focal adhesions to promote enteric neural crest cell migration and differentiation *in vivo*.

N-cadherin expression is also essential for the formation of sympathetic ganglia along the rostrocaudal axis of the trunk.[22] Upon completion of migration, neural crest cells begin a reiterative process of filopodial extension and retraction to contact neighboring cells, leading to cell sorting, coalescing, and condensation into ganglia.[117] N-cadherin expression is induced at the onset of coalescence[8–10,22] and is required for proper aggregation, as injection of dominant-negative N-cadherin expression constructs or N-cadherin-blocking antibodies at ganglia target sites significantly increases aggregate length along the rostrocaudal axis.[22] Alternatively, N-cadherin overexpression decreases the area but not the length of the forming ganglion. Neural crest cells overexpressing N-cadherin were also observed to extend significantly more filopodia out into the interganglionic regions than controls, thereby establishing cell–cell contacts between sympathetic ganglia. These experiments highlight the important contribution of N-cadherin in facilitating proper cell adhesion during sympathetic ganglia formation.

4.2. Heart formation

Cardiac neural crest cells originate from the postotic hindbrain neural fold corresponding to the segmented rhombomeres of the hindbrain neural tube to the third somite level.[118] Cardiac neural crest cells follow trajectories into the third, fourth, and sixth caudal pharyngeal arches on their way to the cardiac outflow tract, where they differentiate into cardiac ganglia and condensed mesenchymal cells of the aorticopulmonary septation complex.[118–121] As mouse cardiac neural crest cells migrate into the outflow tract, N-cadherin is faintly detected in membrane regions at sites of intercellular contact (see also Chapter 12). As differentiation is initiated between E11.5 and E12.5, N-cadherin levels are noticeably increased in neural crest cells undergoing elongation and condensation.[23]

The *N-cadherin* knockout mouse exhibits early heart defects and is embryonic lethal around E10, before cardiac differentiation is well underway.[19] In conditional knockout mice in which *N-cadherin* is lost from premigratory neural crest cells, cranial defects are observed by E10.5, yet no defects are obvious in the developing outflow tract at that time.[23] At E11.75, when cardiac neural crest cells of wild-type embryos begin to express N-cadherin, conditional knockout embryos exhibit abnormal aorticopulmonary septation defects, including truncus arteriosus and abnormal septation or loss of septation. Cardiac neural crest cells residing in the pharyngeal arteries at E11.75, however, are normal in number and differentiate into smooth muscle. In the outflow tract region at E12.5, *N-cadherin*-deficient cardiac neural crest cells fail to elongate and possess fewer intercellular contacts than wild-type cells, thereby abrogating differentiation. By E13, these embryos die, potentially due to neural crest cell-specific defects in septation. Interestingly, N-cadherin is not essential prior to condensation within the outflow tract,[23] as *N-cadherin*-deficient cardiac neural crest cells are able to reach the outflow tract, most likely due to Cad11 expression.[26,27] These phenotypes highlight a role for N-cadherin in maintaining cell–cell contacts that are necessary for proper cardiac neural crest cell organization and differentiation during outflow tract remodeling.

4.3. Other derivatives

Neural crest cells emigrating at all axial levels differentiate into melanocytes or pigment cells, which express E-cadherin, N-cadherin, or P-cadherin at various levels depending on their target destination.[122] In addition, trunk neural crest cells differentiate into neurons, glia, and Schwann cells of the peripheral nervous system. Differentiated chick Schwann cells and their precursors express N-cadherin, Cad19, and Cad7.[14,122,123] Mouse Schwann cells express Cad6 and N-cadherin, and rat Schwann cells express Cad19 but not Cad7.[29,92,124]

5. CONCLUSIONS

The dynamic regulation, expression, and usage of multiple members of the cadherin superfamily in neural crest cells as they undergo major morphogenetic changes reflect the extreme importance of cadherins throughout all aspects of neural crest ontogeny. The significance of these phenomena is only beginning to be unraveled, as continuing research elucidates new roles for cadherins beyond their classical adhesion and tissue segregation functions. In this regard, neural crest cells serve as an excellent model for studying the universal roles cadherins play in modulating cellular identities, behaviors (migration), and tissue/organ formation. Importantly, ongoing research on cadherins during neural crest cell development also directly translates to understanding the involvement of cadherins in human disorders, cancer metastasis, and various neural crest-related diseases.[125]

REFERENCES

1. Le Douarin NM, Kalcheim C. *The neural crest*. New York: Cambridge University Press; 1999.
2. Betancur P, Bronner-Fraser M, Sauka-Spengler T. Assembling neural crest regulatory circuits into a gene regulatory network. *Annu Rev Cell Dev Biol* 2010;**26**:581–603.
3. Sauka-Spengler T, Bronner-Fraser M. A gene regulatory network orchestrates neural crest formation. *Nat Rev Mol Cell Biol* 2008;**9**:557–68.
4. Duband JL, Monier F, Delannet M, Newgreen D. Epithelium-mesenchyme transition during neural crest development. *Acta Anat (Basel)* 1995;**154**:63–78.
5. Thiery JP, Sleeman JP. Complex networks orchestrate epithelial-mesenchymal transitions. *Nat Rev Mol Cell Biol* 2006;**7**:131–42.

6. Duband JL. Diversity in the molecular and cellular strategies of epithelium-to-mesenchyme transitions: insights from the neural crest. *Cell Adh Migr* 2010;**4**:458–82.
7. Theveneau E, Mayor R. Neural crest delamination and migration: from epithelium-to-mesenchyme transition to collective cell migration. *Dev Biol* 2012;**366**:34–54.
8. Duband JL, Volberg T, Sabanay I, Thiery JP, Geiger B. Spatial and temporal distribution of the adherens-junction-associated adhesion molecule A-CAM during avian embryogenesis. *Development* 1988;**103**:325–44.
9. Hatta K, Takeichi M. Expression of N-cadherin adhesion molecules associated with early morphogenetic events in chick development. *Nature* 1986;**320**:447–9.
10. Akitaya T, Bronner-Fraser M. Expression of cell adhesion molecules during initiation and cessation of neural crest cell migration. *Dev Dyn* 1992;**194**:12–20.
11. Nakagawa S, Takeichi M. Neural crest emigration from the neural tube depends on regulated cadherin expression. *Development* 1998;**125**:2963–71.
12. Hatta K, Takagi S, Fujisawa H, Takeichi M. Spatial and temporal expression pattern of N-cadherin cell adhesion molecules correlated with morphogenetic processes of chicken embryos. *Dev Biol* 1987;**120**:215–27.
13. Shoval I, Ludwig A, Kalcheim C. Antagonistic roles of full-length N-cadherin and its soluble BMP cleavage product in neural crest delamination. *Development* 2007;**134**:491–501.
14. Nakagawa S, Takeichi M. Neural crest cell-cell adhesion controlled by sequential and subpopulation-specific expression of novel cadherins. *Development* 1995;**121**:1321–32.
15. Taneyhill LA, Coles EG, Bronner-Fraser M. Snail2 directly represses cadherin6B during epithelial-to-mesenchymal transitions of the neural crest. *Development* 2007;**134**:1481–90.
16. Park KS, Gumbiner BM. Cadherin 6B induces BMP signaling and de-epithelialization during the epithelial mesenchymal transition of the neural crest. *Development* 2010;**137**:2691–701.
17. Prasad MS, Paulson AF. A combination of enhancer/silencer modules regulates spatially restricted expression of cadherin-7 in neural epithelium. *Dev Dyn* 2011;**240**:1756–68.
18. Chalpe AJ, Prasad M, Henke AJ, Paulson AF. Regulation of cadherin expression in the chicken neural crest by the Wnt/β-catenin signaling pathway. *Cell Adh Migr* 2010;**4**:431–8.
19. Radice GL, Rayburn H, Matsunami H, Knudsen KA, Takeichi M, Hynes RO. Developmental defects in mouse embryos lacking N-cadherin. *Dev Biol* 1997;**181**:64–78.
20. Breau MA, Pietri T, Eder O, Blanche M, Brakebusch C, Fässler R, et al. Lack of beta1 integrins in enteric neural crest cells leads to a Hirschsprung-like phenotype. *Development* 2006;**133**:1725–34.
21. Broders-Bondon F, Paul-Gilloteaux P, Carlier C, Radice GL, Dufour S. N-Cadherin and β1-integrins cooperate during the development of the enteric nervous system. *Dev Biol* 2012;**364**:178–91.
22. Kasemeier-Kulesa JC, Bradley R, Pasquale EB, Lefcort F, Kulesa PM. Eph/ephrins and N-cadherin coordinate to control the pattern of sympathetic ganglia. *Development* 2006;**133**:4839–47.
23. Luo Y, High FA, Epstein JA, Radice GL. N-Cadherin is required for neural crest remodeling of the cardiac outflow tract. *Dev Biol* 2006;**299**:517–28.
24. Inoue T, Chisaka O, Matsunami H, Takeichi M. Cadherin-6 expression transiently delineates specific rhombomeres, other neural tube subdivisions, and neural crest subpopulations in mouse embryos. *Dev Biol* 1997;**183**:183–94.
25. Moore R, Champeval D, Denat L, Tan SS, Faure F, Julien-Grille S, et al. Involvement of cadherins 7 and 20 in mouse embryogenesis and melanocyte transformation. *Oncogene* 2004;**23**:6726–35.

26. Hoffmann I, Balling R. Cloning and expression analysis of a novel mesodermally expressed cadherin. *Dev Biol* 1995;**169**:337–46.
27. Kimura Y, Matsunami H, Inoue T, Shimamura K, Uchida N, Ueno T, et al. Cadherin-11 expressed in association with mesenchymal morphogenesis in the head, somite, and limb bud of early mouse embryos. *Dev Biol* 1995;**169**:347–58.
28. Takahashi M, Osumi N. Expression study of cadherin7 and cadherin20 in the embryonic and adult rat central nervous system. *BMC Dev Biol* 2008;**8**:87.
29. Takahashi M, Osumi N. Identification of a novel type II classical cadherin: rat cadherin19 is expressed in the cranial ganglia and Schwann cell precursors during development. *Dev Dyn* 2005;**232**:200–8.
30. Simonneau L, Kitagawa M, Suzuki S, Thiery JP. Cadherin 11 expression marks the mesenchymal phenotype: towards new functions for cadherins? *Cell Adhes Commun* 1995;**3**:115–30.
31. Davidson LA, Keller RE. Neural tube closure in Xenopus laevis involves medial migration, directed protrusive activity, cell intercalation and convergent extension. *Development* 1999;**126**:4547–56.
32. Hadeball B, Borchers A, Wedlich D. Xenopus cadherin-11 (Xcadherin-11) expression requires the Wg/Wnt signal. *Mech Dev* 1998;**72**:101–13.
33. Vallin J, Girault JM, Thiery JP, Broders F. Xenopus cadherin-11 is expressed in different populations of migrating neural crest cells. *Mech Dev* 1998;**75**:171–4.
34. Kashef J, Köhler A, Kuriyama S, Alfandari D, Mayor R, Wedlich D. Cadherin-11 regulates protrusive activity in Xenopus cranial neural crest cells upstream of Trio and the small GTPases. *Genes Dev* 2009;**23**:1393–8.
35. Borchers A, David R, Wedlich D. Xenopus cadherin-11 restrains cranial neural crest migration and influences neural crest specification. *Development* 2001;**128**:3049–60.
36. Piloto S, Schilling TF. Ovo1 links Wnt signaling with N-cadherin localization during neural crest migration. *Development* 2010;**137**:1981–90.
37. Ezin AM, Fraser SE, Bronner-Fraser M. Fate map and morphogenesis of presumptive neural crest and dorsal neural tube. *Dev Biol* 2009;**330**:221–36.
38. Schuh R, Vestweber D, Riede I, Ringwald M, Rosenberg UB, Jäckle H, et al. Molecular cloning of the mouse cell adhesion molecule uvomorulin: cDNA contains a B1-related sequence. *Proc Natl Acad Sci USA* 1986;**83**:1364–8.
39. Edelman GM, Gallin WJ, Delouvée A, Cunningham BA, Thiery JP. Early epochal maps of two different cell adhesion molecules. *Proc Natl Acad Sci USA* 1983;**80**:4384–8.
40. Thiery JP, Delouvée A, Gallin WJ, Cunningham BA, Edelman GM. Ontogenetic expression of cell adhesion molecules: L-CAM is found in epithelia derived from the three primary germ layers. *Dev Biol* 1984;**102**:61–78.
41. Kemler R, Babinet C, Eisen H, Jacob F. Surface antigen in early differentiation. *Proc Natl Acad Sci USA* 1977;**74**:4449–52.
42. Bronner-Fraser M, Wolf JJ, Murray BA. Effects of antibodies against N-cadherin and N-CAM on the cranial neural crest and neural tube. *Dev Biol* 1992;**153**:291–301.
43. Ruhrberg C, Schwarz Q. In the beginning: generating neural crest cell diversity. *Cell Adh Migr* 2010;**4**:622–30.
44. Coles EG, Taneyhill LA, Bronner-Fraser M. A critical role for Cadherin6B in regulating avian neural crest emigration. *Dev Biol* 2007;**312**:533–44.
45. Ahlstrom JD, Erickson CA. New views on the neural crest epithelial-mesenchymal transition and neuroepithelial interkinetic nuclear migration. *Commun. Integr. Biol.* 2009;**2**:489–93.
46. Ahlstrom JD, Erickson CA. The neural crest epithelial-mesenchymal transition in 4D: a 'tail' of multiple non-obligatory cellular mechanisms. *Development* 2009;**136**:1801–12.
47. Viebahn C. Epithelio-mesenchymal transformation during formation of the mesoderm in the mammalian embryo. *Acta Anat (Basel)* 1995;**154**:79–97.

48. Nichols DH. Mesenchyme formation from the trigeminal placodes of the mouse embryo. *Am J Anat* 1986;**176**:19–31.
49. Zolessi FR, Poggi L, Wilkinson CJ, Chien CB, Harris WA. Polarization and orientation of retinal ganglion cells in vivo. *Neural Dev* 2006;**1**:2.
50. Kalcheim C, Burstyn-Cohen T. Early stages of neural crest ontogeny: formation and regulation of cell delamination. *Int J Dev Biol* 2005;**49**:105–16.
51. Alfandari D, Cousin H, Marsden M. Mechanism of Xenopus cranial neural crest cell migration. *Cell Adh Migr* 2010;**4**:553–60.
52. Sadaghiani B, Thiébaud CH. Neural crest development in the Xenopus laevis embryo, studied by interspecific transplantation and scanning electron microscopy. *Dev Biol* 1987;**124**:91–110.
53. Theveneau E, Marchant L, Kuriyama S, Gull M, Moepps B, Parsons M, et al. Collective chemotaxis requires contact-dependent cell polarity. *Dev Cell* 2010;**19**:39–53.
54. Nichols DH. Ultrastructure of neural crest formation in the midbrain/rostral hindbrain and preotic hindbrain regions of the mouse embryo. *Am J Anat* 1987;**179**:143–54.
55. Nichols DH. Neural crest formation in the head of the mouse embryo as observed using a new histological technique. *J Embryol Exp Morphol* 1981;**64**:105–20.
56. Duband JL, Thiery JP. Distribution of fibronectin in the early phase of avian cephalic neural crest cell migration. *Dev Biol* 1982;**93**:308–23.
57. Knecht AK, Bronner-Fraser M. Induction of the neural crest: a multigene process. *Nat Rev Genet* 2002;**3**:453–61.
58. Halloran MC, Berndt JD. Current progress in neural crest cell motility and migration and future prospects for the zebrafish model system. *Dev Dyn* 2003;**228**:497–513.
59. Berndt JD, Clay MR, Langenberg T, Halloran MC. Rho-kinase and myosin II affect dynamic neural crest cell behaviors during epithelial to mesenchymal transition in vivo. *Dev Biol* 2008;**324**:236–44.
60. Clay MR, Halloran MC. Control of neural crest cell behavior and migration: insights from live imaging. *Cell Adh Migr* 2010;**4**:586–94.
61. Erickson CA, Weston JA. An SEM analysis of neural crest migration in the mouse. *J Embryol Exp Morphol* 1983;**74**:97–118.
62. Sela-Donenfeld D, Kalcheim C. Regulation of the onset of neural crest migration by coordinated activity of BMP4 and Noggin in the dorsal neural tube. *Development* 1999;**126**:4749–62.
63. Sela-Donenfeld D, Kalcheim C. Inhibition of noggin expression in the dorsal neural tube by somitogenesis: a mechanism for coordinating the timing of neural crest emigration. *Development* 2000;**127**:4845–54.
64. Sela-Donenfeld D, Kalcheim C. Localized BMP4-noggin interactions generate the dynamic patterning of noggin expression in somites. *Dev Biol* 2002;**246**:311–28.
65. Hall RJ, Erickson CA. ADAM 10: an active metalloprotease expressed during avian epithelial morphogenesis. *Dev Biol* 2003;**256**:146–59.
66. Uemura K, Kihara T, Kuzuya A, Okawa K, Nishimoto T, Ninomiya H, et al. Characterization of sequential N-cadherin cleavage by ADAM10 and PS1. *Neurosci Lett* 2006;**402**:278–83.
67. Reiss K, Maretzky T, Ludwig A, Tousseyn T, de Strooper B, Hartmann D, et al. ADAM10 cleavage of N-cadherin and regulation of cell-cell adhesion and beta-catenin nuclear signalling. *EMBO J* 2005;**24**:742–52.
68. Monsonego-Ornan E, Kosonovsky J, Bar A, Roth L, Fraggi-Rankis V, Simsa S, et al. Matrix metalloproteinase 9/gelatinase B is required for neural crest cell migration. *Dev Biol* 2012;**364**:162–77.
69. Monier-Gavelle F, Duband JL. Control of N-cadherin-mediated intercellular adhesion in migrating neural crest cells in vitro. *J Cell Sci* 1995;**108**(Pt. 12):3839–53.

70. Mackay DR, Hu M, Li B, Rhéaume C, Dai X. The mouse Ovol2 gene is required for cranial neural tube development. *Dev Biol* 2006;**291**:38–52.
71. Anderson CB, Meier S. The influence of the metameric pattern in the mesoderm on migration of cranial neural crest cells in the chick embryo. *Dev Biol* 1981;**85**:385–402.
72. Tosney KW. The segregation and early migration of cranial neural crest cells in the avian embryo. *Dev Biol* 1982;**89**:13–24.
73. Lumsden A, Sprawson N, Graham A. Segmental origin and migration of neural crest cells in the hindbrain region of the chick embryo. *Development* 1991;**113**:1281–91.
74. Kos R, Reedy MV, Johnson RL, Erickson CA. The winged-helix transcription factor FoxD3 is important for establishing the neural crest lineage and repressing melanogenesis in avian embryos. *Development* 2001;**128**:1467–79.
75. Liu JP, Jessell TM. A role for rhoB in the delamination of neural crest cells from the dorsal neural tube. *Development* 1998;**125**:5055–67.
76. Taneyhill LA, Bronner-Fraser M. Dynamic alterations in gene expression after Wnt-mediated induction of avian neural crest. *Mol Biol Cell* 2005;**16**:5283–93.
77. Théveneau E, Duband JL, Altabef M. Ets-1 confers cranial features on neural crest delamination. *PLoS One* 2007;**2**:e1142.
78. Cheung M, Chaboissier MC, Mynett A, Hirst E, Schedl A, Briscoe J. The transcriptional control of trunk neural crest induction, survival, and delamination. *Dev Cell* 2005;**8**:179–92.
79. Dottori M, Gross MK, Labosky P, Goulding M. The winged-helix transcription factor Foxd3 suppresses interneuron differentiation and promotes neural crest cell fate. *Development* 2001;**128**:4127–38.
80. Jhingory S, Wu CY, Taneyhill LA. Novel insight into the function and regulation of alphaN-catenin by Snail2 during chick neural crest cell migration. *Dev Biol* 2010;**344**:896–910.
81. Cano A, Pérez-Moreno MA, Rodrigo I, Locascio A, Blanco MJ, del Barrio MG, et al. The transcription factor snail controls epithelial-mesenchymal transitions by repressing E-cadherin expression. *Nat Cell Biol* 2000;**2**:76–83.
82. Batlle E, Sancho E, Francí C, Domínguez D, Monfar M, Baulida J, et al. The transcription factor snail is a repressor of E-cadherin gene expression in epithelial tumour cells. *Nat Cell Biol* 2000;**2**:84–9.
83. Bolós V, Peinado H, Pérez-Moreno MA, Fraga MF, Esteller M, Cano A. The transcription factor Slug represses E-cadherin expression and induces epithelial to mesenchymal transitions: a comparison with Snail and E47 repressors. *J Cell Sci* 2003;**116**:499–511.
84. Côme C, Arnoux V, Bibeau F, Savagner P. Roles of the transcription factors snail and slug during mammary morphogenesis and breast carcinoma progression. *J Mammary Gland Biol Neoplasia* 2004;**9**:183–93.
85. LaBonne C, Bronner-Fraser M. Snail-related transcriptional repressors are required in Xenopus for both the induction of the neural crest and its subsequent migration. *Dev Biol* 2000;**221**:195–205.
86. Carl TF, Dufton C, Hanken J, Klymkowsky MW. Inhibition of neural crest migration in Xenopus using antisense slug RNA. *Dev Biol* 1999;**213**:101–15.
87. David R, Wedlich D. Xenopus cadherin-6 is expressed in the central and peripheral nervous system and in neurogenic placodes. *Mech Dev* 2000;**97**:187–90.
88. Locascio A, Manzanares M, Blanco MJ, Nieto MA. Modularity and reshuffling of Snail and Slug expression during vertebrate evolution. *Proc Natl Acad Sci USA* 2002;**99**:16841–6.
89. Murray SA, Gridley T. Snail family genes are required for left-right asymmetry determination, but not neural crest formation, in mice. *Proc Natl Acad Sci USA* 2006;**103**:10300–4.

90. Jiang R, Lan Y, Norton CR, Sundberg JP, Gridley T. The Slug gene is not essential for mesoderm or neural crest development in mice. *Dev Biol* 1998;**198**:277–85.
91. Murray SA, Oram KF, Gridley T. Multiple functions of Snail family genes during palate development in mice. *Development* 2007;**134**:1789–97.
92. Inoue T, Inoue YU, Asami J, Izumi H, Nakamura S, Krumlauf R. Analysis of mouse Cdh6 gene regulation by transgenesis of modified bacterial artificial chromosomes. *Dev Biol* 2008;**315**:506–20.
93. Inoue YU, Asami J, Inoue T. Genetic labeling of mouse rhombomeres by Cadherin-6:: EGFP-BAC transgenesis underscores the role of cadherins in hindbrain compartmentalization. *Neurosci Res* 2009;**63**:2–9.
94. Espeseth A, Marnellos G, Kintner C. The role of F-cadherin in localizing cells during neural tube formation in Xenopus embryos. *Development* 1998;**125**:301–12.
95. Espeseth A, Johnson E, Kintner C. Xenopus F-cadherin, a novel member of the cadherin family of cell adhesion molecules, is expressed at boundaries in the neural tube. *Mol Cell Neurosci* 1995;**6**:199–211.
96. Liu B, Joel Duff R, Londraville RL, Marrs JA, Liu Q. Cloning and expression analysis of cadherin7 in the central nervous system of the embryonic zebrafish. *Gene Expr Patterns* 2007;**7**:15–22.
97. Liu Q, Marrs JA, Londraville RL, Wilson AL. Cadherin-7 function in zebrafish development. *Cell Tissue Res* 2008;**334**:37–45.
98. Dufour S, Beauvais-Jouneau A, Delouvée A, Thiery JP. Differential function of N-cadherin and cadherin-7 in the control of embryonic cell motility. *J Cell Biol* 1999;**146**:501–16.
99. Cheung M, Briscoe J. Neural crest development is regulated by the transcription factor Sox9. *Development* 2003;**130**:5681–93.
100. McKeown SJ, Lee VM, Bronner-Fraser M, Newgreen DF, Farlie PG. Sox10 overexpression induces neural crest-like cells from all dorsoventral levels of the neural tube but inhibits differentiation. *Dev Dyn* 2005;**233**:430–44.
101. Kools P, Van Imschoot G, van Roy F. Characterization of three novel human cadherin genes (CDH7, CDH19, and CDH20) clustered on chromosome 18q22-q23 and with high homology to chicken cadherin-7. *Genomics* 2000;**68**:283–95.
102. Luo J, Wang H, Lin J, Redies C. Cadherin expression in the developing chicken cochlea. *Dev Dyn* 2007;**236**:2331–7.
103. McCusker C, Cousin H, Neuner R, Alfandari D. Extracellular cleavage of cadherin-11 by ADAM metalloproteases is essential for Xenopus cranial neural crest cell migration. *Mol Biol Cell* 2009;**20**:78–89.
104. McCusker CD, Alfandari D. Life after proteolysis: exploring the signaling capabilities of classical cadherin cleavage fragments. *Commun Integr Biol* 2009;**2**:155–7.
105. Inuzuka H, Redies C, Takeichi M. Differential expression of R- and N-cadherin in neural and mesodermal tissues during early chicken development. *Development* 1991;**113**:959–67.
106. Packer AI, Elwell VA, Parnass JD, Knudsen KA, Wolgemuth DJ. N-Cadherin protein distribution in normal embryos and in embryos carrying mutations in the homeobox gene Hoxa-4. *Int J Dev Biol* 1997;**41**:459–68.
107. Redies C, Inuzuka H, Takeichi M. Restricted expression of N- and R-cadherin on neurites of the developing chicken CNS. *J Neurosci* 1992;**12**:3525–34.
108. Liu Q, Kerstetter AE, Azodi E, Marrs JA. Cadherin-1, -2, and -11 expression and cadherin-2 function in the pectoral limb bud and fin of the developing zebrafish. *Dev Dyn* 2003;**228**:734–9.
109. Kerstetter AE, Azodi E, Marrs JA, Liu Q. Cadherin-2 function in the cranial ganglia and lateral line system of developing zebrafish. *Dev Dyn* 2004;**230**:137–43.

110. Simpson MJ, Zhang DC, Mariani M, Landman KA, Newgreen DF. Cell proliferation drives neural crest cell invasion of the intestine. *Dev Biol* 2007;**302**:553–68.
111. Simpson MJ, Landman KA, Hughes BD, Newgreen DF. Looking inside an invasion wave of cells using continuum models: proliferation is the key. *J Theor Biol* 2006;**243**:343–60.
112. Heanue TA, Pachnis V. Enteric nervous system development and Hirschsprung's disease: advances in genetic and stem cell studies. *Nat Rev Neurosci* 2007;**8**:466–79.
113. Gaidar YA, Lepekhin EA, Sheichetova GA, Witt M. Distribution of N-cadherin and NCAM in neurons and endocrine cells of the human embryonic and fetal gastroenteropancreatic system. *Acta Histochem* 1998;**100**:83–97.
114. Hackett-Jones EJ, Landman KA, Newgreen DF, Zhang D. On the role of differential adhesion in gangliogenesis in the enteric nervous system. *J Theor Biol* 2011;**287**:148–59.
115. Young HM, Bergner AJ, Anderson RB, Enomoto H, Milbrandt J, Newgreen DF, et al. Dynamics of neural crest-derived cell migration in the embryonic mouse gut. *Dev Biol* 2004;**270**:455–73.
116. Young HM, Jones BR, McKeown SJ. The projections of early enteric neurons are influenced by the direction of neural crest cell migration. *J Neurosci* 2002;**22**:6005–18.
117. Kasemeier-Kulesa JC, Kulesa PM, Lefcort F. Imaging neural crest cell dynamics during formation of dorsal root ganglia and sympathetic ganglia. *Development* 2005;**132**:235–45.
118. Kirby ML, Hutson MR. Factors controlling cardiac neural crest cell migration. *Cell Adh Migr* 2010;**4**:609–21.
119. Waldo K, Miyagawa-Tomita S, Kumiski D, Kirby ML. Cardiac neural crest cells provide new insight into septation of the cardiac outflow tract: aortic sac to ventricular septal closure. *Dev Biol* 1998;**196**:129–44.
120. Bajolle F, Zaffran S, Meilhac SM, Dandonneau M, Chang T, Kelly RG, et al. Myocardium at the base of the aorta and pulmonary trunk is prefigured in the outflow tract of the heart and in subdomains of the second heart field. *Dev Biol* 2008;**313**:25–34.
121. Kirby ML, Gale TF, Stewart DE. Neural crest cells contribute to normal aorticopulmonary septation. *Science* 1983;**220**:1059–61.
122. Pla P, Moore R, Morali OG, Grille S, Martinozzi S, Delmas V, et al. Cadherins in neural crest cell development and transformation. *J Cell Physiol* 2001;**189**:121–32.
123. Lin J, Luo J, Redies C. Cadherin-19 expression is restricted to myelin-forming cells in the chicken embryo. *Neuroscience* 2010;**165**:168–78.
124. Padilla F, Broders F, Nicolet M, Mege RM. Cadherins M, 11, and 6 expression patterns suggest complementary roles in mouse neuromuscular axis development. *Mol Cell Neurosci* 1998;**11**:217–33.
125. Acloque H, Adams MS, Fishwick K, Bronner-Fraser M, Nieto MA. Epithelial-mesenchymal transitions: the importance of changing cell state in development and disease. *J Clin Invest* 2009;**119**:1438–49.

CHAPTER FOURTEEN

Cadherins and Epithelial-to-Mesenchymal Transition

Alexander Gheldof[*,†], **Geert Berx**[*,†]
[*]Department for Molecular Biomedical Research, Unit of Molecular and Cellular Oncology, VIB, Ghent, Belgium
[†]Department of Biomedical Molecular Biology, Ghent University, Ghent, Belgium

Contents

Abstract

Epithelial–mesenchymal transition (EMT) is a process whereby epithelial cells are transcriptionally reprogrammed, resulting in decreased adhesion and enhanced migration or invasion. EMT occurs during different stages of embryonic development, including gastrulation and neural crest cell delamination, and is induced by a panel of specific transcription factors. These factors comprise, among others, members of the Snail, ZEB, and Twist families, and are all known to modulate cadherin expression and, in particular, E-cadherin. By regulating expression of the cadherin family of proteins, EMT-inducing transcription factors dynamically modulate cell adhesion, allowing many developmental processes to take place. However, during cancer progression EMT can be utilized by cancer cells to contribute to malignancy. This is also reflected at the level of the cadherins, where the cadherin switch between E- and N-cadherins is a classical example seen in cancer-related EMT. In this chapter, we give a detailed overview of the entanglement between EMT-inducing transcription factors and cadherin

Progress in Molecular Biology and Translational Science, Volume 116
ISSN 1877-1173
http://dx.doi.org/10.1016/B978-0-12-394311-8.00014-5

modulation during embryonic development and cancer progression. We describe how classical cadherins such as E- and N-cadherins are regulated during EMT, as well as cadherin 7, -6B, and −11.

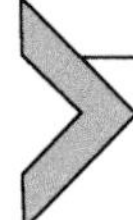

1. INTRODUCTION

Formation of multicellular organisms depends on stable contacts between cells. As multicellular life became more complex during evolution, dynamic regulation and tissue-specific development of cell–cell interactions became essential. The spatiotemporal expression of adhesion complexes not only contributes to the formation and integrity of different tissues during embryogenesis but also plays an important role in cell signaling.[1] During embryogenesis, distinct cell types can migrate as cohesive cell sheets or as individual cells. These two modes of migration must not be seen as dichotomous, since different intermediate modes of migration have been shown to exist. For instance, cephalic neural crest cells (NCCs) migrate as multicellular sheets toward their destination, whereas trunk NCCs move as single chains.[2] Furthermore, cells can change their migratory state depending on the developmental stage and cellular context. Melanocyte precursors initially migrate in groups through mesodermal tissue, but at the dermis–epidermis border, they migrate as single cells into the hair follicles.[3] Nevertheless, all types of cell migration involve alteration of cell polarity and dynamic regulation of cell–cell adhesion complexes.

One of the major classes of proteins involved in tissue morphogenesis and coordinated cell movement is the cadherin family.[4] Cadherins are transmembrane glycoproteins that mediate Ca^{2+}-dependent cell–cell adhesion.[5] The classical example of a cadherin family member is E-cadherin (CDH1), which is classified in the type-I group. The existence of E-cadherin was first proposed in the late 1970s and was soon found to be a crucial mediator of epithelial cell–cell adhesion.[5] Other classical cadherins that will be discussed below are N- and P-cadherins (CDH2 and CDH3, respectively).

The complex tissue reorganizations that occur during morphogenesis can also be regarded as a dynamic interchange or interaction between cells arranged in two dimensions (epithelia) and cells organized in three dimensions (mesenchyme).[6] During gastrulation, epithelial ectodermal cells in the primitive streak region undergo epithelial mesenchymal transition (EMT), which leads to their detachment from their original layer and to enhanced motility.[7] This is needed for formation of the two other germ layers (mesoderm and endoderm). NCCs are generated in the region between the neural plate and the ectoderm, where they undergo EMT and then delaminate and migrate to their destination.[8] EMT is induced by a group of transcription factors, including members of the Snail, ZEB, and Twist families. For example, the endoderm of Snail-deficient mice is morphologically abnormal, and antisense-mediated inhibition of Slug in the chicken embryo leads to defects in delamination of the mesoderm and NCCs.[9,10] During EMT, expression of epithelial proteins is strongly repressed or their functionality is severely compromised. One of the main targets of EMT-inducing transcription factors is E-cadherin, emphasizing its role in suppression of cell migration. Snail, Slug, ZEB1, and ZEB2 have all been demonstrated to repress transcription by binding directly to the E-box sequences of the E-cadherin promoter.[5] Strikingly, while downregulation of E-cadherin seems to be necessary for epithelial cell movement, execution of the EMT program leads to enhanced expression of alternative cadherins. During formation of the primitive streak and subsequent gastrulation, N-cadherin is strongly expressed in premigrating cells and at the same time E-cadherin expression is diminished.[11] These findings, together with the increased migration of epithelial cells overexpressing N-cadherin, point to the importance and specificity of cell–cell adhesion complexes during migration. In this chapter, we will focus on the interplay between modulation of cadherins, in particular, type-I cadherins, and the activation of the EMT program during development and in disease.

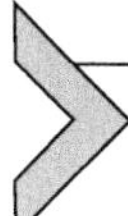

2. EMT AND THE MODULATION OF CADHERINS DURING DEVELOPMENT

2.1. Cadherin modulation during initial embryonic stages

During mammalian embryogenesis, E-cadherin can be detected from the initial stages on in blastomere cells where it is necessary for correct intercellular adhesion. Throughout this early phase, a transition takes place as maternally provided proteins are replaced by proteins produced by the embryonic cells. In mouse oocytes lacking maternal E-cadherin expression, the blastomeres do not adhere; adherence occurs only in the morula, in which embryonically produced E-cadherin compensates for the absent maternal form.[12] The first epithelialized structures become apparent during the formation of the trophectoderm, where functional E-cadherin expression is essential for successful development. When E-cadherin is replaced by N-cadherin by a knock-in strategy, the outer cells of the inner cell mass can still undergo epithelial differentiation, but trophectoderm formation is compromised.[13] Trophectodermal cells lacking E-cadherin cannot correctly express the tight junction protein ZO1 and the organization of their actin cytoskeleton is defective. Furthermore, the typical localization of ezrin at the apical side is lost, suggesting that N-cadherin cannot support a fully polarized epithelial phenotype.[13] Interestingly, in the developing heart, expression of E-cadherin can compensate for the absence of N-cadherin, which is essential for the formation of the heart tube[14] (see also Chapter 12). One hypothesis for this discrepancy is the fact that the involvement of N-cadherin in cell signaling is less pronounced during cardiac development, since the myocytes of N-cadherin knockout mice can beat synchronously.[15] Furthermore, E-cadherin-based AJ are significantly stronger than junctions with N-cadherin, which indicates that replacement of E- by N-cadherin in tissues subjected to strong tension would not provide enough mechanical rigidity.[16]

2.2. Implantation

During implantation of the developing embryo in the decidua, trophectodermal cells prepare for the formation of the placenta. Trophoblasts differentiate into invasive extravillous cytotrophoblasts, which migrate through the endometrium and further differentiate into multinucleated placental bed giant cells. It is presumed that this process can be classified as EMT, although there is limited experimental evidence for that.[17] Indeed, at the embryo–decidua junction, cytotrophoblasts initially form a coherent, epithelial structure with cells showing a typical apicobasal polarization. Before invasion into the decidua, intercellular adhesion is negatively modulated, resulting in loosely organized clusters of migrating cells. The trophoblasts that have migrated the longest distance have considerably less E-cadherin.[18] It is known that reduced levels of fibroblast growth factor 4 (FGF4) trigger trophoblast migration. Removal of FGF4 from the medium of cultured trophoblasts induces a shift toward a mesenchymal morphology together with downregulation of E-cadherin and induction of Snail.[19] Other EMT-inducing transcription factors, such as Slug, ZEB1, and Twist1, remain unaffected. Though P-cadherin, another classical cadherin, is present in murine placental tissue, where it is involved in adhesion of the placenta to the uterine wall, its expression has not been detected in human placenta.[17]

2.3. Gastrulation

During gastrulation, the organization of the developing embryo undergoes drastic changes. In order to form a complex body plan, the single-layered blastula segregates into three different cell layers: ectoderm, mesoderm, and endoderm. In most amniote embryos, formation of both the mesoderm and endoderm starts with epiblast cells that undergo EMT in the region of the primitive streak. Formation of the primitive streak cannot be generalized among amniotes, as large differences exist.[20] In rabbit and avian embryos, the primitive streak is formed as a consequence of large-scale rotational movements that bring streak precursor cells into the correct position. However, in the mouse embryo, no such cell movements can be detected. Instead, the primitive streak develops *in situ* in the posterior region of the epiblast and elongates due to progressive EMT.[7] This is characterized by a progressive loss of the basal lamina followed by apical constriction of the cells, which allows them to form the underlying mesodermal layer. This is coupled with the loss of the tight junction-associated proteins occludin and aPKC.[7] Which EMT-inducing transcription factors are involved during this process is not clear. Mouse Snail is expressed in the primitive streak region and subsequently in the entire mesodermal layer (Fig. 14.1A).[21] Interestingly, Snail mutant mice can still form a mesodermal layer, but this is morphologically abnormal.[9] At the primitive streak of mutant embryos, the basal lamina is still degraded, suggesting that additional mechanisms besides Snail action drive this initial EMT. For example, Twist1-null mice do go through gastrulation even though they die at midgestation, indicating that the redundancy between Snail and Twist expression is bidirectional.[22] However, it is debatable whether this transition at the beginning of gastrulation can be regarded as a full EMT, as E-cadherin is only downregulated when epiblast cells start to migrate away from the primitive streak region to form the endodermal layer. Downregulation of E-cadherin during more advanced stages of murine gastrulation seems to depend on Snail.[9] Interestingly, Snail-mutant *Drosophila* embryos fail to downregulate E-cadherin at all during mesoderm formation and cannot form a mesoderm with mesenchymal characteristics, indicating that the full functional repertoire of Snail is only necessary during these later steps.[23] These findings were also observed in Twist-mutant *Drosophila* embryos.[24]

In avian embryos, the earliest N-cadherin expression occurs in the early gastrula in the groove of the primitive streak, and specifically in cells that are migrating before forming the mesodermal layer. First, N-cadherin is distributed evenly across the cell membrane. After the mesodermal layer is formed and reepithelialization takes place to form the notochord, somites,

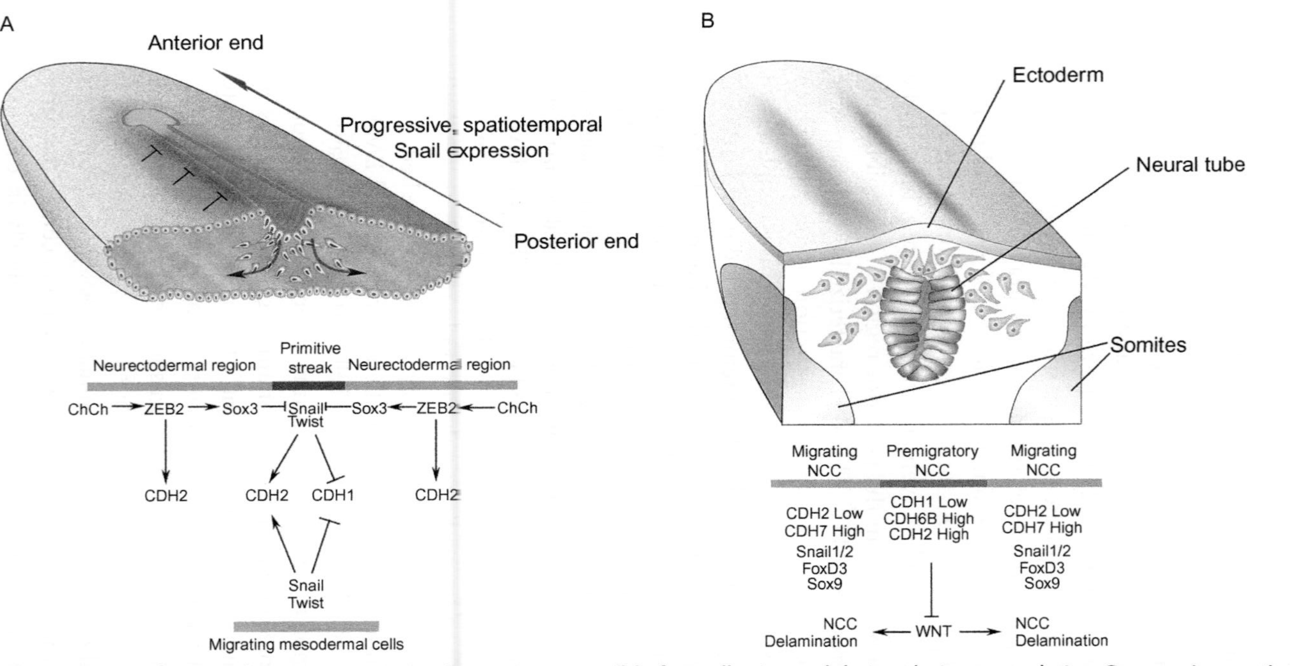

Figure 14.1 (A) Expression of EMT-inducing transcription factors is responsible for cadherin modulation during gastrulation. Progressive spatiotemporal Snail expression is necessary for the formation of the primitive streak in the murine epiblast. Hereby, ectodermal epithelial cells at the posterior side of the embryo undergo an EMT, which is characterized by a switch in classical cadherins yielding high levels of N-cadherin (CDH2) and low expression of E-cadherin (CDH1). Migration of mesoderm forming cells, depicted in green, is dependent on Snail and Twist. At the border of the primitive streak (red) and the neuroectodermal region, streak involution is inhibited by the sequential action of Churchill (ChCh), ZEB2, and Sox3, which counteract the action of Snail. While ZEB2 expression in the neuroectodermal plate leads to the induction of N-cadherin, this is not sufficient to result in an EMT. (B) Different cadherin family members are strongly modulated during neural crest cell delamination. Epithelial cells of the neural tube (red) express high levels of N-cadherin (CDH2) and are highly polarized in the avian embryo prior to neural crest delamination. N-Cadherin inhibits WNT signalization, which is activated during neural crest cell (NCC) delamination. In chicken, premigratory NCC briefly express cadherin 6B (CDH6B), which is eventually lost at the onset of migration. Hereby, cells undergo an EMT whereby N-cadherin is downregulated and cadherin 7 (CDH7) is upregulated (scattered green cells). The transcription factors Snail1/2, FoxD3, and Sox9 lie at the basis of this transition, and expression of all three factors is necessary for a full EMT phenotype. (See Color Insert.)

and lateral plate, the distribution of N-cadherin becomes polarized.[25] During the first stages of gastrulation in *Drosophila*, N-cadherin mRNA is localized in the nucleus of migrating cells.[24] This is consistent with the presence of N-cadherin protein only in more advanced gastrula stages (stage 8 in *Drosophila*). The upregulation of N-cadherin protein during mesoderm formation has been found to depend on both Snail and Twist in *Drosophila*.[24] Indeed, *Drosophila* embryos expressing a mutant form of Twist do not express N-cadherin, whereas embryos mutant for Snail have very low levels of N-cadherin. This observation suggests that Twist is necessary for the initial induction of N-cadherin. While the cadherin switch is a classical hallmark of EMT, N-cadherin does not seem to be crucial for the correct formation of the mesodermal layer. Mice expressing mutant N-cadherin die on day 10 after gestation due to failure in heart formation. However, gastrulation is not affected.[15] Similar findings have been reported for *Drosophila*.[26]

2.4. Neurulation

2.4.1 Role of N-cadherin in neurulation

Neuronal precursors are located at the neuroectodermal plate and display high levels of N-cadherin, which is critical for correct neurulation, as shown by the impaired movement in the anterior neural plate of N-cadherin-lacking zebrafish.[27] Furthermore, expression of N-cadherin is necessary for F-actin assembly and apical constriction in the neural plate of *Xenopus*.[28,29] Neural progenitor cells in the ectodermal epithelium surround the region of the primitive streak and are prevented from undergoing EMT by the transcription factor Sox3, which is a direct transcriptional repressor of Snail2 (Fig. 14.1A).[30] Interestingly, during the subdivision between neural and nonneural tissue, the transcriptional activator Churchill (ChCh) inhibits the expression of mesodermal markers through induction of ZEB2 (SIP1), which is a potent inducer of N-cadherin.[31] These findings highlight the functional differences between the transcription factors of the Snail and the ZEB families.

Cells targeted for neuronal tissue formation will form the neuroectodermal plate and subsequently involute to form the neural tube, from which the brain and spinal cord will develop. Ectodermal cells located at the border of the neuroectodermal plate will develop into the neural crest.[32] N-cadherin is strongly expressed in the epithelium of the neural plate, but E-cadherin expression is not as pronounced. Here, N-cadherin seems to take over the function of E-cadherin, which is to form stable cell–cell contacts in the neural plate and later in the neural tube. It has also been shown that expression of

N-cadherin is essential for inhibition of WNT signaling in the NCCs as long as migration throughout the embryo did not start (Fig. 14.2B).[33] Thus, delamination of the neural crest can only take place if N-cadherin expression is downregulated. The process whereby NCCs become migratory is one of the classical examples of EMT. When the cells derived from the neural crest finally reach their destination and form cellular aggregates, as described for the development of dorsal root ganglia, N-cadherin is again upregulated.[34] The latter event can be defined as mesenchymal-to-epithelial transition (MET) and is associated with an increased differentiation status.[35] Both Snail 1 and 2 have been shown to be potent drivers of NCC delamination in both avian and murine embryos. In mice, Snail1 is expressed in premigratory NCCs while Snail2 is present in migratory NCCs.[36] However, in the chicken embryo, the functionality of Snails 1 and 2 in NCCs is reversed.

2.4.2 Alternative cadherins are involved in neurulation

Cadherin 7 is a type-II cadherin that has been associated with increased NCC migration in avian embryo grafting experiments and is essential for the correct segregation of dorsally and ventrally migrating motor neurons.[37] Perturbing cadherin 7 functionality in chicken embryos prevents motor neuron migration.[38] Downregulation of N-cadherin followed by upregulation of cadherin 7 is reminiscent of the classical cadherin switch during gastrulation. Interestingly, although the same EMT-inducing transcription factors (Snail1 and Snail2) are involved, the outcome of these two cadherin switches is different. This might be explained by a significantly altered transcriptional profile in the NCCs: it has indeed been shown that only a combination of the transcription factors FoxD3, Sox9, and Snail1/2 is sufficient for induction of a full EMT phenotype during neurulation (Fig. 14.2B).[8] More detailed analysis revealed that the upregulation of cadherin 7 is preceded by the induction of cadherin 6B, which is another type-II cadherin present in premigratory NCCs (Fig. 14.2B). Upon Snail2 expression, cadherin 6B is downregulated directly by the binding of Snail2 to the E-box elements in the regulatory region of its promoter[39] (see also Chapter 13). Although the role of cadherin 6B in NCC delamination is not well understood, its downregulation seems to be required for neural crest-associated EMT. This is demonstrated by the ability to rescue the inhibition of NCC migration due to Snail2 knockdown by knocking down cadherin 6B as well.[39] Since cadherin 6B inhibits delamination, it is not logical that its expression would be induced before NCC migration. However, inhibition of cadherin 6B in avian embryos leads to a precocious neural crest

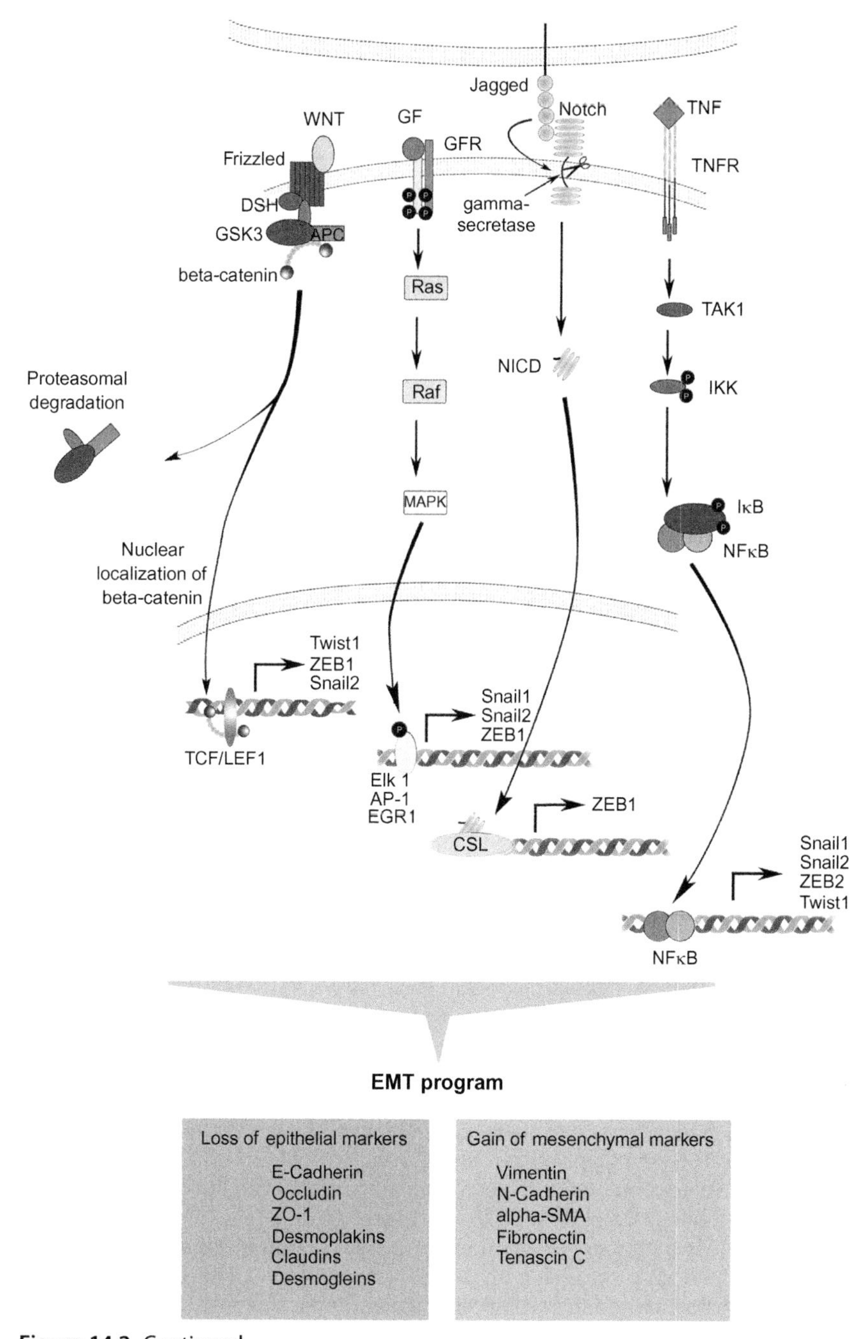

Figure 14.2 Continued

EMT, suggesting that cadherin 6B acts as an additional controlling factor needed for the correct spatiotemporal orchestration of NCC delamination.[40]

Like cadherin 6B, cadherin 11 has also been found to restrain NCC migration in the *Xenopus* embryo. In mice, cadherin 11 is expressed in migrating mesenchymal cells originating from the neural crest and is also present in mesodermal cells during gastrulation. These findings further highlight the observation that cadherin 11 is not expressed in epithelial tissues and why it is generally regarded as a mesenchymal-specific member of the cadherin family.[41] Overexpression of cadherin 11 in cranial NCC in *Xenopus* results in inhibition of migration independent of its cytoplasmic domain.[42] Interestingly, overexpression of cadherin 11 in NCC also leads to decreased levels of Snail and Twist and is associated with the expression of neuronal markers.[42]

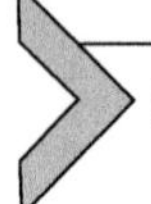

3. TRANSCRIPTIONAL MODULATION OF E-CADHERIN CONTRIBUTES TO MALIGNANT CARCINOMA PROGRESSION

3.1. Cadherin modulation is part of the oncogenic EMT program

Together with its function as an intercellular adhesion protein, a broad spectrum of signaling pathways is modulated by E-cadherin, suggesting that the stability of the adherens junctions is crucial for the maintenance of the epithelial phenotype and for homeostasis of tissues.[5] During the process of malignant cancer formation, several mechanisms have been described that impair the functionality of E-cadherin. These include mutations, epigenetic, and transcriptional silencing, increased endocytosis, and proteasomal degradation.[43]

The first inactivating mutations of E-cadherin were discovered in diffuse-type gastric cancer or cancer with both diffuse or intestinal components (see Chapter 15). A typical mutation is the skipping of exons encoding

Figure 14.2 General overview of several pathways contributing to the EMT phenotype. Apart from the TGF-beta signaling cascade, also the WNT, Notch, and NF-κB pathways have been shown to contribute to the establishment of EMT. In addition, several growth factors (GF) including HGF, EGF, and IGF are able to induce a mesenchymal transition. Activation of these pathways results in upregulation of several EMT-inducing transcription factors. alpha-SMA: alpha smooth muscle actin; CSL, CBF1/RBP-JK/Suppressor of Hairless/LAG-1; DSH, disheveled; GSK3β, glycogen-synthase kinase 3β; GFR, Growth factor receptor; IKK, IκB kinase; MAPK, mitogen-activated protein kinase; NICD, Notch intracellular domain; TAK1, transforming growth factor-β-activated kinase 1; TNF, tumor necrosis factor; TNFR, tumor necrosis factor receptor. (See Color Insert.)

calcium-binding EC repeats, which severely compromises the function of E-cadherin as an adhesion protein.[44] Next to their presence in diffuse gastric carcinoma, E-cadherin mutations have also been characterized in lobular breast carcinomas, which predominantly result in a truncated form of E-cadherin.[45]

Apart from inactivating mutations, methylation of the 5′ CpG islands in the E-cadherin promoter has been described for tumors. More specifically, a progressive increase in E-cadherin promoter hypermethylation has been shown to occur during the development of ductal breast carcinoma.[46] The nature of hypermethylation during carcinogenic progression seems to fit in the frame of a larger change in gene expression. Genome-wide expression analysis in cancer cell lines displaying E-cadherin promoter hypermethylation or carrying inactivating E-cadherin mutations revealed that promoter hypermethylation occurred together with large changes in gene expression, while this was not the case for cells bearing E-cadherin mutations.[47] Moreover, cells with a methylated E-cadherin promoter had a fibroblastic morphology, while cells with mutated E-cadherin still were epithelial like. These findings strongly suggest that hypermethylation of the E-cadherin promoter is part of cellular reprogramming that results in EMT.[47]

Expression of EMT-inducing transcription factors is controlled by growth factor and cytokine signaling. Examples are the WNT, FGF, Notch/delta, and NF-kappaB pathways (Fig. 14.2).[48–51] In parallel with the finding that Snail and Twist are mediators of EMT in the developing embryo came the notion that the transition from a benign tumor toward a cancer with malignant properties was often accompanied by loss of E-cadherin.[43,52] For instance, in human breast carcinomas, expression of Snail was identified in infiltrating ductal carcinomas and was inversely correlated with the differentiation grade of these tumors.[53]

Apart from Snail, additional transcription factors were identified, which are capable of transcriptionally silencing E-cadherin. Both ZEB1 and ZEB2 factors were initially found to repress the delta-crystallin enhancer and the *Xbra2* gene, respectively, in an E-box-dependent manner. Also the E-boxes situated in the E-cadherin promoter were soon identified as active binding sites for both ZEB factors.[54] The first connection that also ZEB2 was able to play a role in malignant tumor progression came from Comijn *et al.*,[55] who showed that binding of ZEB2 to the E-cadherin promoter directly led to transcriptional repression. In addition, conditional expression of ZEB2 in tumor cell lines led to an EMT and a subsequent gain in invasive capacities.[56] Similar conclusions could be made for ZEB1.[57,58] The *in vivo* relevance of these findings was further stipulated by the insight that ZEB1 was expressed in highly aggressive uterine cancers and in invasive cells at the tumor border of advanced colorectal carcinoma.[59,60]

Several alternative EMT-inducing factors have been described, which contribute to malignant cancer progression. These include members of the bHLH family of transcription factors, such as E12/E47, E2-2A/E2-2B, and *TWIST1/2*.[61] Several studies have shown that also these factors are implicated in malignant cancer progression.[62–64] This abundance of EMT-inducing transcription factors might suggest that significant functional overlap exists between these proteins. However, *in vitro* cDNA microarray analysis of MDCK cells overexpressing Snail, Slug, or E12/E47 showed that, apart from some overlap including the downregulation of E-cadherin, almost 70% of the modulated genes seems to be specific to a particular transcription factor.[65] This indicates that different EMT-inducing transcription factors are able to induce common but also specific EMT-related programs and hence might be responsible for a differential role in malignant cancer progression.

3.2. Involvement of alternative cadherins in cancer progression

While reduction in E-cadherin expression is one of the characteristics of advanced carcinoma, the opposite holds true for N-cadherin which is in line with the cadherin switch observed during gastrulation. Transfection of N-cadherin into various cancer cell lines enhances their invasion and metastasis in nude mice.[66] Interestingly, this effect occurs even in the presence of E-cadherin, which demonstrates the dominant effect of N-cadherin. How can enhanced migration take place when a protein that mediates adhesion is upregulated? In one of the proposed mechanisms, N-cadherin physically interacts with the FGFR1 (fibroblast growth factor receptor 1), leading to continuous activation of the MAPK-ERK pathway (Fig. 14.3).[67] Alternatively, N-cadherin expression might enhance interaction with cells of the stromal and endothelial compartment. This stimulates tumor cell survival and promotes access in and out of the circulation. It has indeed been shown that melanoma cells expressing N-cadherin are more easily captured in the vasculature.[68] Modulation of stromal interaction is further exemplified by the fact that myeloma cells displaying aberrant N-cadherin expression are more strongly retained in the bone marrow, where they interact with N-cadherin-positive osteoblasts. This has been shown to facilitate the onset and progression of osteolytic lesions and to inhibit osteogenic differentiation.[69] N-Cadherin-mediated heterotypic contacts of tumor cells with

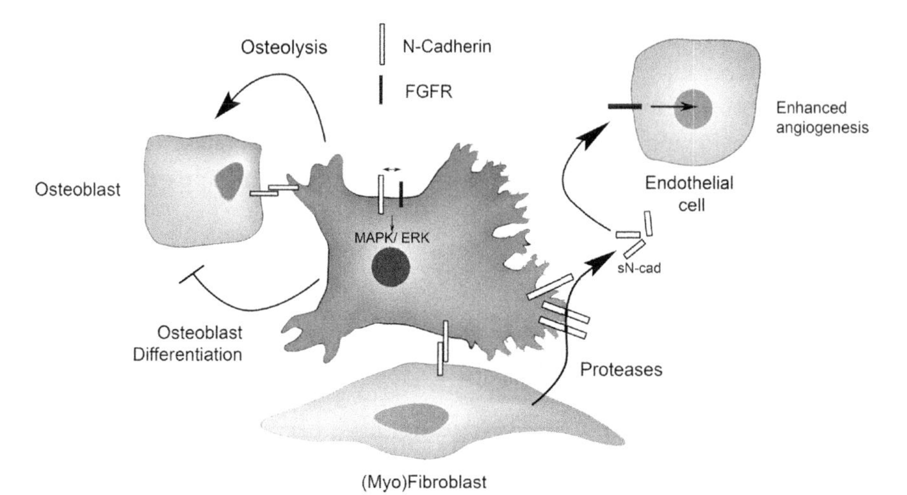

Figure 14.3 Aberrant N-cadherin expression in cancer cells contributes to tumor malignancy N-cadherin interacts with the FGF-receptor (FGFR), thereby triggering activation of the MAPK/ERK signaling cascade. This may lead to enhanced tumor cell proliferation as well as the expression of EMT-inducing transcription factors. Furthermore, N-cadherin stimulates interaction of the tumor cells with stromal cells. For example, enhanced interaction with (myo) fibroblasts, which secrete different proteases, leads to the generation of soluble N-cadherin fragments (sN-cad). sN-cad enhances FGF signaling in endothelial cells, which results in enhanced angiogenesis. Interaction with osteoblasts via N-cadherin leads to inhibition of osteoblast differentiation and therefore enhanced osteolysis. See text for details and references. (See Color Insert.)

myofibroblasts have been suggested as a mechanism for enhanced migration due to the high motility of (myo)fibroblasts.[70] Myofibroblasts have also been shown to secrete many different proteases into the tumor microenvironment. Accumulation of soluble N-cadherin ectodomains due to proteolytic cleavage results in enhanced angiogenesis by stimulating FGF signaling in endothelial cells (Fig. 14.3). Taken together, these findings demonstrate the important role of N-cadherin in tumor progression and are in line with the cadherin switch that occurs in response to EMT-inducing transcription factors.

P-Cadherin has also been associated with tumor progression and increased migration and invasion.[71] However, this cannot be generalized because adverse effects have also been reported.[71] For example, expression of P-cadherin in ductal breast carcinoma has been associated with reduced survival.[72] Tissue microarray analysis revealed that enhanced P-cadherin expression correlates with the occurrence of HER2 positivity and breast carcinoma subtypes of the basal B phenotype. The basal B subgroup is characterized by increased expression of the EMT-inducing transcription factor

ZEB1.[73] In accordance with these findings, suppression of P-cadherin in bladder cancer cell lines leads to decreased *in vitro* invasion or migration and to increased intercellular adhesion.[74] Interestingly, an inverse correlation between P- and E-cadherin has been uncovered in colon cancer cell lines: knockdown of P-cadherin resulted in a marked upregulation of E-cadherin, which subsequently led to downregulation of β-catenin and to reduced formation of liver metastases in a nude mouse model.[75] In contrast to these findings, loss of P-cadherin was associated with a higher tumor grade in colorectal adenocarcinomas.[76] In line with the latter observations, knockdown of P-cadherin in colorectal cancer cell lines results in increased invasion and migration, indicating that in these cases P-cadherin stabilizes the differentiated phenotype.[74] Similar findings have been done in melanoma: overexpression of P-cadherin in melanoma cell lines reduced tumor growth and invasion in xenograft mouse models.[77] Progression of melanoma has been found to be closely associated with onset of EMT. In particular, the transcription factors ZEB1 and Snail2 are dominant players in the repression of E-cadherin in several melanoma cell lines, although this is not the case for Twist1 and Snail1.[78]

3.3. Role of cadherin 11 in cancer progression

Expression of cadherin 11 leads to different outcomes in terms of EMT in cancer settings. Recent findings have shown that cadherin 11 is frequently inactivated by promoter methylation in several cancer cell lines of hepatocellular, colorectal, and breast origin.[79] Overexpression of cadherin 11 in nasopharyngeal cancer cells leads to MET and reduced expression of stem cell markers. In cell lines derived from primary head and neck tumors, cadherin 11 could still be detected, but this was not the case in cell lines derived from lymph nodes and metastases. Overexpression of cadherin 11 in invasive breast cancer cells resulted in decreased migration *in vitro* and in reduced formation of lymph node metastasis after their orthotopic injection in nude mice. In head and neck carcinoma, modulation of cadherin 11 expression was also linked to promoter methylation, suggesting that transcriptional repressors such as EMT inducers are responsible for its downregulation.[80] However, contradictory data have been published about this matter. For instance, cadherin 11 is strongly expressed in invasive breast cancer cell lines, but not in noninvasive lines.[81] This agrees with the initial finding that cadherin 11 is mainly expressed in mesenchymal tissue and is hardly found in epithelial structures.[82] Moreover, cadherin 11 expression has been correlated to the basal B group of breast cancer and is increased in sparsely seeded MCF-10A breast cancer cells.[83] These specific seeding conditions favor the establishment of an EMT-related phenotype, which exhibits high levels of vimentin and low levels of E- and P-cadherin. An additional connection between EMT and cadherin 11 expression is found during wound healing and in fibrosis. Cadherin 11 is expressed more strongly in fibrotic lung tissue than in healthy lung.[84] This increased cadherin 11 expression is correlated with large depositions of collagen and with local accumulation of myofibroblasts. Interestingly, the induction of fibrosis is severely hampered in the lungs of cadherin-11-deficient mice, indicating that cadherin 11 plays a critical role in this process. Further *in vitro* analysis revealed that cadherin 11 is necessary for TGFβ-induced expression of Snail2 and the subsequent establishment of a mesenchymal morphology in A549 cells.[84] The mechanism of the interplay between cadherin 11 and TGFβ-induced EMT is still unknown. However, we can speculate that, like N-cadherin assisted activation of the FGFR, cadherin 11 is necessary for complete activation of the TGFβ receptor.

4. CONCLUDING REMARKS

Proteins of the cadherin family are crucial mediators of cell–cell adhesion and their correct function is thus essential for proper embryonic development and homeostasis. This knowledge has become firmly established during the past 30 years, during which it has also become increasingly clear that modulation of cadherin expression is entangled with EMT. The E-cadherin gene is a direct target of several of EMT-inducing transcription factors and its downregulation is required for efficient cell migration. Although this paradigm holds true for E-cadherin, it certainly does not apply to all cadherin members. For example, the expression of N-cadherin in the neural tube is necessary for tight cell–cell adhesion, while N-cadherin upregulation is observed in migrating mesodermal cells. These findings demonstrate that the cellular context is a crucial determinant of the outcome of cadherin modulation. Moreover, the view that cadherins are implicated only in cell–cell adhesion is too narrow because of the accumulating evidence for their roles in cell signaling. Due to this complexity, we are only beginning to uncover a more detailed relation between EMT, cadherins, and cellular migration in the developing embryo. However, it has become increasingly clear that modulation of cadherin expression is important in the onset and progression of cancer. Adding to the already complex regulation of and response to cadherin modulation is the altered microenvironment in cancer. Infiltration of immune cells, stimulation of angiogenesis, and activation of fibroblasts all contribute to a dynamic and diverse expression of cadherins to which the tumor cells respond. Moreover, these infiltrating cells secrete many cytokines such as TGFβ and members of the FGF family, which modulate several EMT-inducing transcription factors. Feedback loops are established for cadherin expression to modulate the transcription of EMT inducers and *vice versa*. Since different cadherin members are simultaneously expressed in cancer cells and cancer-associated cells, new opportunities might be found in investigating how different cadherin family members influence each other's expression and functionality. New insights might aid in the development of novel cadherin-based signatures for prognosis and eventually in more refined anticancer treatments.

REFERENCES

1. Takeichi M. Cadherins: a molecular family important in selective cell-cell adhesion. *Annu Rev Biochem* 1990;**59**:237–52.
2. Theveneau E, Mayor R. Collective cell migration of the cephalic neural crest: the art of integrating information. *Genesis* 2010;**49**:164–76.
3. Theveneau E, Mayor R. Neural crest delamination and migration: from epithelium-to-mesenchyme transition to collective cell migration. *Dev Biol* 2011;**366**:34–54.
4. Halbleib JM, Nelson WJ. Cadherins in development: cell adhesion, sorting, and tissue morphogenesis. *Genes Dev* 2006;**20**:3199–214.
5. van Roy F, Berx G. The cell-cell adhesion molecule E-cadherin. *Cell Mol Life Sci* 2008;**65**:3756–88.
6. Nakaya Y, Sheng G. Epithelial to mesenchymal transition during gastrulation: an embryological view. *Dev Growth Differ* 2008;**50**:755–66.
7. Williams M, Burdsal C, Periasamy A, Lewandoski M, Sutherland A. Mouse primitive streak forms in situ by initiation of epithelial to mesenchymal transition without migration of a cell population. *Dev Dyn* 2012;**241**:270–83.
8. Cheung M, Chaboissier MC, Mynett A, Hirst E, Schedl A, Briscoe J. The transcriptional control of trunk neural crest induction, survival, and delamination. *Dev Cell* 2005;**8**:179–92.
9. Carver EA, Jiang R, Lan Y, Oram KF, Gridley T. The mouse snail gene encodes a key regulator of the epithelial-mesenchymal transition. *Mol Cell Biol* 2001;**21**:8184–8.
10. Nieto MA. The snail superfamily of zinc-finger transcription factors. *Nat Rev Mol Cell Biol* 2002;**3**:155–66.
11. Wheelock MJ, Shintani Y, Maeda M, Fukumoto Y, Johnson KR. Cadherin switching. *J Cell Sci* 2008;**121**:727–35.
12. De Vries WN, Evsikov AV, Haac BE, Fancher KS, Holbrook AE, Kemler R, et al. Maternal beta-catenin and E-cadherin in mouse development. *Development* 2004;**131**:4435–45.
13. Kan NG, Stemmler MP, Junghans D, Kanzler B, de Vries WN, Dominis M, et al. Gene replacement reveals a specific role for E-cadherin in the formation of a functional trophectoderm. *Development* 2007;**134**:31–41.
14. Luo Y, Ferreira-Cornwell M, Baldwin H, Kostetskii I, Lenox J, Lieberman M, et al. Rescuing the N-cadherin knockout by cardiac-specific expression of N- or E-cadherin. *Development* 2001;**128**:459–69.

15. Radice GL, Rayburn H, Matsunami H, Knudsen KA, Takeichi M, Hynes RO. Developmental defects in mouse embryos lacking N-cadherin. *Dev Biol* 1997;**181**:64–78.
16. Chu YS, Eder O, Thomas WA, Simcha I, Pincet F, Ben-Ze'ev A, et al. Prototypical type I E-cadherin and type II cadherin-7 mediate very distinct adhesiveness through their extracellular domains. *J Biol Chem* 2006;**281**:2901–10.
17. Kokkinos MI, Murthi P, Wafai R, Thompson EW, Newgreen DF. Cadherins in the human placenta–epithelial-mesenchymal transition (EMT) and placental development. *Placenta* 2010;**31**:747–55.
18. Zhou Y, Damsky CH, Fisher SJ. Preeclampsia is associated with failure of human cytotrophoblasts to mimic a vascular adhesion phenotype. One cause of defective endovascular invasion in this syndrome? *J Clin Invest* 1997;**99**:2152–64.
19. Abell AN, Jordan NV, Huang W, Prat A, Midland AA, Johnson NL, et al. MAP3K4/CBP-regulated H2B acetylation controls epithelial-mesenchymal transition in trophoblast stem cells. *Cell Stem Cell* 2011;**8**:525–37.
20. Halacheva V, Fuchs M, Donitz J, Reupke T, Puschel B, Viebahn C. Planar cell movements and oriented cell division during early primitive streak formation in the mammalian embryo. *Dev Dyn* 2011;**240**:1905–16.
21. Smith DE, Franco del Amo F, Gridley T. Isolation of Sna, a mouse gene homologous to the Drosophila genes snail and escargot: its expression pattern suggests multiple roles during postimplantation development. *Development* 1992;**116**:1033–9.
22. Chen ZF, Behringer RR. Twist is required in head mesenchyme for cranial neural tube morphogenesis. *Genes Dev* 1995;**9**:686–99.
23. Grau Y, Carteret C, Simpson P. Mutations and chromosomal rearrangements affecting the expression of Snail, a gene involved in embryonic patterning in Drosophila melanogaster. *Genetics* 1984;**108**:347–60.
24. Oda H, Tsukita S, Takeichi M. Dynamic behavior of the cadherin-based cell-cell adhesion system during Drosophila gastrulation. *Dev Biol* 1998;**203**:435–50.
25. Hatta K, Takagi S, Fujisawa H, Takeichi M. Spatial and temporal expression pattern of N-cadherin cell adhesion molecules correlated with morphogenetic processes of chicken embryos. *Dev Biol* 1987;**120**:215–27.
26. Iwai Y, Usui T, Hirano S, Steward R, Takeichi M, Uemura T. Axon patterning requires DN-cadherin, a novel neuronal adhesion receptor, in the Drosophila embryonic CNS. *Neuron* 1997;**19**:77–89.
27. Hong E, Brewster R. N-cadherin is required for the polarized cell behaviors that drive neurulation in the zebrafish. *Development* 2006;**133**:3895–905.
28. Morita H, Nandadasa S, Yamamoto TS, Terasaka-Iioka C, Wylie C, Ueno N. Nectin-2 and N-cadherin interact through extracellular domains and induce apical accumulation of F-actin in apical constriction of Xenopus neural tube morphogenesis. *Development* 2010;**137**:1315–25.
29. Nandadasa S, Tao Q, Menon NR, Heasman J, Wylie C. N- and E-cadherins in Xenopus are specifically required in the neural and non-neural ectoderm, respectively, for F-actin assembly and morphogenetic movements. *Development* 2009;**136**:1327–38.
30. Acloque H, Ocana OH, Matheu A, Rizzoti K, Wise C, Lovell-Badge R, et al. Reciprocal repression between Sox3 and snail transcription factors defines embryonic territories at gastrulation. *Dev Cell* 2011;**21**:546–58.
31. Vandewalle C, Van Roy F, Berx G. The role of the ZEB family of transcription factors in development and disease. *Cell Mol Life Sci* 2009;**66**:773–87.
32. Theveneau E, Mayor R. Can mesenchymal cells undergo collective cell migration? The case of the neural crest. *Cell Adh Migr* 2011;**5**:490–8.
33. Shoval I, Ludwig A, Kalcheim C. Antagonistic roles of full-length N-cadherin and its soluble BMP cleavage product in neural crest delamination. *Development* 2007;**134**:491–501.

34. Nakagawa S, Takeichi M. Neural crest cell-cell adhesion controlled by sequential and subpopulation-specific expression of novel cadherins. *Development* 1995;**121**:1321–32.
35. Acloque H, Adams MS, Fishwick K, Bronner-Fraser M, Nieto MA. Epithelial-mesenchymal transitions: the importance of changing cell state in development and disease. *J Clin Invest* 2009;**119**:1438–49.
36. Sefton M, Sanchez S, Nieto MA. Conserved and divergent roles for members of the Snail family of transcription factors in the chick and mouse embryo. *Development* 1998;**125**:3111–21.
37. Dufour S, Beauvais-Jouneau A, Delouvee A, Thiery JP. Differential function of N-cadherin and cadherin-7 in the control of embryonic cell motility. *J Cell Biol* 1999;**146**:501–16.
38. Bello SM, Millo H, Rajebhosale M, Price SR. Catenin-dependent cadherin function drives divisional segregation of spinal motor neurons. *J Neurosci* 2012;**32**:490–505.
39. Taneyhill LA, Coles EG, Bronner-Fraser M. Snail2 directly represses cadherin6B during epithelial-to-mesenchymal transitions of the neural crest. *Development* 2007;**134**:1481–90.
40. Coles EG, Taneyhill LA, Bronner-Fraser M. A critical role for Cadherin6B in regulating avian neural crest emigration. *Dev Biol* 2007;**312**:533–44.
41. Hadeball B, Borchers A, Wedlich D. Xenopus cadherin-11 (Xcadherin-11) expression requires the Wg/Wnt signal. *Mech Dev* 1998;**72**:101–13.
42. Borchers A, David R, Wedlich D. Xenopus cadherin-11 restrains cranial neural crest migration and influences neural crest specification. *Development* 2001;**128**:3049–60.
43. Berx G, van Roy F. Involvement of members of the cadherin superfamily in cancer. *Cold Spring Harb Perspect Biol* 2009;**1**:a003129.
44. Becker KF, Atkinson MJ, Reich U, Huang HH, Nekarda H, Siewert JR, et al. Exon skipping in the E-cadherin gene transcript in metastatic human gastric carcinomas. *Hum Mol Genet* 1993;**2**:803–4.
45. Berx G, Becker KF, Höfler H, van Roy F. Mutations of the human E-cadherin (CDH1) gene. *Hum Mutat* 1998;**12**:226–37.
46. Nass SJ, Herman JG, Gabrielson E, Iversen PW, Parl FF, Davidson NE, et al. Aberrant methylation of the estrogen receptor and E-cadherin 5′ CpG islands increases with malignant progression in human breast cancer. *Cancer Res* 2000;**60**:4346–8.
47. Lombaerts M, van Wezel T, Philippo K, Dierssen JWF, Zimmerman RME, Oosting J, et al. E-Cadherin transcriptional downregulation by promoter methylation but not mutation is related to epithelial-to-mesenchymal transition in breast cancer cell lines. *Br J Cancer* 2006;**94**:661–71.
48. Ding W, You H, Dang H, LeBlanc F, Galicia V, Lu SC, et al. Epithelial-to-mesenchymal transition of murine liver tumor cells promotes invasion. *Hepatology* 2010;**52**:945–53.
49. Grotegut S, Von Schweinitz D, Christofori G, Lehembre F. Hepatocyte growth factor induces cell scattering through MAPK/Egr-1-mediated upregulation of Snail. *EMBO J* 2006;**25**:3534–45.
50. Lee MY, Chou CY, Tang MJ, Shen MR. Epithelial-mesenchymal transition in cervical cancer: correlation with tumor progression, epidermal growth factor receptor overexpression, and snail up-regulation. *Clin Cancer Res* 2008;**14**:4743.
51. Leroy P, Mostov KE. Slug is required for cell survival during partial epithelial-mesenchymal transition of HGF-induced tubulogenesis. *Mol Biol Cell* 2007;**18**:1943–52.
52. Vleminckx K, Vakaet Jr L, Mareel M, Fiers W, van Roy F. Genetic manipulation of E-cadherin expression by epithelial tumor cells reveals an invasion suppressor role. *Cell* 1991;**66**:107–19.
53. Blanco MJ, Moreno-Bueno G, Sarrio D, Locascio A, Cano A, Palacios J, et al. Correlation of Snail expression with histological grade and lymph node status in breast carcinomas. *Oncogene* 2002;**21**:3241–6.

54. Gheldof A, Hulpiau P, van Roy F, De Craene B, Berx G. Evolutionary functional analysis and molecular regulation of the ZEB transcription factors. *Cell Mol Life Sci* 2012;**69**:2527–41.
55. Comijn J, Berx G, Vermassen P, Verschueren K, van Grunsven L, Bruyneel E, et al. The two-handed E box binding zinc finger protein SIP1 downregulates E-cadherin and induces invasion. *Mol Cell* 2001;**7**:1267–78.
56. Vandewalle C, Comijn J, De Craene B, Vermassen P, Bruyneel E, Andersen H, et al. SIP1/ZEB2 induces EMT by repressing genes of different epithelial cell-cell junctions. *Nucleic Acids Res* 2005;**33**:6566–78.
57. Aigner K, Dampier B, Descovich L, Mikula M, Sultan A, Schreiber M, et al. The transcription factor ZEB1 (deltaEF1) promotes tumour cell dedifferentiation by repressing master regulators of epithelial polarity. *Oncogene* 2007;**26**:6979–88.
58. Eger A, Stockinger A, Schaffhauser B, Beug H, Foisner R. Epithelial mesenchymal transition by c-Fos estrogen receptor activation involves nuclear translocation of beta-catenin and upregulation of beta-catenin/lymphoid enhancer binding factor-1 transcriptional activity. *J Cell Biol* 2000;**148**:173–88.
59. Spaderna S, Schmalhofer O, Hlubek F, Berx G, Eger A, Merkel S, et al. A transient, EMT-linked loss of basement membranes indicates metastasis and poor survival in colorectal cancer. *Gastroenterology* 2006;**131**:830–40.
60. Spoelstra NS, Manning NG, Higashi Y, Darling D, Singh M, Shroyer KR, et al. The transcription factor ZEB1 is aberrantly expressed in aggressive uterine cancers. *Cancer Res* 2006;**66**:3893–902.
61. Peinado H, Olmeda D, Cano A. Snail, Zeb and bHLH factors in tumour progression: an alliance against the epithelial phenotype? *Nat Rev Cancer* 2007;**7**:415–28.
62. Slattery C, Ryan MP, McMorrow T. E2A proteins: regulators of cell phenotype in normal physiology and disease. *Int J Biochem Cell Biol* 2008;**40**:1431–6.
63. Sobrado VR, Moreno-Bueno G, Cubillo E, Holt LJ, Nieto MA, Portillo F, et al. The class I bHLH factors E2-2A and E2-2B regulate EMT. *J Cell Sci* 2009;**122**:1014–24.
64. Yang J, Mani SA, Donaher JL, Ramaswamy S, Itzykson RA, Come C, et al. Twist, a master regulator of morphogenesis, plays an essential role in tumor metastasis. *Cell* 2004;**117**:927–39.
65. Moreno-Bueno G, Cubillo E, Sarrió D, Peinado H, Rodríguez-Pinilla SM, Villa S, et al. Genetic profiling of epithelial cells expressing E-cadherin repressors reveals a distinct role for Snail, Slug, and E47 factors in epithelial-mesenchymal transition. *Cancer Res* 2006;**66**:9543.
66. Hazan RB, Phillips GR, Qiao RF, Norton L, Aaronson SA. Exogenous expression of N-cadherin in breast cancer cells induces cell migration, invasion, and metastasis. *J Cell Biol* 2000;**148**:779–90.
67. Kim JB, Islam S, Kim YJ, Prudoff RS, Sass KM, Wheelock MJ, et al. N-Cadherin extracellular repeat 4 mediates epithelial to mesenchymal transition and increased motility. *J Cell Biol* 2000;**151**:1193–206.
68. Qi J, Chen N, Wang J, Siu CH. Transendothelial migration of melanoma cells involves N-cadherin-mediated adhesion and activation of the beta-catenin signaling pathway. *Mol Biol Cell* 2005;**16**:4386–97.
69. Groen RW, de Rooij MF, Kocemba KA, Reijmers RM, de Haan-Kramer A, Overdijk MB, et al. N-Cadherin-mediated interaction with multiple myeloma cells inhibits osteoblast differentiation. *Haematologica* 2011;**96**:1653–61.
70. Derycke L, Morbidelli L, Ziche M, De Wever O, Bracke M, Van Aken E. Soluble N-cadherin fragment promotes angiogenesis. *Clin Exp Metastasis* 2006;**23**:187–201.
71. Paredes J, Albergaria A, Oliveira JT, Jeronimo C, Milanezi F, Schmitt FC. P-Cadherin overexpression is an indicator of clinical outcome in invasive breast carcinomas and is associated with CDH3 promoter hypomethylation. *Clin Cancer Res* 2005;**11**:5869–77.

72. Turashvili G, McKinney SE, Goktepe O, Leung SC, Huntsman DG, Gelmon KA, et al. P-Cadherin expression as a prognostic biomarker in a 3992 case tissue microarray series of breast cancer. *Mod Pathol* 2011;**24**:64–81.
73. May CD, Sphyris N, Evans KW, Werden SJ, Guo W, Mani SA. Epithelial-mesenchymal transition and cancer stem cells: a dangerously dynamic duo in breast cancer progression. *Breast Cancer Res* 2011;**13**:202–11.
74. Van Marck V, Stove C, Jacobs K, Van den Eynden G, Bracke M. P-Cadherin in adhesion and invasion: opposite roles in colon and bladder carcinoma. *Int J Cancer* 2011;**128**:1031–44.
75. Sun L, Hu H, Peng L, Zhou Z, Zhao X, Pan J, et al. P-Cadherin promotes liver metastasis and is associated with poor prognosis in colon cancer. *Am J Pathol* 2011;**179**:380–90.
76. Koehler A, Bataille F, Schmid C, Ruemmele P, Waldeck A, Blaszyk H, et al. Gene expression profiling of colorectal cancer and metastases divides tumours according to their clinicopathological stage. *J Pathol* 2004;**204**:65–74.
77. Jacobs K, Feys L, Vanhoecke B, Van Marck V, Bracke M. P-Cadherin expression reduces melanoma growth, invasion, and responsiveness to growth factors in nude mice. *Eur J Cancer Prev* 2011;**20**:207–16.
78. Wels C, Joshi S, Koefinger P, Bergler H, Schaider H. Transcriptional activation of ZEB1 by Slug leads to cooperative regulation of the epithelial-mesenchymal transition-like phenotype in melanoma. *J Invest Dermatol* 2011;**131**:1877–85.
79. Li L, Ying J, Li H, Zhang Y, Shu X, Fan Y, et al. The human cadherin 11 is a pro-apoptotic tumor suppressor modulating cell stemness through Wnt/beta-catenin signaling and silenced in common carcinomas. *Oncogene* 2012;**31**:3901–12.
80. Carmona FJ, Villanueva A, Vidal A, Munoz C, Puertas S, Penin RM, et al. Epigenetic disruption of cadherin-11 in human cancer metastasis. *J Pathol* 2012;**228**:230–40.
81. Pishvaian MJ, Feltes CM, Thompson P, Bussemakers MJ, Schalken JA, Byers SW. Cadherin-11 is expressed in invasive breast cancer cell lines. *Cancer Res* 1999;**59**:947–52.
82. Hoffmann I, Balling R. Cloning and expression analysis of a novel mesodermally expressed cadherin. *Dev Biol* 1995;**169**:337–46.
83. Sarrio D, Rodriguez-Pinilla SM, Hardisson D, Cano A, Moreno-Bueno G, Palacios J. Epithelial-mesenchymal transition in breast cancer relates to the basal-like phenotype. *Cancer Res* 2008;**68**:989–97.
84. Schneider DJ, Wu M, Le TT, Cho SH, Brenner MB, Blackburn MR, et al. Cadherin-11 contributes to pulmonary fibrosis: potential role in TGF-beta production and epithelial to mesenchymal transition. *FASEB J* 2012;**26**:503–12.

CHAPTER FIFTEEN

E-Cadherin Alterations in Hereditary Disorders with Emphasis on Hereditary Diffuse Gastric Cancer

Carla Oliveira[*,†], **Hugo Pinheiro**[*], **Joana Figueiredo**[‡], **Raquel Seruca**[†,‡], **Fátima Carneiro**[†,‡,§]

[*]Expression Regulation in Cancer Group, Institute of Molecular Pathology and Immunology of the University of Porto (IPATIMUP), Porto, Portugal
[†]Faculty of Medicine, University of Porto, Porto, Portugal
[‡]Cancer Genetics Group, Institute of Molecular Pathology and Immunology of the University of Porto (IPATIMUP), Porto, Portugal
[§]Department of Pathology, Centro Hospitalar de São João, Porto, Portugal

Contents

Abstract

The only gastric cancer (GC) syndrome with a proven inherited defect is designated as hereditary diffuse gastric cancer (HDGC) and is caused by germline E-cadherin/*CDH1* alterations. Other E-cadherin-associated hereditary disorders have been identified, encompassing HDGC families with or without cleft-lip/palate involvement, isolated early-onset diffuse GCs, and lobular breast cancer families without GC. To date, 141 probands harboring more than 100 different germline *CDH1* alterations, mainly point

Progress in Molecular Biology and Translational Science, Volume 116
ISSN 1877-1173
http://dx.doi.org/10.1016/B978-0-12-394311-8.00015-7

mutations and large deletions, have been described in these different settings. A third of all HDGC families described so far carry recurrent *CDH1* alterations. Full screening of *CDH1* is recommended in patients fulfilling the HDGC criteria and total prophylactic gastrectomy is the only reliable intervention for carriers of pathogenic alterations. In this chapter, we discuss *CDH1*-associated syndromes, frequency and type of *CDH1* germline alterations, clinical criteria, and guidelines for genetic counseling, molecular pathology, and available animal/cell line models of the disease.

1. GENERAL INTRODUCTION

Gastric cancer (GC) is the sixth most common cancer and the third leading cause of cancer-associated death worldwide (Globocan, assessed 6 March 2012). According to Laurén's classification,[1] two major GC types can be identified, diffuse (DGC) and intestinal (IGC), with distinct epidemiological, morphological, and molecular features.

IGC is more prevalent in elderly males, whereas DGC tends to occur in younger individuals, mainly females, and frequently represents hereditary conditioning. The IGC incidence is steadily decreasing in most countries, in contrast to DGC, which is quite stable or even increasing.[2] The majority of GC is sporadic. GC clustering is a feature in about 10% of the cases, and 1–3% are genetically determined.[3]

2. E-CADHERIN ALTERATION IN HEREDITARY DISORDERS

The human *CDH1* gene [MIM + 192090] is localized in the long arm of chromosome 16, comprises 16 exons transcribed into a 4.5-kb mRNA, and encodes E-cadherin[4] (Fig. 15.1). E-cadherin is a transmembrane

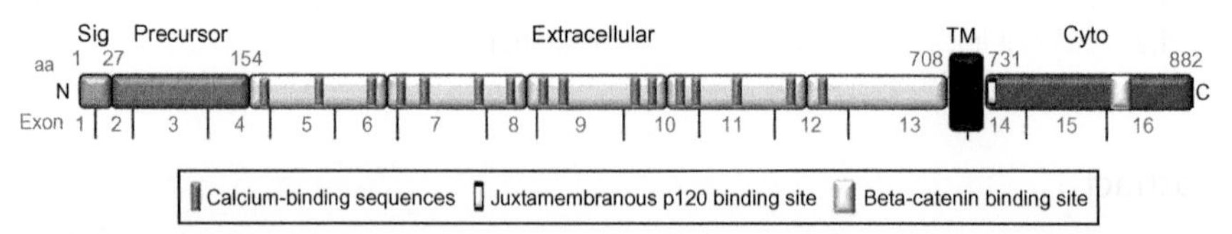

Figure 15.1 Schematic structure of the *CDH1* mRNA and protein. The immature form of E-cadherin is 882 amino acids (aa) long and is encoded by 16 exons. After cleavage of the signal (Sig) and precursor peptides, the 728-aa mature form of the protein is generated. This protein comprises a long extracellular domain with five E-cadherin repeats (EC repeats), a single transmembrane (TM) domain, and a cytoplasmic (Cyto) domain that includes a juxtamembranous p120 binding site and a beta-catenin binding site. The aa numbering is marked on top of the image and the exon numbering (Exon) is marked underneath.

calcium-dependent protein that is predominantly expressed at the basolateral membrane of epithelial cells, where its functions are primarily cell–cell adhesion and suppression of invasion.[5]

E-cadherin dysfunction is a major contributor to cancer development and progression and may occur through several molecular mechanisms, including *CDH1* mutations,[6–8] epigenetic silencing by promoter hypermethylation,[9,10] loss of heterozygosity (LOH),[7] transcriptional silencing by a variety of transcriptional repressors that target the *CDH1* promoter,[11,12] and microRNAs that regulate E-cadherin expression.[13,14]

Germline *CDH1* alterations, although rare, are the cause of several inherited diseases that are addressed in the next sections.

2.1. Hereditary diffuse GC (HDGC)

2.1.1 Definition

Familial clustering of DGC was recognized almost 50 years ago[15] and its molecular basis was identified in 1998. Guilford *et al.*[8] reported three Maori kindred with early-onset, multigenerational DGC, in which germline *CDH1* mutations were detected. These findings led to the identification of a new inherited cancer syndrome designated as hereditary diffuse GC (HDGC) [MIM #137215].[8] Families from other ethnicities and sharing similar features were later identified.[16–18]

2.1.2 Clinical criteria

In 1999, the International GC Linkage Consortium (IGCLC) proposed the following criteria to define HGDC: (1) Any family with two documented cases of DGC in first or second degree relatives, of whom at least one is under the age of 50. (2) Three or more documented cases of DGC in first or second degree relatives of any age. Instances of families with aggregation of GC and an index case with DGC, but not fulfilling the criteria for HDGC, were called familial diffuse GC (FDGC).[19]

Later, Brooks-Wilson *et al.*[20] expanded the criteria for full *CDH1* screening and, in 2010, the IGCLC endorsed the original definition of HDGC and suggested a broader set of clinical criteria as indications for genetic testing for *CDH1* mutations, namely, relaxing the restriction for histopathological DGC confirmation of one family member, including individuals with DGC before the age of 40 without a family history, and including individuals and families with diagnosis of both DGC and lobular breast cancer (LBC) (one diagnosis below 50 years).[21]

An alternative genetically based nomenclature was proposed by the New Zealand group, in which the term HDGC is restricted to families with *CDH1* germline mutations.[18] The IGCLC definition for HDGC is used in the remainder of this chapter.

2.1.3 CDH1 *germline alterations*

Heterozygous *CDH1* alterations remain the only germline genetic event underlying HDGC.[8,21]

2.1.3.1 Causal alterations

2.1.3.1.1 Frequency, site, and type Germline *CDH1* alterations are not restricted to specific sites of the *CDH1* gene or specific E-cadherin protein domains, as they are distributed throughout the coding regions and include splice-site sequences and UTRs (5′- and 3′-untranslated regions) of the gene, as well as throughout all protein functional domains (Fig. 15.2A and B).

Thus far, 122 HDGC families have been reported to harbor such alterations (for a review, see Ref. 22). Truncating *CDH1* germline alterations, predicted to generate premature termination codons, occur in 77% of HDGC families[23–25] (Table 15.1). These alterations are small frameshifts (28.7%), splice-site mutations (27.0%), nonsense (19.7%), and 3′-end large deletions (1.6%). "Nonexpressing" mutations (predicted to lead to complete ablation of mRNA expression) account for 4.1% of the cases. A single family has been reported to carry promoter methylation (0.8%) and four families (3.3%) to carry 5′-end deletions that impair transcription initiation. Missense mutations affect 18.0% of all HDGC families and a single family has been reported to carry an in-frame deletion (0.8%) (Table 15.1).

Penetrance in proven mutation carriers is incomplete, with an estimated lifetime risk for DGC of >80% in both men and women by age 80, and of 60% for LBC in women by the same age.[21] The combined risk of GC and breast cancer in women has been calculated to be 90% at 80 years.[26]

2.1.3.1.2 Different clinical settings The first study comparing different settings of families with FDGC/HDGC was published in 2002.[27] It reported *CDH1* germline mutations in 36.4% of HDGC families, while no mutations were identified in FDGC families. So far, the most complete study analyzed the occurrence of *CDH1* point or small frameshift mutations and large *CDH1* locus rearrangements in 283 families (160 HDGC and 123 FDGC, selected by clinical criteria).[23] Among HDGC

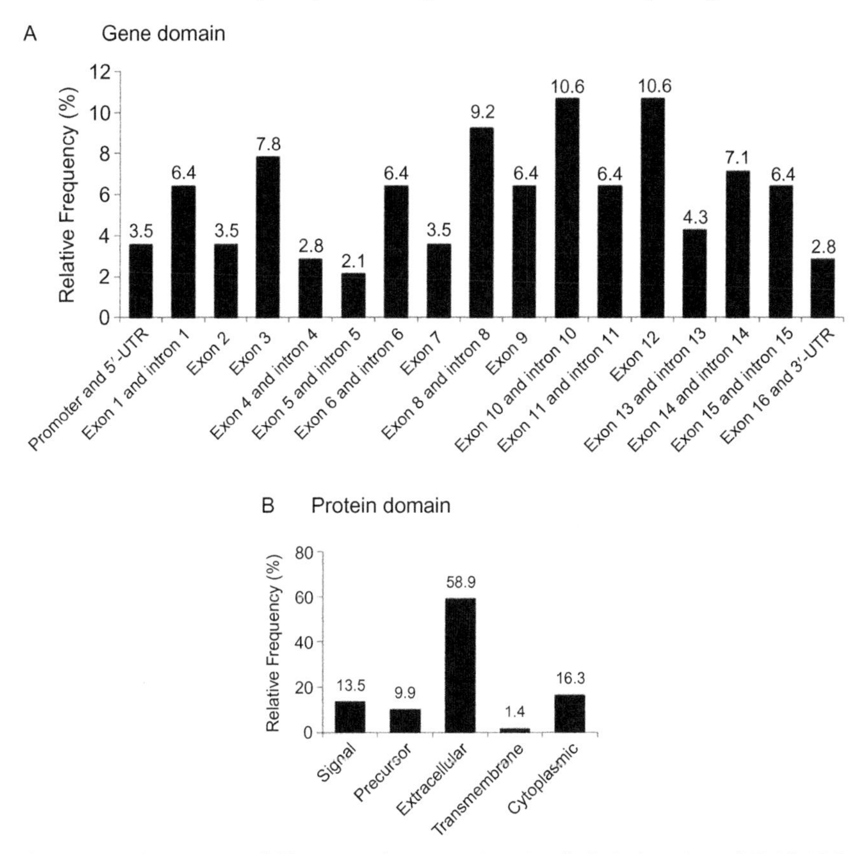

Figure 15.2 Frequency of *CDH1* germline mutations in all clinical settings (HDGC, LBC, and EOGC) together. Alterations for (A) each gene domain and (B) each protein domain. Frequencies are represented as percentages above the bars.

probands, 45.6% (73/160) displayed *CDH1* germline alterations (41.8% point or small frameshift mutations and 3.8% large deletions), while among FDGC probands, 5.7% (7/123) carried *CDH1* mutations and none presented large deletions.

2.1.3.1.3 Different populations The frequency of *CDH1* germline mutations in families with GC clustering varies greatly between geographic regions, and the overall GC incidence also varies.[23,28–33] Specifically, it was verified that in low incidence regions (North America, Canada, and UK), the frequency of germline *CDH1* mutations in such families was approximately 50%, while in moderate (Germany) and high

Table 15.1 Frequency of families/probands regarding mutation type and mutation effect in each E-cadherin-associated disease setting

Mutation type[a]	**HDGC (*n*=122)**	**LBC (*n*=10)**	**Cleft-lip**[b] **(*n*=5)**	**EODGC (*n*=9)**	**Total (*n*=141)**
Frameshift (*n*=42)	35 (28.7%)	2 (20%)	1 (20%)	5 (55.6%)	42 (29.8%)
Splice-site (*n*=34)	33 (27.0%)	1 (10.0%)	3 (60.0%)	0 (0.0%)	34 (24.1%)
Nonsense (*n*=26)	24 (19.7%)	1 (10%)	0 (0.0%)	1 (11.1%)	26 (18.4%)
3′-end large deletions (*n*=2)	2 (1.6%)	0 (0.0%)	0 (0.0%)	0 (0.0%)	2 (1.4%)
Total truncating (*n*=104)	94 (77.0%)	4 (40%)	4 (80%)	6 (66.7%)	104 (73.8%)
Methylation (*n*=1)	1 (0.8%)	0 (0.0%)	0 (0.0%)	0 (0.0%)	1 (0.7%)
5′-end large deletions (*n*=4)	4 (3.3%)	0 (0.0%)	0 (0.0%)	0 (0.0%)	4 (2.8%)
Total nonexpressing (*n*=5)	5 (4.1%)	0 (0.0%)	0 (0.0%)	0 (0.0%)	5 (3.5%)
Missense (*n*=31)	22 (18.0%)	6 (60.0%)	1 (20.0%)	3 (33.3%)	31 (22.0%)
In-frame deletion (*n*=1)	1 (0.8%)	0 (0.0%)	0 (0.0%)	0 (0.0%)	1 (0.7%)

[a]HDGC, hereditary diffuse gastric cancer; LBC, lobular breast cancer; Cleft-lip, HDGC-associated cleft-lip; EODGC, early-onset diffuse gastric cancer.
[b]The figures in this column do not account for the total number of cases as cleft lip/palate cases are already included within the HDGC figures.

(Portugal and Italy) incidence countries, the frequency of alterations was 25% and 22%, respectively.[23] Moreover, in countries with a high incidence of sporadic GC (Japan, China, and Korea), the frequency of germline mutations in families with GC aggregation drops to frequencies near 10%.[34]

In high incidence countries, most families may well represent GC clustering due to the cumulative effect of environmental risk factors (virulent *Helicobacter pylori* strains) and genetic susceptibility of the individuals associated with low penetrance genes,[35] rather than families carrying a *CDH1* germline defect.[34]

2.1.3.1.4 The relevance of *CDH1* germline missense mutations Missense mutations result in full-length E-cadherin molecules containing amino acid substitutions, which can be deleterious. For evaluation of the pathogenicity of missense mutations, it is recommended to evaluate the following parameters: cosegregation of the mutations in families; mutation recurrence in different families; and mutation frequency in healthy control population.[36] However, in most cases, this type of analysis is not possible due to the lack of biological material or to the small size of the families. To circumvent this limitation and improve genetic counseling, functional *in vitro* assays and an *in silico* model were developed to characterize HDGC-associated *CDH1* germline missense mutations.[36]

This *in silico* model estimates the degree of conservation within species of the mutated site and its effect on splicing and protein structure.[36] The functional *in vitro* assays discriminate between pathogenic *CDH1* missense mutations (impairing cell adhesion and leading to invasion) and those that do not affect the phenotype.[20,28,31,37–41]

2.1.3.1.5 *CDH1* germline allelic imbalance Tumors from HDGC families, carrying *CDH1* germline alterations or not, similarly present abnormal or absent E-cadherin protein expression.[42] These findings suggest the existence of germline *CDH1* genetic or epigenetic defects in HDGC families lacking currently recognized *CDH1* germline alterations. Germline *CDH1* allele specific expression imbalance (AI) was advanced as a marker to detect *CDH1* germline impairment in HDGC.[24] Monoallelic *CDH1* expression or high AI was observed in RNA extracted from blood lymphocytes of 80% of *CDH1* mutation carriers and 70.6% of *CDH1* mutation-negative HDGC probands, but not of cancer-free individuals.[24] Germline large deletions and promoter hypermethylation were found in 25% of *CDH1* mutation-negative probands displaying high AI.[24]

Importantly, when considering *CDH1* AI as part of the overall array of germline alterations affecting *CDH1*, the percentage of HDGC families caused by *CDH1* alterations rises to approximately 80%, instead of the previously mentioned 40% (see previous sections).

2.1.3.2 Noncausal alterations: *CDH1* polymorphisms

Two *CDH1* polymorphisms located in the *CDH1* promoter region, known as −160 (rs16260) and −347 (rs5030625),[43] have been tested for *in vitro* functionality. In *CDH1* −160C > A, the A allele has been shown to decrease transcription efficiency by 68%, when compared to the C-allele.[44] In a case-control study performed on an Italian population, the *CDH1* −160C > A variant was associated with an increased susceptibility to DGC, especially in AA homozygotes (95% CI 2.89–21.24).[45] Nevertheless, these results were not confirmed in 899 GC patients and control populations from Portugal, Canada, and Germany.[46]

The second example, $-347\text{G}>\text{GA}$, has also been reported to affect *CDH1* transcriptional activity. The GA allele decreased *CDH1* transcriptional efficiency by 10-fold ($p<0.001$) and had a weak transcription factor binding compared to the G allele.[47]

Despite the large number of studies published so far, the association of these polymorphisms with GC, including HDGC, remains controversial, as an association with disease has not been consistently observed.

2.1.4 Genetic counseling and screening

Genetic counseling is the first step in HDGC management. The process should be complemented by interviews with experts from distinct clinical areas, namely gastroenterology, gastric surgery, and nutrition, with the solid support of molecular pathology and advanced imaging.[21] Mutation carriers should be informed about the limitations of genetic testing and screening methods in this type of cancer. Whenever genetic testing is recommended, informed consent is obligatory.

The youngest age at which testing should be offered is not well established. Rare cases of DGC have been reported before the age of 18, but the overall risk before the age of 20 is very low.[26,48] Genetic testing should be initiated in an affected proband. In the case of missense mutations, functional *in vitro* assays and/or *in silico* studies should be performed in order to evaluate the pathogenicity of the mutation. If alterations are not found by conventional direct sequencing methods, it is indicated to complete the genetic testing with a search for large deletions and AI.[23,24] Total gastrectomy is recommended in at-risk family members aged >20 who have a *CDH1* mutation.[21]

2.1.5 Pathology of E-cadherin-related HDGC

2.1.5.1 Macroscopy

Macroscopic features of the stomach of asymptomatic *CDH1* mutation carriers nearly always appear normal to the naked eye; there is no mass lesion, and slicing shows normal mucosal thickness.[49] Most index HDGC cases present cancers that are indistinguishable from sporadic DGC and can involve all topographic regions of the stomach.

2.1.5.2 Microscopy

Complete mapping of total gastrectomies from asymptomatic carriers of *CDH1* mutations shows microscopic, usually multiple, foci of intramucosal (T1a) signet-ring cell (diffuse) carcinoma in almost all cases.[2,21] Individual

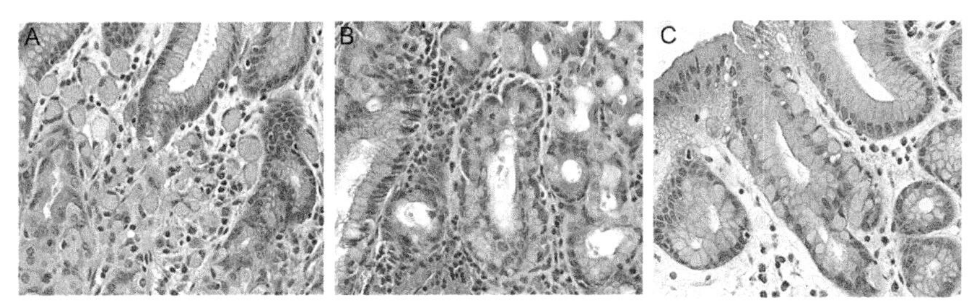

Figure 15.3 Microscopic features of HDGC: (A) Invasive focus of T1a intramucosal signet-ring cell (diffuse) carcinoma (original magnification 400 ×). (B) Signet-ring cell carcinoma *in situ* (original magnification 400 ×). (C) Pagetoid spread of signet-ring cells (original magnification 400 ×). (See Color Insert.)

foci of intramucosal (T1a) signet-ring cell (diffuse) carcinoma are small, ranging from 0.1 to 10 mm (Fig. 15.3A) and every region of the gastric mucosa can be affected.

Two distinct types of lesions were identified in prophylactic gastrectomies as precursors of the invasive cancers: (i) *in situ* signet-ring cell carcinoma (SRCC), corresponding to the presence of signet-ring cells within the basal membrane, generally with hyperchromatic and depolarized nuclei (Fig. 15.3B) and (ii) pagetoid spread of signet-ring cells below the preserved epithelium of glands/foveolae (Fig. 15.3C).[49] E-cadherin immunoexpression was shown to be reduced or absent in early invasive gastric carcinomas (T1a), contrasting with the normal membranous expression in adjacent nonneoplastic mucosa. However, one should be aware that E-cadherin may be expressed at the cell membrane of neoplastic cells (with reduced intensity and/or in a dotted pattern) as well as in the cytoplasm.[6]

2.2. Inherited E-cadherin-associated LBC

2.2.1 Definition

Overall, LBC and its variants represent 5–15% of all invasive breast cancers.[50] The first case of histologically defined LBC in association with HDGC, caused by germline *CDH1* mutations, was described in 1999.[51] Since then, several other cases have been reported. LBC is the second most frequent type of neoplasia in HDGC families and is widely accepted as part of the tumor spectrum of this disease.[26–28,31,51] Penetrance data based on 11 *CDH1*-positive HDGC families estimated the cumulative risk of LBC for female mutation carriers at 39% (95% CI, 12–84%) by 80 years of age.[26] More recently, the estimated cumulative risk of breast cancer for females by the age of 75 was reported as 52% (95% CI, 29–94%), based on the analysis of four pedigrees from Newfoundland with the 2398delC *CDH1* founder mutation.[28]

Several studies were performed to discover whether *CDH1* is a breast cancer susceptibility gene. In LBC families lacking *BRCA1* and *BRCA2* gene mutations, *CDH1* germline mutations were identified in patients having breast cancer as the predominant cancer.[52–54]

2.2.2 Clinical criteria

Germline mutations in *BRCA1* and *BRCA2* are responsible for approximately one-third of hereditary breast cancers among young women with the disease. Mutations in other genes such as *TP53*, *PTEN*, *STK11*, *CHEK2*, and *ATM* account for a small proportion of hereditary breast cancer syndromes, often with distinct clinical features.[55] Despite the involvement of these genes in patients with early-onset breast cancer and/or with a family history of this disease, a proportion of young women with LBC still remain without a molecular diagnosis. The identification of germline *CDH1* mutations in early-onset LBC patients and in LBC families without DGC, signposted *CDH1* as a novel LBC-susceptibility gene. Therefore, women who are diagnosed before the age of 45 or have a family history of LBC and lack *BRCA1* and *BRCA2* germline mutations are eligible for *CDH1* genetic testing.[54]

2.2.3 CDH1 *germline alterations*

Thus far, 10 LBC probands/families have been reported to harbor *CDH1* germline mutations[41,52–54,56] (Table 15.1). As in HDGC families, germline *CDH1* alterations affecting LBC patients are not restricted to specific *CDH1* gene sites or E-cadherin protein domains. In contrast to HDGC, missense mutations are the most frequent alterations, accounting for 60% of the cases described so far. Truncating mutations account for the remaining 40%.

From the mutations reported in LBC families, one of the missense mutations (2494G>A)[54] and a nonsense mutation (283C>T)[56] have been previously reported in HDGC families without LBC involvement.[28,32,57] This indicates that similar *CDH1* mutations may have different clinical outcomes.

2.2.4 Genetic counseling and screening

Besides *BRCA1* and *BRCA2* mutation carriers, three other groups of women are at risk of development of LBC: carriers of *CDH1* mutations in the setting of HDGC, carriers of *CDH1* mutations not belonging to HDGC families, women from families with clustering of DGC in whom no *CDH1* mutation has been identified. It is considered premature to recommend genetic testing of women with a family history of breast cancer unless that at least one of the breast cancers is shown to be lobular.[53] A recent work supports this opinion and reports a 1.3% prevalence of potentially pathogenic *CDH1* variants in patients with early-onset or familial LBC, without DGC family history.[54]

For the surveillance of women carrying germline *CDH1* mutations or untested women from *CDH1*-positive families, it is recommended to perform breast self-examination, annual mammograms, and semiannual clinical breast examination, beginning at the age of 30 at the latest.[53] The importance of regular bilateral breast magnetic ressonance imaging (MRI) is also recognized, as LBCs are known to frequently elude mammographic detection.[53] Prophylactic mastectomy may also be considered for some *CDH1* mutation-positive women, particularly those who have been previously diagnosed with breast cancer in one breast and those who underwent multiple biopsies for abnormal clinical findings.[53]

2.2.5 Pathology of E-cadherin-related LBC

2.2.5.1 Macroscopy

Tumors are often multicentric, bilateral, and may be difficult to define macroscopically.[50]

2.2.5.2 Microscopy

The proliferation of small cells lacking cohesion and appearing individually dispersed through a fibrous connective tissue or arranged in single-file linear cords that invade the stroma is the hallmark of invasive LBC. Histological variants of invasive lobular carcinoma (ILC) include solid, alveolar, pleomorphic, and mixed variants.[50] In a minority of invasive breast cancers, both ductal and lobular features are present[58] and E-cadherin immunohistochemistry may help to differentiate them, although the immunophenotype remains ambiguous in a marginal percentage of cases.[50]

2.3. E-Cadherin-related cleft-lip with or without cleft palate

2.3.1 Definition

In 2006, several members of two HDGC families carrying two different germline mutations of *CDH1* presented cleft-lip, with or without cleft palate (CL/P).[59] CL/P are congenital developmental disorders which result in abnormal facial structures, namely cleft-lip, cleft-lip and palate, or cleft palate alone.[60]

Environmental factors such as maternal smoking, maternal alcohol intake, poor nutrition, viral infection, medical drugs, and teratogens have been identified as risk factors for CL/P. Nevertheless, it has been demonstrated that genetic alterations are important clues for these types of abnormalities.[60]

Approximately 70% of CL/P cases are nonsyndromic, occurring as isolated entities with no other apparent cognitive or craniofacial structural abnormalities. However, and supporting its genetic component, the remaining cases are associated with several malformation syndromes in which the underlying genetic mutation has been identified.[60]

2.3.2 CDH1 *germline alterations*

An association of *CDH1* and CL/P was reported for the first time in two families with HDGC with *CDH1* splicing mutations that generated aberrant transcripts with an in-frame deletion, both affecting the E-cadherin extracellular domain.[59]

Recently, another study reported 10 Dutch *CDH1* mutation families, 3 of them with CL/P ($n=7$). Four individuals were mutation carriers and three were not tested.[41] *CDH1* polymorphisms were also described in nonsyndromic CL/P.[61] The specific single nucleotide polymorphism rs16260 (C>A), located upstream of the transcriptional start site of the *CDH1* promoter, was found to be associated with an increased risk of cleft palate alone in a Chinese population.[62]

All these studies are in agreement with the finding that, in human embryos, *CDH1* is strongly expressed at fourth and fifth weeks of embryogenesis in the frontonasal prominence and at sixth week in the lateral and medial nasal prominences[59] (which correspond to the critical stages of lip and palate development[60]). These findings strongly support the hypothesis that *CDH1* alterations are involved in the disturbed lip and palate closure.

In summary, CL/P has been reported so far in five HDGC families, encompassing four truncating mutations (three splice-site and one frameshift) and one pathogenic missense mutation (Table 15.1). Of these, a single mutation was recurrent (1137G>A) in three HDGC families, although only one of these families was reported to display CL/P in mutation carriers.

2.3.3 *Genetic counseling and screening of HDGC-associated CL/P*

In the opinion of the Dutch Working Group on Hereditary GC, there is no reason for informing future parents about the risk of CL/P in offspring as an integral part of genetic counseling in all HDGC families. However, the occurrence of CL/P abnormalities in family members should be reported in families with HDGC history. In CL/P families, current knowledge should be carefully communicated with counseling, since the exact risks of CL/P remain unknown.[41]

2.4. Early-onset diffuse GC

2.4.1 Definition

Early-onset GC (EOGC) is defined as any GC presenting at the age of 45 or earlier and represents approximately 10% of all patients with stomach cancer.[63] GC occurrence before the age of 30 is very rare (1.1–1.6%) and patients diagnosed before 20 years are exceptional.[64–66] Many cases display diffuse histology (EODGC) and occur in HDGC families.[8,23,37]

2.4.2 CDH1 *germline alterations*

2.4.2.1 Causal alterations

CDH1 germline inactivating mutations and deletions are a well-documented genetic factor associated with EODGC within HDGC families.[8,23,37] The largest series of EODGC screened for *CDH1* germline mutations was reported by Bacani *et al.*,[67] who identified 8 germline sequence variants in 81 (9.9%) patients younger than 50 years. Suriano *et al.*[37] described a large series of 54 patients with diffuse or mixed type EOGC, for whom 5 (9.3%) germline sequence variants were reported.

To date, more than 250 patients harboring apparently sporadic diffuse or mixed GCs, and aged 51 years or less, have been screened for the presence of *CDH1* germline alterations.[68] Of these, less than 10% carried *CDH1* constitutional germline sequence variants, and even less presented variants with a proven deleterious effect *in vitro*. From all EODGC patients described to carry potentially deleterious germline *CDH1* alterations, only 8 have been proved to carry causative *CDH1* germline mutations, 66.7% (6/9) were truncating (5 frameshifts and 1 nonsense) and 33.3% (3/9) were missense, following the trend observed for HDGC families (Table 15.1) (detailed review in Ref. 22).

2.4.3 Genetic counseling and screening

The age of onset has been used in several types of hereditary cancer syndromes to recruit patients without a family history for genetic screening, as these isolated patients represent putative carriers of *de novo* germline mutations.

The approach, in these cases, was similar to the one described in the HDGC section and the age for screening of individuals without a family history of DGC or LBC was set at 40 years. In practical terms, after the direct sequencing of all exons and intron–exon boundaries of the *CDH1* gene, the functional assessment of any mutation that has not been previously described as causative is considered mandatory.

2.5. Final remarks on E-cadherin-associated diseases

CDH1 germline mutations can generate different inherited disorders that can either cluster in the HDGC spectrum, such as LBC, EODGC, and CL/P, or appear as independent clinical presentations.

Numerous germline *CDH1* alterations have been described in 141 families/probands in all clinical settings, dispersed in all gene and protein domains, and with different predicted effects (Fig. 15.4A and B). Of these 141 probands, 86.5% were from HDGC families with or without cleft-lip/palate, 7.1% belonged to LBC families, and 6.4% were EODGC patients (Fig. 15.4C).

About 15% of all *CDH1* mutations recurrently appeared in several families, suggesting that *CDH1*-associated disorders can either arise from a common ancestor[23,28] or be the result of a mutation hotspot. Specifically, a third of all families (51/141; 36.2%) described so far in the literature either display a shared ancestor, as proved for at least 12 of them (4 mutations and a large deletion),[23,28] or carry a hotspot mutation (reviewed in Ref. 22).

Importantly, the same *CDH1* mutation can be observed in families with different E-cadherin-associated diseases, meaning that other coinherited factors may also be involved in these different clinical presentations, highlighting the role of genetic modifiers of the genotype–phenotype correlation.

In summary, *CDH1* germline alterations cause a heterogeneous set of diseases that may be modulated by coinherited factors yet to be identified.

3. MOLECULAR PATHOLOGY IN *CDH1* GERMLINE MUTATION CARRIERS

3.1. Somatic alterations in HDGC tumors

3.1.1 CDH1 *second-hit inactivating mechanisms*

Heterozygous carriers of *CDH1* germline mutations possess a single functional *CDH1* allele in their genomes. This allele apparently produces a sufficient amount of protein in the stomach, to preclude the negative effects of the nonfunctional allele for at least two decades of life. The inactivation of the wild-type allele, by a second-hit molecular mechanism, leads to biallelic inactivation of the *CDH1* gene and determines DGC development.[6,9,69] The initial reports addressing the type and frequency of second *CDH1* hits

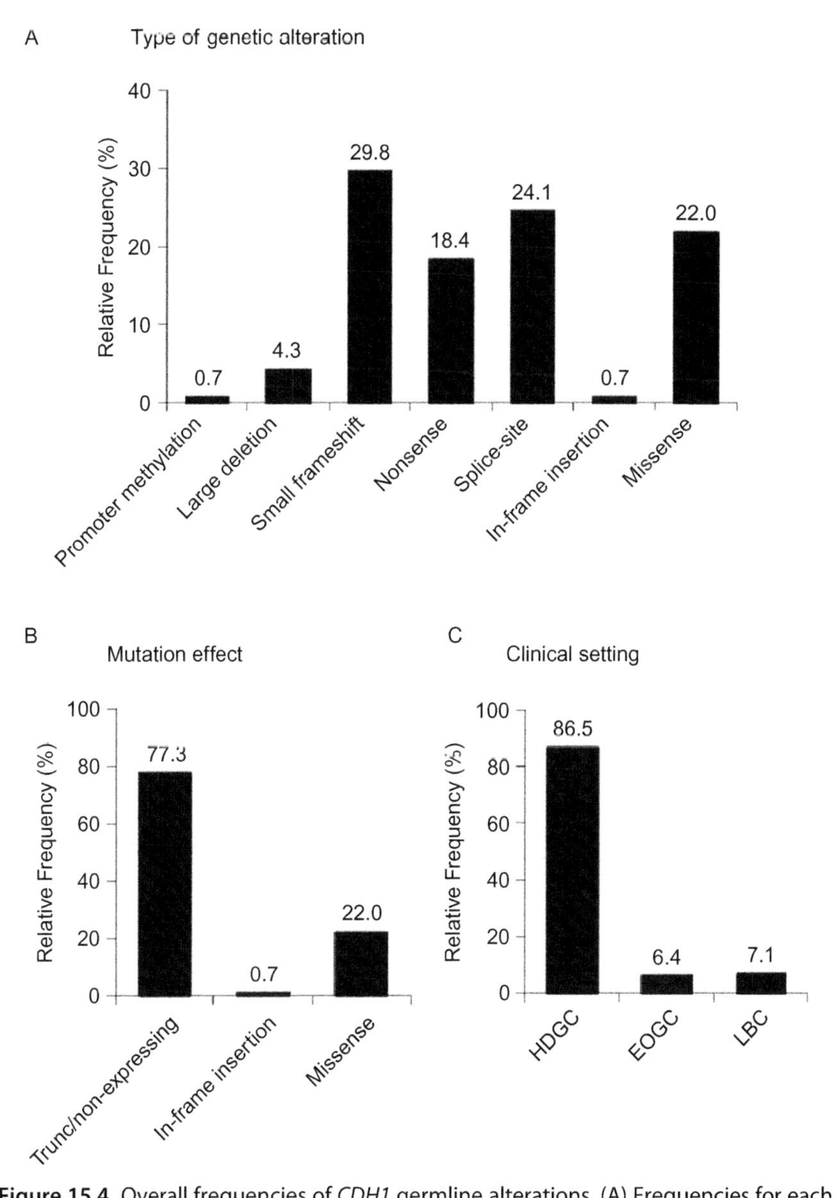

Figure 15.4 Overall frequencies of *CDH1* germline alterations. (A) Frequencies for each type of genetic alteration in all clinical settings (HDGC, LBC, and EOGC) together. (B) Frequencies for each mutation effect in all clinical settings together. (C) Overall frequencies per clinical setting. Frequencies are represented as percentages above the bars.

in HDGC tumors indicated *CDH1* promoter hypermethylation as the most common second-hit mechanism of inactivation,[9,69] while a second mutation or deletion (LOH or intragenic deletions)[6,9,69] was less frequently identified. Consequently, *CDH1* promoter hypermethylation has been suggested as the basis for development of early detection tools as well as for chemoprophylaxis in unaffected *CDH1* mutation carriers.[70] However, such a strategy assumes hypermethylation as the unique mechanism in different neoplastic lesions developing in the same patient, which is not the case.

Oliveira *et al.*[42] performed a systematic analysis of somatic second *CDH1* hits occurring in independent primary cancer foci and lymph node metastases from different *CDH1* germline mutation carriers belonging to HDGC families (in some carriers, several cancer foci were analyzed). Somatic *CDH1* promoter methylation and/or LOH at the *CDH1* locus were detected in 80% of HDGC families and in 75% of all lesions analyzed. Promoter hypermethylation was found in 32.1%, LOH in 25%, both alterations in 17.9%, and no alterations in 25% of the lesions analyzed. In primary cancer, half of the second *CDH1* hits were epigenetic modifications. In metastases, the most frequent second-hit was LOH (58.3%). Different neoplastic lesions from the same patient frequently displayed distinct second-hit mechanisms, and different second-hit mechanisms were also detected in the same neoplastic lesion.[42] Taking into account the heterogeneity of these alterations in HDGC neoplastic lesions and the plasticity of hypermethylated promoters during tumor initiation and progression, drugs targeting only epigenetic alterations may be less effective than initially predicted, particularly in patients with metastatic HDGC.

There has been no systematic analysis of second *CDH1* hit inactivating mechanisms in tumors from isolated cases of EODGC caused by germline *CDH1* mutations.

3.1.2 c-Src kinase activation in early HDGC

To understand the general mechanisms underlying early HDGC development, Humar *et al.*[71] analyzed samples from various stages of hereditary and sporadic DGC, assuming that the progression from early to advanced HDGC was mediated by the acquisition of epithelial-to-mesenchymal transition (EMT) features. The c-Src kinase, an EMT inducer,[72] was not expressed in the small early HDGC foci, while strong expression of the protein was observed in large intramucosal HDGC lesions, as well as in cells invading the *muscularis mucosae*.[71] Moreover, fibronectin (a downstream target of c-Src kinase) was also strongly expressed in carcinoma cells invading beyond the mucosa, consistent with the acquisition of EMT features. Furthermore, active Fak and Stat3 were expressed in c-Src expressing neoplastic cells.[71] Thus, the prevention of EMT by c-Src antagonists seems to be a promising therapeutic strategy for patients with early HDGC.[71]

In keeping with these findings, it was found that cells expressing mutants affecting the extracellular E-cadherin domain exhibit increased motility *in vitro*, concomitantly with increased activation of c-Src kinase.[73]

3.2. Somatic alterations in inherited E-cadherin-associated LBC

3.2.1 CDH1 *second-hit inactivating mechanisms*

The nature and frequency of second-hit *CDH1* mechanisms have not been extensively explored so far in the LBC context. A germline mutation was found in a woman whose LBC was diagnosed at the age of 42 and whose mother had developed LBC at the age of 28.[52] By immunohistochemistry, the proband breast cancer was negative for E-cadherin, indicating the presence of a second molecular event leading to complete inactivation of *CDH1*. LOH downstream of the gene promoter region and encompassing at least exon 4 was detected using intragenic markers.[52]

In sporadic LBC cases, somatic *CDH1* mutations, LOH, and methylation have been proved to inactivate E-cadherin.[74,75] Therefore, it is possible that a similar scenario exists in larger series of E-cadherin-associated inherited LBC.

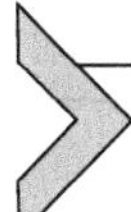

4. ANIMAL AND CELL CULTURE MODELS FOR STUDYING *CDH1*-ASSOCIATED DISEASES

4.1. *Cdh1*$^{+/-}$ mice

In 1995, Riethmacher *et al.*[76] introduced a targeted mutation into the E-cadherin gene (*Cdh1*) in mouse embryonic stem cells. Embryos carrying the homozygous mutation showed severe abnormalities before implantation and were not viable. The heterozygous mutant animals (*Cdh1*$^{+/-}$) were apparently normal[76] and were further used to establish the first mouse model of DGC.[77]

Wild-type and *Cdh1*$^{+/-}$ mice were treated with a carcinogen (*N*-methyl-*N*-nitrosourea) to promote gastric carcinogenesis. When compared to wild-type mice, *Cdh1*$^{+/-}$ mice developed intramucosal SRCC with an 11-times higher frequency.[77] Murine SRCCs were histologically similar to their human counterpart, all lesions displaying decreased E-cadherin expression, low proliferative activity, and absence of nuclear β-catenin accumulation, suggesting that there is no activation of the Wnt pathway.[77]

Another transgenic mouse model for DGC was tried without success. In this model, the specific loss of E-cadherin in the parietal cell lineage strongly affected gastric epithelial cell polarity, growth and differentiation, but this was not sufficient for the development of tumors.[78] In further research, a double conditional knockout mouse line was genetically engineered in which E-cadherin and p53 were specifically inactivated by Cre-mediated recombination under the control of the *Atp4b* gene promoter.[79] DGC developed in all these mice within 12 months (100% penetrance) and cancers were mainly composed of poorly differentiated cells and partly of SRCCs, as in human HDGC.[79] Gene expression patterns of DGC in double knockout mice showed increased expression of mesenchymal markers and EMT-related genes, resembling the human condition.[79] Similar models were used to unravel the role of E-cadherin loss in ILC.[80,81]

4.2. CHO cell lines stably expressing *CDH1* mutations

An *in vitro* model was developed to study the pathogenicity of *CDH1* missense mutations.[37] This model consists of transfection of cadherin-negative cell lines with cDNAs encoding either the wild-type E-cadherin or the mutant forms, to evaluate the two main functions of E-cadherin, namely, cell–cell adhesion and invasion suppression.[37] Among several cadherin-negative cell lines, CHO (Chinese hamster ovary) cells showed the best results. When transfected with wild-type E-cadherin, CHO cells gained the capacity to form aggregates on soft agar and became less able to invade through a Matrigel matrix. On the contrary, pathogenic E-cadherin mutants abrogated cell–cell adhesion and increased invasion ability.[37]So far, 28 *CDH1* missense mutations have been received at IPATIMUP (Porto), a reference laboratory of the IGCLC. All mutations were cloned and their effects on cell invasion and adhesion were fully characterized. The majority was classified as pathogenic and this information was communicated to the initial requesters and used in mutation carrier's management.[28,31,37–41,82] These cell lines are unique in the world and constitute a powerful tool to identify the E-cadherin-mediated signaling pathways pivotal for cancer development.

REFERENCES

1. Lauren P. The two histological main types of gastric carcinoma: diffuse and so-called intestinal-type carcinoma. An attempt at a histo-clinical classification. *Acta Pathol Microbiol Scand* 1965;**64**:31–49.

2. Lauwers GY, Carneiro F, Graham DY, Curado MP, Franceschi S, Montgomery E, et al. Gastric carcinoma. In: Bosman FT, Carneiro F, Hruban RH, Theise ND, editors. *WHO classification of tumors of the digestive system.* Lyon: IARC Press; 2010. p. 48–68.
3. Palli D, Galli M, Caporaso NE, Cipriani F, Decarli A, Saieva C, et al. Family history and risk of stomach cancer in Italy. *Cancer Epidemiol Biomarkers Prev* 1994;**3**:15–8.
4. Berx G, Cleton-Jansen AM, Nollet F, de Leeuw WJ, van de Vijver M, Cornelisse C, et al. E-cadherin is a tumour/invasion suppressor gene mutated in human lobular breast cancers. *EMBO J* 1995;**14**:6107–15.
5. Vleminckx K, Vakaet Jr L, Mareel M, Fiers W, van Roy F. Genetic manipulation of E-cadherin expression by epithelial tumor cells reveals an invasion suppressor role. *Cell* 1991;**66**:107–19.
6. Oliveira C, de Bruin J, Nabais S, Ligtenberg M, Moutinho C, Nagengast FM, et al. Intragenic deletion of CDH1 as the inactivating mechanism of the wild-type allele in an HDGC tumour. *Oncogene* 2004;**23**:2236–40.
7. Berx G, Becker KF, Hofler H, van Roy F. Mutations of the human E-cadherin (CDH1) gene. *Hum Mutat* 1998;**12**:226–37.
8. Guilford P, Hopkins J, Harraway J, McLeod M, McLeod N, Harawira P, et al. E-cadherin germline mutations in familial gastric cancer. *Nature* 1998;**392**:402–5.
9. Grady WM, Willis J, Guilford PJ, Dunbier AK, Toro TT, Lynch H, et al. Methylation of the CDH1 promoter as the second genetic hit in hereditary diffuse gastric cancer. *Nat Genet* 2000;**26**:16–7.
10. Machado JC, Oliveira C, Carvalho R, Soares P, Berx G, Caldas C, et al. E-cadherin gene (CDH1) promoter methylation as the second hit in sporadic diffuse gastric carcinoma. *Oncogene* 2001;**20**:1525–8.
11. Peinado H, Olmeda D, Cano A. Snail, Zeb and bHLH factors in tumour progression: an alliance against the epithelial phenotype? *Nat Rev Cancer* 2007;**7**:415–28.
12. Herranz N, Pasini D, Diaz VM, Franci C, Gutierrez A, Dave N, et al. Polycomb complex 2 is required for E-cadherin repression by the Snail1 transcription factor. *Mol Cell Biol* 2008;**28**:4772–81.
13. Carvalho J, van Grieken NC, Pereira PM, Sousa S, Tijssen M, Buffart TE, et al. Lack of microRNA-101 causes E-cadherin functional deregulation through EZH2 upregulation in intestinal gastric cancer. *J Pathol* 2012;**228**:31–44.
14. Hurteau GJ, Carlson JA, Spivack SD, Brock GJ. Overexpression of the microRNA hsa-miR-200c leads to reduced expression of transcription factor 8 and increased expression of E-cadherin. *Cancer Res* 2007;**67**:7972–6.
15. Jones EG. Familial gastric cancer. *N Z Med J* 1964;**63**:287–96.
16. Gayther SA, Gorringe KL, Ramus SJ, Huntsman D, Roviello F, Grehan N, et al. Identification of germ-line E-cadherin mutations in gastric cancer families of European origin. *Cancer Res* 1998;**58**:4086–9.
17. Richards FM, McKee SA, Rajpar MH, Cole TR, Evans DG, Jankowski JA, et al. Germline E-cadherin gene (CDH1) mutations predispose to familial gastric cancer and colorectal cancer. *Hum Mol Genet* 1999;**8**:607–10.
18. Guilford PJ, Hopkins JB, Grady WM, Markowitz SD, Willis J, Lynch H, et al. E-cadherin germline mutations define an inherited cancer syndrome dominated by diffuse gastric cancer. *Hum Mutat* 1999;**14**:249–55.
19. Caldas C, Carneiro F, Lynch HT, Yokota J, Wiesner GL, Powell SM, et al. Familial gastric cancer: overview and guidelines for management. *J Med Genet* 1999;**36**:873–80.
20. Brooks-Wilson AR, Kaurah P, Suriano G, Leach S, Senz J, Grehan N, et al. Germline E-cadherin mutations in hereditary diffuse gastric cancer: assessment of 42 new families and review of genetic screening criteria. *J Med Genet* 2004;**41**:508–17.
21. Fitzgerald RC, Hardwick R, Huntsman D, Carneiro F, Guilford P, Blair V, et al. Hereditary diffuse gastric cancer: updated consensus guidelines for clinical management and directions for future research. *J Med Genet* 2010;**47**:436–44.

22. Corso G, Marrelli D, Pascale V, Vindigni C, Roviello F. Frequency of CDH1 germline mutations in gastric carcinoma coming from high- and low-risk areas: metanalysis and systematic review of the literature. *BMC Cancer* 2012;**12**:8.
23. Oliveira C, Senz J, Kaurah P, Pinheiro H, Sanges R, Haegert A, et al. Germline CDH1 deletions in hereditary diffuse gastric cancer families. *Hum Mol Genet* 2009;**18**:1545–55.
24. Pinheiro H, Bordeira-Carrico R, Seixas S, Carvalho J, Senz J, Oliveira P, et al. Allele-specific CDH1 downregulation and hereditary diffuse gastric cancer. *Hum Mol Genet* 2010;**19**:943–52.
25. Guilford P, Humar B, Blair V. Hereditary diffuse gastric cancer: translation of CDH1 germline mutations into clinical practice. *Gastric Cancer* 2010;**13**:1–10.
26. Pharoah PD, Guilford P, Caldas C. Incidence of gastric cancer and breast cancer in CDH1 (E-cadherin) mutation carriers from hereditary diffuse gastric cancer families. *Gastroenterology* 2001;**121**:1348–53.
27. Oliveira C, Bordin MC, Grehan N, Huntsman D, Suriano G, Machado JC, et al. Screening E-cadherin in gastric cancer families reveals germline mutations only in hereditary diffuse gastric cancer kindred. *Hum Mutat* 2002;**19**:510–7.
28. Kaurah P, MacMillan A, Boyd N, Senz J, De Luca A, Chun N, et al. Founder and recurrent CDH1 mutations in families with hereditary diffuse gastric cancer. *J Am Med Assoc* 2007;**297**:2360–72.
29. Oliveira C, Seruca R, Carneiro F. Genetics, pathology, and clinics of familial gastric cancer. *Int J Surg Pathol* 2006;**14**:21–33.
30. Oliveira C, Ferreira P, Nabais S, Campos L, Ferreira A, Cirnes L, et al. E-Cadherin (CDH1) and p53 rather than SMAD4 and Caspase-10 germline mutations contribute to genetic predisposition in Portuguese gastric cancer patients. *Eur J Cancer* 2004;**40**:1897–903.
31. Suriano G, Yew S, Ferreira P, Senz J, Kaurah P, Ford JM, et al. Characterization of a recurrent germ line mutation of the E-cadherin gene: implications for genetic testing and clinical management. *Clin Cancer Res* 2005;**11**:5401–9.
32. Yabuta T, Shinmura K, Tani M, Yamaguchi S, Yoshimura K, Katai H, et al. E-cadherin gene variants in gastric cancer families whose probands are diagnosed with diffuse gastric cancer. *Int J Cancer* 2002;**101**:434–41.
33. Wang Y, Song JP, Ikeda M, Shinmura K, Yokota J, Sugimura H. Ile-Leu substitution (I415L) in germline E-cadherin gene (CDH1) in Japanese familial gastric cancer. *Jpn J Clin Oncol* 2003;**33**:17–20.
34. Carneiro F, Oliveira C, Suriano G, Seruca R. Molecular pathology of familial gastric cancer, with an emphasis on hereditary diffuse gastric cancer. *J Clin Pathol* 2008;**61**:25–30.
35. Figueiredo C, Machado JC, Pharoah P, Seruca R, Sousa S, Carvalho R, et al. Helicobacter pylori and interleukin 1 genotyping: an opportunity to identify high-risk individuals for gastric carcinoma. *J Natl Cancer Inst* 2002;**94**:1680–7.
36. Suriano G, Seixas S, Rocha J, Seruca R. A model to infer the pathogenic significance of CDH1 germline missense variants. *J Mol Med (Berl)* 2006;**84**:1023–31.
37. Suriano G, Oliveira C, Ferreira P, Machado JC, Bordin MC, De Wever O, et al. Identification of CDH1 germline missense mutations associated with functional inactivation of the E-cadherin protein in young gastric cancer probands. *Hum Mol Genet* 2003;**12**:575–82.
38. Keller G, Vogelsang H, Becker I, Plaschke S, Ott K, Suriano G, et al. Germline mutations of the E-cadherin(CDH1) and TP53 genes, rather than of RUNX3 and HPP1, contribute to genetic predisposition in German gastric cancer patients. *J Med Genet* 2004;**41**:e89.
39. More H, Humar B, Weber W, Ward R, Christian A, Lintott C, et al. Identification of seven novel germline mutations in the human E-cadherin (CDH1) gene. *Hum Mutat* 2007;**28**:203.

40. Corso G, Roviello F, Paredes J, Pedrazzani C, Novais M, Correia J, et al. Characterization of the P373L E-cadherin germline missense mutation and implication for clinical management. *Eur J Surg Oncol* 2007;**33**:1061–7.
41. Kluijt I, Siemerink EJ, Ausems MG, van Os TA, de Jong D, Simoes-Correia J, et al. CDH1-related hereditary diffuse gastric cancer syndrome: Clinical variations and implications for counseling. *Int J Cancer* 2012;**131**:367–76.
42. Oliveira C, Sousa S, Pinheiro H, Karam R, Bordeira-Carrico R, Senz J, et al. Quantification of epigenetic and genetic 2nd hits in CDH1 during hereditary diffuse gastric cancer syndrome progression. *Gastroenterology* 2009;**136**:2137–48.
43. Medina-Franco H, Ramos-De la Medina A, Vizcaino G, Medina-Franco JL. Single nucleotide polymorphisms in the promoter region of the E-cadherin gene in gastric cancer: case-control study in a young Mexican population. *Ann Surg Oncol* 2007;**14**: 2246–9.
44. Li LC, Chui RM, Sasaki M, Nakajima K, Perinchery G, Au HC, et al. A single nucleotide polymorphism in the E-cadherin gene promoter alters transcriptional activities. *Cancer Res* 2000;**60**:873–6.
45. Humar B, Graziano F, Cascinu S, Catalano V, Ruzzo AM, Magnani M, et al. Association of CDH1 haplotypes with susceptibility to sporadic diffuse gastric cancer. *Oncogene* 2002;**21**:8192–5.
46. Pharoah PD, Oliveira C, Machado JC, Keller G, Vogelsang H, Laux H, et al. CDH1 c-160a promotor polymorphism is not associated with risk of stomach cancer. *Int J Cancer* 2002;**101**:196–7.
47. Shin Y, Kim IJ, Kang HC, Park JH, Park HR, Park HW, et al. The E-cadherin −347G-> GA promoter polymorphism and its effect on transcriptional regulation. *Carcinogenesis* 2004;**25**:895–9.
48. Blair V, Martin I, Shaw D, Winship I, Kerr D, Arnold J, et al. Hereditary diffuse gastric cancer: diagnosis and management. *Clin Gastroenterol Hepatol* 2006;**4**:262–75.
49. Carneiro F, Huntsman DG, Smyrk TC, Owen DA, Seruca R, Pharoah P, et al. Model of the early development of diffuse gastric cancer in E-cadherin mutation carriers and its implications for patient screening. *J Pathol* 2004;**203**:681–7.
50. Tavassoli FA, Devilee P. *World Health Organization Classification of Tumours. Pathology and genetics of tumours of the breast and female genital organs*. Lyon: IARC Press; 2003. pp. 13–62.
51. Keller G, Vogelsang H, Becker I, Hutter J, Ott K, Candidus S, et al. Diffuse type gastric and lobular breast carcinoma in a familial gastric cancer patient with an E-cadherin germline mutation. *Am J Pathol* 1999;**155**:337–42.
52. Masciari S, Larsson N, Senz J, Boyd N, Kaurah P, Kandel MJ, et al. Germline E-cadherin mutations in familial lobular breast cancer. *J Med Genet* 2007;**44**:726–31.
53. Schrader KA, Masciari S, Boyd N, Wiyrick S, Kaurah P, Senz J, et al. Hereditary diffuse gastric cancer: association with lobular breast cancer. *Fam Cancer* 2008;**7**:73–82.
54. Schrader KA, Masciari S, Boyd N, Salamanca C, Senz J, Saunders DN, et al. Germline mutations in CDH1 are infrequent in women with early-onset or familial lobular breast cancers. *J Med Genet* 2011;**48**:64–8.
55. Garber JE, Offit K. Hereditary cancer predisposition syndromes. *J Clin Oncol* 2005;**23**:276–92.
56. Xie ZM, Li LS, Laquet C, Penault-Llorca F, Uhrhammer N, Xie XM, et al. Germline mutations of the E-cadherin gene in families with inherited invasive lobular breast carcinoma but no diffuse gastric cancer. *Cancer* 2011;**117**:3112–7.
57. Dussaulx-Garin L, Blayau M, Pagenault M, Le Berre-Heresbach N, Raoul JL, Campion JP, et al. A new mutation of E-cadherin gene in familial gastric linitis plastica cancer with extra-digestive dissemination. *Eur J Gastroenterol Hepatol* 2001;**13**:711–5.
58. Martinez V, Azzopardi JG. Invasive lobular carcinoma of the breast: incidence and variants. *Histopathology* 1979;**3**:467–88.

59. Frebourg T, Oliveira C, Hochain P, Karam R, Manouvrier S, Graziadio C, et al. Cleft lip/palate and CDH1/E-cadherin mutations in families with hereditary diffuse gastric cancer. *J Med Genet* 2006;**43**:138–42.
60. Dixon MJ, Marazita ML, Beaty TH, Murray JC. Cleft lip and palate: understanding genetic and environmental influences. *Nat Rev Genet* 2011;**12**:167–78.
61. Letra A, Menezes R, Granjeiro JM, Vieira AR. AXIN2 and CDH1 polymorphisms, tooth agenesis, and oral clefts. *Birth Defects Res A Clin Mol Teratol* 2009;**85**:169–73.
62. Song Y, Zhang S. Association of CDH1 promoter polymorphism and the risk of non-syndromic orofacial clefts in a Chinese Han population. *Arch Oral Biol* 2011;**56**:68–72.
63. Koea JB, Karpeh MS, Brennan MF. Gastric cancer in young patients: demographic, clinicopathological, and prognostic factors in 92 patients. *Ann Surg Oncol* 2000;**7**:346–51.
64. Mori M, Sugimachi K, Ohiwa T, Okamura T, Tamura S, Inokuchi K. Early gastric carcinoma in Japanese patients under 30 years of age. *Br J Surg* 1985;**72**:289–91.
65. Nakamura T, Yao T, Niho Y, Tsuneyoshi M. A clinicopathological study in young patients with gastric carcinoma. *J Surg Oncol* 1999;**71**:214–9.
66. McGill TW, Downey EC, Westbrook J, Wade D, de la Garza J. Gastric carcinoma in children. *J Pediatr Surg* 1993;**28**:1620–1.
67. Bacani JT, Soares M, Zwingerman R, di Nicola N, Senz J, Riddell R, et al. CDH1/E-cadherin germline mutations in early-onset gastric cancer. *J Med Genet* 2006;**43**:867–72.
68. Corso G, Pedrazzani C, Pinheiro H, Fernandes E, Marrelli D, Rinnovati A, et al. E-cadherin genetic screening and clinico-pathologic characteristics of early onset gastric cancer. *Eur J Cancer* 2011;**47**:631–9.
69. Barber ME, Save V, Carneiro F, Dwerryhouse S, Lao-Sirieix P, Hardwick RH, et al. Histopathological and molecular analysis of gastrectomy specimens from hereditary diffuse gastric cancer patients has implications for endoscopic surveillance of individuals at risk. *J Pathol* 2008;**216**:286–94.
70. Humar B, Guilford P. Hereditary diffuse gastric cancer and lost cell polarity: a short path to cancer. *Future Oncol* 2008;**4**:229–39.
71. Humar B, Fukuzawa R, Blair V, Dunbier A, More H, Charlton A, et al. Destabilized adhesion in the gastric proliferative zone and c-Src kinase activation mark the development of early diffuse gastric cancer. *Cancer Res* 2007;**67**:2480–9.
72. Avizienyte E, Frame MC. Src and FAK signalling controls adhesion fate and the epithelial-to-mesenchymal transition. *Curr Opin Cell Biol* 2005;**17**:542–7.
73. Mateus AR, Simoes-Correia J, Figueiredo J, Heindl S, Alves CC, Suriano G, et al. E-cadherin mutations and cell motility: a genotype-phenotype correlation. *Exp Cell Res* 2009;**315**:1393–402.
74. Droufakou S, Deshmane V, Roylance R, Hanby A, Tomlinson I, Hart IR. Multiple ways of silencing E-cadherin gene expression in lobular carcinoma of the breast. *Int J Cancer* 2001;**92**:404–8.
75. Graff JR, Herman JG, Lapidus RG, Chopra H, Xu R, Jarrard DF, et al. E-cadherin expression is silenced by DNA hypermethylation in human breast and prostate carcinomas. *Cancer Res* 1995;**55**:5195–9.
76. Riethmacher D, Brinkmann V, Birchmeier C. A targeted mutation in the mouse E-cadherin gene results in defective preimplantation development. *Proc Natl Acad Sci USA* 1995;**92**:855–9.
77. Humar B, Blair V, Charlton A, More H, Martin I, Guilford P. E-cadherin deficiency initiates gastric signet-ring cell carcinoma in mice and man. *Cancer Res* 2009;**69**:2050–6.
78. Mimata A, Fukamachi H, Eishi Y, Yuasa Y. Loss of E-cadherin in mouse gastric epithelial cells induces signet ring-like cells, a possible precursor lesion of diffuse gastric cancer. *Cancer Sci* 2011;**102**:942–50.

79. Shimada S, Mimata A, Sekine M, Mogushi K, Akiyama Y, Fukamachi H, et al. Synergistic tumour suppressor activity of E-cadherin and p53 in a conditional mouse model for metastatic diffuse-type gastric cancer. *Gut* 2012;**61**:344–53.
80. Derksen PW, Liu X, Saridin F, van der Gulden H, Zevenhoven J, Evers B, et al. Somatic inactivation of E-cadherin and p53 in mice leads to metastatic lobular mammary carcinoma through induction of anoikis resistance and angiogenesis. *Cancer Cell* 2006; **10**:437–49.
81. Derksen PW, Braumuller TM, van der Burg E, Hornsveld M, Mesman E, Wesseling J, et al. Mammary-specific inactivation of E-cadherin and p53 impairs functional gland development and leads to pleomorphic invasive lobular carcinoma in mice. *Dis Model Mech* 2011;**4**:347–58.
82. Simoes-Correia J, Figueiredo J, Oliveira C, van Hengel J, Seruca R, van Roy F, et al. Endoplasmic reticulum quality control: a new mechanism of E-cadherin regulation and its implication in cancer. *Hum Mol Genet* 2008;**17**:3566–76.

CHAPTER SIXTEEN

Cadherin Defects in Inherited Human Diseases

Aziz El-Amraoui[*,†,‡], **Christine Petit**[*,†,‡,§]
[*]Institut Pasteur, Unité de Génétique et Physiologie de l'Audition, Paris, France
[†]Inserm UMRS 1120, Paris, France
[‡]UPMC, Paris, France
[§]Collège de France, Paris, France

Contents

Abstract

The tight control of cell–cell connectivity mediated by cadherins is a key issue in human health and disease. The human genome contains over 115 genes encoding cadherins and cadherin-like proteins. Defects in about 21 of these proteins (8 classical, 5 desmosomal, 8 atypical cadherins) have been linked to inherited disorders in humans, including skin and hair disorders, cardiomyopathies, sensory defects associated with deafness and blindness, and psychiatric disorders. With the advent of exome and genome sequencing techniques, we can anticipate the discovery of yet more evidence for the involvement of additional cadherins. Elucidation of the related physiopathological mechanisms underlying these conditions should help to clarify the roles played by these cadherins in tissues and the ways in which defects in different cadherins cause such a wide spectrum of associated phenotypes. These disorders also constitute disparate model systems for investigations of the relative contributions of mechanical adhesive strength and intracellular signaling pathways to the pathogenic process for a given cadherin.

Progress in Molecular Biology and Translational Science, Volume 116
ISSN 1877-1173
http://dx.doi.org/10.1016/B978-0-12-394311-8.00016-9

1. INTRODUCTION

The coordinated control of cell–cell connectivity by various signaling pathways plays an essential role in fundamental physiological and pathological processes, ranging from morphogenesis, tissue differentiation and renewal, selective axon targeting and connectivity to neurite elongation, immune and inflammatory responses, and tumor progression and metastasis.[1–3] Each cell senses its neighborhood by adapting its intracellular signaling pathways to various environmental cues. Several different types of intercellular junction complexes mediate cell–cell adhesion in vertebrates: tight and adherens junctions, desmosomes, and gap junctions.[4] Adherens junctions and desmosomes are cadherin-based protein complexes responsible for cell–cell adhesion in epithelia. The Ca^{2+}-dependent adhesion molecules of the cadherin superfamily are key actors in cell–cell adhesion, often acting in concert with other intercellular or matrix-associated complexes including nectins, tight-junction proteins, gap-junction proteins, and integrins.[3,5]

Mammalian genomes contain over 115 genes encoding cadherins and cadherin-related proteins. Cadherins are transmembrane proteins with an extracellular region consisting of extracellular cadherin-specific repeats (EC domains) (Fig. 16.1), each of which is composed of a seven-stranded β-barrel forming a Greek key topological protein module.[2,3,6] The functions of cadherins involve their homophilic or heterophilic interactions, usually at cell–cell interfaces, but also between plasma membrane extensions in the same cell. Given their highly divergent cytodomains, cadherins may differ in their abilities to sense changes in the mechanical environment of the cell, resulting in the activation of cadherin subclass-specific and appropriate intracellular signaling pathways.[1,3] Cadherins can be classified into 6 or 12 distinct families, depending on the criteria used: sequence characteristics or overall structure and function (see Refs. 3,7,8). Figure 16.1 summarizes the domain organization of representative subfamilies of these proteins in humans: classical and desmosomal cadherins, PCDH15/CDHR15 and CDH23/CDHR23, 7D-cadherins, protocadherins, fat and dachsous, flamingo/celsr, calsyntenins, Ret, and T-cadherin (see Ref. 3; also Chapter 4).

Any given cell generally expresses multiple cadherin subtypes, resulting in a "cadherin code" that may govern molecular cell fate specification and participate in various morphogenetic events, from the maintenance of tissue integrity to promoting dynamic cellular rearrangements. Despite the

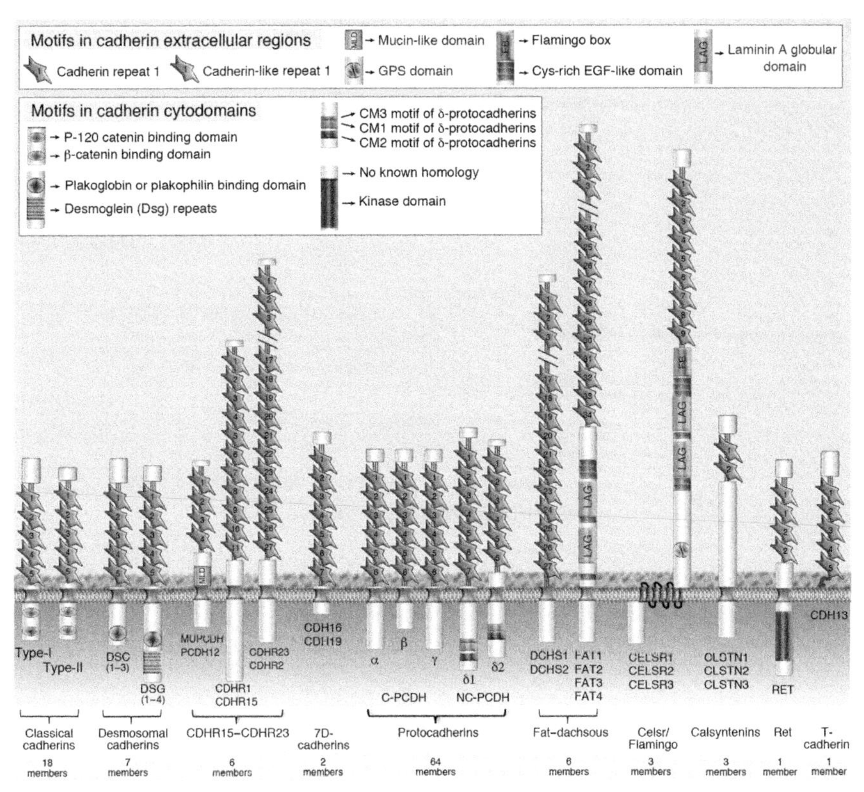

Figure 16.1 Domain organization of various representative families of cadherins from humans, as they have been classified in Ref. 3. Cadherins have extracellular cadherin-specific (EC) domains, which are organized in a tandem fashion (depicted as numbered motifs). With the exception of T-cadherin (CDH13), which lacks a cytodomain, the cadherins have cytodomains that vary in length, sequence, and interacting partners. (See Color Insert.)

functional redundancy of cell–cell junction proteins, some tissues are more prone to cadherin defects than others, depending on the importance of biomechanical determinants of tissue development and functioning. Defects in cell–cell adhesion and/or the associated signaling pathways are common in tumor growth, malignant transformation, and metastases (see Chapter 14). We focus here on cadherin-linked inherited disorders such as defects of the skin and/or hair follicles, cardiomyopathies, sensory defects leading to deafness and blindness, and psychiatric disorders. Defects of 21 cadherins (see Table 16.1; 8 classical, 5 desmosomal, and 8 atypical cadherins) have been implicated in inherited diseases in humans (see also Refs. 3,9). Elucidation of the pathogenesis of these conditions should help to clarify the

Table 16.1 Cadherins involved in inherited human diseases

Classical type-I and desmosomal cadherins in skin, hair follicle, and heart diseases					
Skin and hair diseases				Heart diseases	
***CDH3*, P-cadherin**	***DSG1*, Desmoglein-1**	***DSG4*, Desmoglein-4**	***DSC3*, Desmocolin-3**	***DSG2*, Desmoglein-2**	***DSC2*, Desmocolin-2**
Hypotrichosis, juvenile macular dystrophy (HJMD, OMIM 601553) & Ectodermal dysplasia, ectrodactyly, macular dystrophy syndrome (EEM, OMIM 225280)	Striate palmoplantar keratoderma (SPPK, OMIM 125670 and 148700)	Localized recessive hypotrichosis (LAH, OMIM 607892 and 607903) & Recessive monilethrix (OMIM 158000 and 252200)	Hypotrichosis and recurrent skin vesicles (HRSV; OMIM 613102)	Arrhythmogenic right ventricular cardiomyopathy, ARVC10 (or AC10) (OMIM 125671 and 610193)	Arrhythmogenic right ventricular cardiomyopathy, ARVC11 (or AC11) (OMIM 125645 and 610476) +/- palmoplantar keratoderma and woolly hair (Naxos disease, OMIM601214)

Cadherins and cognitive disorders			
***PCDH19*, Protocadherin-19** X-linked epileptic encephalopathy, early infantile, 9 (EIEE9, or Juberg-Hellman syndrome, OMIM 300088) and Dravet syndrome (OMIM 607208)	***CDH7*** (OMIM 605806) Bipolar disorders ***CDH8***, (OMIM 603008) Learning disabilities, and/or autism ***CDH15*** (OMIM 114019) Intellectual disability	***CDH10*** (OMIM 604555) and ***CDH9*** (OMIM 609974) **loci** Autism susceptibility 1 (OMIM 209850) ***PCDH10*** (OMIM 608286) and ***PCDH18*** (OMIM 608287) **loci** Autism	***CDH12*** (OMIM 600562) and ***CDH18*** (OMIM 603019) **loci** Schizophrenia ***PCDH11X*** (OMIM 300246) and ***PCDH11Y*** (OMIM 400022) **loci** Severe language delay
	No mutations of these cadherins were detected in the patients analyzed so far		

Cadherins and neurosensory diseases		
Inner ear and retinal diseases		Retinal dystrophy
***CDH23*, Cadherin-23 (OMIM 605516)** Sensorineural deafness, DFNB12 (OMIM 601386) and Usher syndrome type ID (OMIM 601067)	***PCDH15*, Protocadherin-15 (OMIM 605514)** Sensorineural deafness, DFNB23 (OMIM 609533) and Usher syndrome type IF (OMIM 602083)	***PCDH21*, Protocadherin-21 (OMIM 609502)** Autosomal recessive one-rod dystrophy 15 (OMIM 613660)

The OMIM entries indicated provide much information about the links between cadherin molecules and diseases.

precise role of cadherins in tissues and the ways in which defects of different cadherins result in such a wide spectrum of associated phenotypes.

2. CLASSICAL TYPE-I AND DESMOSOMAL CADHERINS IN THE SKIN, HAIR FOLLICLES, AND HEART

The classical and desmosomal cadherins have five EC domains, a single transmembrane segment, and a cytoplasmic domain that contains highly conserved binding sites for catenin proteins (Fig. 16.1), providing indirect links to the cytoskeleton: actin, microtubules, and/or intermediate filaments.[1]

2.1. Classical type-I cadherins in the skin

P-cadherin and E-cadherin are classical cadherins expressed in large amounts in the hair follicle. During hair follicle morphogenesis, E-cadherin levels decrease markedly and P-cadherin levels then increase.[10] The notion that this change in cadherin profile is essential for correct hair follicle formation is supported by the identification of deleterious mutations in the P-cadherin gene (*CDH3*) that cause hypotrichosis with juvenile macular dystrophy

(HJMD; OMIM 601553), an autosomal recessive disorder characterized by sparse hair and retinal cone–rod dystrophy.[11,12] Mutations in *CDH3* are also known to cause another disorder with recessive inheritance: ectodermal dysplasia, ectrodactyly, and macular dystrophy syndrome (EEM; OMIM 225280).[13] Interestingly, the overlap between the phenotypes of EEM and another autosomal recessive disease, ectodermal dysplasia, ectrodactyly, and cleft lip/palate (EEC, OMIM 129900), which is caused by mutations in the gene encoding p63,[13] led to the conclusion that *CDH3* is a direct transcriptional target gene of p63.[14,15] *CDH3* and *P63* transcripts have similar distributions in the skin and limbs, but not in the retina, consistent with the absence of retinal defects in individuals with p63 defects.[15]

2.2. Desmosomal cadherins in the skin, hair, and heart

The desmosomal cadherins, desmogleins (Dsgs), and desmocollins (Dscs) form the adhesive core of desmosomes, specialized cell–cell junctions that play a crucial role in maintaining the mechanical integrity of tissues exposed to high levels of mechanical stress, such as the skin and heart.[16] Seven desmosomal cadherins have been described in humans: four desmogleins (DSG1–4) and three desmocollins (DSC1–3, each of which has two splice variants). These molecules have five EC domains, a single transmembrane domain, and a cytoplasmic region that interacts with the intermediate filament-based cytoskeleton through the plakoglobin, plakophilin, and desmoplakin proteins (Fig. 16.2A and B) (see also Chapter 5). Different cell types express different types of desmosomal cadherins and armadillo proteins, accounting for the tissue-specificity of desmosomes. Thus, defects in desmosomal proteins manifest clinically as cardiac and/or cutaneous diseases, depending on the tissue-specific distribution of the mutant protein and the severity of the mutation (see Fig. 16.2C).[17,18]

2.2.1 Desmosomal cadherins in skin and hair diseases

Mutations in *DSG1*, *DSG4*, and *DSC3* have been associated with different hair loss and/or inherited skin disorders in humans.[19–21] Desmoglein-1 was the first cadherin to be identified as the target of a human autosomal dominant hereditary disease, striate palmoplantar keratoderma (SPPK), which is characterized by a thickening of the skin on the palms of the hands and the soles of the feet.[19] More than 16 different heterozygous mutations have been identified in *DSG1*, most of them resulting in the introduction of a premature stop codon.[22,23] Ultrastructural analysis has shown that the palmar skin of SPPK patients contains very few desmosomes, and that those present in

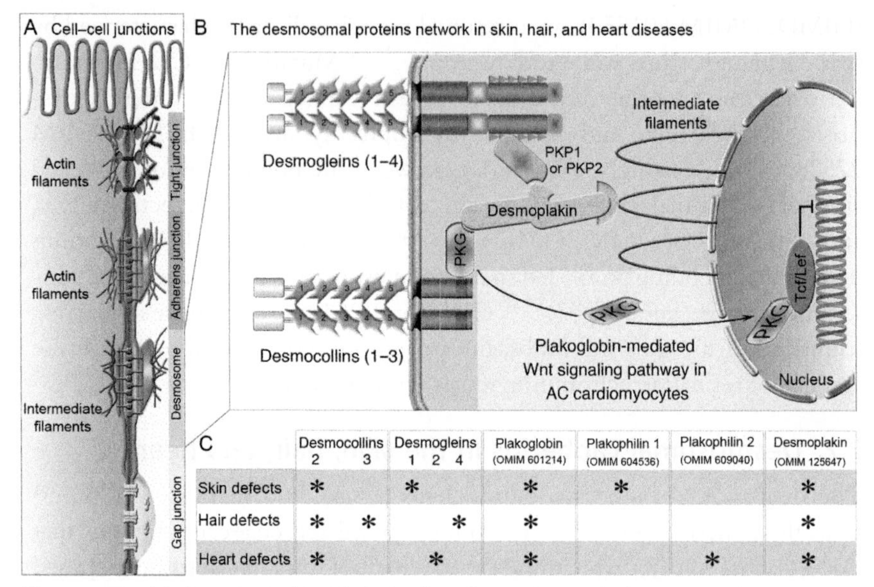

	Desmocollins 2	3	Desmogleins 1	2	4	Plakoglobin (OMIM 601214)	Plakophilin 1 (OMIM 604536)	Plakophilin 2 (OMIM 609040)	Desmoplakin (OMIM 125647)
Skin defects	*		*			*	*		*
Hair defects	*	*			*	*			*
Heart defects	*			*		*		*	*

Figure 16.2 (A) Overview of epithelial cell–cell junctions, which contain distinct intercellular junctional complexes: tight and adherens junctions, desmosomes, and gap junctions. (B) Architecture of the desmosomal molecular network. PKP1, plakophilin 1; PKP2, plakophilin 2; PKG, plakoglobin. (C) List of the desmosomal proteins involved in human diseases; asterisks indicate the affected tissues. For more details on the proteins and/or associated human diseases, please refer to the indicated OMIM numbers. *Panels A,B: Redrawn and modified from Ref. 9.* (See Color Insert.)

the suprabasal layers are abnormally small and poorly formed.[24] Diverse mutations of *DSG4* (intragenic deletions, missense and null mutations) have been reported in two autosomal recessive human hair disorders with overlapping clinical features: localized autosomal recessive hypotrichosis (LAH; OMIM 607903)[20,25] and the autosomal recessive monilethrix-like form (OMIM 607903). LAH patients lack hair principally on the scalp, trunk, and extremities,[25] whereas monilethrix is characterized by a beaded appearance of the scalp hair, due to periodic thinning of the hair shaft.[26,27] These abnormalities are consistent with strong *DSG4* expression in the hair follicle shaft cortex.[28] Major differences in clinical features (age at onset, severity, and progression) have been reported for individuals carrying *DSG4* mutations, both within and between families, suggesting that there may be modifier genes and that different types and combinations of mutations may have different effects in affected patients. A homozygous nonsense mutation in *DSC3* was recently detected in an autosomal recessive disease, hypotrichosis, and recurrent skin vesicles (HRSV; OMIM 613102).[21] However, the presence of "skin vesicles" in the patients has since been disputed,[29] further highlighting the similarities between *DSC3*- and *DSG4*-associated phenotypes. How cadherin defects lead to different hair and skin phenotypes, and genotype–phenotype correlations that might account for variability in affected individuals, are yet to be determined.[27,30]

2.2.2 Desmosomal proteins in heart diseases

Myocardial desmosomes consist essentially of desmoglein-2 and desmocollin-2. Mutations in the genes encoding desmoglein-2 (*DSG2*) and desmocollin-2 (*DSC2*) and three other desmosomal proteins, plakoglobin (*JUP*; OMIM 601214; skin, hair, heart), plakophilin-2 (*PKP2*; OMIM 609040; heart), and desmoplakin (*DSP*; OMIM 125647; skin, hair, heart) (see Fig. 16.2C) are found in approximately 50% of patients with arrhythmogenic right ventricular cardiomyopathy (ARVC, OMIM 107970).[17,31,32] ARVC patients have a high incidence of serious ventricular tachyarrhythmia and sudden death, essentially due to a progressive loss of cardiomyocytes, which are replaced by fibro-fatty tissue.[33–37] There has been increasing evidence in mutant mice, and in humans, for biventricular involvement in the disease, and the name arrhythmogenic cardiomyopathy (AC) may thus be more appropriate.[38]

Analyses of the cardiomyocyte intercalated discs in biopsy specimens from AC patients have revealed the presence of very small numbers of desmosomes, desmosomal gap widening, and inappropriately located desmosomes.[36,39] Mice with null mutations for desmosomal proteins often die before birth, but the advent of technologies for producing transgenic mice and mice with a cardiac tissue-specific conditional knockout has opened up new possibilities for dissection of the molecular and cellular events underlying the onset and progression of myocardial dystrophy.[38]

The specific overexpression of a dominant-negative form of *DSG2* in the heart causes a biventricular cardiomyopathy with aneurysms, ventricular arrhythmia, and sudden death, together with fibrous replacement of the ventricular myocardium, mimicking the human AC phenotype.[40] Myocyte necrosis is the initial event in these mice, leading to progressive myocardial dystrophy, as observed in individuals with *DSG2* mutations. The almost total absence of adipose tissue in AC transgenic mice, unlike in AC patients, is remarkable and remains unexplained. Krusche *et al.*[41] recently produced a homozygous *DSG2* mutant mouse lacking EC domains in the DSG2 extracellular region ($DSG2^{mt/mt}$) in which disease progression was found to be correlated with an increase in mRNA levels for several markers of cardiac stress, remodeling, and heart failure: *c-myc*, *ANF*, *BNF*, *CTGF*, and *GDF15*. The physiological relevance of these changes in terms of disease progression remains to be established.

Intercalated discs, which consist of hybrid adhering junctions or area composite, made up of three distinct junctional complexes, adherens junctions, desmosomes, and gap junctions, connect adjacent cardiomyocytes.[42,43] Recent studies have indicated that appropriate interactions between the N-cadherin/catenin complex at the adherens junctions, the connexin 43-based gap junctions, and the desmosomal–plakoglobin complex in the intercalated discs are essential for normal mechanical and electrical coupling between cardiomyocytes.[44] Further studies are required to unravel the relative contributions of each of these junctional complexes to AC. A model that involves the Wnt/β–catenin signaling pathway, which has been shown to be important in heart myogenesis, has been proposed to account for the pathogenesis of AC (see Refs. 34,36,37). Translocation of the armadillo protein plakoglobin (γ-catenin) to the nucleus seems to inhibit Wnt signaling in cardiac progenitor cells (see Fig. 16.2B).[45] The resulting decrease in the level of transcription of β-catenin target genes, such as the c-Myc and cyclin-D1 encoding genes, ultimately promotes the expression of adipogenic genes. The degenerating cardiomyocytes are then gradually replaced by fibro-fatty scar tissue, resulting in a predisposition to ventricular tachycardia and sudden death, as observed in AC patients.[34,36,37]

Efforts have been made to revise and to improve the molecular diagnosis of AC, with the aim of reducing the risk of sudden arrhythmia/death and slowing disease progression in AC patients.[46] The immunoreactive signal for plakoglobin in the intercalated discs is weaker than normal in most AC patients.[47] Immunohistochemical analysis of endomyocardial biopsies has therefore been proposed as a tool for AC diagnosis. The combination of this approach with a genetic test (even though this test is informative in only 50% of affected families) should facilitate the accurate, early detection of affected individuals and carriers, with implications for genetic counseling and clinical and lifestyle counseling for the patients (e.g., type of surgery, contraindication of intense physical activity).[48] However, genetic counseling remains complicated, due to the considerable variability in disease expression and severity, even among family members harboring the same disease-causing mutation.[36,37] Studies of the genetics, epigenetics, and/or environmental modifiers of the disease and of the interactions between them should provide new insight into the pathogenesis of sudden death, and could lead to the development of targeted therapies.

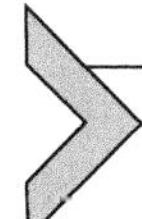

3. CADHERINS IN THE CENTRAL NERVOUS SYSTEM

3.1. Classical type-I cadherins in the central nervous system

In recent years, cadherins have been considered possible targets for cognitive disorders because of their involvement not only in various aspects of brain development, morphogenesis, and maturation, but also in synapse formation, differentiation, signaling, and plasticity.[3] Multiple and distinct cohorts of patients with autism spectrum disorder, schizophrenia, or bipolar disorder have been analyzed worldwide, in studies making use of the recent progress in whole-exome and genome-sequencing techniques. Several cadherin and cadherin-related genes have been identified as susceptibility genes or risk factors for various cognitive disorders (see also Table 16.1): a deletion of *CDH15* has been found in patients with intellectual disability (ID)[49]; single nucleotide polymorphisms between the *CDH9* and *CDH10* loci have been implicated in susceptibility to autism[50]; a deletion between the *CDH12* and *CDH18* loci has been linked to schizophrenia[51]; common variants located close to *CDH7* have been linked to bipolar disorders[52,53]; several deletions involving *CDH8* have been implicated in susceptibility to autism and/or learning disability[54]; a deletion between the *PCDH10* and *PCDH18* loci and recurrent and overlapping copy number variations encompassing *PCDH9* have been identified in a number of patients with autism (Ref. 55 and references therein); a loss of the *PCDH11X* and *PCDH11Y* loci has been reported in a boy with severe language delay[56]; and, finally, an increasing number of mutations in *PCHD19* have been detected in patients with epilepsy-associated phenotypes[57,58] (see Section 3.2 and Chapter 8). However, with the exception of *PCDH19* (see below), compelling evidence for these cadherin genes being the causal genes is lacking, despite the demonstration of links to the disease. Additional studies are therefore required to determine the physiological roles of these cadherins and the corresponding pathophysiological mechanisms.

3.2. Protocadherin-19 is associated with epilepsy and ID

Heterozygous mutations in *PCDH19*, encoding a δ2-protocadherin (see Fig. 16.1), were initially identified in an X-linked disorder characterized by early-onset seizures and ID limited to female patients (EFMR), hemizygous males being asymptomatic.[57,58] This disorder is now referred to as epileptic encephalopathy, early infantile, 9 (EIEE9 or Juberg–Hellman syndrome, OMIM 300088).[59,60] More than 60 different mutations of *PCDH19* have been identified, including missense mutations, nonsense mutations, small nucleotide deletions and insertions, mutations altering the splice sites, and intragenic or whole-gene deletions (Ref. 61, see also http://www.lovd.nl/PCDH19). *PCDH19* is now considered the second most frequently identified gene involved in epilepsy, after *SCN1A* (sodium channel, neuronal type-I, alpha subunit) (OMIM 182389).

Based on the unusual X-linked pattern of inheritance, with *PCDH19* mutations selectively affecting females, occurrence of "cellular interference" has been suggested. Males carrying a deleterious *PCDH19* mutation display normal cognitive functioning, indicating that the loss of PCDH19 function is not pathogenic. By contrast, females heterozygous for a *PCDH19* mutation display tissue mosaicism for this mutation due to random X inactivation. Thus, some cells contain the normal PCDH19 and others contain the mutant form. The association of these two types of cells ultimately impedes cell–cell communication.[57,61] The recent identification of an affected male patient, with fibroblasts displaying mosaicism for a *PCDH19* deletion, provides strong support for a cellular interference mechanism caused by *PCDH19* mutations.[58] However, studies of mutant mice are required to confirm or refute this possibility and to determine the precise function of PCDH19.

4. THE EAR AND THE EYE: TWO SENSORY ORGANS UNDER MECHANICAL CONSTRAINTS

4.1. Hearing and the inner ear

4.1.1 Anatomy and pathology of the inner ear

Hearing results from the processing of sound pressure waves by the auditory sensory cells, which are of two types (Fig. 16.3A): the outer hair cells (OHCs), which amplify the mechanical stimulus of the auditory organ, and the inner hair cells (IHCs), the genuine sensory cells of the cochlea, which transmit sound information to the brain. The hair bundle, which crowns the apical surface of hair cells, is where the mechanoelectrical transduction process takes place. The hair bundle consists of 50–300 F-actin-filled stereocilia organized into three rows of increasing heights (Fig. 16.3A and B), and connected by different fibrous links, specifically, the top connectors and the tip-links (Fig. 16.3B). There is a large body of evidence to suggest that the tip-links control the gating of the mechanoelectrical transduction ion channels, which are located at the

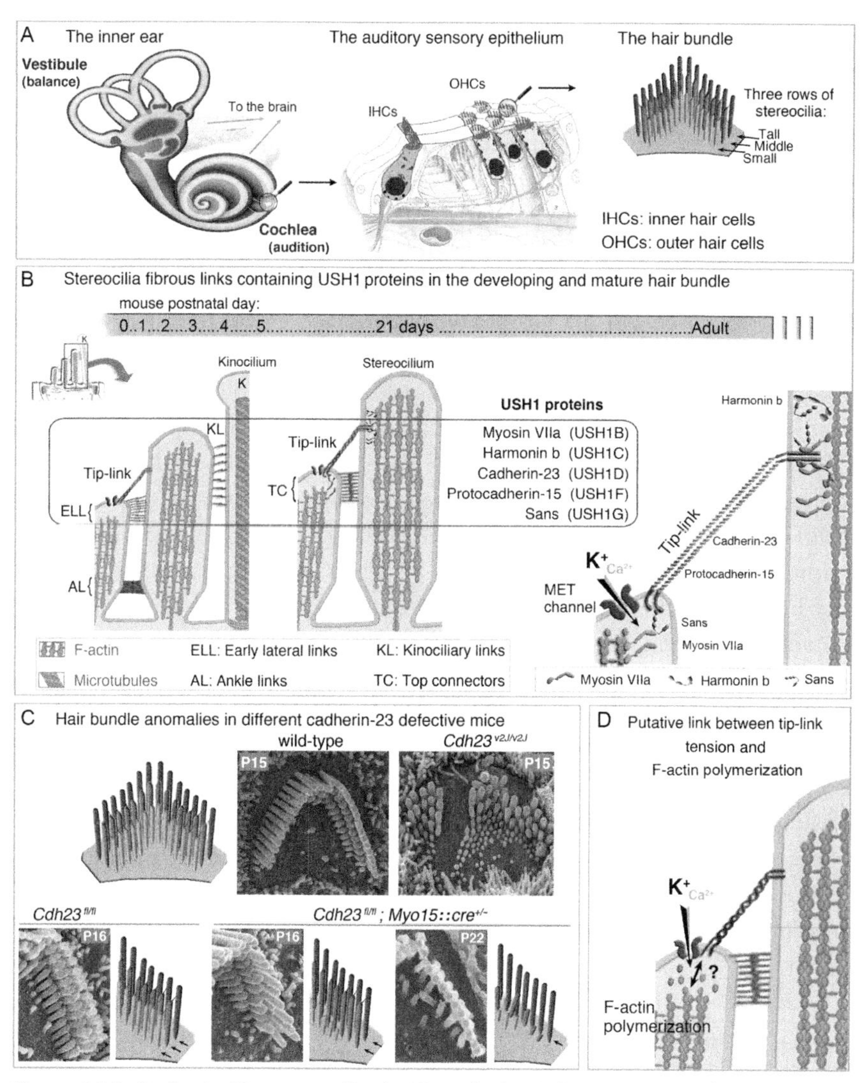

Figure 16.3 Cadherin-23, protocadherin-15, and other USH1 proteins in the hair bundle of auditory hair cells. (A) The inner ear (left panel) contains the vestibule (balance organ) and the cochlea (the auditory organ). All along the cochlea, in the auditory sensory epithelium, sensory IHCs and OHCs are organized in a single medial-side row and three lateral-side rows, respectively. A mechanosensitive hair bundle crowns the apical surface of every hair cell (right panel). (B) The architecture and number of the intersterociliary links vary according to the developmental stage of the hair bundles. A single transient cilium, the kinocilium (K) is located toward the periphery of the hair bundle, attached to stereocilia in the tallest row. In the developing hair bundle (time scale on top represents postnatal mouse development), stereocilia are connected by the early lateral links (ELL), the ankle links (AL), and the kinociliary links (KL), later

(Continued)

tips of stereocilia from the small and middle rows of the hair bundle (Fig. 16.3B, right panel). During development in mammals, the hair bundle also contains the transient early lateral links (ELL), and the ankle links (AL), connecting the growing stereocilia, and includes a single transient genuine cilium, the kinocilium, which is connected to the adjacent stereocilia from the tallest row by fibrous kinociliary links (KL) (Fig. 16.3B, left panel).[63–65]

Studies of one form of syndromic deafness, Usher syndrome of type-I (USH1), over the last 10 years have provided considerable insights into the development and functioning of the hair bundle.[64–66] Usher syndrome (USH) is an autosomal recessive disorder characterized by combined deafness–blindness. It affects approximately one out of 25,000 children. USH has three clinical subtypes, USH1–3, of which type-I (USH1) is the most severe form. It is characterized by congenital, severe to profound deafness, constant vestibular dysfunction, and retinitis pigmentosa beginning before puberty and rapidly leading to blindness.[64–66] At least seven USH1 loci have been characterized, and five causal genes have been identified, encoding the actin-based motor protein myosin VIIa, the PDZ domain-containing protein harmonin, the scaffolding protein sans and the Ca^{2+}-dependent adhesion proteins cadherin-23 and protocadherin-15.[65]

Deaf mutant mice with mutations in each of the five USH1 genes identified to date have been reported and studies of the molecular, morphological (stereocilia formation and cohesiveness), and functional (mechanoelectrical transduction) properties of the hair bundle in wild type and Ush1 mutant mice have greatly advanced our understanding of the cellular and molecular mechanisms accounting for hearing impairment in this syndrome.[63–65]

replaced by the top connectors (TC). At mature stages (right panel), cadherin-23 and protocadherin-15 form the tip-link; they interact with other USH1 proteins at both insertion points of the tip-link. MET channel, mechanoelectrical transducer channel. (C) Viewed from the top of the hair cells, scanning electron microscopy images from P15 mice illustrate the fragmentation of the hair bundle in complete cadherin-23 knockout mice, whereas wild-type mice have a well-organized hair bundle structure (top panels). In conditional knockout mice with delayed cadherin-23 ablation (bottom panels), hair bundles are not fragmented, but at P16 and P22, most of the stereocilia in the small and middle rows are much shorter, and many have disappeared in comparison to control $Cdh23^{fl/fl}$ mice (small arrows). (D) Magnification of the tips of the stereocilia from the middle and tall rows of a hair bundle, illustrating the putative link between F-actin polymerization and tip-link tension and/or Ca^{2+} flow through the MET channel. *C: Adapted from Refs. 9,74.* (See Color Insert.)

4.1.2 USH1 proteins are required for the formation of correctly shaped hair bundles

All five USH1 null mutant mice form fragmented hair bundles[67] (see Fig. 16.3C). A defect in the formation of USH1 protein containing links, ELL between stereocilia and KL between stereocilia and kinocilium, has been identified as the principal cause of hearing impairment in USH1 patients (Refs. 67–71; reviewed in Refs. 62–66). The cohesive forces applied to the transient ELL and KL (see Fig. 16.3B), both consisting of cadherin-23 and protocadherin-15, in the developing hair bundle seem to be necessary for the formation of unitary and properly shaped hair bundles.[67,68,70–74] Analysis of cadherin-23 and protocadherin-15 sequences has led to their classification in the cadherin-related 2 (Cr-2) and Cr-1 subfamilies, respectively. The two proteins have been renamed accordingly, CDHR23 for cadherin-23, and CDHR15 for protocadherin-15 (see Ref. 7 and Chapter 4). Over the years, evidence has been obtained for multiple *in vitro* and *in vivo* interactions between USH1 proteins.[63,68,69,75,76] In the hair bundle, the cadherin-mediated links are anchored to the stereocilia actin filaments by direct interactions with myosin VIIa, harmonin b, and sans (see Fig. 16.3B, right panel).[62,68,69,75]

4.1.3 USH1 Proteins at the core of the mechanoelectrical transduction machinery

USH1 proteins also play a role in mechanoelectrical transduction in mature hair cells. The tip-link was recently shown to have an asymmetric structure and to consist of heterodimers of cadherin-23 and protocadherin-15.[77] Harmonin b (the largest harmonin isoform), which binds to actin filaments[68] and cadherin-23,[68,78,79] operates as an intracellular link, limiting adaptation of mechanoelectrical transduction and engaging adaptation motors at the upper insertion point of the tip-link on stereocilia.[78,79] A tripartite complex, consisting of myosin VIIa, harmonin b, and cadherin-23, has also been shown to bind to phosphatidylinositol 4,5-bisphosphate (PIP2).[80] This may explain the role of this membrane phospholipid in the process of mechanoelectrical transduction adaptation in the hair bundle.[80]

Delayed conditional knockout mice with ablation of either sans or cadherin-23 in the hair cells after the period of hair bundle maturation have been engineered. In these mice, the hair bundle forms and differentiates normally, at least until postnatal day 15 for cadherin-23 deficient mice (see Fig. 16.3C).[74] Morphological and electrophysiological analyses of sans deficient mice have shown that sans, which interacts with cadherin-23, is also a critical component of the tip-link complex.[74] In mice with a delayed conditional knockout of either sans or cadherin-23, stereocilia from the small and middle rows, but not from the tall row, shorten considerably (Fig. 16.3C), suggesting that the tip-link tension or Ca^{2+} flowing through the mechanoelectrical transducer channel is coupled to F-actin polymerization at the tips of stereocilia in the small and middle rows of cochlear hair bundles (see Fig. 16.3D).[62,74] Additional studies are required to elucidate the molecular relationship between the tip-link, mechanoelectrical transducer channels, and the F-actin polymerization machinery involved in stereocilia.[62]

4.1.4 Structural basis of the adhesive interactions between cadherins and the force applied to the tip-link

At epithelial cell–cell contacts, an intermembrane space approximately 22 nm across, cadherin domains from apposed cells dimerize via their membrane-distal EC1 domains to form *trans*-adhesive homodimers.[4,6] Hundreds of classic cadherin homodimers in adherens junctions, thousands of desmosomal cadherins in desmosomes, bring the cells into contact with the required strength for adhesion. However, in the hair bundle, the tip-links, which are about 150- to 200-nm long, stretch from the tips of the stereocilia in the small and middle rows to the sides of the neighboring, taller stereocilia. It has been suggested that these links are instead formed by one cadherin-23 homodimer interacting in *trans* with one protocadherin 15 homodimer.[77] It remains unclear how this cadherin-mediated link structure withstands the mechanical force to which it is subjected during the stimulation of hair cells with sound. This force is estimated to range from 1 to 10 pN per tip-link for a stimulation intensity of 80 dB SPL (reviewed in Ref. 81). The major mechanism of cadherin *trans*-dimerization, particularly that between classical or desmosomal cadherins, is known as strand-swap binding and involves the exchange of N-terminal β-strands of EC1 domains between partner molecules. Strand-swap binding requires the tryptophan residue at position 2 (Trp2) in the EC1 of a type-I cadherin molecule, or Trp2 and Trp4 in type-II cadherins. These residues dock into a complementary hydrophobic pocket in the EC1 of the interacting partner.[82] However, these Trp residues are not conserved in the EC1 domains of several nonclassical cadherins, including members of the Cr-2 cadherin subfamily [cadherin-23 (CDHR23), protocadherin-21 (CDHR1), and protocadherin-24 (CDHR2)],[7] suggesting that these cadherins may make use of different binding mechanisms (see below). The three-dimensional (3D) structures of the cadherin-23 and protocadherin-15 EC1 domains are similar but not identical to those of other cadherins as they show several unusual features.[83,84] As in classical cadherins, each EC domain has three Ca^{2+} ion-binding sites (referred to as sites 1, 2, and 3), which render the cadherin extracellular domains more rigid and promote *trans* junctional interactions. However, the EC1 domains of CDHR23 and CDHR15 each have an N-terminal extension that is conserved across species and comprises several polar amino acid residues. These polar residues, which are also present at the N-terminus of CDHR1, CDHR2, and CDHR23[83] have an additional Ca^{2+} ion-binding site (referred to as site 0) at the very tip of EC1, which may bind to Ca^{2+} ions with a lower affinity.[83,84] Ca^{2+} concentration may therefore have a significant effect on the formation and stability of adhesion complexes of inner ear hair cells, particularly for the low concentrations of Ca^{2+} (about 20 μM) present in the endolymph surrounding the tip-links.[83,84] The first structural view of a *trans* heterophilic cadherin complex has recently been obtained from the interaction between the two amino-terminal cadherin repeats EC1 and EC2 of CDHR15 and CDHR23, which has revealed an overlapped, antiparallel heterodimer that matched the well-known Greek key motif of classical cadherins.[85] The CDHR15 and CDHR23 heterophilic interface, however, exploits unique structural protrusions within strand A of each cadherin EC1 repeat, which in turn are stabilized by a disulfide bond and the additional Ca^{2+}-binding site of CDHR23; key interactions and salt-bridges (such as those between R113 in CDHR15 and E77 and Q98 in CDHR23) have been identified, which seems to provide the selectivity that prevents the same antiparallel binding for homophilic complexes; and the EC2 repeat of each cadherin also is involved in the interaction further extending the heterodimer interface.[85] Finally, molecular dynamics simulation has predicted the heterodimer bond between CDHR15 and CDHR23 to be mechanically strong enough to withstand the forces produced on the tip-links by sounds of moderate intensity.[85]

4.2. Vision and the retina

4.2.1 Vision in USH1 patients and Ush1 mutant mice

A complete picture of the molecular or cellular pathways involving USH1 proteins in the sensory organs is required before therapeutic options for preventing or delaying the onset of retinitis pigmentosa in USH1 can be developed. As described above, mutations of the cadherin-23 and protocadherin-15 genes affect both the inner ear and the retina in humans. Analyses of Ush1 mutant mice, which have hearing defects mimicking those observed in USH1 patients, have been instrumental in deciphering the molecular and cellular mechanisms underlying hearing loss in USH1. By contrast, progress in clarifying the critical role of USH1 proteins in the retina has been very slow, due largely to the lack of a visual defect in any of the Ush1 mutant mice generated (see Refs. 63,64,86,87). Several explanations have been put forward for this phenotypic discrepancy, including the shorter lifespan of the mouse, interspecies differences in light exposure, functional redundancy of USH1 proteins in mouse (but not in human retina), and intrinsic differences in the photoreceptor cell architecture and/or functioning between humans and mice.[63,88]

4.2.2 Cadherin-23, protocadherin-15, and USH1 proteins in the retina

In vivo analyses of human USH1B patients, based on optical imaging of the retina and visual function, have shown that the first detectable defect affects photoreceptor cells.[89] The retinopathy in all USH1 patients is classified as a rod–cone dystrophy, and electroretinogram abnormalities are already detectable in 2-year-old affected children. Rod dysfunction is detected first and this rapidly worsens, to be followed by cone failure, which progresses more slowly.[89–92] No such electroretinogram anomalies or abnormal visual phenotypes have been observed in the mutant mice defective for cadherins or the other USH1 proteins.[93–95] Nonetheless, some minor retinal abnormalities have been reported in mice with a myosin VIIa defect.[96–98] In the absence of functional myosin VIIa, opsin concentration increases moderately in the connecting cilia of photoreceptor cells, suggesting a defect in the trafficking of this protein between the inner and outer segments of the photoreceptor.[98] The consequences of this delayed trafficking for retinal degeneration are unknown. It also remains unclear whether other Ush1 mutant mice display such a delay in opsin transport. The pathogenesis of the USH1 hearing impairment was elucidated by looking for holistic explanations for the pathogenesis of deafness in all five genetic forms of USH1, which are clinically indistinguishable in terms of the hearing loss phenotype.[64–66] We would expect there to be common mechanisms in the retina too, because it is not possible to distinguish between USH1 genetic forms in terms of the clinical presentation of retinitis pigmentosa.[89–92] USH1 proteins are present in different types of retinal cells, including photoreceptor cells and nonsensory cells, and in various subcellular compartments of the photoreceptor cells. The photoreceptor synaptic region has been shown to contain the five USH1 proteins analyzed, but no significant retinal synaptic defect was observed in any of the corresponding Ush1 mutant mice (see Refs. 63,64,86,87). Comparative studies determining the subcellular distribution of USH1 proteins in several species have shown that the five proteins analyzed colocalize to the junction between the inner and outer segment of retinal photoreceptors in humans, macaques, and frogs but not in mice.[99] Ultrastructural and immunogold analyses have revealed that USH1 proteins, including CDHR15 and CDHR23, are present in the photoreceptor calyceal processes, F-actin-rich long microvilli that crown the inner segment and wrap around the basolateral region of the outer segment.[99] Together, these findings led to the proposal that USH1 proteins form an adhesion belt around the basolateral region of the photoreceptor outer segment in humans, and that defects in this structure probably cause the retinal degeneration in USH1 patients. By contrast, mouse photoreceptors lack the calyceal processes and do not have USH1 proteins at the inner–outer segment interface, which would account for the observation that Ush1 mouse models do not develop retinitis pigmentosa.[99]

4.2.3 *Cdhr1 (protocadherin-21) and prominin-1 at the base of photoreceptor cells*

CDHR1, previously named *PCDH21*, is a gene encoding another member of the Cr-2 cadherin subfamily (see Ref. 7), which has been implicated in human recessive retinal dystrophies.[100,101] The encoded cadherin-related protein CDHR1 has six EC repeats and a single cytoplasmic domain, and is abundant in the retina, particularly at the junction between the inner and outer segments of rod and cone photoreceptor cells. Unlike cadherin-23 (Cdhr23) and protocadherin-15 (Cdhr15) defects, *Cdhr1* inactivation in mice does cause retinal degeneration, because 6-month-old animals have only 50% of the normal number of photoreceptor cells, and these cells have abnormally short outer segments with imperfectly stacked disks.[102] Cdhr1 has been shown to interact with prominin-1, a pentaspan transmembrane glycoprotein that is also located at the base of the outer segment.[103] *PROM1* mutations have also been implicated in autosomal dominant cone–rod dystrophy and autosomal recessive retinitis pigmentosa.[104,105] Furthermore, transgenic mice carrying the human R373C *PROM1* mutation found in blind patients have overly large outer disks aligned perpendicular to the normal orientation of the disks,[103] reminiscent of the phenotype observed in *Cdhr1* knockout mice.[102] In these prominin-1-transgenic mice,

Cdhr1 and prominin-1 were found to be distributed throughout the photoreceptors rather than targeted to the base of the photoreceptor outer segment as would normally be the case. Together, these findings suggest that Cdhr1 and prominin-1 cooperate at the junction between the inner and outer segments, to ensure the correct establishment and/or maintenance of the topological properties of the outer segment.[102,103]

5. CONCLUDING REMARKS

Progress in next-generation sequencing will probably rapidly extend the list of known inherited human disorders involving cadherin defects. The diversity and variability of abnormal phenotypes, even among members of the same family with the same cadherin defect, highlights the importance of as yet unexplored factors modulating the expression of cadherin genes, such as modifier genes, and environmental and epigenetic factors (allelic exclusion imprinting). The generation of appropriate animal models (in mice or other species) faithfully mimicking the corresponding human phenotypes remains, however, essential, if we are to achieve a profound understanding of cadherin functions/dysfunctions. Molecular, cellular, and subcellular differences between species should be considered, to determine whether they are responsible for the phenotypic differences between humans and the corresponding animal models. Genetically modified animals with more elaborate cognitive functions than those of the mouse are required for confirmation of the role of cadherins in cognitive disorders. Correlations between the protein composition of the junctional complexes and the intra- and intercellular forces involved in the morphogenesis and maintenance of a given tissue are of critical importance. Cell–cell junctional networks are generally considered as complete, single structures, but detailed analyses of the individual proteins constituting these networks are now required (How many copies of each protein are there? How are they positioned? What are their dynamics?). Building on this knowledge, we should be able to define the differences in network properties between developmental contexts or disease states, and to identify the determinants of these differences.

ACKNOWLEDGMENTS

We are grateful to Jean-Pierre Hardelin for critically reading the manuscript and Jacques Boutet de Monvel for his comments. We apologize for not citing all the relevant references because of space limitation. The work in the authors' laboratory is funded by the European Union

Seventh Framework Programme, under grant agreement HEALTH-F2-2010-242013 (TREATRUSH), LHW-Stiftung, Fondation Raymonde & Guy Strittmatter, FAUN Stiftung (Suchert Foundation), and Conny Maeva Charitable Foundation (CP) and an ANR-07-MRAR E-009-01 grant to A. E.

REFERENCES

1. Harris TJ, Tepass U. Adherens junctions: from molecules to morphogenesis. *Nat Rev Mol Cell Biol* 2010;**11**:502–14.
2. Gomez GA, McLachlan RW, Yap AS. Productive tension: force-sensing and homeostasis of cell-cell junctions. *Trends Cell Biol* 2011;**21**:499–505.
3. Hirano S, Takeichi M. Cadherins in brain morphogenesis and wiring. *Physiol Rev* 2012;**92**:597–634.
4. Oda H, Takeichi M. Evolution: structural and functional diversity of cadherin at the adherens junction. *J Cell Biol* 2011;**193**:1137–46.
5. Weber GF, Bjerke MA, DeSimone DW. Integrins and cadherins join forces to form adhesive networks. *J Cell Sci* 2011;**124**:1183–93.
6. Brasch J, Harrison OJ, Honig B, Shapiro L. Thinking outside the cell: how cadherins drive adhesion. *Trends Cell Biol* 2012;**22**:299–310.
7. Hulpiau P, van Roy F. Molecular evolution of the cadherin superfamily. *Int J Biochem Cell Biol* 2009;**41**:349–69.
8. Hulpiau P, van Roy F. New insights into the evolution of metazoan cadherins. *Mol Biol Evol* 2011;**28**:647–57.
9. El-Amraoui A, Petit C. Cadherins as targets for genetic diseases. *Cold Spring Harb Perspect Biol* 2010;**2**:a003095.
10. Jamora C, DasGupta R, Kocieniewski P, Fuchs E. Links between signal transduction, transcription and adhesion in epithelial bud development. *Nature* 2003;**422**:317–22.
11. Sprecher E, Bergman R, Richard G, Lurie R, Shalev S, Petronius D, et al. Hypotrichosis with juvenile macular dystrophy is caused by a mutation in CDH3, encoding P-cadherin. *Nat Genet* 2001;**29**:134–6.
12. Indelman M, Eason J, Hummel M, Loza O, Suri M, Leys MJ, et al. Novel CDH3 mutations in hypotrichosis with juvenile macular dystrophy. *Clin Exp Dermatol* 2007;**32**:191–6.
13. Kjaer KW, Hansen L, Schwabe GC, Marques-de-Faria AP, Eiberg H, Mundlos S, et al. Distinct CDH3 mutations cause ectodermal dysplasia, ectrodactyly, macular dystrophy (EEM syndrome). *J Med Genet* 2005;**42**:292–8.
14. Shimomura Y, Wajid M, Shapiro L, Christiano AM. P-cadherin is a p63 target gene with a crucial role in the developing human limb bud and hair follicle. *Development* 2008;**135**:743–53.
15. Shimomura Y. Congenital hair loss disorders: rare, but not too rare. *J Dermatol* 2012;**39**:3–10.
16. Delva E, Tucker DK, Kowalczyk AP. The desmosome. *Cold Spring Harb Perspect Biol* 2009;**1**:a002543.
17. McGrath JA. Inherited disorders of desmosomes. *Australas J Dermatol* 2005;**46**:221–9.
18. Brooke MA, Nitoiu D, Kelsell DP. Cell-cell connectivity: desmosomes and disease. *J Pathol* 2012;**226**:158–71.
19. Rickman L, Simrak D, Stevens HP, Hunt DM, King IA, Bryant SP, et al. N-terminal deletion in a desmosomal cadherin causes the autosomal dominant skin disease striate palmoplantar keratoderma. *Hum Mol Genet* 1999;**8**:971–6.
20. Kljuic A, Bazzi H, Sundberg JP, Martinez-Mir A, O'Shaughnessy R, Mahoney MG, et al. Desmoglein 4 in hair follicle differentiation and epidermal adhesion: evidence from inherited hypotrichosis and acquired pemphigus vulgaris. *Cell* 2003;**113**:249–60.

21. Ayub M, Basit S, Jelani M, Ur Rehman F, Iqbal M, Yasinzai M, et al. A homozygous nonsense mutation in the human desmocollin-3 (DSC3) gene underlies hereditary hypotrichosis and recurrent skin vesicles. *Am J Hum Genet* 2009;**85**:515–20.
22. Dua-Awereh MB, Shimomura Y, Kraemer L, Wajid M, Christiano AM. Mutations in the desmoglein 1 gene in five Pakistani families with striate palmoplantar keratoderma. *J Dermatol Sci* 2009;**53**:192–7.
23. Hershkovitz D, Lugassy J, Indelman M, Bergman R, Sprecher E. Novel mutations in DSG1 causing striate palmoplantar keratoderma. *Clin Exp Dermatol* 2009;**34**:224–8.
24. Wan H, Dopping-Hepenstal PJ, Gratian MJ, Stone MG, Zhu G, Purkis PE, et al. Striate palmoplantar keratoderma arising from desmoplakin and desmoglein 1 mutations is associated with contrasting perturbations of desmosomes and the keratin filament network. *Br J Dermatol* 2004;**150**:878–91.
25. Wajid M, Bazzi H, Rockey J, Lubetkin J, Zlotogorski A, Christiano AM. Localized autosomal recessive hypotrichosis due to a frameshift mutation in the desmoglein 4 gene exhibits extensive phenotypic variability within a Pakistani family. *J Invest Dermatol* 2007;**127**:1779–82.
26. Schweizer J. More than one gene involved in monilethrix: intracellular but also extracellular players. *J Invest Dermatol* 2006;**126**:1216–9.
27. Shimomura Y, Sakamoto F, Kariya N, Matsunaga K, Ito M. Mutations in the desmoglein 4 gene are associated with monilethrix-like congenital hypotrichosis. *J Invest Dermatol* 2006;**126**:1281–5.
28. Bazzi H, Getz A, Mahoney MG, Ishida-Yamamoto A, Langbein L, Wahl 3rd JK, et al. Desmoglein 4 is expressed in highly differentiated keratinocytes and trichocytes in human epidermis and hair follicle. *Differentiation* 2006;**74**:129–40.
29. Payne AS. No evidence of skin blisters with human desmocollin-3 gene mutation. *Am J Hum Genet* 2010;**86**:292 (author reply 292).
30. Amagai M, Stanley JR. Desmoglein as a target in skin disease and beyond. *J Invest Dermatol* 2012;**132**:776–84.
31. Syrris P, Ward D, Evans A, Asimaki A, Gandjbakhch E, Sen-Chowdhry S, et al. Arrhythmogenic right ventricular dysplasia/cardiomyopathy associated with mutations in the desmosomal gene desmocollin-2. *Am J Hum Genet* 2006;**79**:978–84.
32. Awad MM, Calkins H, Judge DP. Mechanisms of disease: molecular genetics of arrhythmogenic right ventricular dysplasia/cardiomyopathy. *Nat Clin Pract Cardiovasc Med* 2008;**5**:258–67.
33. Saffitz JE. Arrhythmogenic cardiomyopathy: advances in diagnosis and disease pathogenesis. *Circulation* 2011;**124**:e390–e392.
34. Saffitz JE. The pathobiology of arrhythmogenic cardiomyopathy. *Annu Rev Pathol* 2011;**6**:299–321.
35. Watkins H, Ashrafian H, Redwood C. Inherited cardiomyopathies. *N Engl J Med* 2011;**364**:1643–56.
36. Basso C, Bauce B, Corrado D, Thiene G. Pathophysiology of arrhythmogenic cardiomyopathy. *Nat Rev Cardiol* 2012;**9**:223–33.
37. Jacoby D, McKenna WJ. Genetics of inherited cardiomyopathy. *Eur Heart J* 2012;**33**:296–304.
38. Pilichou K, Bezzina CR, Thiene G, Basso C. Arrhythmogenic cardiomyopathy: transgenic animal models provide novel insights into disease pathobiology. *Circ Cardiovasc Genet* 2011;**4**:318–26.
39. Basso C, Czarnowska E, Della Barbera M, Bauce B, Beffagna G, Wlodarska EK, et al. Ultrastructural evidence of intercalated disc remodelling in arrhythmogenic right ventricular cardiomyopathy: an electron microscopy investigation on endomyocardial biopsies. *Eur Heart J* 2006;**27**:1847–54.

40. Pilichou K, Remme CA, Basso C, Campian ME, Rizzo S, Barnett P, et al. Myocyte necrosis underlies progressive myocardial dystrophy in mouse dsg2-related arrhythmogenic right ventricular cardiomyopathy. *J Exp Med* 2009;**206**:1787–802.
41. Krusche CA, Holthofer B, Hofe V, van de Sandt AM, Eshkind L, Bockamp E, et al. Desmoglein 2 mutant mice develop cardiac fibrosis and dilation. *Basic Res Cardiol* 2011;**106**:617–33.
42. Goossens S, Janssens B, Bonne S, De Rycke R, Braet F, van Hengel J, et al. A unique and specific interaction between alphaT-catenin and plakophilin-2 in the area composita, the mixed-type junctional structure of cardiac intercalated discs. *J Cell Sci* 2007;**120**:2126–36.
43. Li J, Goossens S, van Hengel J, Gao E, Cheng L, Tyberghein K, et al. Loss of alphaT-catenin alters the hybrid adhering junctions in the heart and leads to dilated cardiomyopathy and ventricular arrhythmia following acute ischemia. *J Cell Sci* 2012;**125**:1058–67.
44. Swope D, Cheng L, Gao E, Li J, Radice GL. Loss of cadherin-binding proteins beta-catenin and plakoglobin in the heart leads to gap junction remodeling and arrhythmogenesis. *Mol Cell Biol* 2012;**32**:1056–67.
45. Lombardi R, da Graca Cabreira-Hansen M, Bell A, Fromm RR, Willerson JT, Marian AJ. Nuclear plakoglobin is essential for differentiation of cardiac progenitor cells to adipocytes in arrhythmogenic right ventricular cardiomyopathy. *Circ Res* 2011;**109**:1342–53.
46. Marcus FI, McKenna WJ, Sherrill D, Basso C, Bauce B, Bluemke DA, et al. Diagnosis of arrhythmogenic right ventricular cardiomyopathy/dysplasia: proposed modification of the Task Force Criteria. *Eur Heart J* 2010;**31**:806–14.
47. Asimaki A, Tandri H, Huang H, Halushka MK, Gautam S, Basso C, et al. A new diagnostic test for arrhythmogenic right ventricular cardiomyopathy. *N Engl J Med* 2009;**360**:1075–84.
48. Murray B. Arrhythmogenic right ventricular dysplasia/cardiomyopathy (ARVD/C): a review of molecular and clinical literature. *J Genet Couns* 2012;**21**:494–504.
49. Bhalla K, Luo Y, Buchan T, Beachem MA, Guzauskas GF, Ladd S, et al. Alterations in CDH15 and KIRREL3 in patients with mild to severe intellectual disability. *Am J Hum Genet* 2008;**83**:703–13.
50. Wang K, Zhang H, Ma D, Bucan M, Glessner JT, Abrahams BS, et al. Common genetic variants on 5p14.1 associate with autism spectrum disorders. *Nature* 2009;**459**:528–33.
51. Singh SM, Castellani C, O'Reilly R. Autism meets schizophrenia via cadherin pathway. *Schizophr Res* 2010;**116**:293–4.
52. Soronen P, Ollila HM, Antila M, Silander K, Palo OM, Kieseppa T, et al. Replication of GWAS of bipolar disorder: association of SNPs near CDH7 with bipolar disorder and visual processing. *Mol Psychiatry* 2010;**15**:4–6.
53. Willemsen MH, Fernandez BA, Bacino CA, Gerkes E, de Brouwer AP, Pfundt R, et al. Identification of ANKRD11 and ZNF778 as candidate genes for autism and variable cognitive impairment in the novel 16q24.3 microdeletion syndrome. *Eur J Hum Genet* 2010;**18**:429–35.
54. Pagnamenta AT, Khan H, Walker S, Gerrelli D, Wing K, Bonaglia MC, et al. Rare familial 16q21 microdeletions under a linkage peak implicate cadherin 8 (CDH8) in susceptibility to autism and learning disability. *J Med Genet* 2011;**48**:48–54.
55. Kim SY, Yasuda S, Tanaka H, Yamagata K, Kim H. Non-clustered protocadherin. *Cell Adh Migr* 2011;**5**:97–105.
56. Speevak MD, Farrell SA. Non-syndromic language delay in a child with disruption in the Protocadherin11X/Y gene pair. *Am J Med Genet B Neuropsychiatr Genet* 2011;**156B**:484–9.
57. Dibbens LM, Tarpey PS, Hynes K, Bayly MA, Scheffer IE, Smith R, et al. X-linked protocadherin 19 mutations cause female-limited epilepsy and cognitive impairment. *Nat Genet* 2008;**40**:776–81.

58. Depienne C, Bouteiller D, Keren B, Cheuret E, Poirier K, Trouillard O, et al. Sporadic infantile epileptic encephalopathy caused by mutations in PCDH19 resembles Dravet syndrome but mainly affects females. *PLoS Genet* 2009;**5**:e1000381.
59. Depienne C, Trouillard O, Bouteiller D, Gourfinkel-An I, Poirier K, Rivier F, et al. Mutations and deletions in PCDH19 account for various familial or isolated epilepsies in females. *Hum Mutat* 2011;**32**:E1959–E1975.
60. Dibbens LM, Kneen R, Bayly MA, Heron SE, Arsov T, Damiano JA, et al. Recurrence risk of epilepsy and mental retardation in females due to parental mosaicism of PCDH19 mutations. *Neurology* 2011;**76**:1514–9.
61. Depienne C, LeGuern E. PCDH19-related infantile epileptic encephalopathy: an unusual X-linked inheritance disorder. *Hum Mutat* 2012;**33**:627–34.
62. Caberlotto E, Michel V, de Monvel JB, Petit C. Coupling of the mechanotransduction machinery and F-actin polymerization in the cochlear hair bundles. *Bioarchitecture* 2011;**1**:169–74.
63. El-Amraoui A, Petit C. Usher I syndrome: unravelling the mechanisms that underlie the cohesion of the growing hair bundle in inner ear sensory cells. *J Cell Sci* 2005; **118**:4593–603.
64. Petit C, Richardson G. Linking deafness genes to hair-bundle development and function. *Nat Neurosci* 2009;**12**:703–10.
65. Bonnet C, El-Amraoui A. Usher syndrome (sensorineural deafness and retinitis pigmentosa): pathogenesis, molecular diagnosis and therapeutic approaches. *Curr Opin Neurol* 2012;**25**:42–9.
66. Richardson GP, de Monvel JB, Petit C. How the genetics of deafness illuminates auditory physiology. *Annu Rev Physiol* 2011;**73**:311–34.
67. Lefèvre G, Michel V, Weil D, Lepelletier L, Bizard E, Wolfrum U, et al. A core cochlear phenotype in USH1 mouse mutants implicates fibrous links of the hair bundle in its cohesion, orientation and differential growth. *Development* 2008;**135**:1427–37.
68. Boëda B, El-Amraoui A, Bahloul A, Goodyear R, Daviet L, Blanchard S, et al. Myosin VIIa, harmonin and cadherin 23, three Usher I gene products that cooperate to shape the sensory hair cell bundle. *EMBO J* 2002;**21**:6689–99.
69. Siemens J, Kazmierczak P, Reynolds A, Sticker M, Littlewood-Evans A, Muller U. The Usher syndrome proteins cadherin 23 and harmonin form a complex by means of PDZ-domain interactions. *Proc Natl Acad Sci U S A* 2002;**99**:14946–51.
70. Michel V, Goodyear RJ, Weil D, Marcotti W, Perfettini I, Wolfrum U, et al. Cadherin 23 is a component of the transient lateral links in the developing hair bundles of cochlear sensory cells. *Dev Biol* 2005;**280**:281–94.
71. Lagziel A, Ahmed ZM, Schultz JM, Morell RJ, Belyantseva IA, Friedman TB. Spatiotemporal pattern and isoforms of cadherin 23 in wild type and waltzer mice during inner ear hair cell development. *Dev Dyn* 2005;**205**:295–306.
72. Michalski N, Michel V, Bahloul A, Lefèvre G, Chardenoux S, Yagi H, et al. Molecular characterization of the ankle link complex in cochlear hair cells and its role in the hair bundle functioning. *J Neurosci* 2007;**27**:6478–88.
73. Webb SW, Grillet N, Andrade LR, Xiong W, Swarthout L, Della Santina CC, et al. Regulation of PCDH15 function in mechanosensory hair cells by alternative splicing of the cytoplasmic domain. *Development* 2011;**138**:1607–17.
74. Caberlotto E, Michel V, Foucher I, Bahloul A, Goodyear RJ, Pepermans E, et al. Usher type 1 G protein sans is a critical component of the tip-link complex, a structure controlling actin polymerization in stereocilia. *Proc Natl Acad Sci U S A* 2011;**108**: 5825–30.
75. Adato A, Lefevre G, Delprat B, Michel V, Michalski N, Chardenoux S, et al. Usherin, the defective protein in Usher syndrome type IIA, is likely to be a component of

interstereocilia ankle links in the inner ear sensory cells. *Hum Mol Genet* 2005;**14**: 3921–32.
76. Pan L, Zhang M. Structures of usher syndrome 1 proteins and their complexes. *Physiology (Bethesda)* 2012;**27**:25–42.
77. Kazmierczak P, Sakaguchi H, Tokita J, Wilson-Kubalek EM, Milligan RA, Muller U, et al. Cadherin 23 and protocadherin 15 interact to form tip-link filaments in sensory hair cells. *Nature* 2007;**449**:87–91.
78. Michalski N, Michel V, Caberlotto E, Lefèvre GM, van Aken AFJ, Tinevez J-Y, et al. Harmonin-b, an actin-binding scaffold protein, is involved in the adaptation of mechanoelectrical transduction by sensory hair cells. *Pflugers Arch* 2009;**459**:115–30.
79. Grillet N, Xiong W, Reynolds A, Kazmierczak P, Sato T, Lillo C, et al. Harmonin mutations cause mechanotransduction defects in cochlear hair cells. *Neuron* 2009;**62**:375–87.
80. Bahloul A, Michel V, Hardelin JP, Nouaille S, Hoos S, Houdusse A, et al. Cadherin-23, myosin VIIa and harmonin, encoded by Usher syndrome type I genes, form a ternary complex and interact with membrane phospholipids. *Hum Mol Genet* 2010;**19**:3557–65.
81. Fettiplace R, Hackney CM. The sensory and motor roles of auditory hair cells. *Nat Rev Neurosci* 2006;**7**:19–29.
82. Vendome J, Posy S, Jin X, Bahna F, Ahlsen G, Shapiro L, et al. Molecular design principles underlying beta-strand swapping in the adhesive dimerization of cadherins. *Nat Struct Mol Biol* 2011;**18**:693–700.
83. Elledge HM, Kazmierczak P, Clark P, Joseph JS, Kolatkar A, Kuhn P, et al. Structure of the N terminus of cadherin 23 reveals a new adhesion mechanism for a subset of cadherin superfamily members. *Proc Natl Acad Sci U S A* 2010;**107**:10708–12.
84. Sotomayor M, Weihofen WA, Gaudet R, Corey DP. Structural determinants of cadherin-23 function in hearing and deafness. *Neuron* 2010;**66**:85–100.
85. Sotomayor M, Weihofen WA, Gaudet R, Corey DP. Structure of a force-conveying cadherin bond essential for inner-ear mechanotransduction. *Nature* 2012;**492**:128–32.
86. Reiners J, Nagel-Wolfrum K, Jurgens K, Marker T, Wolfrum U. Molecular basis of human Usher syndrome: deciphering the meshes of the Usher protein network provides insights into the pathomechanisms of the Usher disease. *Exp Eye Res* 2006;**83**:97–119.
87. Williams DS. Usher syndrome: animal models, retinal function of Usher proteins, and prospects for gene therapy. *Vision Res* 2008;**48**:433–41.
88. Petit C. Usher syndrome: from genetics to pathogenesis. *Annu Rev Genomics Hum Genet* 2001;**2**:271–97.
89. Jacobson SG, Cideciyan AV, Aleman TS, Sumaroka A, Roman AJ, Gardner LM, et al. Usher syndromes due to MYO7A, PCDH15, USH2A or GPR98 mutations share retinal disease mechanism. *Hum Mol Genet* 2008;**17**:2405–15.
90. Jacobson SG, Cideciyan AV, Gibbs D, Sumaroka A, Roman AJ, Aleman TS, et al. Retinal disease course in Usher syndrome 1B due to MYO7A mutations. *Invest Ophthalmol Vis Sci* 2011;**52**:7924–36.
91. Malm E, Ponjavic V, Moller C, Kimberling WJ, Andreasson S. Phenotypes in defined genotypes including siblings with Usher syndrome. *Ophthalmic Genet* 2011;**32**:65–74.
92. Malm E, Ponjavic V, Moller C, Kimberling WJ, Stone ES, Andreasson S. Alteration of rod and cone function in children with Usher syndrome. *Eur J Ophthalmol* 2011;**21**:30–8.
93. Libby RT, Kitamoto J, Holme RH, Williams DS, Steel KP. Cdh23 mutations in the mouse are associated with retinal dysfunction but not retinal degeneration. *Exp Eye Res* 2003;**77**:731–9.
94. Ball SL, Bardenstein D, Alagramam KN. Assessment of retinal structure and function in Ames waltzer mice. *Invest Ophthalmol Vis Sci* 2003;**44**:3986–92.
95. Haywood-Watson 2nd RJ, Ahmed ZM, Kjellstrom S, Bush RA, Takada Y, Hampton LL, et al. Ames Waltzer deaf mice have reduced electroretinogram

amplitudes and complex alternative splicing of Pcdh15 transcripts. *Invest Ophthalmol Vis Sci* 2006;**47**:3074–84.

96. El-Amraoui A, Schonn J-S, Küssel-Andermann P, Blanchard S, Desnos C, Henry J-P, et al. MyRIP, a novel Rab effector, enables myosin VIIa recruitment to retinal melanosomes. *EMBO Rep* 2002;**3**:463–70.
97. Gibbs D, Azarian SM, Lillo C, Kitamoto J, Klomp AE, Steel KP, et al. Role of myosin VIIa and Rab27a in the motility and localization of RPE melanosomes. *J Cell Sci* 2004;**117**:6473–83.
98. Liu X, Udovichenko IP, Brown SD, Steel KP, Williams DS. Myosin VIIa participates in opsin transport through the photoreceptor cilium. *J Neurosci* 1999;**19**:6267–74.
99. Sahly I, Dufour E, Schietroma C, Michel V, Bahloul A, Perfettini I, et al. Localization of Usher 1 proteins to the photoreceptor calyceal processes, which are absent from mice. *J Cell Biol* 2012;**199**:381–99.
100. Henderson RH, Li Z, Abd El Aziz MM, Mackay DS, Eljinini MA, Zeidan M, et al. Biallelic mutation of protocadherin-21 (PCDH21) causes retinal degeneration in humans. *Mol Vis* 2010;**16**:46–52.
101. Ostergaard E, Batbayli M, Duno M, Vilhelmsen K, Rosenberg T. Mutations in PCDH21 cause autosomal recessive cone-rod dystrophy. *J Med Genet* 2010;**47**:665–9.
102. Rattner A, Smallwood PM, Williams J, Cooke C, Savchenko A, Lyubarsky A, et al. A photoreceptor-specific cadherin is essential for the structural integrity of the outer segment and for photoreceptor survival. *Neuron* 2001;**32**:775–86.
103. Yang Z, Chen Y, Lillo C, Chien J, Yu Z, Michaelides M, et al. Mutant prominin 1 found in patients with macular degeneration disrupts photoreceptor disk morphogenesis in mice. *J Clin Invest* 2008;**118**:2908–16.
104. Michaelides M, Gaillard MC, Escher P, Tiab L, Bedell M, Borruat FX, et al. The PROM1 mutation p.R373C causes an autosomal dominant bull's eye maculopathy associated with rod, rod-cone, and macular dystrophy. *Invest Ophthalmol Vis Sci* 2010;**51**:4771–80.
105. Permanyer J, Navarro R, Friedman J, Pomares E, Castro-Navarro J, Marfany G, et al. Autosomal recessive retinitis pigmentosa with early macular affectation caused by premature truncation in PROM1. *Invest Ophthalmol Vis Sci* 2010;**51**:2656–63.

Cadherin-Associated Proteins

Beta-Catenin Versus the Other Armadillo Catenins: Assessing Our Current View of Canonical Wnt Signaling

Rachel K. Miller[*], **Ji Yeon Hong**[†,1], **William A. Muñoz**[*,‡,1], **Pierre D. McCrea**[*,‡]

[*]Department of Biochemistry and Molecular Biology, The University of Texas MD Anderson Cancer Center, Houston, Texas, USA
[†]The Catholic University of Korea, Catholic High Performance Cell Therapy Center, 505 Banpo-dong, Seocho-gu, Seoul, South Korea
[‡]Program in Genes and Development, The University of Texas Graduate School of Biomedical Sciences, Houston, Texas, USA
[1]These authors contributed equally to this work

Contents

Abstract

The prevailing view of canonical Wnt signaling emphasizes the role of beta-catenin acting downstream of Wnt activation to regulate transcriptional activity. However, emerging

Progress in Molecular Biology and Translational Science, Volume 116
ISSN 1877-1173
http://dx.doi.org/10.1016/B978-0-12-394311-8.00017-0

evidence indicates that other armadillo catenins in vertebrates, such as members of the p120 subfamily, convey parallel signals to the nucleus downstream of canonical Wnt pathway activation. Their study is thus needed to appreciate the networked mechanisms of canonical Wnt pathway transduction, especially as they may assist in generating the diversity of Wnt effects observed in development and disease. In this chapter, we outline evidence of direct canonical Wnt effects on p120 subfamily members in vertebrates and speculate upon these catenins' roles in conjunction with or aside from beta-catenin.

1. OVERVIEW OF CATENINS

The catenin proteins contribute to a number of fundamental processes in multicellular animals. The best known roles of beta-catenin (Armadillo in *Drosophila*) are in cell–cell adhesion and in canonical Wnt signaling. Vertebrate catenins were first isolated in association with cadherin proteins at cell–cell junctions.[1] Catenin proteins, which were named after the Latin word for chain, "catena," provide among other things an indirect link between cadherin proteins and the contractile cortical actin cytoskeleton.[2] Beta-catenin is also a central transducer of canonical Wnt signals, as it can enter the nucleus to modulate expression of gene targets, such as through the relief of TCF/LEF-mediated transcriptional repression.[3] Although the current Wnt literature focuses on beta-catenin's role, recent evidence has begun to suggest that other catenins comprising a central armadillo domain (see later) and that likewise bind to cadherins, also act downstream of canonical Wnt or other signals to modulate gene activity (Fig. 17.1).[5,8,9]

Despite its name and (indirect) association with cadherins, alpha-catenin lacks armadillo repeats and instead bears homology to the actin-binding protein vinculin.[10] Intriguingly, alpha-catenin has been implicated in mechanosensing within the cadherin complex[11] (see Chapter 2) and has also been reported in the nucleus.[12] However, it is not discussed further here, given its lack of homology with armadillo catenins.

2. CADHERIN–CATENIN COMPLEXES

By definition catenins bind cadherins, generally, but not exclusively, at adherens or desmosomal cell–cell junctions. Most catenins are distinguished by a central domain composed of 9 or 12 armadillo repeats.[13,14] In vertebrates, there are three groups of armadillo-repeat catenins, namely, the beta-catenin, p120-catenin, and plakophilin-catenin subfamilies (Fig. 17.2). Within

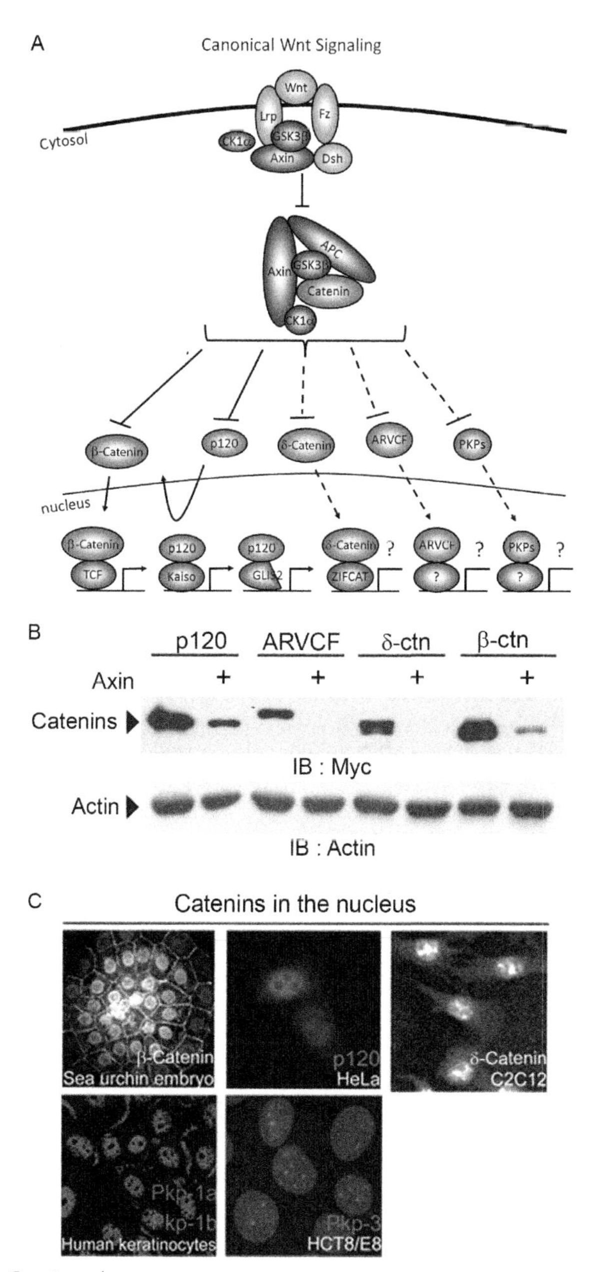

Figure 17.1 Continued

adherens junctions, the cytoplasmic domains of classic cadherins, such as C-cadherin, E-cadherin, and N-cadherin, contain two catenin binding sites.[15] Nearest to the membrane is the juxtamembrane or membrane-proximal domain of cadherins, where members of the p120-catenin subfamily (vertebrate p120-, ARVCF-, delta-, or p0071-catenin) use their armadillo domains to bind in a mutually exclusive fashion (Fig. 17.3). In contrast, the beta-catenin subfamily in vertebrates has two members, beta- and gamma-catenin (plakoglobin), which employ their armadillo domains to bind in a mutually exclusive manner to the C-terminal, membrane-distal region of cadherins. Although the functions of catenins at adherens junctions remain under active investigation, beta-catenin subfamily members indirectly link cadherins to the contractile actin–myosin cytoskeleton through their associations with proteins including alpha-catenin, vinculin, and EPLIN.[2,16] When bound, p120 subfamily members reduce cadherin endocytosis, promoting cadherin stabilization, association with the cortical actin cytoskeleton, and the clustering of cadherins at adhesive sites[17,18] (see Chapter 18).

Related but distinct from the classic cadherins are the desmosomal cadherins (see Chapter 5). For example, at desmosomal junctions, desmocollin- and desmoglein-cadherins associate with plakoglobin, a member of the beta-catenin subfamily, and further bind members of the plakophilin-catenin subfamily (plakophilin-1, -2, and -3) (Figs. 17.2 and 17.3).[19–22] Within desmosomes, plakoglobin, plakophilins, and additional proteins assist in linking desmosomal cadherins to the intermediate filament cytoskeleton. These catenin interactions appear roughly analogous to those of beta-catenin in complex with classic cadherins at adherens junctions in that they bridge the desmosomal cadherin complex within this case, the intermediate filament cytoskeleton.[23,24] Also, the functions of plakophilin subfamily members within desmosomes may parallel those of p120 subfamily members at adherens junctions, in that they promote desmosomal cadherin stability and clustering.[25,26]

Figure 17.1 *Catenins and canonical Wnt signaling.* (A) Canonical Wnt signaling inhibits the destruction complex (purple), preventing degradation of armadillo catenin proteins (red). Upon stabilization, catenin proteins may enter the nucleus to bind transcription factors (blue), regulating downstream gene expression. Canonical regulation of specific catenin isoforms of the p120- and plakophilin-catenin subfamilies is distinct from but possibly overlapping with the noncanonical Wnt signaling trajectories (not shown). Dotted lines represent relationships that are currently uncertain. (B) In *Xenopus* embryos, overexpression of Axin, a destruction complex component, reduces the levels of p120-catenin subfamily members. Beta-catenin is also destabilized by Axin (positive control). (C) Members of all subfamilies of armadillo catenins localize to the nucleus in certain contexts.[4] *Panel (B): reproduced with permission from Ref. 5. Panel (C): Reproduced with permission from Refs. 4,6,7,96,97.* (See Color Insert.)

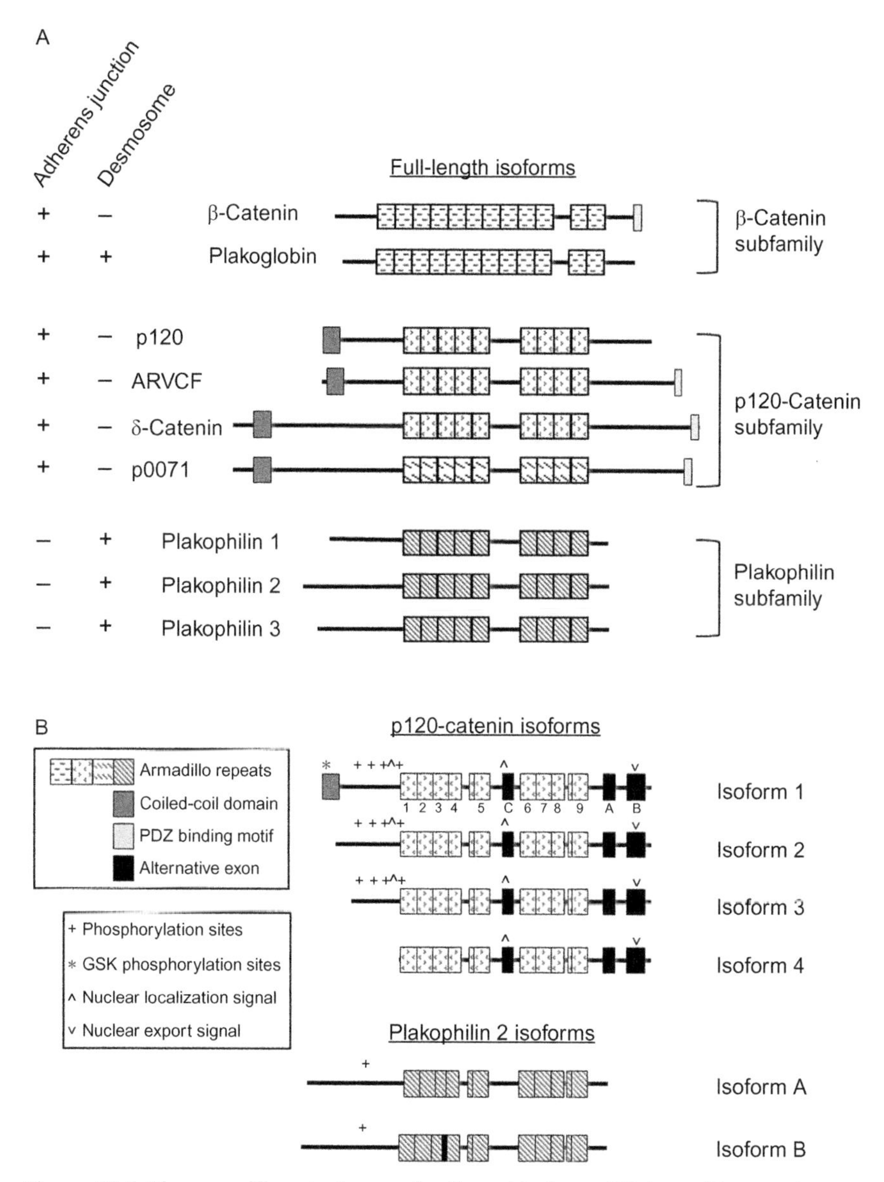

Figure 17.2 *The armadillo catenin superfamily and isoforms.* (A) Armadillo catenin proteins are characterized by a central domain comprising 9 or 12 armadillo repeats. They are classified into three distinct subfamilies, the beta-catenin subfamily, the p120-catenin subfamily, and the plakophilin subfamily. (B) The best characterized member of the p120 subfamily, p120-catenin, has numerous isoforms. These isoforms are generated by a combination of differential translational start sites (e.g., isoforms 1–4) in humans and differentially spliced exons (e.g., exons A, B, and C in black). The isoforms vary in their regulation. For example, only the longest isoform of p120-catenin (isoform 1) possesses a "destruction box" with conserved GSK3-beta (and CK1-alpha) phosphorylation sites at its N-terminus (*). Plakophilin isoforms arise through alternatively spliced exons. For example, plakophilin 2B has one additional exon (black) as compared with plakophilin 2A. (See Color Insert.)

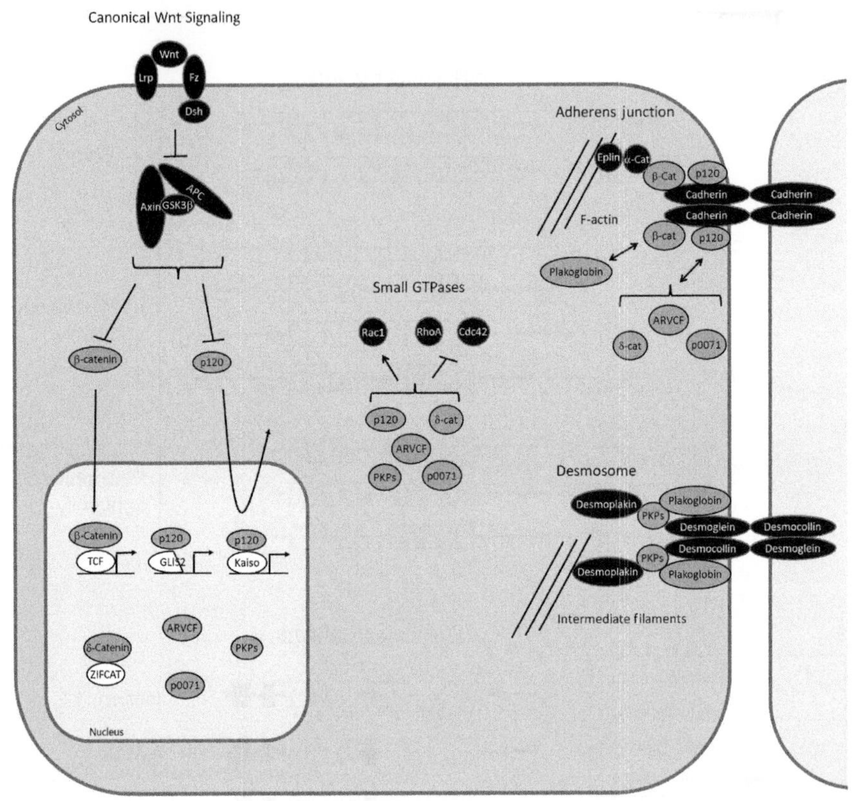

Figure 17.3 *Cellular roles of catenins.* Beta-catenin and p120-catenin have confirmed roles in Wnt signaling, although certain isoforms of all catenins (gray) localize to the nucleus and may interact with transcription factors (white). p120-catenin and plakophilin subfamily members (gray) also regulate small GTPases (Rac1, RhoA, Cdc42—black). All catenins associate with cadherins at junctional complexes, with beta-catenin and p120 subfamily members residing exclusively at adherens junctions, and with plakophilins residing exclusively at desmosomes. Plakoglobin is present at both adherens and desmosomal junctions. (See Color Insert.)

3. THE "CANONICAL" VIEW OF CANONICAL Wnt SIGNALING

In addition to their roles in cell–cell interactions via the adherens and desmosomal junctions, catenins are essential to cellular communication via their function in Wnt signaling.[27] In fact, a recent study suggests that the Wnt-related functions of beta-catenin might have evolved earlier (are "more fundamental") than their adhesive roles[28] (see Chapter 1).

3.1. Discovery of Wnt signaling components

Wnt signaling has been investigated over the past 30 years,[3] beginning with a genetic screen conducted by Nusslein-Volhard and Wieschaus to identify genes required for embryonic segment polarity in *Drosophila*.[29] The *wingless* (*Drosophila* Wnt ligand) gene was identified, and soon thereafter, Nusse and Varmus observed activation of the *int-1* gene upon nearby integration of mouse mammary tumor virus.[30] The *Drosophila* and mammalian gene products were shown to be orthologous, and the gene names of *wingless* and *int-1* were combined to form Wnt1.[31]

Since the discovery of genes encoding Wnt ligands, numerous components within the signaling cascade have been identified and characterized (Fig. 17.1), such as *armadillo*, the *Drosophila* ortholog of vertebrate beta-catenin.[32] Experiments in *Xenopus laevis* (frog) embryos demonstrated a fundamental role for Wnt1 in body axis formation.[33] *Xenopus* beta-catenin was cloned based on its association with cadherins,[34] and then, unexpectedly, found to participate in Wnt signaling/body axis formation.[35] Key studies in *Drosophila* later examined the epistatic relationships of Wnt pathway components, including (among others) *porcupine*, *dishevelled*, *armadillo* (beta-catenin), and *zeste-white 3* (GSK3beta), followed later by the discovery of the plasma membrane-associated Frizzled receptors.[3] LEF/TCF (HMG-box transcription factor) was discovered early on, followed by characterization of its binding with, and regulation by, beta-catenin.[36,37,98] Components of the canonical Wnt signaling pathway have continued to be discovered, such as the Wnt-ligand coreceptors LRP5/6 (*Drosophila arrow*), inhibitory secreted antagonists such as Dkk1, and the augmentation of low-level Wnt signaling via activation of LGR receptors by R-spondin ligand.[3,38]

3.2. The basics of canonical Wnt signaling

Canonical Wnt signaling has been extensively studied given its relationship to development and disease.[27] It is initiated when secreted Wnt ligands bind a Frizzled–LRP receptor–coreceptor complex (Fig. 17.1). Conversely, in the absence of appropriate Wnt ligands, the "signaling pool" of beta-catenin (i.e., noncadherin bound), is diminished by its interaction with a "destruction complex" composed of the GSK3beta and CK1-alpha kinases, and the APC (adenomatous polyposis coli tumor suppressor) and axin scaffolds, among other components. The phosphorylation of beta-catenin by GSK3beta and CK1-alpha leads to ubiquitination by beta-TRCP (E3-ubiquitin ligase), followed by degradation in the proteasome. Since beta-catenin's signaling pool is thereby lowered (in the absence of canonical Wnt signals), it is less likely that beta-catenin will enter the nucleus to relieve repression conferred by TCF/LEF, which reside in association with repressive factors such as Groucho.[3]

Conversely, when secreted Wnt ligands bind the Frizzled–LRP receptor–coreceptor complex, LRP's intracellular domain becomes phosphorylated by CK1-alpha. In one model, this leads to LRP-mediated sequestration of the axin–GSK3beta complex away from the destruction complex,[39] promoting beta-catenin's accumulation within a growing cytoplasmic "signaling pool." This signaling pool is distinct from the usually much larger pool of beta-catenin bound to cadherins at cell–cell junctions, which is frequently viewed as being Wnt signaling-inactive. It should be pointed out, however, that evidence exists for beta-catenin's dissociation from cadherins in response to other stimuli (e.g., RTKs or Src),[40] resulting in activation of canonical Wnt reporters. In any case, a portion of the cytoplasmic pool of beta-catenin that arises following canonical Wnt-ligand activation translocates to the nucleus, where it binds TCF/LEF and relieves their repression of gene targets. Upon binding beta-catenin, the corepressors in association with TCF/LEF (e.g., Groucho and CtBP)[41] are shed in favor of coactivators/chromatin remodelers (e.g., p300 and Pygo)[42,43] to promote gene-target activation.

4. THE "OTHER" CATENINS IN THE NUCLEUS

To this point, research has largely focused on the adhesion-related roles of the catenin family, in addition to the transcriptional roles of beta-catenin, in particular, within the Wnt signaling pathway. However, catenins within all subfamilies localize to the nucleus (Fig. 17.3), and a growing body of evidence suggests that they have transcriptional roles.[4] Thus, further assessment of the nuclear roles of the "other" catenins is necessary.

4.1. The beta-catenin subfamily

Like beta-catenin, plakoglobin (gamma-catenin) utilizes its armadillo repeats to interact with binding partners (Fig. 17.2), including cadherins and nuclear transcription factors such as TCF/LEF.[44] Both plakoglobin and beta-catenin regulate transcription within the nucleus, and although they modulate some of the same genes, beta-catenin's role in the canonical Wnt pathway appears more direct. For example, although plakoglobin activates some TCF/LEF-controlled gene targets[45] and shares with beta-catenin mechanisms regulating its protein stability, the means by which it executes transcriptional effects may be distinct from those of beta-catenin.[44] Surprisingly, for example, cytoplasmic more than nuclear accumulation of plakoglobin appears to promote Wnt signaling,[44] and its nuclear presence inhibits TCF/LEF gene-target activation in the copresence of beta-catenin.[46] It has been proposed that the cytoplasmic sequestration of plakoglobin results in activation of Wnt targets by retention of TCF/LEF outside the nucleus, which contrasts with the mechanism generally noted for beta-catenin.[44] Noteworthy, however, is that in some developmental contexts, beta-catenin has likewise been found to derepress TCF by means that involve TCF dissociation from DNA.[47]

4.2. The p120 subfamily

Increasing evidence shows that p120 subfamily members, like p120-, ARVCF-, delta-, and p0071-catenin (Fig. 17.2), regulate gene activity. In the late 1990s, p120 was found to bind the transcriptional repressor Kaiso.[48,49] By competitively displacing Kaiso from its sequence-specific gene targets and/or sequestering Kaiso in the cytoplasm, p120 is thought to derepress (activate) transcription.[48] p120 also interacts with the transcriptional repressor GLIS2, facilitating GLIS2 cleavage and altering GLIS2's selection of gene targets.[50] More recently, studies have uncovered nuclear activities of delta-catenin.[51,52]

In the context of the nucleus, some catenins appear to have partially overlapping transcriptional partners. For example, both p120 and delta-catenin bind and modulate Kaiso, albeit with different outcomes,[6,48] and both beta-catenin and plakoglobin bind TCF/LEF. The shared versus exclusive transcription-factor associations of catenins may facilitate both networked and differential effects. One possibility, yet to be tested, is that different Wnt-ligand:Wnt-receptor pairings may preferentially stabilize certain catenins more than others, partially accounting for distinctive downstream nuclear outputs. Recent studies have identified components such as Frodo (presumed scaffold) and Dyrk1A (Ser/Thr kinase) as binding and facilitating the stabilization of p120-catenin, thus enhancing its signaling capacity. Since Frodo and Dyrk1A may preferentially bind and stabilize p120-catenin,[8,53] they serve as examples of the type of molecules that could facilitate selective catenin activity downstream of particular receptor–ligand pairings (e.g., canonical Wnt pathways).

4.3. The plakophilin subfamily

Plakophilin-1, -2, and -3 (Fig. 17.2) have also been reported in the nucleus (Fig. 17.3), and several nuclear functions are proposed. For example, plakophilin-2 has been reported to associate with beta-catenin and modulate

Wnt gene activation.[54] Plakophilin-1 binds single-strand DNA, affecting cell survival following DNA damage.[55] Finally, our group has preliminarily resolved plakophilin-3's association with two established transcription factors (unpublished results). Clearly, the continued study of plakophilin family members is required to broaden our perspective of their nuclear roles.

5. ISOFORM DIVERSITY IN THE "OTHER" CATENINS

As described, catenins other than beta-catenin modulate different aspects of cellular behavior, including cell adhesion, transcription, and GTPase activity (described later) (Fig. 17.3). Regulation of these processes may be influenced through specific isoform usage within different contexts. For example, p120-catenin exists as multiple isoforms that arise due to alternative translational initiation and/or RNA splicing (Fig. 17.2).[9] The same cell can express more than one isoform. Isoform 1 (longest) has been associated somewhat more with motile cell states, and isoform 3 (shorter) with sessile behaviors.[56,57] Such associations are thought to result in part from differential p120 isoform effects on small GTPases,[58] as addressed later (see also Chapter 18). Alternative isoform usage by p120 subfamily members has repeatedly been observed across tissues and developmental or disease stages, but its basis and functional distinctions require further study.[59] As the organization of exons within the p120 subfamily is conserved,[19] it is expected that at least some of the alternative splice isoforms of p120 may additionally exist in other subfamily members, including ARVCF-, delta-, and p0071-catenin (Fig. 17.2). The exon organizations of plakophilins resemble each other, as well as the organization of p120 subfamily members, but are not similar to those of beta-catenin subfamily members.[19] However, whereas p120 subfamily members appear to have multiple isoforms, only two have been identified for plakophilin-1 and −2 (Fig. 17.2).[7,60,61] Differential usage and regulation of catenin isoforms may achieve specificity of cellular responses within the contexts of cell adhesion, small-GTPase activity, and transcription.

6. Wnt SIGNALING AND "THE OTHER" CATENINS

6.1. Beta-catenin: A paradigm

As noted earlier, canonical Wnt signaling is currently defined as the pathway in which signal transduction occurs via beta-catenin, overlooking the potential roles of multiple similarly structured catenins in vertebrates (Fig. 17.1).

This is understandable because beta-catenin is the dominant player, with its loss or gain often resulting in graphic phenotypic and molecular effects. When manipulating members of the p120 subfamily, the effects on Wnt signaling are more modest. For example, to resolve external Wnt phenotypes in *Xenopus* embryos, p120 effects are best tested in the context of a sensitized background, such as in conjunction with altered beta-catenin levels.[62] Thus, in contrast to beta-catenin, p120 cannot induce duplicate axis formation in *Xenopus* when expressed alone in the ventral region, a commonly used readout of ectopic canonical Wnt signaling. Yet, if the derepression of TCF/LEF occurs coordinately with the depletion of Kaiso (the transcriptional repressor that p120-catenin analogously relieves), partial duplicate axes are observed.[62] The p120 subfamily effects on Wnt signaling may further be masked by known p120-subfamily effects on small GTPases,[8] which regulate cytoskeletal structures and cell junctions. Finally, unlike beta-catenin, each p120 subfamily member has multiple isoforms generated from alternative translational start sites and/or by RNA splicing (Fig. 17.2). Thus, as discussed later, owing to the resulting presence or absence of a "destruction box" targeted by the destruction components (e.g., GSK3beta, CK1-alpha, APC, axin, and beta-TRCP), different isoforms of the same catenin are not equally responsive to canonical Wnt signals.[5]

In conclusion, the greater focus on beta-catenin is justified in view of its dominant role within the canonical Wnt pathway. However, the fact that some isoforms of other catenins, such as p120 isoform1, likewise respond to canonical Wnt signals, and thereupon enter the nucleus to modulate transcription factors distinct from LEF/TCF, suggests that there is a networked canonical response that requires attention (Fig. 17.1). An especially pertinent point is that canonical Wnt signals generate numerous distinct outcomes depending on the organism, tissue or temporal, or disease setting. While this is well known to be accounted for in part by the contributions of other distinct signals, some of which cross talk with the Wnt pathway (e.g., Notch, FGF, TGF-beta),[63] it is equally interesting to consider that the effects on other catenins may contribute to context-specific outcomes.

6.2. Cross talk between cell adhesion and Wnt signaling

Similar to beta- and gamma-catenin/plakoglobin (both members of the beta-catenin subfamily), most catenins of the p120-catenin and plakophilin subfamilies have been observed in the nucleus (Fig. 17.1). While evidence is still being accumulated, one possibility is that the repertoire of catenin components within adherens junctions and desmosomes facilitates cross talk between these adhesive structures and the nucleus (Fig. 17.3). In addition to highly polarized, mature junctions participating in contact inhibition (e.g., via modulation of hippo signaling),[64–66] the transit of catenins between cell junctions and the nucleus might assist in regulating more dynamic cadherin contacts, as is needed during morphogenesis or wound healing.

In common with beta-catenin, the nuclear localization of p120-catenin and plakophilin subfamily members is context dependent. In most instances, their junctional localization is much more prominent, to the extent that their presence in the nucleus might understandably be overlooked. Relative to beta-catenin, which is often likewise difficult to visualize in the nucleus, there has been relatively scant attention paid to the nuclear roles of p120 and plakophilin family members, despite the potential or known relevance of the p120 and plakophilin subfamilies in development and disease[67] (see Chapter 16).

Under some conditions, release of catenins from adherens junctions may precede their nuclear translocation to engage in gene regulation. For example, even though it is not the generally discussed route, beta-catenin's dissociation from cadherin in response to RTK or Src activity has been associated in certain contexts with higher Wnt-reporter activity.[40] Kinases and phosphatases are enriched at cell junctions, and some modulate the retention versus release of catenins from cadherins.[68,69] Likewise, cadherin absence has been associated in certain studies with higher canonical Wnt/beta-catenin signaling. This might be ascribed to beta-catenin no longer being sequestered by cadherin, enhancing its potential signaling pool (especially in cells with a less than fully active destruction complex).[70] Additionally, cadherin-associated p120 has recently been tied to the assembly of the canonical Wnt signalosome, resulting in greater beta-catenin signaling activity.[71] However, cell adhesion has also been implicated in destabilizing the signaling pool of beta-catenin by promoting destruction complex activity.[72]

As for many complexes, it is accepted that catenin protein associations and localizations respond to secondary protein modifications. For example, upon Wnt pathway activation, not only is beta-catenin stability enhanced (its destruction box remains unphosphorylated), but the active signaling pool of beta-catenin is modified such that it binds TCF but not cadherin.[73] Likewise, findings suggest that the localization of catenins to junctions, cytoplasm or nucleus is responsive to their modification, or modifications upon their direct binding partners.[40] In addition to the responsiveness of the cadherin–catenin complex to RTKs, Src, or Wnt signals, the complex has been implicated in functional interactions with the Hippo and other pathways.[64–66] Concerning the roles of catenins in particular, considerably more work remains to address the extent to which catenins that are cadherin-bound represent a latent signaling pool, which upon release in response to an appropriate stimulus allows them to fulfill a nuclear function.

Further relationships indicate links of the canonical Wnt pathway with the cadherin–catenin complex.[74] Two recent papers, for example, reported biochemical and/or functional associations of Wnt pathway components with the cadherin–catenin complex.[71,72] In addition, beta-catenin's transcriptional targets include Twist, Snail, and Slug,[75,76] which repress the expression of E-cadherin (see Chapter 14).[77,78] Interestingly, the destruction complex regulating the stability of beta-catenin (and those other catenin isoforms encoding destruction boxes) additionally acts upon Snail, which is likewise a GSK3beta target and is degraded by the proteasome. Indeed, Wnt1 has been shown to stabilize Snail.[79,80] Thus, as a consequence of the nuclear entry of beta-catenin (perhaps in concert with other catenins), canonical Wnt pathway activity may lead to reduced cadherin levels and enhanced cell migration in some morphogenic or wound-healing contexts. A reduced level of E-cadherin may also favor the formation of larger cytoplasmic/signaling pools of p120-subfamily catenins to more fully modulate small GTPases or to enhance the catenins' nuclear impact.

6.3. Roles of the "other" catenins in Wnt signaling

As was noted, a potentially significant but understudied role of the p120-subfamily catenins is in canonical Wnt signaling (Fig. 17.1). While certain work has drawn into question the extent to which gene control exercised by Kaiso (and by inference p120) is pertinent to Wnt signaling,[81] growing evidence indicates that p120 subfamily members act in parallel with beta-catenin in Wnt-mediated transcriptional regulation. Initial clues came from findings that p120 relieves Kaiso-mediated repression of *Xenopus siamois*[62] and *wnt11*,[82] both established canonical Wnt target genes and thus responsive to beta-catenin relief of TCF-mediated repression. It was found that the combined and respective p120 and beta-catenin relief of Kaiso and TCF is additive in activating target gene promoters and has the predicted developmental consequences.[62] This was evident in both molecular assays (e.g., endogenous gene or luciferase reporter readouts), and in gross phenotypes, such as the formation of ectopic dorsal axes in early *Xenopus* embryos.[62] Further, as predicted, exogenously expressed Kaiso inhibits beta-catenin mediated axis duplication in embryos.[62] Together, these data support the role of p120/Kaiso, acting together with beta-catenin/TCF, in some canonical Wnt developmental contexts.

Thus far, the regulation of Wnt target genes has been reported for beta-catenin, plakoglobin (gamma-catenin), p120-catenin, and plakophilin-2.[45,54,62] We expect that additional p120 and plakophilin subfamily members are involved. For example, delta-catenin has also been reported to bind and modulate Kaiso.[6] Also quite suggestive is that axin lowers the levels not only of beta-catenin, plakoglobin, and p120-catenin but also of delta-catenin and ARVCF-catenin (Fig. 17.1).[5] Further, work from our group has shown that isoform-1 of p120 (longest) possesses a conserved destruction box at its amino-terminus and responds as predicted to both Wnt signals and the pathway's destruction machinery (Fig. 17.2).[5] Independent results indicate the sensitivity of delta-catenin to GSK3-beta, likewise leading to ubiquitin/proteasome-mediated destruction.[83] Together, these data suggest that Wnt signaling initiates multiple responses through various catenins, likely modulating specific cellular responses within different biological contexts.

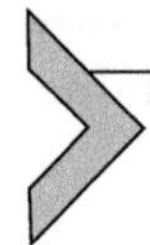

7. CANONICAL Wnt SIGNALING IN THE CONTEXT OF CATENIN–SMALL-GTPase EFFECTS

All three catenin subfamilies act at cell junctions, but the p120 and plakophilin subfamily members additionally regulate small GTPases within the cytoplasmic space (Fig. 17.3). Through their association with GEFs and GAPs[84] or via direct small-GTPase interactions,[85] catenins modulate the activity of RhoA, Rac1, and Cdc42 to modulate cytoskeletal dynamics[57] (see also Chapter 18). For example, in a context-dependent manner, both p120 and plakophilin subfamily members can activate Rac1 and inhibit RhoA.[57] Thus, upon release from cadherin, these catenins are expected to regulate small GTPases, either distal (e.g., cell-free edge) or in the immediate junctional region. Catenin–small-GTPase communication is would likewise be predicted to arise upon canonical Wnt pathway stabilization of p120 isoform1, and/or Wnt-mediated stabilization of other p120 subfamily or plakophilin-catenins. Consistent with this possibility are initial findings that a little understood component of the Wnt pathway, Frodo, enhances p120 stabilization to produce an increased modulatory effect on small GTPases.[86] It is also intriguing that catenins further associate with the microtubule cytoskeleton, kinesins and centrosomes.[87–89] Since p120 and plakophilin subfamily catenins bind and regulate classic and/or desmosomal cadherins, as well as small GTPases, these proteins appear well positioned to contribute to outcomes including motile versus sessile cell behaviors (Fig. 17.3).[90,91] Indeed, full-length and shorter p120 isoforms have distinguishable effects on RhoA,[58,90] such that the stabilization of certain p120 isoforms would be expected to alter cell behavior (Fig. 17.2). As the field progresses, the role of catenin stabilization in response to canonical Wnt signals needs to take into consideration catenin relationships with small GTPases (cell architecture, trafficking, and motility), in addition to effects on nuclear gene activity. Thus, while they require further study, the p120- and plakophilin-catenins potentially assist in coordinating cellular outcomes involving the canonical Wnt pathway, cadherin–catenin and cytoskeletal complexes, and gene regulation.

8. REDEFINING "CANONICAL" Wnt SIGNALING

As evidence for cross talk between Wnt signaling and adherens and desmosomal junctions increases, it may become more difficult to define the boundaries of "canonical" Wnt signaling. An underdeveloped area of study focuses on whether Wnt signaling affects cell adhesion via direct mechanisms independent of downstream transcription. Wnts have been connected with endosomal internalization of cadherins in gastrulation.[92] Further, studies in *Drosophila* have shown that Wnt signaling induces reduction of membrane-associated beta-catenin and concomitant destabilization of cadherins, thereby reducing cell adhesion.[93] Reduction in cell adhesion is then followed by Wnt-induced expression of DE-cadherin.[93,94] In other contexts, Wnt signaling is also associated with increased transcription of Twist, Slug, and Snail, which has been reported to repress E-cadherin (see Chapter 14).[77] Further, Wnt pathway components physically interact with cadherins and protocadherins within adherens junctions.[95] Given that members of the p120- and plakophilin subfamilies stabilize cadherins when bound to them within cell–cell junctions,[14,15] Wnt signaling effects upon catenin levels may also modulate cell adhesion. In addition, such "other" catenins, namely the p120- and plakophilin-catenin family members, may regulate small GTPases differentially as a consequence of canonical Wnt signaling. Thus, as more is learned in the years ahead, the boundaries between "noncanonical" Wnt trajectories, which also regulate these small GTPases, and "canonical" Wnt signaling may not appear that separable in many contexts.

9. CONCLUSION AND GAPS IN KNOWLEDGE

Following its stabilization, beta-catenin is the central signal transducer of the canonical Wnt pathway entering the nucleus to relieve repression conferred by TCF/LEF. However, evidence is beginning to suggest a broader effect of canonical signals through additional stabilizing effects on p120 isoform 1 and likely additional p120 or plakophilin subfamily members. Given that p120-catenin family members regulate small GTPases and the stability of cadherins, canonical Wnt signals may produce a number of rather direct catenin-mediated effects beyond those that catenins are known, or conjectured, to execute in the nucleus.

Numerous questions remain regarding the role of these "other" catenins in Wnt signaling. First, it will be interesting to identify the transcription-factor partners and the genes that are directly regulated by each catenin. This helps us understand each catenin's distinct versus overlapping functions in response to Wnt or other signals. Further, if multiple catenins respond to Wnt signals, are they differentially regulated in accordance with a particular Wnt-ligand:Frizzled-receptor pairing? What factors might then assist in transducing such differential effects? One possibility is the involvement of scaffolding proteins with preferences for particular catenins. In all cases, to understand the larger catenin network, we need to identify and functionally characterize many more selective and shared protein partners within junctions and the cytoplasm/cytoskeleton. Likewise, we must resolve direct catenin gene targets and regulatory factors within the nucleus, perhaps using the well-developed model of beta-catenin as a guide. In all cases, there is much that the cadherin–catenin field looks forward to in the coming years, with a primary interest in establishing a deeper understanding of the roles of catenins in promoting and coordinating cellular processes within varied cellular compartments.

ACKNOWLEDGMENTS

This work was supported by an NIH NIGMS R01 Grant (GM052112) (P. D. M.), a postdoctoral fellowship provided by the Odyssey Program and the Theodore N. Law Endowment for Scientific Achievement at the University of Texas MD Anderson Cancer Center (R. K. M.), a postdoctoral fellowship from the National Kidney Foundation (FLB1628) (R. K. M.), an NIH NIDDK K01 Career Development Award (1K01DK092320) (R. K. M.), and a Cancer Prevention and Research Institute of Texas Training Award (RP101502) (W. A. M.).

REFERENCES

1. Ozawa M, Baribault H, Kemler R. The cytoplasmic domain of the cell adhesion molecule uvomorulin associates with three independent proteins structurally related in different species. *EMBO J* 1989;**8**:1711–7.
2. Abe K, Takeichi M. EPLIN mediates linkage of the cadherin catenin complex to F-actin and stabilizes the circumferential actin belt. *Proc Natl Acad Sci USA* 2008;**105**:13–9.
3. Clevers H, Nusse R. Wnt/beta-catenin signaling and disease. *Cell* 2012;**149**:1192–205.
4. McCrea PD, Gu D. The catenin family at a glance. *J Cell Sci* 2010;**123**:637–42.
5. Hong JY, Park JI, Cho K, Gu D, Ji H, Artandi SE, et al. Shared molecular mechanisms regulate multiple catenin proteins: canonical Wnt signals and components modulate p120-catenin isoform-1 and additional p120 subfamily members. *J Cell Sci* 2010;**123**: 4351–65.
6. Rodova M, Kelly KF, VanSaun M, Daniel JM, Werle MJ. Regulation of the rapsyn promoter by Kaiso and {delta}-catenin. *Mol Cell Biol* 2004;**24**:7188–96.
7. Schmidt A, Langbein L, Rode M, Pratzel S, Zimbelmann R, Franke WW. Plakophilins 1a and 1b: widespread nuclear proteins recruited in specific epithelial cells as desmosomal plaque components. *Cell Tissue Res* 1997;**290**:481–99.
8. McCrea PD, Gu D, Balda MS. Junctional music that the nucleus hears: cell-cell contact signaling and the modulation of gene activity. *Cold Spring Harb Perspect Biol* 2009;**1**: a002923.
9. Pieters T, van Roy F, van Hengel J. Functions of p120ctn isoforms in cell-cell adhesion and intracellular signaling. *Front Biosci* 2012;**17**:1669–94.
10. Pokutta S, Drees F, Yamada S, Nelson WJ, Weis WI. Biochemical and structural analysis of alpha-catenin in cell-cell contacts. *Biochem Soc Trans* 2008;**36**:141–7.
11. Lecuit T. alpha-catenin mechanosensing for adherens junctions. *Nat Cell Biol* 2010;**12**:522–4.
12. Scott JA, Yap AS. Cinderella no longer: alpha-catenin steps out of cadherin's shadow. *J Cell Sci* 2006;**119**:4599–605.
13. Huber AH, Nelson WJ, Weis WI. Three-dimensional structure of the armadillo repeat region of beta-catenin. *Cell* 1997;**90**:871–82.
14. Shapiro L, Weis WI. Structure and biochemistry of cadherins and catenins. *Cold Spring Harb Perspect Biol* 2009;**1**:a003053.
15. Yap AS, Crampton MS, Hardin J. Making and breaking contacts: the cellular biology of cadherin regulation. *Curr Opin Cell Biol* 2007;**19**:508–14.
16. Kwiatkowski AV, Maiden SL, Pokutta S, Choi HJ, Benjamin JM, Lynch AM, et al. In vitro and in vivo reconstitution of the cadherin-catenin-actin complex from Caenorhabditis elegans. *Proc Natl Acad Sci USA* 2010;**107**:14591–6.
17. Reynolds AB, Carnahan RH. Regulation of cadherin stability and turnover by p120ctn: implications in disease and cancer. *Semin Cell Dev Biol* 2004;**15**:657–63.
18. Nelson WJ. Regulation of cell-cell adhesion by the cadherin-catenin complex. *Biochem Soc Trans* 2008;**36**:149–55.
19. Hatzfeld M. Plakophilins: multifunctional proteins or just regulators of desmosomal adhesion? *Biochim Biophys Acta* 2007;**1773**:69–77.
20. Bass-Zubek AE, Godsel LM, Delmar M, Green KJ. Plakophilins: multifunctional scaffolds for adhesion and signaling. *Curr Opin Cell Biol* 2009;**21**:708–16.
21. Garrod D, Chidgey M. Desmosome structure, composition and function. *Biochim Biophys Acta* 2008;**1778**:572–87.
22. Franke WW, Rickelt S, Barth M, Pieperhoff S. The junctions that don't fit the scheme: special symmetrical cell-cell junctions of their own kind. *Cell Tissue Res* 2009;**338**:1–17.
23. Delva E, Tucker DK, Kowalczyk AP. The desmosome. *Cold Spring Harb Perspect Biol* 2009;**1**:a002543.

24. Saito M, Tucker DK, Kohlhorst D, Niessen CM, Kowalczyk AP. Classical and desmosomal cadherins at a glance. *J Cell Sci* 2012;**125**:2547–52.
25. Bornslaeger EA, Godsel LM, Corcoran CM, Park JK, Hatzfeld M, Kowalczyk AP, et al. Plakophilin 1 interferes with plakoglobin binding to desmoplakin, yet together with plakoglobin promotes clustering of desmosomal plaque complexes at cell-cell borders. *J Cell Sci* 2001;**114**:727–38.
26. South AP, Wan H, Stone MG, Dopping-Hepenstal PJ, Purkis PE, Marshall JF, et al. Lack of plakophilin 1 increases keratinocyte migration and reduces desmosome stability. *J Cell Sci* 2003;**116**:3303–14.
27. Nusse R. Wnt signaling. *Cold Spring Harb Perspect Biol* 2012;**4**:a011163.
28. Dickinson DJ, Nelson WJ, Weis WI. A polarized epithelium organized by beta- and alpha-catenin predates cadherin and metazoan origins. *Science* 2011;**331**:1336–9.
29. Nusslein-Volhard C, Wieschaus E. Mutations affecting segment number and polarity in Drosophila. *Nature* 1980;**287**:795–801.
30. Nusse R, Varmus HE. Many tumors induced by the mouse mammary tumor virus contain a provirus integrated in the same region of the host genome. *Cell* 1982;**31**: 99–109.
31. Rijsewijk F, van Deemter L, Wagenaar E, Sonnenberg A, Nusse R. Transfection of the *int*-1 mammary oncogene in cuboidal RAC mammary cell line results in morphological transformation and tumorigenicity. *EMBO J* 1987;**6**:127–31.
32. Wieschaus E, Riggleman R. Autonomous requirements for the segment polarity gene armadillo during Drosophila embryogenesis. *Cell* 1987;**49**:177–84.
33. McMahon AP, Moon RT. Ectopic expression of the proto-oncogene *int-1* in *Xenopus* embryos leads to duplication of the embryonic axis. *Cell* 1989;**58**:1075–84.
34. McCrea PD, Turck CW, Gumbiner B. A homolog of the armadillo protein in Drosophila (plakoglobin) associated with E-cadherin. *Science* 1991;**254**:1359–61.
35. McCrea PD, Brieher WM, Gumbiner BM. Induction of a secondary body axis in Xenopus by antibodies to beta-catenin. *J Cell Biol* 1993;**123**:477–84.
36. Nusse R. A versatile transcriptional effector of Wingless signaling. *Cell* 1997;**89**:321–3.
37. Porfiri E, Rubinfeld B, Albert I, Hovanes K, Waterman M, Polakis P. Induction of a beta-catenin-LEF-1 complex by wnt-1 and transforming mutants of beta-catenin. *Oncogene* 1997;**15**:2833–9.
38. Haegebarth A, Clevers H. Wnt signaling, lgr5, and stem cells in the intestine and skin. *Am J Pathol* 2009;**174**:715–21.
39. Zeng X, Huang H, Tamai K, Zhang X, Harada Y, Yokota C, et al. Initiation of Wnt signaling: control of Wnt coreceptor Lrp6 phosphorylation/activation via frizzled, dishevelled and axin functions. *Development* 2008;**135**:367–75.
40. Nelson WJ, Nusse R. Convergence of Wnt, beta-catenin, and cadherin pathways. *Science* 2004;**303**:1483–7.
41. Polakis P. Wnt signaling in cancer. *Cold Spring Harb Perspect Biol* 2012;**4**:a008052.
42. Miyagishi M, Fujii R, Hatta M, Yoshida E, Araya N, Nagafuchi A, et al. Regulation of Lef-mediated transcription and p53-dependent pathway by associating beta-catenin with CBP/p300. *J Biol Chem* 2000;**275**:35170–5.
43. Thompson B, Townsley F, Rosin-Arbesfeld R, Musisi H, Bienz M. A new nuclear component of the Wnt signalling pathway. *Nat Cell Biol* 2002;**4**:367–73.
44. Zhurinsky J, Shtutman M, Ben-Ze'ev A. Plakoglobin and beta-catenin: protein interactions, regulation and biological roles. *J Cell Sci* 2000;**113**(Pt 18):3127–39.
45. Maeda O, Usami N, Kondo M, Takahashi M, Goto H, Shimokata K, et al. Plakoglobin (gamma-catenin) has TCF/LEF family-dependent transcriptional activity in beta-catenin-deficient cell line. *Oncogene* 2004;**23**:964–72.
46. Garcia-Gras E, Lombardi R, Giocondo MJ, Willerson JT, Schneider MD, Khoury DS, et al. Suppression of canonical Wnt/beta-catenin signaling by nuclear plakoglobin

recapitulates phenotype of arrhythmogenic right ventricular cardiomyopathy. *J Clin Invest* 2006;**116**:2012–21.
47. Hikasa H, Ezan J, Itoh K, Li X, Klymkowsky MW, Sokol SY. Regulation of TCF3 by Wnt-dependent phosphorylation during vertebrate axis specification. *Dev Cell* 2010;**19**: 521–32.
48. Daniel JM. Dancing in and out of the nucleus: p120(ctn) and the transcription factor Kaiso. *Biochim Biophys Acta* 2007;**1773**:59–68.
49. Daniel JM, Reynolds AB. The catenin p120(ctn) interacts with Kaiso, a novel BTB/POZ domain zinc finger transcription factor. *Mol Cell Biol* 1999;**19**:3614–23.
50. Hosking CR, Ulloa F, Hogan C, Ferber EC, Figueroa A, Gevaert K, et al. The transcriptional repressor Glis2 is a novel binding partner for p120 catenin. *Mol Biol Cell* 2007;**18**:1918–27.
51. Gu D, Tonthat NK, Lee M, Ji H, Bhat KP, Hollingsworth F, et al. Caspase-3 cleavage links delta-catenin to the novel nuclear protein ZIFCAT. *J Biol Chem* 2011;**286**:23178–88.
52. Koutras C, Lessard CB, Levesque G. A nuclear function for the presenilin 1 neuronal partner NPRAP/delta-catenin. *J Alzheimers Dis* 2011;**27**:307–16.
53. Hong JY, Park JI, Lee M, Munoz WA, Miller RK, Ji H, et al. Down's-syndrome-related kinase Dyrk1A modulates the p120-catenin-Kaiso trajectory of the Wnt signaling pathway. *J Cell Sci* 2012;**125**:561–9.
54. Chen X, Bonne S, Hatzfeld M, van Roy F, Green KJ. Protein binding and functional characterization of plakophilin 2. Evidence for its diverse roles in desmosomes and beta-catenin signaling. *J Biol Chem* 2002;**277**:10512–22.
55. Sobolik-Delmaire T, Reddy R, Pashaj A, Roberts BJ, Wahl 3rd JK. Plakophilin-1 localizes to the nucleus and interacts with single-stranded DNA. *J Invest Dermatol* 2010;**130**:2638–46.
56. Slorach EM, Chou J, Werb Z. Zepp o1 is a novel metastasis promoter that represses E-cadherin expression and regulates p120-catenin isoform expression and localization. *Genes Dev* 2011;**25**:471–84.
57. Anastasiadis PZ. p120-ctn: a nexus for contextual signaling via Rho GTPases. *Biochim Biophys Acta* 2007;**1773**:34–46.
58. Yanagisawa M, Huveldt D, Kreinest P, Lohse CM, Cheville JC, Parker AS, et al. A p120 catenin isoform switch affects Rho activity, induces tumor cell invasion, and predicts metastatic disease. *J Biol Chem* 2008;**283**:18344–54.
59. Pieters T, van Hengel J, van Roy F. Functions of p120ctn in development and disease. *Front Biosci* 2012;**17**:760–83.
60. Bonne S, Gilbert B, Hatzfeld M, Chen X, Green KJ, van Roy F. Defining desmosomal plakophilin-3 interactions. *J Cell Biol* 2003;**161**:403–16.
61. Gandjbakhch E, Charron P, Fressart V, Lorin de la Grandmaison G, Simon F, Gary F, et al. Plakophilin 2A is the dominant isoform in human heart tissue: consequences for the genetic screening of arrhythmogenic right ventricular cardiomyopathy. *Heart* 2011;**97**: 844–9.
62. Park JI, Kim SW, Lyons JP, Ji H, Nguyen TT, Cho K, et al. Kaiso/p120-catenin and TCF/beta-catenin complexes coordinately regulate canonical Wnt gene targets. *Dev Cell* 2005;**8**:843–54.
63. Katoh M. Networking of WNT, FGF, Notch, BMP, and Hedgehog signaling pathways during carcinogenesis. *Stem Cell Rev* 2007;**3**:30–8.
64. Kim NG, Koh E, Chen X, Gumbiner BM. E-cadherin mediates contact inhibition of proliferation through Hippo signaling-pathway components. *Proc Natl Acad Sci USA* 2011;**108**:11930–5.
65. Cordenonsi M, Zanconato F, Azzolin L, Forcato M, Rosato A, Frasson C, et al. The Hippo transducer TAZ confers cancer stem cell-related traits on breast cancer cells. *Cell* 2011;**147**:759–72.

66. Varelas X, Miller BW, Sopko R, Song S, Gregorieff A, Fellouse FA, et al. The Hippo pathway regulates Wnt/beta-catenin signaling. *Dev Cell* 2010;**18**:579–91.
67. McCrea PD, Park JI. Developmental functions of the P120-catenin sub-family. *Biochim Biophys Acta* 2007;**1773**:17–33.
68. Andl CD, Rustgi AK. No one-way street: cross-talk between e-cadherin and receptor tyrosine kinase (RTK) signaling: a mechanism to regulate RTK activity. *Cancer Biol Ther* 2005;**4**:28–31.
69. Lilien J, Balsamo J. The regulation of cadherin-mediated adhesion by tyrosine phosphorylation/dephosphorylation of beta-catenin. *Curr Opin Cell Biol* 2005;**17**:459–65.
70. Jeanes A, Gottardi CJ, Yap AS. Cadherins and cancer: how does cadherin dysfunction promote tumor progression? *Oncogene* 2008;**27**:6920–9.
71. Casagolda D, Del Valle-Perez B, Valls G, Lugilde E, Vinyoles M, Casado-Vela J, et al. A p120-catenin-CK1epsilon complex regulates Wnt signaling. *J Cell Sci* 2010;**123**:2621–31.
72. Maher MT, Flozak AS, Stocker AM, Chenn A, Gottardi CJ. Activity of the beta-catenin phosphodestruction complex at cell-cell contacts is enhanced by cadherin-based adhesion. *J Cell Biol* 2009;**186**:219–28.
73. Gottardi CJ, Gumbiner BM. Distinct molecular forms of beta-catenin are targeted to adhesive or transcriptional complexes. *J Cell Biol* 2004;**167**:339–49.
74. Janssens S, Crabbe E, Vleminckx K. The role of Wnt in cell signaling and cell adhesion during early vertebrate development. *Front Biosci* 2012;**17**:2352–66.
75. Vallin J, Thuret R, Giacomello E, Faraldo MM, Thiery JP, Broders F. Cloning and characterization of three Xenopus slug promoters reveal direct regulation by Lef/beta-catenin signaling. *J Biol Chem* 2001;**276**:30350–8.
76. Tan C, Costello P, Sanghera J, Dominguez D, Baulida J, de Herreros AG, et al. Inhibition of integrin linked kinase (ILK) suppresses beta-catenin-Lef/Tcf-dependent transcription and expression of the E-cadherin repressor, snail, in APC −/− human colon carcinoma cells. *Oncogene* 2001;**20**:133–40.
77. Heuberger J, Birchmeier W. Interplay of cadherin-mediated cell adhesion and canonical Wnt signaling. *Cold Spring Harb Perspect Biol* 2010;**2**:a002915.
78. Foubert E, De Craene B, Berx G. Key signalling nodes in mammary gland development and cancer. The Snail1-Twist1 conspiracy in malignant breast cancer progression. *Breast Cancer Res* 2010;**12**:206.
79. Yook JI, Li XY, Ota I, Fearon ER, Weiss SJ. Wnt-dependent regulation of the E-cadherin repressor snail. *J Biol Chem* 2005;**280**:11740–8.
80. Zhou BP, Deng J, Xia W, Xu J, Li YM, Gunduz M, et al. Dual regulation of Snail by GSK-3beta-mediated phosphorylation in control of epithelial-mesenchymal transition. *Nat Cell Biol* 2004;**6**:931–40.
81. Ruzov A, Savitskaya E, Hackett JA, Reddington JP, Prokhortchouk A, Madej MJ, et al. The non-methylated DNA-binding function of Kaiso is not required in early Xenopus laevis development. *Development* 2009;**136**:729–38.
82. Kim SW, Park JI, Spring CM, Sater AK, Ji H, Otchere AA, et al. Non-canonical Wnt signals are modulated by the Kaiso transcriptional repressor and p120-catenin. *Nat Cell Biol* 2004;**6**:1212–20.
83. Oh M, Kim H, Yang I, Park JH, Cong WT, Baek MC, et al. GSK-3 phosphorylates delta-catenin and negatively regulates its stability via ubiquitination/proteosome-mediated proteolysis. *J Biol Chem* 2009;**284**:28579–89.
84. Niessen CM, Yap AS. Another job for the talented p120-catenin. *Cell* 2006;**127**:875–7.
85. Anastasiadis PZ, Moon SY, Thoreson MA, Mariner DJ, Crawford HC, Zheng Y, et al. Inhibition of RhoA by p120 catenin. *Nat Cell Biol* 2000;**2**:637–44.
86. Park JI, Ji H, Jun S, Gu D, Hikasa H, Li L, et al. Frodo links Dishevelled to the p120-catenin/Kaiso pathway: distinct catenin subfamilies promote Wnt signals. *Dev Cell* 2006;**11**:683–95.

87. Bahmanyar S, Kaplan DD, Deluca JG, Giddings Jr TH, O'Toole ET, Winey M, et al. beta-Catenin is a Nek2 substrate involved in centrosome separation. *Genes Dev* 2008;**22**:91–105.
88. Chen X, Kojima S, Borisy GG, Green KJ. p120 catenin associates with kinesin and facilitates the transport of cadherin-catenin complexes to intercellular junctions. *J Cell Biol* 2003;**163**:547–57.
89. Ichii T, Takeichi M. p120-catenin regulates microtubule dynamics and cell migration in a cadherin-independent manner. *Genes Cells* 2007;**12**:827–39.
90. Reynolds AB, Roczniak-Ferguson A. Emerging roles for p120-catenin in cell adhesion and cancer. *Oncogene* 2004;**23**:7947–56.
91. Yanagisawa M, Anastasiadis PZ. p120 catenin is essential for mesenchymal cadherin-mediated regulation of cell motility and invasiveness. *J Cell Biol* 2006;**174**:1087–96.
92. Ulrich F, Krieg M, Schotz EM, Link V, Castanon I, Schnabel V, et al. Wnt11 functions in gastrulation by controlling cell cohesion through Rab5c and E-cadherin. *Dev Cell* 2005;**9**:555–64.
93. Wodarz A, Stewart DB, Nelson WJ, Nusse R. Wingless signaling modulates cadherin-mediated cell adhesion in Drosophila imaginal disc cells. *J Cell Sci* 2006;**119**:2425–34.
94. Yanagawa S, Lee JS, Haruna T, Oda H, Uemura T, Takeichi M, et al. Accumulation of Armadillo induced by Wingless, Dishevelled, and dominant-negative Zeste-White 3 leads to elevated DE-cadherin in Drosophila clone 8 wing disc cells. *J Biol Chem* 1997;**272**:25243–51.
95. Hay E, Buczkowski T, Marty C, Da Nascimento S, Sonnet P, Marie PJ. Peptide-based mediated disruption of N-cadherin-LRP5/6 interaction promotes Wnt signalling and bone formation. *J Bone Miner Res* 2012;**27**:1852–63.
96. Logan CY, Miller JR, Ferkowicz MJ, McClay DR. Nuclear beta-catenin is required to specify vegetal cell fates in the sea urchin embryo. *Development* 1999;**126**:345–57.
97. Bonne S, van Hengel J, Nollet F, Kools P, van Roy F. Plakophilin-3, a novel armadillo-like protein present in nuclei and desmosomes of epithelial cells. *J Cell Sci* 1999;**112** (Pt 14):2265–76.
98. Waterman ML, Jones KA. Purification of TCF-1 alpha, a T-cell-specific transcription factor that activates the T-cell receptor C alpha gene enhancer in a context-dependent manner. *New Biol* 1990;**2**(7):621–36.

CHAPTER EIGHTEEN

p120 Catenin: An Essential Regulator of Cadherin Stability, Adhesion-Induced Signaling, and Cancer Progression

Antonis Kourtidis, Siu P. Ngok, Panos Z. Anastasiadis
Department of Cancer Biology, Mayo Clinic, Jacksonville, Florida, USA

Contents

Abstract

p120 catenin is the best studied member of a subfamily of proteins that associate with the cadherin juxtamembrane domain to suppress cadherin endocytosis. p120 also recruits the minus ends of microtubules to the cadherin complex, leading to junction maturation. In addition, p120 regulates the activity of Rho family GTPases through multiple interactions with Rho GEFs, GAPs, Rho GTPases, and their effectors. Nuclear signaling is affected by the interaction of p120 with Kaiso, a transcription factor regulating Wnt-responsive genes as well as transcriptionally repressing methylated promoters. Multiple alternatively spliced p120 isoforms and complex phosphorylation events affect

Progress in Molecular Biology and Translational Science, Volume 116
ISSN 1877-1173
http://dx.doi.org/10.1016/B978-0-12-394311-8.00018-2

these p120 functions. In cancer, reduced p120 expression correlates with reduced E-cadherin function and with tumor progression. In contrast, in tumor cells that have lost E-cadherin expression, p120 promotes cell invasion and anchorage-independent growth. Furthermore, p120 is required for Src-induced oncogenic transformation and provides a potential target for future therapeutic interventions.

1. INTRODUCTION: HISTORY OF p120

The p120 catenin protein (p120ctn; p120 herein) was originally identified in a screen for substrates of the Src tyrosine kinase as a 120-kDa protein, the phosphorylation of which on tyrosine residues correlated with cellular transformation (Ref. 1, reviewed in Ref. 2). A number of additional Src substrates, including focal adhesion kinase, p130cas, talin, and cortactin, were also identified in the screen and subsequently shown to regulate cytoskeletal remodeling and focal adhesion signaling.[3] Unlike the other Src substrates, p120 was shown to interact with classical cadherins at the adherens junctions (AJs),[4–6] where it suppresses cadherin endocytosis.[7–9] p120 regulates cytoskeletal reorganization by affecting the activities of Rho GTPases,[10–12] or by mediating the association of microtubules with the cell junctions.[13–16] In addition, p120 interacts with the transcription factor Kaiso to regulate nuclear signaling events.[17,18] Recent studies reevaluating the role of p120 in cellular transformation indicate that p120 is required for the anchorage-independent growth (AIG) of tumor cells lacking E-cadherin, and of cells overexpressing Src.[19,20]

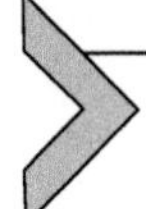

2. STRUCTURE

2.1. Family members

p120 is the best studied member of a subfamily of proteins containing armadillo repeats (ARMs) (see also Chapter 4) and sharing a common genetic and protein structure; the other members are also found at cell–cell junctions and most, if not all, can participate in Rho signaling[2]: ARVCF and δ-catenin, like p120, are found at AJs; plakophilins are at desmosomes; and p0071 is found at both (reviewed in Refs. 21–23). p120 is ubiquitously expressed, while the expression of the other family members is more restricted, suggesting specialized cellular functions. One difference between p120 and the other three family members found at AJs is that p120 lacks a C-terminal PDZ (postsynaptic density 95/disc large/zonula occludens-1)-binding motif with high affinity for Erbin and hScrib.[24,25]

p120 is also superficially similar to β-catenin, both in structure, being a member of the ARM family of proteins, and in function, interacting with cadherins at AJs and regulating transcription. p120's functions, however, are substantially different from those of β-catenin.[2] p120 and β-catenin associate with the cadherin juxtamembrane domain (JMD) and the catenin-binding domain (CBD), respectively. Surprisingly, the mechanism is similar, with both proteins utilizing basic grooves within their ARM domains to recognize charged and hydrophobic residues in the JMD or CBD.[26] A distinct set of surface-exposed residues is responsible for the differential ability of these proteins to bind either the JMD or the CBD. Through its binding to the JMD, p120 selectively regulates cadherin endocytosis and stability. Another important difference is that p120 has a well-established role in regulating members of the Rho family of small GTPases, which are key mediators of cytoskeletal dynamics and cadherin-mediated cell–cell adhesion (reviewed in Ref. 27).

2.2. Isoforms

Cloning and analysis of the *p120* gene revealed four alternatively spliced exons and four transcriptional start sites that together can potentially produce 64 different isoforms of p120 (Refs. 28,29, reviewed in Refs. 21,30). p120 isoforms are named after the transcriptional start site used (1–4), and the alternatively spliced exons they express (A–D). The complexity of p120's exon–intron structure and regulation as well as the potential competition for cadherin binding from other family members[31,32] suggests that p120 and the JMD play a key regulatory function in the cadherin complex. This hypothesis is further supported by evidence that p120 can be phosphorylated in multiple serine, threonine, and tyrosine residues that affect its function.[33–36] The potential of this variation in protein sequence and phosphorylation status makes p120 an interesting but also a complex member of AJs.

The functions of particular p120 isoforms are largely unexplored. However, a change in the ratio of p120 isoforms is evident in epithelial versus mesenchymal cells.[5,29,37] During epithelial-to-mesenchymal transition (EMT), or after ectopic Snail expression, the larger p120 isoform 1 is induced[38–40] by a process that involves Epithelial Splicing Regulatory Proteins 1 and 2.[41]

2.3. Evolution

Analysis of proteomes from 14 different vertebrate and metazoan species shows that the p120 protein family evolved from a common δ-catenin-like ancestor present in all metazoans.[42,43] Successive rounds of gene duplication and diversification are responsible for the seven p120 family members present in the human genome (see also Chapter 4). Ablation of the single δ-catenin-like p120 family member in invertebrate species (e.g., *Drosophila melanogaster, Caenorhabditis elegans*) produced minor adhesion phenotypic changes but did not affect survival[44–46] (see also Chapter 11). In contrast, p120 knockdown or depletion in vertebrates is embryonic lethal.[47–51] Paradoxically, p120 has evolved in vertebrates into a ubiquitously expressed protein that is essential for life, whereas other family members, that is δ-catenin, ARVCF, and possibly p0071, are more restricted in expression and their loss is better tolerated.[52,53]

3. FUNCTION

3.1. Cadherin stability and function

It is well established that p120 binding to the JMD of the cadherin cytoplasmic tail is directly responsible for stabilizing cadherin expression at the cell surface as well as for inducing cadherin clustering that results in the formation of AJs (reviewed in Refs. 21,54). Dissociation of p120 from the cadherin complex results in endocytic internalization of cadherins[8,9,55] (Fig. 18.1). Endocytosed cadherins can be either degraded in the lysosomes or recycled to the plasma membrane by a process that might involve p120 binding and the Rap1 GTPase.[55,56] Thus, p120 is a master regulator (a rheostat) that controls cadherin stabilization and assembly into AJs.

Insight into the mechanism by which p120 binding prevents cadherin endocytosis was provided recently by the elucidation of the structure of p120 bound to E-cadherin.[26] Both "dynamic" and "static" interactions contribute to the binding of p120's ARM domain to the cadherin JMD. Importantly, these interactions include JMD residues that are implicated in clathrin-mediated endocytosis and Hakai-dependent ubiquitination of E-cadherin[26] (Fig. 18.1). Based on the structural evidence, p120 binding physically impedes the interaction of cadherin molecules with either E3 ligases or the endocytic machinery.

Interestingly, the N-terminal domain of cadherin-bound p120 isoform 1 (the longest p120 isoform) is in close proximity to the CBD and β-catenin. This N-terminal domain includes the first 323 amino acids of p120 isoform 1, is lacking entirely in p120 isoform 4, and is highly regulated by alternative splicing and by phosphorylation. Serine/threonine phosphorylation of this domain is thought to control E-cadherin dynamics at the cell membrane.[35] The data indicate that both cadherin binding, which is mediated by p120's

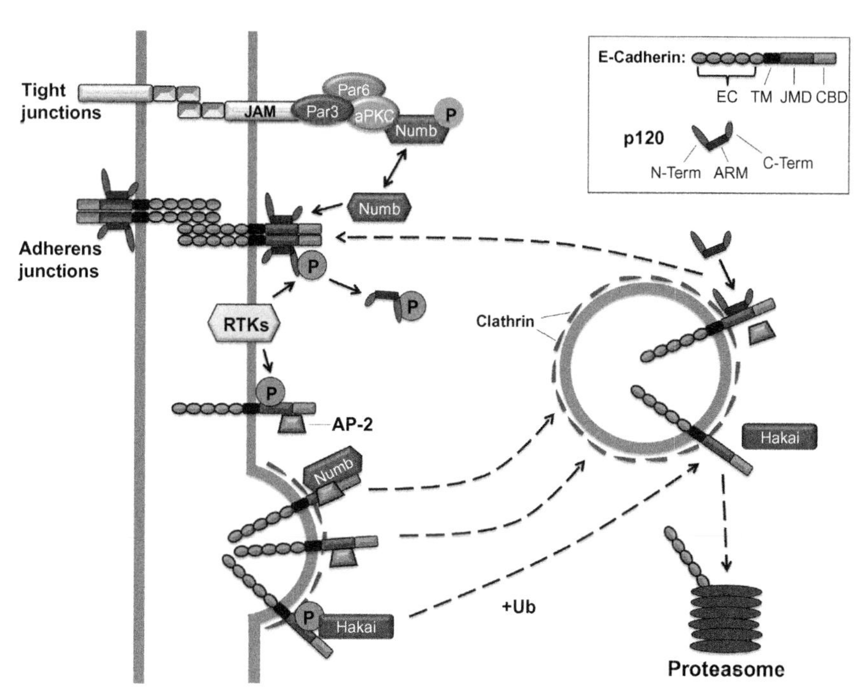

Figure 18.1 E-Cadherin stabilization by p120 catenin. p120 binding to the cadherin juxtamembrane domain (JMD) stabilizes cadherins at the cell membrane and promotes the formation of adherens junctions. Removal of p120 from the cadherin complex can proceed by multiple mechanisms, including phosphorylation of p120 and/or E-cadherin downstream of receptor tyrosine kinases (RTKs), or association of p120 with non-phosphorylated Numb, a protein normally phosphorylated by aPKC. p120 dissociation uncovers an adaptor protein 2 (AP-2) binding motif in the JMD as well as a phosphorylation-dependent motif in E-cadherin for the recruitment of the E3 ligase Hakai. AP-2 binding promotes clathrin-dependent endocytosis of E-cadherin. The endocytosed E-cadherin, upon binding to p120, can be recycled back to the membrane. Alternatively, Hakai-induced ubiquitination of E-cadherin can lead to cadherin degradation in the proteasome. Insert top right: the detailed structure of E-cadherin and p120; EC, extracellular domain; TM, transmembrane domain; JMD, juxtamembrane domain; CBD, catenin-binding domain; N-term, N-terminus of p120 isoform 1; ARM, armadillo repeat domain; C-term, C-terminus of p120. (See Color Insert.)

central ARM domain, and signaling events mediated by p120's N-terminus are required for cadherin stabilization.

Surprisingly, early studies indicated that the JMD has both positive and negative functions in cadherin clustering and cell–cell adhesion (reviewed in Ref. 21). A model whereby p120 is "activated" upon cadherin *trans* dimerization was proposed, whereas intracellular signaling events were suspected to "deactivate" p120 and block its contribution to cell adhesion.[21]

This model is supported by the observation that cadherin activating antibodies induce cadherin conformational changes and promote cell–cell adhesion via the dephosphorylation of selective N-terminal p120 residues.[57] The positive effect of dephosphorylated "activated" p120 might be due to increased cadherin clustering, possibly via direct p120 dimerization.[26] Alternatively, p120 might serve as a scaffold for additional proteins, enabling them to directly affect adhesion or to regulate the function of other cadherin/catenin complex proteins. One such candidate is PLEKHA7, a novel protein interacting with the p120 N-terminal that is required for mature junction formation.[16,58] In addition, p120 can recruit a number of kinases and phosphatases that could influence adhesion by affecting the entire cadherin–catenin complex. p120 "deactivation" might also relate to signaling events leading to cadherin endocytosis. Recently, p120 was shown to mediate the interaction of E-cadherin complexes with a protein called Numb. Numb is normally phosphorylated by aPKC, a main component of tight junctions. However, upon dephosphorylation, Numb can bind either to the C-terminus of p120 or to non-p120-bound E-cadherin, mediating in both cases clathrin-dependent endocytosis of E-cadherin[59] (Fig. 18.1).

Finally, the signaling that regulates the interaction of p120 with cadherins likely includes dynamic phosphorylation of p120 and cadherin (reviewed in Ref. 60). Tyrosine phosphorylation of E-cadherin at Y755/756 or VE-cadherin at Y658 disrupts p120 binding[26,61,62] (Fig. 18.1). Additional kinases, such as PAK (p21-associated kinase) or CK1 epsilon, can reportedly also phosphorylate cadherins and disrupt p120 association.[63,64] Clearly, the JMD plays a critical role in cadherin stability and cell adhesion, and its function is highly regulated by phosphorylation and by interaction with a host of p120-related proteins (including isoforms and family members).

3.2. Regulation of Rho GTPase signaling

The Rho GTPases RhoA, Rac1, and Cdc42 are molecular switches that regulate cell migration, as well as cadherin-mediated adhesion and cell–cell junction formation, through the manipulation of the actin cytoskeleton.[65–69] Interestingly, cadherin ligation can reciprocally modulate the signaling of Rho GTPases, and such regulation is thought to require p120 (reviewed in Ref. 27). Experiments to date suggest the existence of several different mechanisms by which p120 can affect Rho GTPases:

1. The first mechanism involves the direct association of p120 with Rho GTPases. When not bound to cadherins, p120 acts as a guanine-nucleotide

dissociation inhibitor, and directly binds to and suppresses RhoA activity.[36,70] At least two p120 domains are essential for a stable interaction with RhoA and inhibition of its activity: a central polybasic region termed "ΔRho" (amino acids 622–628 in human p120 isoform 1A)[36,71] and an N-terminal region (including amino acids 131–156).[72] The interaction of p120 with RhoA is further stabilized by tyrosine phosphorylation of p120 at Y217 and Y228 by Src, or destabilized by phosphorylation at Y112 by Fyn.[36] In addition to RhoA, p120 can also associate with Rac1 and especially Rac1b, an alternatively spliced constitutively active Rac1 isoform and this promotes cell motility (see below).[73] Unlike the inhibition of RhoA, this interaction does not interfere with the intrinsic GTPase activity of either Rac1 or Rac1b.

2. The second mechanism involves the association of p120 with Rho GEFs or Rho GAPs, which regulate the activation or inactivation of downstream Rho GTPases. At nascent AJs, cadherin-associated p120 interacts locally with Vav2 to activate Rac1 and Cdc42 to promote reorganization of the actin cytoskeleton.[11,50] It is now clear that association with Rho regulatory proteins in addition to Rho GTPases is a general feature of the p120 family of proteins. The p120 family member p0071 has been shown to associate with the Rho GEF Ect-2 to regulate cytokinesis,[74] while δ-catenin binds p190 RhoGEF.[75] p120 is also responsible for the cortical localization of p190 RhoGAP, which is activated by Src signaling and mediates Rac-mediated suppression of RhoA.[76]
3. A third mechanism of action is p120 interaction with Rho GTPase effector proteins. p120 can associate with and recruit Rho-associated protein kinase 1 to sites of cell–cell contact.[77] Additionally, cytosolic p120 associates directly with and inhibits the activity of myosin phosphatase Rho-interacting protein (Mrip), an antagonist of Rho/Rock function and ROCK activation.[78] While questions remain, the data suggest that p120 family members act as scaffolds to promote the association of positive or negative regulators with Rho GTPases and their subsequent effector proteins at particular subcellular sites depending on contextual signals. Consistent with this hypothesis, the recruitment of Rac1 to newly formed cell–cell contacts and the accumulation of the Rac1 effector PAK at these sites require the cadherin JMD and p120 association.[79]

Three things make the picture more complicated: (1) p120 isoforms differ in their ability to regulate Rho GTPases. Expression of the mesenchymal p120 isoform 1 inhibits Rho activity; expression of the epithelial p120 isoform 3 has no effect, while isoform 4 activates RhoA.[71] (2) Since the stability of microtubules can affect the activities of Rho GTPases via microtubule-binding proteins such as GEF-H1,[80] the ability of p120 to associate with and stabilize microtubules[13–16] can indirectly influence Rho GTPases. (3) The effects of p120 on Rho GTPase activation/inhibition might depend on the cell context, and specifically the type of cadherins that a cell expresses.

3.3. Nuclear signaling

Another signaling function of p120 is the regulation of gene transcription, particularly the convergent regulation of canonical Wnt/β-catenin target genes such as cyclin D1 and Wnt-11. Regulation of transcription by p120 occurs through its interaction with the transcriptional repressor Kaiso (reviewed in Refs. 81,82). Kaiso recognizes two DNA motifs: a sequence-specific DNA consensus site (5′CTGCNA3′) known as a Kaiso binding sequence or KBS[83] and methylated CpG dinucleotides.[18,83] Many Wnt/β-catenin target genes contain the sequence-specific KBS in their promoters and Kaiso acts as a transcriptional repressor of these genes, antagonizing β-catenin-mediated transcriptional activation.[84,85] p120 interacts with Kaiso's zinc finger domain to prevent Kaiso–DNA binding. Thus, p120 antagonizes Kaiso's transcriptional repression, facilitating in this way Wnt/β-catenin signaling. However, Kaiso exhibits even higher affinity for the methylated 5′TCTCGCGAGA3′ palindromic motif, which is responsible for the recruitment of the SMRT (silencing mediator of retinoic acid and thyroid hormone receptor) corepressor to promoter regions of proadipogenic genes,[86] raising the possibility that p120 binding to Kaiso may also affect adipogenesis.

The signals that govern the p120–Kaiso interaction and their combined regulation of transcription are largely unclear. However, a recent study in *Xenopus* indicates that p120 is stabilized by Frodo, a downstream target of Wnt/Dsh signaling.[85] This stabilization resulted in the inhibition of Kaiso-mediated repression of Wnt/β-catenin target genes. Consistent with this, Wnt3a signaling controls the transcriptional repressor activity of Kaiso via the CK1ε-mediated phosphorylation of p120.[87] Thus, p120's transcriptional activity might be regulated by Wnt signaling: another point of signaling convergence between p120 and β-catenin. The signals that govern the cytoplasmic versus nuclear localization of Kaiso are also not well characterized, although it has been suggested that the cell's microenvironment plays a role.[88] As with p120's other roles in mediating cadherin stabilization and Rho GTPase signaling, cell context undoubtedly affects the ability of p120 to regulate transcription; the details regarding how remain to be elucidated. Finally, an interaction of nuclear p120 with the transcriptional repressor Glis2 (Gli-similar 2) was also reported, although its functional significance is still unclear.[89]

3.4. Regulation of cell motility by p120-Rho-GTPase signaling

Low-level p120 overexpression by a retroviral delivery system promotes cell motility.[12] This effect is mediated by Rho GTPases, and possibly their subsequent effects on integrin-mediated signaling after adhesion to the extracellular matrix (ECM).[90–92] The effects of p120 on cell migration depend on the specific cadherins expressed in the cells.[93] How mesenchymal cadherins promote cell migration is currently unclear, but it might involve the p120-mediated activation of Rac1 signaling and subsequent actin reorganization.[93–95] Inhibition of RhoA by cytoplasmic p120 in cells expressing mesenchymal cadherins might also promote migration by allowing the reorganization of the actin cytoskeleton.[93,96] In contrast, in cells expressing E-cadherin, p120 is important for the promigratory signaling of EGF or HGF, via a mechanism that involves the N-terminus of p120 isoform 1 and RhoA activation.[97] In breast cancer cells expressing E-cadherin, p120-mediated Rac1 activation was required for HER2/ErbB2-induced cell migration.[94] Therefore, it seems that p120 and Rho GTPase signaling are involved in the motile behavior of cells expressing mesenchymal cadherins or activated receptor tyrosine kinases (RTKs), but the specific mechanisms are not well defined.

3.5. p120 and microtubules

It has been shown that p120 interacts with the microtubule network in several ways, both directly and indirectly. This interaction supports a bidirectional functional role regulating (i) p120 trafficking in the cell, and (ii) microtubule bundling and tethering to the junctions. The ARM repeats of p120 interact directly with microtubules, and this interaction is mutually exclusive to E-cadherin binding.[13] p120 can also bind to microtubules indirectly via N-terminal sequences present in isoforms 1–3.[13,14,16] An interaction with conventional kinesin heavy chain controls p120 trafficking[13,14,98] and nuclear accumulation.[13,99] Dynein-mediated tethering of microtubules to the junctions enables delivery of p120 to these sites.[98]

Additionally, p120 promotes microtubule stabilization, bundling, and tethering to the junctions[13,15,100] via a novel binding partner called PLEKHA7.[16,58] The interaction is mediated by sequences within the

N-terminus of p120 isoform 1. p120 and PLEKHA7 form a complex at the AJs that connects to the microtubule network through the microtubule minus end capping protein Nezha and the KIFC3 motor.[16] This p120–PLEKHA7–Nezha interaction links E-cadherin–p120 complexes to the minus ends of noncentrosomal microtubules, thus stabilizing AJs.[16] Importantly, prior studies have shown that noncentrosomal microtubule organization is essential for AJ symmetry and planar polarity in *Drosophila*,[101] while in mammalian MDCK cells, vertically aligned noncentrosomal polyglutamylated microtubules are required for polarized vesicle transport and the establishment of epithelial cell polarity.[102] Interestingly, the interaction of p120 with PLEKHA7 occurs selectively at mature AJs in polarized Caco2 cells, and not at perijunctional cadherin puncta.[16] Overall, p120–microtubule cross talk is essential in mediating both the intracellular localization of p120 and in stabilizing the microtubule–junctional architecture.

4. p120 IN CANCER

Normal adherent cells inhibit their growth and migration when they start to contact each other. These properties are progressively lost in tumor cells, contributing to increased rates of cell proliferation and migration. These findings imply that adhesion-triggered signaling events regulate both cell growth and motility. Based on its ability to regulate E-cadherin function, p120 is expected to act as a tumor suppressor by stabilizing junctions. Surprisingly, recent studies show that signaling events downstream of p120 and cadherins are crucial for the AIG of tumor cells as well as for Src-mediated transformation.[19,20] Therefore, p120 exerts both protumorigenic and antitumorigenic functions, an intriguing behavior (Fig. 18.2). Despite many remaining questions that will form the basis for future investigations, several important factors known to affect p120 behavior in cancer are discussed here.

4.1. p120 and cadherin switch in tumor growth and progression

During gastrulation, ingressing epithelial cells in the ectoderm undergo EMT, which allows them to adopt new cell fates, migrate through the basement membrane away from the epithelial monolayer, and establish new tissues. This normal developmental process overcomes the structural requirements for epithelial cell survival, and is characterized by a switch in the expression of epithelial versus mesenchymal markers, including cell–cell adhesion receptors of the cadherin family. During tumor

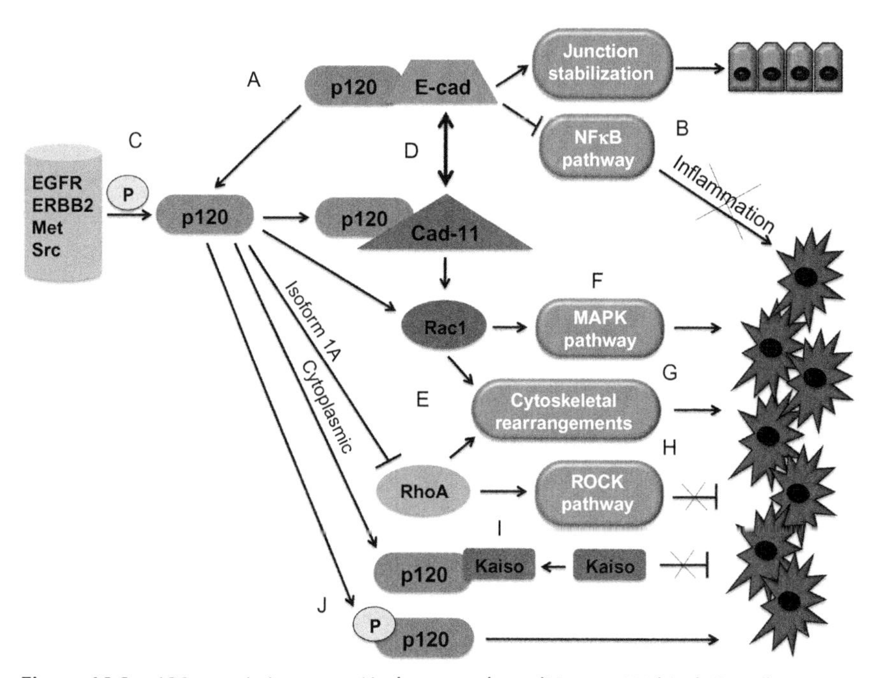

Figure 18.2 p120 catenin in cancer. Under normal conditions, p120 binds E-cadherin, stabilizes junctions (A), and suppresses inflammatory signaling pathways (B). However, upon phosphorylation by any of several tyrosine kinases (C) or after cadherin switch (D), p120 induces protumorigenic pathways, mainly by modulating Rho GTPase activities (E). These protumorigenic events include activation of the MAPK pathway (F), cytoskeletal rearrangements (G), and suppression of the ROCK pathway (H). In addition, p120 cytoplasmic localization results in sequestration of Kaiso in the cytoplasm and its inactivation (I), also relieving transcriptional suppression of oncogenic signals. Finally, p120 phosphorylation at Tyr or Ser sites is well correlated to tumorigenesis (J). (See Color Insert.)

progression, though, differentiated epithelial cells undergo a similar transition, lose their characteristic phenotype and acquire mesenchymal properties, becoming motile and invasive. A major event that influences this transformation is an essential change in the type of junctions that these cells form and the subsequent signaling events that are triggered. This change is due mainly to a switch in the type of cadherins expressed. In particular, while epithelial cells primarily express E-cadherin, which is a well-established tumor suppressor (see Chapter 14), expression of E-cadherin is decreased or diminished during EMT, whereas expression of other cadherins, such as N-cadherin and cadherin-11, is induced. The latter cadherins are collectively described as "mesenchymal" cadherins. This switch results in membrane ruffling, immature junctions, induced cell motility and invasiveness, all of which are hallmarks of tumorigenic transformation. Indeed, it has been well established in a variety of epithelial cancers, such as invasive lobular breast carcinoma and diffuse-type gastric adenocarcinoma, that loss of E-cadherin expression is correlated with increased infiltrative growth and tumor cell motility (reviewed in Refs. 103,104) (see Chapter 15). Several signaling pathways and mediators of this process have been described. For example, E-cadherin engagement during cell–cell adhesion regulates the levels of the cyclin kinase inhibitor p27[105,106] and suppresses signaling from a variety of RTKs, including the EGF receptor (EGFR) and c-Met,[107] resulting in decreased rates of cell growth. In addition, cadherin-mediated cell–cell adhesion suppresses cell migration by regulating the activity of Rho family GTPases via p120.[93]

Since p120 stabilizes E-cadherin complexes, it is expected that it will exert antitumorigenic activities (Fig. 18.2). In several cases of p120 deletion *in vivo*, the main effect was severe epithelial barrier malfunction and induction of inflammation, via RhoA and NFκB activation in the skin,[108] or by accumulation of COX-2-expressing neutrophils in the colon.[47,109] Solid evidence of the tumor suppressor function of p120 *in vivo* came from studies on the salivary gland, where p120 deletion resulted in intraepithelial neoplasia, due to severe adhesion defects and E-cadherin destabilization.[48] The tumor suppressor function of p120 was further established recently: targeted *in vivo* deletion of p120 resulted in invasive squamous neoplastic lesions in the oral cavity, esophagus, and forestomach.[110] p120 deletion was again accompanied by induction of proinflammatory events (Fig. 18.2), such as activation of NFκB, Akt, and Stat-3, and production of the granulocyte macrophage colony-stimulating factor, the macrophage colony-stimulating factor, monocyte chemotactic protein-1, and tumor necrosis factor α from tumor-derived cells. These events were accompanied by desmoplasia and accumulation of immature myeloid cells, both of which are hallmarks of the development of a tumorigenic microenvironment. These findings demonstrate *in vivo* the antitumorigenic and antiinflammatory properties of p120, which are coupled to stabilization of E-cadherin in the epithelial cells.

In agreement with the animal data, several immunohistochemical studies report that p120 expression is downregulated in certain tumors.[111–115] Also, p120 is transcriptionally downregulated by FOXC2 in non-small cell lung cancer.[116] Downregulation of p120 is expected to decrease E-cadherin-mediated adhesion, and possibly relieve inhibition of RTKs. The combined loss of cell adhesiveness, increased RTK signaling, and altered Rho signaling might have profound effects on the aggressiveness of these tumors. Consistent with this, p120 knockdown in lung cancer cell lines enhanced their migration and invasiveness.[117] The correlation of p120 loss with increased tumor aggressiveness[113] also suggests that in a background of other cancer-related mutations, p120 loss might promote tumor progression and metastasis.

On the other hand, the cadherin switch during EMT also affects the subcellular localization of p120, influencing tumorigenic signaling events induced by growth factor receptors or mediated by Rho GTPases (Fig. 18.2). Early studies reported that p120 mislocalizes to the cytoplasm as a result of E-cadherin loss during tumor progression.[118,119] For example, during EMT in colon cancer cells, E-cadherin is gradually lost and p120 is mislocalized to the cytoplasm, where it inhibits Rho activity.[96] In this case, cytoplasmic p120 correlated strongly with later stages of tumor, lymph node metastasis, and reduced survival in a cohort of colorectal cancer cases.[96] Cytoplasmic p120 was also strongly correlated with late-stage pancreatic cancer.[120] Abnormal overexpression and cytoplasmic localization of p120 was associated with poor prognosis in lung squamous cell carcinoma and adenocarcinoma.[114] In ovarian cancer, expression of the gonadotropin-releasing hormone receptor promotes an E- to P-cadherin switch, p120 mislocalization, induction of Rac1 and Cdc42 activities, and tumor invasiveness.[95] A potential mechanism by which p120 can directly affect the cadherin switch was also proposed. Cells that were forced to overexpress R-cadherin had reduced levels of E- and P-cadherin. This was the result of competition between the different cadherins for p120 binding. R-Cadherin overexpression resulted in sequestration of p120 to R-cadherin, which led to E-cadherin destabilization, endocytosis, and degradation.[121]

p120 is not only affected by a cadherin switch and E-cadherin loss, but is also essential in influencing downstream pathways toward induction of cell motility and invasiveness.[19,93] p120-dependent regulation of Rho GTPase signaling is one example (Fig. 18.2). Upon loss of E-cadherin in a murine model of infiltrating lobular carcinoma of the breast (ILC), p120 activated ROCK and induced tumor growth via interaction and inhibition of Mrip, an antagonist of Rho/ROCK function.[78] Experiments on breast and kidney cancer cell lines have revealed that loss of E-cadherin alone is not sufficient for gaining invasiveness. An additional requirement is an increase in mesenchymal cadherin expression and p120/mesenchymal cadherin-associated Rac1 signaling.[93] Therefore, cadherin switching during EMT and deregulated Rho GTPase signaling promote cell migration and tumor growth.

Surprisingly, in some cases, p120 exerts tumorigenic activities even when E-cadherin is present. For example, ERBB2 overexpression in normal mammary epithelial MCF10A cells activates Rac1 and Cdc42 and induces cell migration and invasion in a p120-dependent manner, without any changes in E-cadherin expression.[94] Also in this study, p120 overexpression in an ERBB2-overexpressing but weakly metastatic breast cancer cell line, BT474 cells, dramatically induced its ability to metastasize *in vivo*.[94] Rac1b, a constitutively active isoform of Rac1, required p120 binding in order to promote cell motility in E-cadherin-positive mouse mammary epithelial cells.[73] In another recent example, p120 was shown to be essential in the progression of a highly lethal form of cancer, inflammatory breast cancer (IBC).[122] In IBC, overexpression of the translation initiation factor eIF4GI strongly increases p120 synthesis via its action on a specific IRES in the p120 mRNA. As a result, E-cadherin is stabilized substantially, which favors IBC progression, likely by supporting the formation of the characteristic emboli of IBC. Therefore, in IBC, p120–E-cadherin complexes are overexpressed and protumorigenic. Furthermore, aberrantly cytoplasmic p120 in invasive breast carcinomas that express P-cadherin correlated with poor survival, especially when E-cadherin was also present.[123] Similarly, in cases of ILC in which E-cadherin is still expressed, p120 is cytoplasmic, suggesting misregulation of the cadherin–p120 complex.[124] p120 is also required for the collective migration of A431 cells in a three-dimensional medium (matrigel or collagen) upon EGFR stimulation.[125] This function of p120 is attributed to the cadherin stabilization required for the cells to maintain their junctions in order to collectively migrate and invade.

Overall, although there is solid evidence that p120 exerts both protumorigenic and antitumorigenic functions depending on the type of cadherin present, there are cases where association of p120 with a certain cadherin produces different outcomes, indicating that the full mechanistic details of these functions remain to be elucidated.

4.2. p120 isoform-specific tumorigenic events

An extra layer of complexity in identifying the role of p120 in cancer progression involves the relative expression of p120 isoforms (Fig. 18.2). We have reported that full-length p120 and N-terminally truncated p120 isoforms differentially affect Rho GTPase activities, cell migration, and tumor cell invasion, while a p120 isoform switch during tumor progression predicts metastatic disease.[71] More specifically, the long isoform 1A is capable of promoting invasiveness, whereas isoform 4A cannot, and expression of isoform 1A was found to correlate with renal cancer micrometastasis. Although both isoforms are able to bind RhoA downstream of HGF and Met signaling, p120-mediated RhoA inhibition requires the cooperative binding of RhoA to a central binding domain of p120 and to the N-terminus of the long isoform 1A.[71] By efficiently suppressing the RhoA–ROCK pathway and by activating Rac1, isoform 1A of p120 induces the migration and invasiveness of breast cancer cells. Furthermore, examination of p120 isoform expression across a series of breast cancer cell lines showed that the more invasive, luminal-type cells show strong expression of the large isoforms 1 and 2, whereas isoform 3 is almost universally expressed.[126] Overexpression of isoform 3A attenuated proliferation of colon cancer HT-29 cells,[127] in agreement with the aforementioned observations, although a role for this isoform in aberrant mitosis and induction of polyploidy was also proposed.[127]

A new transcription factor named Zeppo 1 was shown to deregulate junction formation and promote lung metastasis of breast cancer by inducing the expression of the protumorigenic isoform 1 of p120.[128] Other studies also showed that overexpression and mislocalization of p120 isoform 1 strongly correlates with a poor outcome in lung cancer.[129]

Interestingly, both isoforms 1 and 3 are able to induce cell proliferation via different pathways that ultimately affect cyclin D1.[130] In particular, p120 isoform 1A can induce cyclin D1 via increased expression and signaling by β-catenin, whereas isoform 3A does it by decreasing the levels of Kaiso in the nucleus, thereby relieving cyclin D1 repression. This study highlights the complexity in attributing protumorigenic functions to particular p120 isoforms, indicating that these functions might be context-dependent and are clearly not yet fully delineated.

4.3. p120 phosphorylation correlates with cancer occurrence

p120 isoform 1 is phosphorylated by Src family kinases at a number of tyrosines within its N-terminus, including Y96, Y112, Y228, Y257, Y280, Y291, Y296, and Y302.[33] EGFR also phosphorylates p120 at Y228, without Src being the necessary intermediate[131] (Fig. 18.2). A number of studies have shown that p120 phosphorylation correlates with tumorigenic events. In particular, phosphorylation of p120 at Y228 correlates positively with occurrence of oral squamous cancer.[132] Phosphorylation of p120 at several tyrosine and serine sites is inversely related to cadherin activation and adhesion strengthening.[57]

In addition to tyrosine phosphorylation, serine phosphorylation of p120, which is mainly due to PKC action,[60] has been associated with protumorigenic events. Phosphorylation of p120 isoform 3 at S288 increased Kaiso binding and promoted lung cancer cell invasion.[75] Furthermore, Wnt signaling induced phosphorylation of p120 at S268 and S269, dissociating it from E-cadherin and subsequently promoting Kaiso sequestration and activation of downstream Wnt signaling events.[87] Overall, the evidence points toward a protumorigenic role for p120 phosphorylation, but the mechanistic details of this action are still unclear.

4.4. Regulation of AIG by p120

A hallmark of cells that are transformed with oncogenes is their capacity for AIG. The acquisition of AIG correlates strongly with tumorigenicity *in vivo*, and is thought to be a key transition in tumor progression associated with aggressive disease. In essence, AIG allows epithelial cells to grow in the wrong environmental context. Loss of dependence on the basement membrane for survival and progression through the cell cycle is closely related to the induction of tumor growth.

When normal epithelial cells are grown in suspension, they undergo apoptosis by a process termed anoikis. There is evidence that suggests a link between cadherin-mediated adhesion and anoikis. In the colonic epithelium, cadherin-mediated cell–cell junctions prevented anoikis through the activation of Src- and PI3K-dependent signaling pathways.[133,134] Two separate studies argue that (i) anoikis can be prevented by maintaining cell–cell junctions, which are often lost upon cell suspension, and that (ii) anoikis is induced by the EGF-stimulated disruption of enterocytic cell–cell contacts upon detachment from the ECM.[135] However, cellular context and epigenetic modifications might play a critical role in the overall effect of E-cadherin depletion on anoikis. For example, in the context of p53 mutations, loss of E-cadherin in mammary epithelial cells leads to ILC, due in part to anoikis resistance.[78] Overall, it is likely that intercellular adhesion-related signaling events can substitute for focal adhesion signaling and drive the transformed growth of tumor cells. Following up on the original observation that p120 is a transformation-relevant Src substrate, recent work has established that p120 is an essential mediator of the Src- and Rac1-induced AIG[20] (Fig. 18.2). p120-mediated suppression of the Rho–ROCK pathway was necessary for AIG induction driven by the overexpression of either Rac1 or Src.[20] The observation that ROCK is recruited to E-cadherin complexes by binding directly to p120[77] further establishes the connection between p120, ROCK, and AIG.

ECM and focal adhesions cooperate with growth factors to induce sustained activation of ERK/MAP kinase, increased expression of cyclin D1, and cell cycle progression.[136] Interestingly, p120 is essential in bypassing the requirement for ECM in the induction of the ERK/MAP kinase pathway, and for cyclin D1 activation of tumor cells grown in soft agar. This cyclin D1 activation occurred in the presence of cadherin 11 but not of E-cadherin and was mediated by Rac1 activation.[19] The data supported a model in which p120 normally promotes the stabilization of E-cadherin complexes and their ability to suppress Ras and Rac1 signaling, thus blocking cell growth. However, upon E-cadherin loss during EMT or tumor progression, the negative regulation of Ras is relieved, and endogenous p120 induces Rac1 activation, constitutive activation of the ERK/MAP kinase signaling pathway, cell cycle progression, and AIG.[19]

Therefore, it is important to note that anchorage independence in epithelial cells is a complicated process requiring both suppression of apoptosis and progression through the cell cycle via signaling events triggered by cell–cell contact. Despite many remaining questions that will form the basis for future investigations, the current data indicate that intercellular adhesion and p120 can regulate the actin cytoskeleton and the activity of mitogenic pathways under conditions of anchorage independence.

REFERENCES

1. Reynolds AB, Roesel DJ, Kanner SB, Parsons JT. Transformation-specific tyrosine phosphorylation of a novel cellular protein in chicken cells expressing oncogenic variants of the avian cellular *src* gene. *Mol Cell Biol* 1989;**9**:629–38.
2. Reynolds AB. p120-catenin: past and present. *Biochim Biophys Acta* 2007;**1773**:2–7.
3. Kanner SB, Reynolds AB, Vines RR, Parsons JT. Monoclonal antibodies to individual tyrosine-phosphorylated protein substrates of oncogene-encoded tyrosine kinases. *Proc Natl Acad Sci USA* 1990;**87**:3328–32.
4. Reynolds AB, Daniel J, McCrea PD, Wheelock MJ, Wu J, Zhang Z. Identification of a new catenin: the tyrosine kinase substrate p120cas associates with E-cadherin complexes. *Mol Cell Biol* 1994;**14**:8333–42.
5. Staddon JM, Smales C, Schulze C, Esch FS, Rubin LL. p120, a p120-related protein (p100), and the cadherin/catenin complex. *J Cell Biol* 1995;**130**:369–81.
6. Shibamoto S, Hayakawa M, Takeuchi K, Hori T, Miyazawa K, Kitamura N, et al. Association of p120, a tyrosine kinase substrate, with E-cadherin/catenin complexes. *J Cell Biol* 1995;**128**:949–57.
7. Ireton RC, Davis MA, van Hengel J, Mariner DJ, Barnes K, Thoreson MA, et al. A novel role for p120 catenin in E-cadherin function. *J Cell Biol* 2002;**159**:465–76.
8. Davis MA, Ireton RC, Reynolds AB. A core function for p120-catenin in cadherin turnover. *J Cell Biol* 2003;**163**:525–34.
9. Xiao K, Allison DF, Buckley KM, Kottke MD, Vincent PA, Faundez V, et al. Cellular levels of p120 catenin function as a set point for cadherin expression levels in microvascular endothelial cells. *J Cell Biol* 2003;**163**:535–45.

10. Anastasiadis PZ, Moon SY, Thoreson MA, Mariner DJ, Crawford HC, Zheng Y, et al. Inhibition of RhoA by p120 catenin. *Nat Cell Biol* 2000;**2**:637–44.
11. Noren NK, Liu BP, Burridge K, Kreft B. p120 catenin regulates the actin cytoskeleton via Rho family GTPases. *J Cell Biol* 2000;**150**:567–80.
12. Grosheva I, Shtutman M, Elbaum M, Bershadsky AD. p120 catenin affects cell motility via modulation of activity of Rho-family GTPases: a link between cell-cell contact formation and regulation of cell locomotion. *J Cell Sci* 2001;**114**:695–707.
13. Yanagisawa M, Kaverina IN, Wang A, Fujita Y, Reynolds AB, Anastasiadis PZ. A novel interaction between kinesin and p120 modulates p120 localization and function. *J Biol Chem* 2004;**279**:9512–21.
14. Chen X, Kojima S, Borisy GG, Green KJ. p120 catenin associates with kinesin and facilitates the transport of cadherin-catenin complexes to intercellular junctions. *J Cell Biol* 2003;**163**:547–57.
15. Franz CM, Ridley AJ. p120 catenin associates with microtubules: inverse relationship between microtubule binding and Rho GTPase regulation. *J Biol Chem* 2004;**279**:6588–94.
16. Meng W, Mushika Y, Ichii T, Takeichi M. Anchorage of microtubule minus ends to adherens junctions regulates epithelial cell-cell contacts. *Cell* 2008;**135**:948–59.
17. Daniel JM, Reynolds AB. The catenin p120(ctn) interacts with Kaiso, a novel BTB/POZ domain zinc finger transcription factor. *Mol Cell Biol* 1999;**19**:3614–23.
18. Prokhortchouk A, Hendrich B, Jorgensen H, Ruzov A, Wilm M, Georgiev G, et al. The p120 catenin partner Kaiso is a DNA methylation-dependent transcriptional repressor. *Genes Dev* 2001;**15**:1613–8.
19. Soto E, Yanagisawa M, Marlow LA, Copland JA, Perez EA, Anastasiadis PZ. p120 catenin induces opposing effects on tumor cell growth depending on E-cadherin expression. *J Cell Biol* 2008;**183**:737–49.
20. Dohn MR, Brown MV, Reynolds AB. An essential role for p120-catenin in Src- and Rac1-mediated anchorage-independent cell growth. *J Cell Biol* 2009;**184**:437–50.
21. Anastasiadis PZ, Reynolds AB. The p120 catenin family: complex roles in adhesion, signaling and cancer. *J Cell Sci* 2000;**113**(Pt. 8):1319–34.
22. Hatzfeld M. The armadillo family of structural proteins. *Int Rev Cytol* 1999;**186**:179–224.
23. Hatzfeld M. Plakophilins: multifunctional proteins or just regulators of desmosomal adhesion? *Biochim Biophys Acta* 2007;**1773**:69–77.
24. Laura RP, Witt AS, Held HA, Gerstner R, Deshayes K, Koehler MF, et al. The Erbin PDZ domain binds with high affinity and specificity to the carboxyl termini of delta-catenin and ARVCF. *J Biol Chem* 2002;**277**:12906–14.
25. Zhang Y, Yeh S, Appleton BA, Held HA, Kausalya PJ, Phua DC, et al. Convergent and divergent ligand specificity among PDZ domains of the LAP and zonula occludens (ZO) families. *J Biol Chem* 2006;**281**:22299–311.
26. Ishiyama N, Lee SH, Liu S, Li GY, Smith MJ, Reichardt LF, et al. Dynamic and static interactions between p120 catenin and E-cadherin regulate the stability of cell-cell adhesion. *Cell* 2010;**141**:117–28.
27. Anastasiadis PZ. p120-ctn: a nexus for contextual signaling via Rho GTPases. *Biochim Biophys Acta* 2007;**1773**:34–46.
28. Aho S, Rothenberger K, Uitto J. Human p120ctn catenin: tissue-specific expression of isoforms and molecular interactions with BP180/type XVII collagen. *J Cell Biochem* 1999;**73**:390–9.
29. Keirsebilck A, Bonne S, Staes K, van Hengel J, Nollet F, Reynolds A, et al. Molecular cloning of the human p120ctn catenin gene (CTNND1): expression of multiple alternatively spliced isoforms. *Genomics* 1998;**50**:129–46.

30. Pieters T, van Roy F, van Hengel J. Functions of p120ctn isoforms in cell-cell adhesion and intracellular signaling. *Front Biosci* 2012;**17**:1669–94.
31. Mariner DJ, Wang J, Reynolds AB. ARVCF localizes to the nucleus and adherens junction and is mutually exclusive with p120(ctn) in E cadherin complexes. *J Cell Sci* 2000;**113**:1481–90.
32. Lu Q, Paredes M, Medina M, Zhou JH, Cavallo R, Peifer M, et al. Delta-catenin, an adhesive junction-associated protein which promotes cell scattering. *J Cell Biol* 1999;**144**:519–32.
33. Mariner DJ, Anastasiadis P, Keilhack H, Bohmer FD, Wang J, Reynolds AB. Identification of Src phosphorylation sites in the catenin p120ctn. *J Biol Chem* 2001;**276**:28006–13.
34. Xia X, Mariner DJ, Reynolds AB. Adhesion-associated and PKC-modulated changes in serine/threonine phosphorylation of p120-catenin. *Biochemistry* 2003;**42**:9195–204.
35. Fukumoto Y, Shintani Y, Reynolds AB, Johnson KR, Wheelock MJ. The regulatory or phosphorylation domain of p120 catenin controls E-cadherin dynamics at the plasma membrane. *Exp Cell Res* 2008;**314**:52–67.
36. Castano J, Solanas G, Casagolda D, Raurell I, Villagrasa P, Bustelo XR, et al. Specific phosphorylation of p120-catenin regulatory domain differently modulates its binding to RhoA. *Mol Cell Biol* 2007;**27**:1745–57.
37. Mo YY, Reynolds AB. Identification of murine p120 isoforms and heterogeneous expression of p120cas isoforms in human tumor cell lines. *Cancer Res* 1996;**56**:2633–40.
38. Ohkubo T, Ozawa M. The transcription factor Snail downregulates the tight junction components independently of E-cadherin downregulation. *J Cell Sci* 2004;**117**:1675–85.
39. Shapiro IM, Cheng AW, Flytzanis NC, Balsamo M, Condeelis JS, Oktay MH, et al. An EMT-driven alternative splicing program occurs in human breast cancer and modulates cellular phenotype. *PLoS Genet* 2011;**7**:e1002218.
40. Eger A, Stockinger A, Schaffhauser B, Beug H, Foisner R. Epithelial mesenchymal transition by c-Fos estrogen receptor activation involves nuclear translocation of beta-catenin and upregulation of beta-catenin/lymphoid enhancer binding factor-1 transcriptional activity. *J Cell Biol* 2000;**148**:173–88.
41. Warzecha CC, Sato TK, Nabet B, Hogenesch JB, Carstens RP. ESRP1 and ESRP2 are epithelial cell-type-specific regulators of FGFR2 splicing. *Mol Cell* 2009;**33**:591–601.
42. Carnahan RH, Rokas A, Gaucher EA, Reynolds AB. The molecular evolution of the p120-catenin subfamily and its functional associations. *PLoS One* 2010;**5**:e15747.
43. Zhao ZM, Reynolds AB, Gaucher EA. The evolutionary history of the catenin gene family during metazoan evolution. *BMC Evol Biol* 2011;**11**:198.
44. Myster SH, Cavallo R, Anderson CT, Fox DT, Peifer M. Drosophila p120catenin plays a supporting role in cell adhesion but is not an essential adherens junction component. *J Cell Biol* 2003;**160**:433–49.
45. Pettitt J, Cox EA, Broadbent ID, Flett A, Hardin J. The Caenorhabditis elegans p120 catenin homologue, JAC-1, modulates cadherin-catenin function during epidermal morphogenesis. *J Cell Biol* 2003;**162**:15–22.
46. Magie CR, Pinto-Santini D, Parkhurst SM. Rho1 interacts with p120ctn and alpha-catenin, and regulates cadherin-based adherens junction components in Drosophila. *Development* 2002;**129**:3771–82.
47. Smalley-Freed WG, Efimov A, Burnett PE, Short SP, Davis MA, Gumucio DL, et al. p120-catenin is essential for maintenance of barrier function and intestinal homeostasis in mice. *J Clin Invest* 2010;**120**:1824–35.
48. Davis MA, Reynolds AB. Blocked acinar development, E-cadherin reduction, and intraepithelial neoplasia upon ablation of p120-catenin in the mouse salivary gland. *Dev Cell* 2006;**10**:21–31.

49. Oas RG, Xiao K, Summers S, Wittich KB, Chiasson CM, Martin WD, et al. p120-catenin is required for mouse vascular development. *Circ Res* 2010;**106**:941–51.
50. Fang X, Ji H, Kim SW, Park JI, Vaught TG, Anastasiadis PZ, et al. Vertebrate development requires ARVCF and p120 catenins and their interplay with RhoA and Rac. *J Cell Biol* 2004;**165**:87–98.
51. Elia LP, Yamamoto M, Zang K, Reichardt LF. p120 catenin regulates dendritic spine and synapse development through Rho-family GTPases and cadherins. *Neuron* 2006;**51**:43–56.
52. Puech A, Saint-Jore B, Merscher S, Russell RG, Cherif D, Sirotkin H, et al. Normal cardiovascular development in mice deficient for 16 genes in 550 kb of the velocardiofacial/DiGeorge syndrome region. *Proc Natl Acad Sci USA* 2000;**97**:10090–5.
53. Israely I, Costa RM, Xie CW, Silva AJ, Kosik KS, Liu X. Deletion of the neuron-specific protein delta-catenin leads to severe cognitive and synaptic dysfunction. *Curr Biol* 2004;**14**:1657–63.
54. Xiao K, Oas RG, Chiasson CM, Kowalczyk AP. Role of p120-catenin in cadherin trafficking. *Biochim Biophys Acta* 2007;**1773**:8–16.
55. Hoshino T, Sakisaka T, Baba T, Yamada T, Kimura T, Takai Y. Regulation of E-cadherin endocytosis by nectin through afadin, Rap1, and p120ctn. *J Biol Chem* 2005;**280**:24095–103.
56. Balzac F, Avolio M, Degani S, Kaverina I, Torti M, Silengo L, et al. E-Cadherin endocytosis regulates the activity of Rap1: a traffic light GTPase at the crossroads between cadherin and integrin function. *J Cell Sci* 2005;**118**:4765–83.
57. Petrova YI, Spano MM, Gumbiner BM. Conformational epitopes at cadherin calcium-binding sites and p120-catenin phosphorylation regulate cell adhesion. *Mol Biol Cell* 2012;**23**:2092–108.
58. Pulimeno P, Bauer C, Stutz J, Citi S. PLEKHA7 is an adherens junction protein with a tissue distribution and subcellular localization distinct from ZO-1 and E-cadherin. *PLoS One* 2010;**5**:e12207.
59. Sato K, Watanabe T, Wang S, Kakeno M, Matsuzawa K, Matsui T, et al. Numb controls E-cadherin endocytosis through p120 catenin with aPKC. *Mol Biol Cell* 2011;**22**:3103–19.
60. Alema S, Salvatore AM. p120 catenin and phosphorylation: mechanisms and traits of an unresolved issue. *Biochim Biophys Acta* 2007;**1773**:47–58.
61. Fujita Y, Krause G, Scheffner M, Zechner D, Leddy HE, Behrens J, et al. Hakai, a c-Cbl-like protein, ubiquitinates and induces endocytosis of the E-cadherin complex. *Nat Cell Biol* 2002;**4**:222–31.
62. Potter MD, Barbero S, Cheresh DA. Tyrosine phosphorylation of VE-cadherin prevents binding of p120- and beta-catenin and maintains the cellular mesenchymal state. *J Biol Chem* 2005;**280**:31906–12.
63. Gavard J, Patel V, Gutkind JS. Angiopoietin-1 prevents VEGF-induced endothelial permeability by sequestering Src through mDia. *Dev Cell* 2008;**14**:25–36.
64. Casagolda D, Del Valle-Perez B, Valls G, Lugilde E, Vinyoles M, Casado-Vela J, et al. A p120-catenin-CK1epsilon complex regulates Wnt signaling. *J Cell Sci* 2010;**123**:2621–31.
65. Hall A. Rho GTPases and the actin cytoskeleton. *Science* 1998;**279**:509–14.
66. Braga VM, Machesky LM, Hall A, Hotchin NA. The small GTPases Rho and Rac are required for the establishment of cadherin-dependent cell-cell contacts. *J Cell Biol* 1997;**137**:1421–31.
67. Jou TS, Nelson WJ. Effects of regulated expression of mutant RhoA and Rac1 small GTPases on the development of epithelial (MDCK) cell polarity. *J Cell Biol* 1998;**142**:85–100.

68. Takaishi K, Sasaki T, Kotani H, Nishioka H, Takai Y. Regulation of cell-cell adhesion by rac and rho small G proteins in MDCK cells. *J Cell Biol* 1997;**139**:1047–59.
69. Kodama A, Takaishi K, Nakano K, Nishioka H, Takai Y. Involvement of Cdc42 small G protein in cell-cell adhesion, migration and morphology of MDCK cells. *Oncogene* 1999;**18**:3996–4006.
70. Anastasiadis PZ, Reynolds AB. Regulation of Rho GTPases by p120-catenin. *Curr Opin Cell Biol* 2001;**13**:604–10.
71. Yanagisawa M, Huveldt D, Kreinest P, Lohse CM, Cheville JC, Parker AS, et al. A p120 catenin isoform switch affects Rho activity, induces tumor cell invasion, and predicts metastatic disease. *J Biol Chem* 2008;**283**:18344–54.
72. Lee SH, Peng IF, Ng YG, Yanagisawa M, Bamji SX, Elia LP, et al. Synapses are regulated by the cytoplasmic tyrosine kinase Fer in a pathway mediated by p120catenin, Fer, SHP-2, and beta-catenin. *J Cell Biol* 2008;**183**:893–908.
73. Orlichenko L, Geyer R, Yanagisawa M, Khauv D, Radisky ES, Anastasiadis PZ, et al. The 19-amino acid insertion in the tumor-associated splice isoform Rac1b confers specific binding to p120 catenin. *J Biol Chem* 2010;**285**:19153–61.
74. Wolf A, Keil R, Gotzl O, Mun A, Schwarze K, Lederer M, et al. The armadillo protein p0071 regulates Rho signalling during cytokinesis. *Nat Cell Biol* 2006;**8**:1432–40.
75. Kim H, Han JR, Park J, Oh M, James SE, Chang S, et al. Delta-catenin-induced dendritic morphogenesis. An essential role of p190RhoGEF interaction through Akt1-mediated phosphorylation. *J Biol Chem* 2008;**283**:977–87.
76. Wildenberg GA, Dohn MR, Carnahan RH, Davis MA, Lobdell NA, Settleman J, et al. p120-catenin and p190RhoGAP regulate cell-cell adhesion by coordinating antagonism between Rac and Rho. *Cell* 2006;**127**:1027–39.
77. Smith AL, Dohn MR, Brown MV, Reynolds AB. Association of Rho-associated protein kinase 1 with E-cadherin complexes is mediated by p120-catenin. *Mol Biol Cell* 2012;**23**:99–110.
78. Schackmann RC, van Amersfoort M, Haarhuis JH, Vlug EJ, Halim VA, Roodhart JM, et al. Cytosolic p120-catenin regulates growth of metastatic lobular carcinoma through Rock1-mediated anoikis resistance. *J Clin Invest* 2011;**121**:3176–88.
79. Goodwin M, Kovacs EM, Thoreson MA, Reynolds AB, Yap AS. Minimal mutation of the cytoplasmic tail inhibits the ability of E-cadherin to activate Rac but not phosphatidylinositol 3-kinase: direct evidence of a role for cadherin-activated Rac signaling in adhesion and contact formation. *J Biol Chem* 2003;**278**:20533–9.
80. Krendel M, Zenke FT, Bokoch GM. Nucleotide exchange factor GEF-H1 mediates cross-talk between microtubules and the actin cytoskeleton. *Nat Cell Biol* 2002;**4**:294–301.
81. Daniel JM. Dancing in and out of the nucleus: p120(ctn) and the transcription factor Kaiso. *Biochim Biophys Acta* 2007;**1773**:59–68.
82. van Roy FM, McCrea PD. A role for Kaiso-p120ctn complexes in cancer? *Nat Rev Cancer* 2005;**5**:956–64.
83. Daniel JM, Spring CM, Crawford HC, Reynolds AB, Baig A. The p120(ctn)-binding partner Kaiso is a bi-modal DNA-binding protein that recognizes both a sequence-specific consensus and methylated CpG dinucleotides. *Nucleic Acids Res* 2002;**30**:2911–9.
84. Kim SW, Park JI, Spring CM, Sater AK, Ji H, Otchere AA, et al. Non-canonical Wnt signals are modulated by the Kaiso transcriptional repressor and p120-catenin. *Nat Cell Biol* 2004;**6**:1212–20.
85. Park JI, Ji H, Jun S, Gu D, Hikasa H, Li L, et al. Frodo links Dishevelled to the p120-catenin/Kaiso pathway: distinct catenin subfamilies promote Wnt signals. *Dev Cell* 2006;**11**:683–95.

86. Raghav SK, Waszak SM, Krier I, Gubelmann C, Isakova A, Mikkelsen TS, et al. Integrative genomics identifies the corepressor SMRT as a gatekeeper of adipogenesis through the transcription factors C/EBPbeta and KAISO. *Mol Cell* 2012;**46**:335–50.
87. Del Valle-Perez B, Casagolda D, Lugilde E, Valls G, Codina M, Dave N, et al. Wnt controls the transcriptional activity of Kaiso through CK1epsilon-dependent phosphorylation of p120-catenin. *J Cell Sci* 2011;**124**:2298–309.
88. Soubry A, van Hengel J, Parthoens E, Colpaert C, Van Marck E, Waltregny D, et al. Expression and nuclear location of the transcriptional repressor Kaiso is regulated by the tumor microenvironment. *Cancer Res* 2005;**65**:2224–33.
89. Hosking CR, Ulloa F, Hogan C, Ferber EC, Figueroa A, Gevaert K, et al. The transcriptional repressor Glis2 is a novel binding partner for p120 catenin. *Mol Biol Cell* 2007;**18**:1918–27.
90. Clark EA, King WG, Brugge JS, Symons M, Hynes RO. Integrin-mediated signals regulated by members of the rho family of GTPases. *J Cell Biol* 1998;**142**:573–86.
91. Schwartz MA, Shattil SJ. Signaling networks linking integrins and rho family GTPases. *Trends Biochem Sci* 2000;**25**:388–91.
92. Sastry SK, Burridge K. Focal adhesions: a nexus for intracellular signaling and cytoskeletal dynamics. *Exp Cell Res* 2000;**261**:25–36.
93. Yanagisawa M, Anastasiadis PZ. p120 catenin is essential for mesenchymal cadherin-mediated regulation of cell motility and invasiveness. *J Cell Biol* 2006;**174**:1087–96.
94. Johnson E, Seachrist DD, DeLeon-Rodriguez CM, Lozada KL, Miedler J, Abdul-Karim FW, et al. HER2/ErbB2-induced breast cancer cell migration and invasion require p120 catenin activation of Rac1 and Cdc42. *J Biol Chem* 2010;**285**:29491–501.
95. Cheung LW, Leung PC, Wong AS. Cadherin switching and activation of p120 catenin signaling are mediators of gonadotropin-releasing hormone to promote tumor cell migration and invasion in ovarian cancer. *Oncogene* 2010;**29**:2427–40.
96. Bellovin DI, Bates RC, Muzikansky A, Rimm DL, Mercurio AM. Altered localization of p120 catenin during epithelial to mesenchymal transition of colon carcinoma is prognostic for aggressive disease. *Cancer Res* 2005;**65**:10938–45.
97. Cozzolino M, Stagni V, Spinardi L, Campioni N, Fiorentini C, Salvati E, et al. p120 catenin is required for growth factor-dependent cell motility and scattering in epithelial cells. *Mol Biol Cell* 2003;**14**:1964–77.
98. Ligon LA, Holzbaur EL. Microtubules tethered at epithelial cell junctions by dynein facilitate efficient junction assembly. *Traffic* 2007;**8**:808–19.
99. Roczniak-Ferguson A, Reynolds AB. Regulation of p120-catenin nucleocytoplasmic shuttling activity. *J Cell Sci* 2003;**116**:4201–12.
100. Ichii T, Takeichi M. p120-catenin regulates microtubule dynamics and cell migration in a cadherin-independent manner. *Genes Cells* 2007;**12**:827–39.
101. Harris TJ, Peifer M. aPKC controls microtubule organization to balance adherens junction symmetry and planar polarity during development. *Dev Cell* 2007;**12**:727–38.
102. Spiliotis ET, Hunt SJ, Hu Q, Kinoshita M, Nelson WJ. Epithelial polarity requires septin coupling of vesicle transport to polyglutamylated microtubules. *J Cell Biol* 2008;**180**:295–303.
103. Wijnhoven BP, Dinjens WN, Pignatelli M. E-Cadherin-catenin cell-cell adhesion complex and human cancer. *Br J Surg* 2000;**87**:992–1005.
104. Nollet F, Berx G, van Roy F. The role of the E-cadherin/catenin adhesion complex in the development and progression of cancer. *Mol Cell Biol Res Commun* 1999;**2**:77–85.
105. Motti ML, Califano D, Baldassarre G, Celetti A, Merolla F, Forzati F, et al. Reduced E-cadherin expression contributes to the loss of p27kip1-mediated mechanism of contact inhibition in thyroid anaplastic carcinomas. *Carcinogenesis* 2005;**26**:1021–34.

106. St. Croix B, Sheehan C, Rak JW, Florenes VA, Slingerland JM, Kerbel RS. E-Cadherin-dependent growth suppression is mediated by the cyclin- dependent kinase inhibitor p27(KIP1). *J Cell Biol* 1998;**142**:557–71.
107. Qian X, Karpova T, Sheppard AM, McNally J, Lowy DR. E-Cadherin-mediated adhesion inhibits ligand-dependent activation of diverse receptor tyrosine kinases. *EMBO J* 2004;**23**:1739–48.
108. Perez-Moreno M, Davis MA, Wong E, Pasolli HA, Reynolds AB, Fuchs E. p120-catenin mediates inflammatory responses in the skin. *Cell* 2006;**124**:631–44.
109. Smalley-Freed WG, Efimov A, Short SP, Jia P, Zhao Z, Washington MK, et al. Adenoma formation following limited ablation of p120-catenin in the mouse intestine. *PLoS One* 2011;**6**:e19880.
110. Stairs DB, Bayne LJ, Rhoades B, Vega ME, Waldron TJ, Kalabis J, et al. Deletion of p120-catenin results in a tumor microenvironment with inflammation and cancer that establishes it as a tumor suppressor gene. *Cancer Cell* 2011;**19**:470–83.
111. Dillon DA, D'Aquila T, Reynolds AB, Fearon ER, Rimm DL. The expression of p120ctn protein in breast cancer is independent of alpha- and beta-catenin and E-cadherin. *Am J Pathol* 1998;**152**:75–82.
112. Karayiannakis AJ, Syrigos KN, Efstathiou J, Valizadeh A, Noda M, Playford RJ, et al. Expression of catenins and E-cadherin during epithelial restitution in inflammatory bowel disease. *J Pathol* 1998;**185**:413–8.
113. Kallakury BV, Sheehan CE, Winn-Deen E, Oliver J, Fisher HA, Kaufman Jr RP, et al. Decreased expression of catenins (alpha and beta), p120 CTN, and E-cadherin cell adhesion proteins and E-cadherin gene promoter methylation in prostatic adenocarcinomas. *Cancer* 2001;**92**:2786–95.
114. Wang EH, Liu Y, Xu HT, Dai SD, Liu N, Xie CY, et al. Abnormal expression and clinicopathologic significance of p120-catenin in lung cancer. *Histol Histopathol* 2006;**21**:841–7.
115. Bremnes RM, Veve R, Gabrielson E, Hirsch FR, Baron A, Bemis L, et al. High-throughput tissue microarray analysis used to evaluate biology and prognostic significance of the E-cadherin pathway in non-small-cell lung cancer. *J Clin Oncol* 2002;**20**:2417–28.
116. Mortazavi F, An J, Dubinett S, Rettig M. p120-catenin is transcriptionally down-regulated by FOXC2 in non-small cell lung cancer cells. *Mol Cancer Res* 2010;**8**:762–74.
117. Liu Y, Li QC, Miao Y, Xu HT, Dai SD, Wei Q, et al. Ablation of p120-catenin enhances invasion and metastasis of human lung cancer cells. *Cancer Sci* 2009;**100**:441–8.
118. Thoreson MA, Reynolds AB. Altered expression of the catenin p120 in human cancer: implications for tumor progression. *Differentiation* 2002;**70**:583–9.
119. Sarrio D, Perez-Mies B, Hardisson D, Moreno-Bueno G, Suarez A, Cano A, et al. Cytoplasmic localization of p120ctn and E-cadherin loss characterize lobular breast carcinoma from preinvasive to metastatic lesions. *Oncogene* 2004;**23**:3272–83.
120. Mayerle J, Friess H, Buchler MW, Schnekenburger J, Weiss FU, Zimmer KP, et al. Up-regulation, nuclear import, and tumor growth stimulation of the adhesion protein p120 in pancreatic cancer. *Gastroenterology* 2003;**124**:949–60.
121. Maeda M, Johnson E, Mandal SH, Lawson KR, Keim SA, Svoboda RA, et al. Expression of inappropriate cadherins by epithelial tumor cells promotes endocytosis and degradation of E-cadherin via competition for p120(ctn). *Oncogene* 2006;**25**:4595–604.
122. Silvera D, Arju R, Darvishian F, Levine PH, Zolfaghari L, Goldberg J, et al. Essential role for eIF4GI overexpression in the pathogenesis of inflammatory breast cancer. *Nat Cell Biol* 2009;**11**:903–8.

123. Paredes J, Correia AL, Ribeiro AS, Milanezi F, Cameselle-Teijeiro J, Schmitt FC. Breast carcinomas that co-express E- and P-cadherin are associated with p120-catenin cytoplasmic localisation and poor patient survival. *J Clin Pathol* 2008;**61**:856–62.
124. Rakha EA, Patel A, Powe DG, Benhasouna A, Green AR, Lambros MB, et al. Clinical and biological significance of E-cadherin protein expression in invasive lobular carcinoma of the breast. *Am J Surg Pathol* 2010;**34**:1472–9.
125. Macpherson IR, Hooper S, Serrels A, McGarry L, Ozanne BW, Harrington K, et al. p120-catenin is required for the collective invasion of squamous cell carcinoma cells via a phosphorylation-independent mechanism. *Oncogene* 2007;**26**:5214–28.
126. Paredes J, Correia AL, Ribeiro AS, Schmitt F. Expression of p120-catenin isoforms correlates with genomic and transcriptional phenotype of breast cancer cell lines. *Cell Oncol* 2007;**29**:467–76.
127. Chartier NT, Oddou CI, Laine MG, Ducarouge B, Marie CA, Block MR, et al. Cyclin-dependent kinase 2/cyclin E complex is involved in p120 catenin (p120ctn)-dependent cell growth control: a new role for p120ctn in cancer. *Cancer Res* 2007;**67**:9781–90.
128. Slorach EM, Chou J, Werb Z. Zeppo1 is a novel metastasis promoter that represses E-cadherin expression and regulates p120-catenin isoform expression and localization. *Genes Dev* 2011;**25**:471–84.
129. Miao Y, Liu N, Zhang Y, Liu Y, Yu JH, Dai SD, et al. p120ctn isoform 1 expression significantly correlates with abnormal expression of E-cadherin and poor survival of lung cancer patients. *Med Oncol* 2010;**27**:880–6.
130. Jiang G, Wang Y, Dai S, Liu Y, Stoecker M, Wang E, et al. P120-catenin isoforms 1 and 3 regulate proliferation and cell cycle of lung cancer cells via beta-catenin and Kaiso respectively. *PLoS One* 2012;**7**:e30303.
131. Mariner DJ, Davis MA, Reynolds AB. EGFR signaling to p120-catenin through phosphorylation at Y228. *J Cell Sci* 2004;**117**:1339–50.
132. Ma LW, Zhou ZT, He QB, Jiang WW. Phosphorylated p120-catenin expression has predictive value for oral cancer progression. *J Clin Pathol* 2012;**65**:315–9.
133. Hofmann C, Lippert E, Falk W, Scholmerich J, Rogler G, Obermeier F. Primary human colonic epithelial cells are transiently protected from anoikis by a Src-dependent mechanism. *Biochem Biophys Res Commun* 2009;**390**:908–14.
134. Hofmann C, Obermeier F, Artinger M, Hausmann M, Falk W, Schoelmerich J, et al. Cell-cell contacts prevent anoikis in primary human colonic epithelial cells. *Gastroenterology* 2007;**132**:587–600.
135. Lugo-Martinez VH, Petit CS, Fouquet S, Le Beyec J, Chambaz J, Pincon-Raymond M, et al. Epidermal growth factor receptor is involved in enterocyte anoikis through the dismantling of E-cadherin-mediated junctions. *Am J Physiol Gastrointest Liver Physiol* 2009;**296**:G235–G244.
136. Eblen ST, Slack JK, Weber MJ, Catling AD. Rac-PAK signaling stimulates extracellular signal-regulated kinase (ERK) activation by regulating formation of MEK1-ERK complexes. *Mol Cell Biol* 2002;**22**:6023–33.

Afadin/AF-6 and Canoe: Roles in Cell Adhesion and Beyond

Kenji Mandai[*,†], **Yoshiyuki Rikitake**[*,‡,§], **Yohei Shimono**[*], **Yoshimi Takai**[*,†]

[*]Division of Molecular and Cellular Biology, Kobe University Graduate School of Medicine, Kobe, Hyogo, Japan
[†]CREST, Japan Science and Technology Agency, Kobe, Hyogo, Japan
[‡]Division of Cardiovascular Medicine, Kobe University Graduate School of Medicine, Kobe, Hyogo, Japan
[§]Division of Signal Transduction, Kobe University Graduate School of Medicine, Kobe, Hyogo, Japan

Contents

Abstract

Afadin is an actin filament (F-actin) and Rap1 small G protein-binding protein encoded by the *MLLT4/AF-6* gene. It is abundant at cadherin-based adherens junctions in epithelial cells, endothelial cells, and fibroblasts. It contains multiple domains and interacts with many proteins, including cell adhesion molecules and their associated molecules, and signaling molecules. Many lines of evidence show that afadin plays pleiotropic functions not only in the formation of cell junctions but also in cell polarization, migration,

Progress in Molecular Biology and Translational Science, Volume 116
ISSN 1877-1173
http://dx.doi.org/10.1016/B978-0-12-394311-8.00019-4

survival, proliferation, and differentiation. In addition, it is involved in oncogenesis and metastasis. Afadin is evolutionarily conserved from *Caenorhabditis elegans* to *Homo sapiens*. Canoe, the *Drosophila melanogaster* counterpart of afadin, is also localized at adherens junctions and regulates cell adhesion, cytoskeletal organization, planar cell polarity, cell differentiation, and migration. Moreover, canoe regulates asymmetric cell division of *Drosophila* neuroblasts. Thus, afadin/AF-6 and canoe are pivotal regulatory elements in many fundamental signaling cascades in cells.

1. INTRODUCTION

Afadin (*AF-6*, alias *MLLT4*) and Canoe (Cno) are structurally and functionally related proteins (genes). *AF-6* was originally reported as a fusion partner of the *MLL* gene (alias *ALL-1*) in pediatric acute myeloid leukemia with chromosome translocation.[1] However, the physiological function of the *AF-6* gene was not known at that time. Subsequently, its *Drosophila* counterpart, *Cno*, was identified. We biochemically purified an actin filament (F-actin)-binding protein with a molecular mass of 205 kDa (p205) from embryonic rat brains.[2] p205 was copurified with a protein with a molecular mass of 190 kDa (p190) that did not show an F-actin-binding activity. Peptide mapping analysis of p205 and p190 showed that they were almost identical. The amino acid sequences of the peptides common in p205 and p190 showed significant homology to the human *AF-6* gene product and a weaker homology to the *Drosophila Cno* gene product,[2] while the peptides specific to p205 did not show any homology to known gene products. The isolated cDNAs for p205 and p190 clearly showed that they were splice variants and that p190 lacked an F-actin-binding domain, which is located in the C-terminus of p205.[2] We named the p205 protein l-afadin (a large splice variant of *AF-6* protein localized at adherens junctions) and the p190 protein s-afadin (a small splice variant of l-afadin). To facilitate readability, we refer to l-afadin simply as afadin here.

One of the breakthroughs in afadin research came from immunohistochemical analysis with an anti-afadin polyclonal antibody. It revealed that afadin was localized at cell–cell adhesion sites in epithelial cells and fibroblasts.[2] Moreover, the signal for afadin was concentrated at the apical portion of the cell–cell adhesion sites in small intestinal absorptive epithelial cells, while the signal for E-cadherin was more diffusely distributed along the lateral membranes of the epithelial cells, indicating that afadin is localized at adherens junctions (AJs) or tight junctions (TJs). Further analysis of intestinal epithelial cells with immunoelectron microscopy revealed that afadin was localized at AJs, but not at TJs or desmosomes. Thus, afadin is an AJ protein that likely links AJs to the actin cytoskeleton. Afadin has a single PDZ domain (Fig. 19.1), so we hypothesized that afadin binds transmembrane proteins localized at AJs through this domain. Meanwhile, we identified the nectin protein family (nectins) as binding partners of afadin.[3] Nectins are cell adhesion molecules (CAMs) that bind to the PDZ domain of afadin through their C-termini.[3–5] Precise functional characterization of afadin's role in cell–cell adhesion was made possible with the finding of the nectin family. Afadin and nectins cooperatively regulate the formation of cell–cell adhesion, cell polarization, migration, survival, differentiation, and entry of viruses.[3]

Canoe (Cno) is the *Drosophila* counterpart of afadin/AF-6. Its domain composition resembles that of afadin (Fig. 19.1) and its gene was originally characterized on the basis of genetic interaction with the Notch signaling pathway.[6] It is also localized at AJs and regulates the formation of cell–cell adhesion.[7] In addition, Cno regulates cytoskeletal organization, planar cell polarity, cell differentiation, and migration. Therefore, afadin and Cno partly share similar functions.

In this chapter, we summarize the basic properties and functions of afadin/AF-6 and Cno, as well as their implications in pathogenesis.

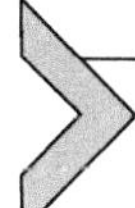

2. GENERAL PROPERTIES OF AFADIN/AF-6 AND Cno

2.1. Biochemical properties

p205 (l-afadin) and p190 (s-afadin) were copurified from rat embryonic brains and appeared in the fractions corresponding to a molecular mass of over 600 kDa on gel-filtration chromatography, which indicated that they formed a multimer.[2] It is not known whether afadin exists as a multimer *in vivo*. Afadin binds along the sides of F-actin and shows an F-actin cross-linking activity, which is weaker than that of α-actinin. The stoichiometry for the binding of the F-actin-binding domain of afadin to actin is approximately 500 actin monomers per one afadin molecule. Its dissociation constant (K_d) is in the order of 10^{-7} M.[2]

2.2. Domain structures

The *MLLT4/AF-6* gene locus encoding afadin consists of more than 30 exons and produces multiple alternatively spliced forms.[8] In mice and humans, the *MLLT4/AF-6* gene presumably produces 9 and 16 splice

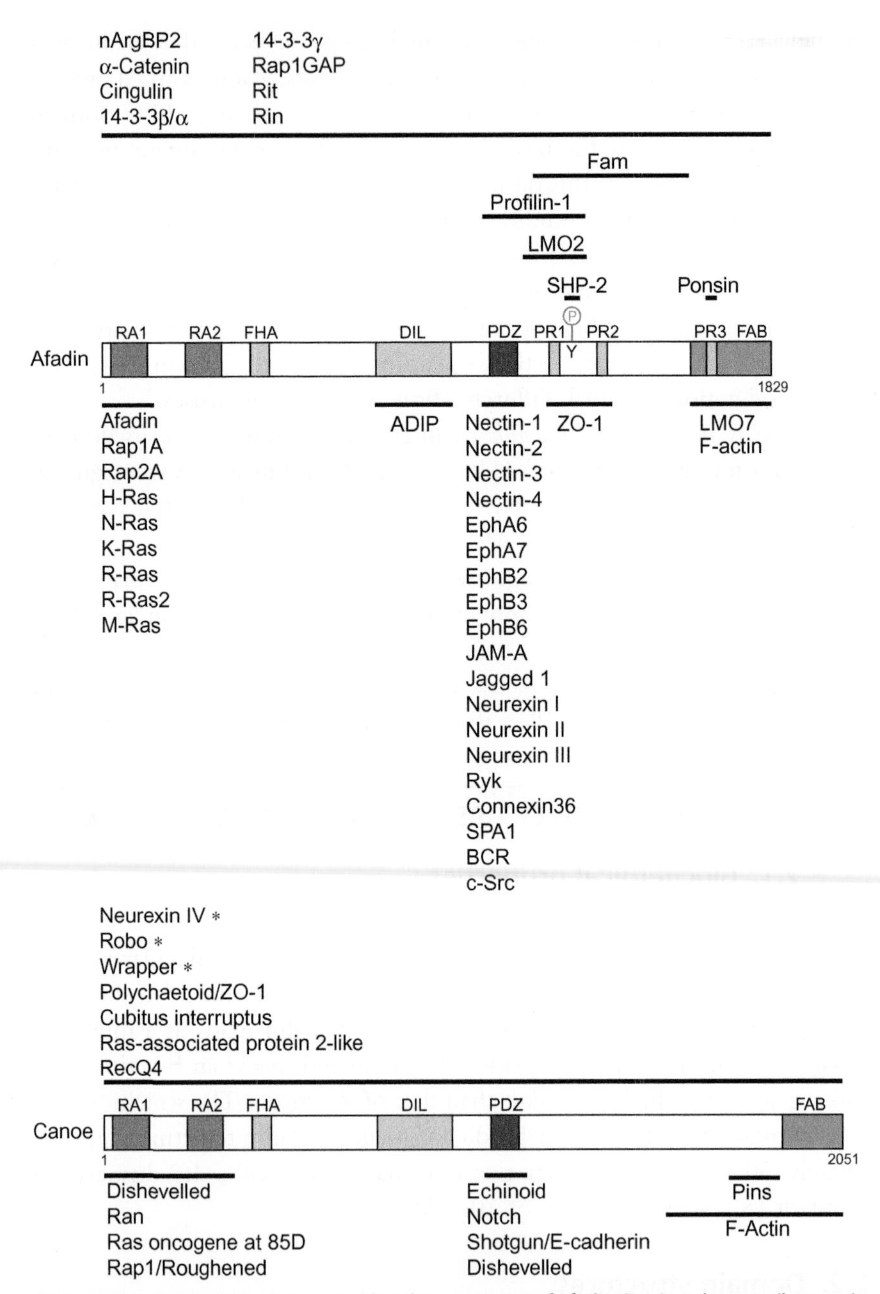

Figure 19.1 Domain structures and binding partners of afadin (top) and canoe (bottom). First and last amino acids are numbered. The following binding domains are shown: RA1, RA2, Ras-associated domain-1 and -2; FHA, forkhead-associated domain; DIL, dilute domain; PDZ, PDZ domain; PR1, PR2, PR3, proline-rich domain-1 to -3; FAB, F-actin-binding domain. *It is not known whether the interaction is direct or indirect.

variants, respectively, according to the Ensembl database. The longest afadin protein, l-afadin, binds F-actin through its F-actin-binding domain located in the C-terminus (Fig. 19.1), but all other variants lack this domain. In the rat brain, both l-afadin and s-afadin are expressed, while in most other rat tissues examined so far, mainly l-afadin is expressed.[2] However, some non-neuronal cells are reported to express both l-afadin and s-afadin.[9,10] Afadin has multiple domains: from the N-terminus to the C-terminus it has two Ras-associated domains, a forkhead-associated domain, a dilute domain, a PDZ domain, three proline-rich domains, and an F-actin-binding domain (Fig. 19.1). s-Afadin lacks the third proline-rich domain and the F-actin-binding domain.

2.3. Binding partners

Many afadin-binding proteins have been identified (Fig. 19.1). They include transmembrane proteins (such as nectins, the Eph receptor tyrosine kinases, the synaptic transmembrane protein neurexins, the Notch receptor ligand jagged-1, the junctional adhesion molecule JAM, and the gap junction protein connexin36), peripheral membrane proteins (such as the TJ protein ZO-1, the afadin- and α-actinin-binding protein ADIP, the actin-binding protein profilin, the vinculin-binding protein ponsin, the cadherin-binding protein α-catenin, and the Arg/Abl-interacting protein nArgBP2), and signaling molecules (such as the Rap1, Ras, Rit, and Rin small G proteins, the Rap1 GTPase-activating protein SPA-1, the Bcr and c-Src protein kinases, the protein tyrosine phosphatase SHP-2, the T-cell oncogene LMO2, and the tumor suppressor LMO7).[3,11–14] The tyrosine kinase Ryk is reported to interact with afadin but this interaction is controversial.[15] Moreover, the first Ras-associated domain of afadin mediates its self-association.[16]

Of these binding partners of afadin, nectins are the most important for explaining the functions of afadin in cell–cell adhesion. Nectin-like molecules (Necls) are proteins structurally similar to nectins. Nectins and Necls comprise families with four and five members, respectively, and all of them are Ca^{2+}-independent immunoglobulin superfamily CAMs.[3] They share three extracellular immunoglobulin-like domains comprising an amino-terminal variable region-like domain and two constant region-like domains, a transmembrane region, and a cytoplasmic region. Nectins and Necls are classified on the basis of their afadin-binding properties. The PDZ domain of afadin binds to the C-terminus of nectins. In contrast to nectins, Necls do not bind afadin.[3] The members of the nectin and Necl families *trans*-interact both homophilically and heterophilically with each other in various combinations.[3] Mutations in *NECTIN-1* and *NECTIN-4* are reported as causes of human hereditary disorders: cleft lip/palate-ectodermal dysplasia[17,18] and ectodermal dysplasia-syndactyly syndrome-1,[19] respectively.

2.4. Cno

Cno, a *Drosophila* counterpart of afadin, is encoded by a gene that presumably produces three splice variants, according to the Ensembl database. The domain organization of Cno resembles that of afadin (Fig. 19.1).[6] Cno interacts physically with many proteins (Fig. 19.1), including transmembrane proteins, such as the immunoglobulin superfamily transmembrane proteins echinoid (ed) and shotgun (shg, *Drosophila* E-cadherin), the peripheral membrane protein polychaetoid (pyd, *Drosophila* ZO-1), signaling molecules, such as small G proteins Rap1 and Ran, Ras-associated protein 2-like (Rap2l), and Ras oncogene at 85D (Ras85D), and other functional molecules, such as the DNA helicase RecQ4, the zinc finger transcription factor cubitus interruptus (ci), Pins (partner of inscuteable) and the transcription factor knot (kn).[7,20–24] In addition, Cno forms complexes with the immunoglobulin superfamily proteins, wrapper and robo, and the transmembrane protein neurexin IV.[25,26] However, it is not known whether Cno binds to them directly. Like afadin, Cno interacts with F-actin through its C-terminus.[20] *Cno* also interacts genetically with the Notch signaling pathway genes *Hairless* (*H*) and *Notch* (*N*), *mushroom body defect* (*Mud*), *Ras oncogene at 64B* (*Ras64B, Drosophila Rras2*), and *basket* (*bsk, Drosophila JNK*).[6,7,24,27–30]

3. FUNCTIONS OF AFADIN/AF-6 IN MAMMALS

3.1. *In vivo* functions

In vivo functions of afadin were assessed using knockout mice.[31,32] Afadin-deficient mice are embryonic lethal and display various developmental defects during and after gastrulation, including disorganized AJs, TJs, and cell polarity of epithelial cells in ectoderm, and impaired migration and differentiation of mesoderm. In addition, afadin-deficient cystic embryoid bodies also show defects resembling those in afadin-deficient embryos.[31] Thus, afadin is essential in fundamental cellular and developmental processes, such as cell adhesion, polarization, differentiation, and migration, which are required for proper ontogenesis of multicellular organisms. In addition, afadin is required for cell survival and proliferation, which is discussed in Section 3.6. Afadin is also required to maintain the barrier function of the small intestinal epithelia independently of its interaction with nectin-2 or nectin-3.[10] Once development of epithelial tissues is completed, molecular machinery containing nectin-free afadin may form and function in small intestinal epithelial cells.

3.2. Formation of AJs

The core molecular components of AJs are the cadherin–catenin and nectin–afadin complexes[3,33] (Fig. 19.2). Afadin and nectins are characterized by their strict localization at AJs in epithelial cells, whereas cadherins, α-catenin, β-catenin, and p120^{ctn} are not strictly limited to AJs, but are widely distributed along the lateral plasma membrane in epithelial cells.[2,3] At AJs, the direct or indirect binding partners of afadin, such as vinculin, ponsin, and PLEKHA7, are also strictly colocalized with afadin and nectins,[34,35] while vinculin and ponsin are also localized at cell–matrix junctions. It was reported that AF-6 is localized at TJs in epithelial MDCK cells.[36] However, in well-polarized MDCK cells in which TJs and AJs are separated, afadin/AF-6 is not localized at TJs.[37] The available evidence indicates that nectins initiate cell–cell adhesion, followed by the recruitment of the complex of E-cadherin and α- and β-catenins to the nectin-based cell–cell adhesion sites through afadin without *trans*-interactions by E-cadherin. Subsequently, the recruited E-cadherin molecules *trans*-interact with each other to form AJs.[38–40] In fact, inhibitors of nectin-1-based cell–cell adhesions (recombinant proteins comprising either the extracellular region of nectin-3 or glycoprotein D, an envelope protein of herpes simplex virus type 1) prevent the formation of E-cadherin-based cell–cell adhesion in epithelial cells and fibroblasts.[39] Moreover, as mentioned above, afadin deficiency inhibits the formation of the cadherin-based AJs and TJs, and afadin-knockdown impairs their formation in MDCK cells.[31,41]

The *trans*-interactions of nectins at the initial cell–cell contact sites first induce the activation of the tyrosine kinase c-Src[40,42] (Fig. 19.2). Activated c-Src then induces the activation of Rap1 through the adaptor proteins Crk and C3G (a guanine nucleotide exchange factor (GEF) for Rap1) and phosphorylates tyrosine residues of FRG (a GEF for the small G protein Cdc42) and Vav2 (a GEF for the small G protein Rac). Activated Cdc42 enhances the activation of phosphorylated Vav2 and eventually induces the activation of Rac and the formation of lamellipodia. Activated Cdc42 and Rac bind their respective downstream effectors that are F-actin-binding proteins, such as IQGAP1, IRSp53, WAVE, N-WASP, and WASP. These signaling events lead to the formation of cell–cell contacts. Besides c-Src, Rap1 is also activated by the *trans*-interactions of nectins at the initial cell–cell contact

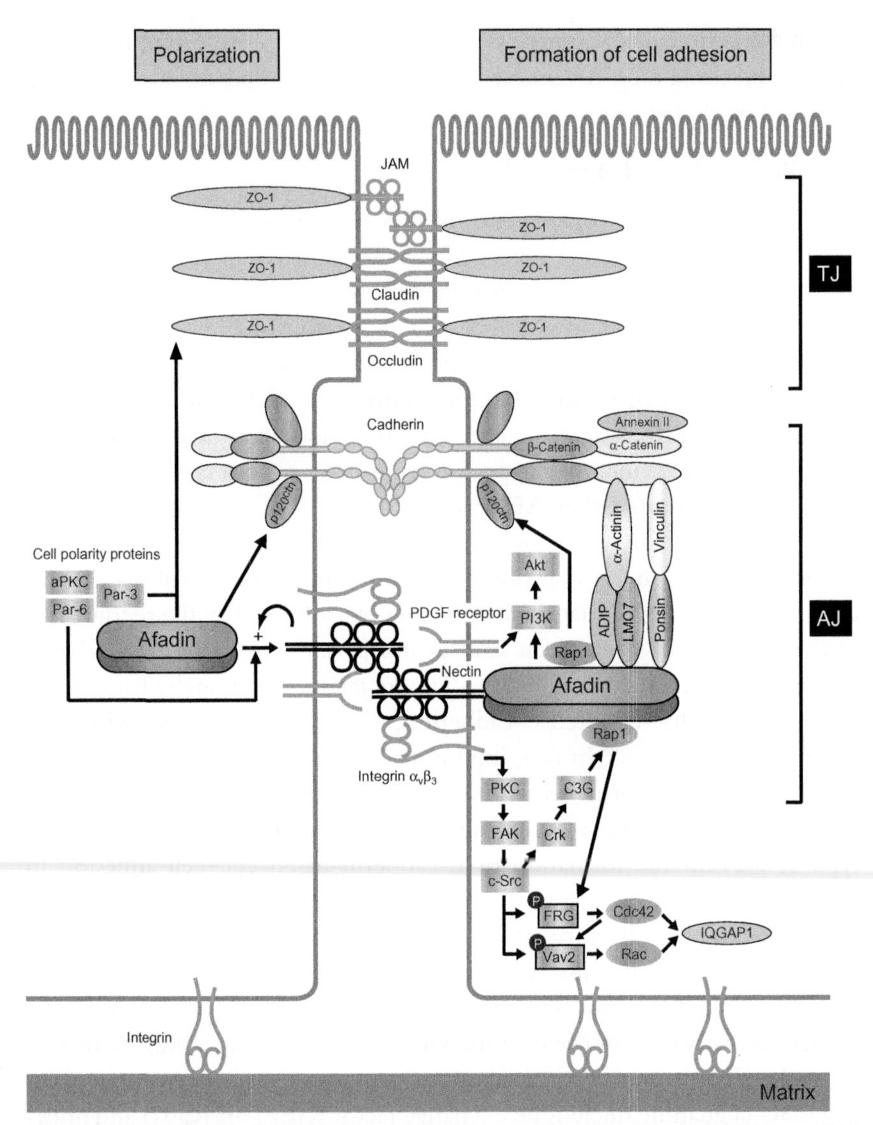

Figure 19.2 Role of afadin in the formation of AJs and TJs. (Right) In the initial step of AJ formation, nectins are associated with integrin $\alpha_V\beta_3$. When two nectin molecules *trans*-interact, they cooperatively induce the activation of c-Src through protein kinase C (PKC) and focal adhesion kinase (FAK). Activated c-Src then induces the activation of Rap1 through Crk and C3G, and phosphorylates tyrosine residues of FRG and Vav2. Activated Rap1 activates phosphorylated FRG, resulting in the activation of Cdc42. Activated Cdc42 enhances the activation of phosphorylated Vav2 and eventually induces the activation of Rac. Activated Cdc42 and Rac reorganize the actin cytoskeleton and recruit the cadherin–catenin complex to the nectin-based cell–cell adhesion sites through many F-actin-binding proteins, such as IQGAP1 and annexin II. These intracellular signaling

sites and subsequently binds to afadin.[43,44] Then, afadin indirectly binds p120^{ctn} that is associated with non-*trans*-interacting E-cadherin.[43] The Rap1-dependent binding of afadin to p120^{ctn} inhibits the endocytosis of E-cadherin, which then enhances both the accumulation of E-cadherin at the nectin-based cell–cell adhesion sites and the cell–cell adhesion activity of E-cadherin, leading to the establishment of AJs.[43,45] Nectins can associate with the actin cytoskeleton through afadin and through its interaction with other actin-binding proteins, such as α-catenin, vinculin, α-actinin, and ZO-1.[40] α-Catenin binds to the reorganized actin cytoskeleton and might recruit the cadherin–β-catenin complex.[43,45]

The nectin-induced intracellular signals and the recruitment of the cadherin–catenin system to the nectin-based cell–cell adhesion sites further enhance the recruitment of the nectin–afadin complex to the cell–cell contact sites. This recruitment then enhances the nectin–nectin interaction, which in turn enhances the nectin–afadin interaction. Thus, the recruitment of afadin to the cell–cell contact sites further facilitates the efficient formation of AJs in a positive feedback manner.[46] In this way, the signaling induced by nectins and cadherins facilitates the dynamic reorganization of the actin cytoskeleton to strengthen the clustering of the CAMs and their cell–cell adhesion activity.

PLEKHA7 is originally identified as a protein of pleckstrin homology domain containing, family A member 7,[35] which is as strictly localized at AJs as afadin and nectins.[35] The interaction of PLEKHA7 with p120^{ctn} and Nezha, which bind, respectively, E-cadherin and the minus ends of

---

pathways are essential for the formation of AJs. The recruitment of the cadherin–catenin complex to the nectin-based cell–cell adhesion sites requires several other connector complexes, ponsin–vinculin, α-actinin–ADIP, and α-actinin–LMO7, which are associated with F-actin. Afadin also interacts directly with α-catenin, but this interaction does not seem to be strong. Finally, activated Rap1 induces the association of afadin with p120^{ctn}, which increases the *trans*-interacting activity of E-cadherin, leading to stabilization of AJs. Interaction of nectin with PDGF receptor and afadin induces the activation of the phosphatidylinositol 3-kinase(PI3K)–Akt signaling pathway, which prevents apoptosis. See text for relevant references. (Left) The Par protein complex is crucial for TJ formation and apicobasal polarization. Afadin and Par-3 cooperatively induce the formation of AJs and TJs. Upon initiation of the formation of cell–cell adhesion, afadin interacts with nectins, and then this complex enhances the nectin–nectin interaction, which further enhances the nectin–afadin interaction in a positive feedback manner. Thus, the nectin–nectin and nectin–afadin interactions cooperatively increase the clustering of the nectin–afadin complex at the cell–cell contact sites, promoting the formation of nectin-based cell–cell adhesion. See text for relevant references. (See Color Insert.)

microtubules, connects the microtubule network to AJs.[35] We found that afadin binds and recruits PLEKHA7 to the nectin-based cell–cell adhesion sites (unpublished observations). Collectively, the direct interaction of afadin with α-catenin and PLEKHA7 together with the recruitment of E-cadherin and the reorganization of the actin cytoskeleton cooperatively contributes to the establishment of AJs.

3.3. Formation of TJs and establishment of apicobasal polarity

In epithelial cells, apicobasal polarity is established by the formation of TJs at the apical side of AJs. The nectin-induced intracellular signaling and the reorganization of the actin cytoskeleton in cooperation with afadin are important for the formation of AJs and TJs as well as for the establishment of apicobasal polarization.[47–49] After or during the formation of AJs, nectins recruit ZO-1 through afadin. The transient interaction of ZO-1 with afadin is necessary for the localization of JAMs and claudins at the apical side of AJs.[50] Afadin and nectins also function with Par-3 to form both AJs and TJs.[51] It was reported that afadin binds to JAM directly,[52] but this finding is controversial.[48] The significance of the binding of afadin to JAM in the formation of TJs remains to be elucidated.

While many CAMs and cell polarity proteins required for apicobasal polarization have been identified, the cooperative roles of these proteins in polarization are not well understood. Intriguingly, exogenous expression of various combinations of CAMs and cell polarity proteins in fibroblasts that have AJs but not TJs leads to the establishment of both epithelial-like AJs and TJs (unpublished observations). In other words, simultaneous expression of several cell polarity proteins, Par-3, aPKC, Par-6, Crb3, Pals1, and Patj, and the CAMs that establish AJs and TJs, is necessary to properly localize TJs at the apical side of AJs (unpublished observations). Among these cell polarity proteins, Crb3, which is localized at the apical membrane, appears to be most important for the localization of TJs at the apical side of AJs. For better understanding of the molecular mechanisms underlying the establishment of apicobasal polarity, how Crb3 is transported to the apical membrane should be clarified.

3.4. Formation of junctions in neurons

Synapses are specialized intercellular junctions that are indispensable for neuronal functions. There are at least two types of junctional structures in synapses: synaptic junctions (SJs) and puncta adherentia junctions (PAJs).[53]

SJs are the places where neurotransmission occurs, whereas PAJs are regarded as mechanical adhesion sites between axon terminals and their targets. In fact, the molecular architecture in PAJs is similar to that of AJs.[54,55] SJs are associated both with synaptic vesicles docked at the presynaptic active zone where Ca^{2+} channels are localized, and with postsynaptic densities where specific neurotransmitter receptors are concentrated. PAJs, in contrast, contain symmetrical paramembranous dense materials and are not associated with synaptic vesicles.

At mossy fiber–CA3 pyramidal cell synapses in the stratum lucidum of the hippocampus, both SJs and PAJs are highly developed as separate clusters consisting of distinct macromolecular complexes. N-Cadherin is localized at PAJs, forming junctions surrounding SJs and plays a role in synaptic plasticity.[55–57] In addition, at PAJs in the mossy fiber–CA3 pyramidal cell synapses, afadin is symmetrically localized at both the presynaptic and the postsynaptic sides, whereas nectin-1 and nectin-3 are asymmetrically localized at, respectively, the presynaptic and the postsynaptic sides.[58] By analogy with AJs in epithelial cells and fibroblasts, the formation of nectin-based cell–cell adhesion may be followed by the recruitment of the N-cadherin–catenin complex to form synapses. Then, SJs and PAJs are segregated. In immature mossy fiber–CA3 pyramidal cell synapses, immunogold signals for αN-catenin, nectin-1, nectin-3, and afadin are homogeneously distributed at synaptic contact sites, whereas SJs in mature synapses are devoid of such signals.[58] In nectin-1-deficient and nectin 3-deficeint mice, the number of PAJs in the CA3 region of the hippocampus of the adult brain is reduced.[59] In the stratum radiatum of the CA1 region of the hippocampus, afadin is necessary for synapse formation.[60] Afadin is also required for the maintenance of synapses in the stratum lucidum of the CA3 region of the hippocampus.[61] Afadin indirectly interacts with a synaptic scaffolding molecule, S-SCAM, which interacts with many synaptic proteins, including neuroligin, NMDA and GluA2-positive AMPA receptors, and β- and δ2-catenins,[54,62] suggesting that afadin might regulate synaptic functions in certain populations of excitatory neurons. Moreover, afadin regulates spine morphology in cooperation with Rap1, which is activated by NMDA receptors.[63] Activated Rap1 recruits afadin to the plasma membrane and induces spine neck elongation, while inactive Rap1 dissociates afadin from the membrane and induces spine enlargement. Thus, afadin could be involved in activity-dependent synaptic plasticity. Afadin may also play a role in electrical synapses formed by gap junctions composed of connexin36. Connexin36 and afadin are colocalized at gap junctions and interact with each other through the C-terminus of connexin36 and the PDZ domain of afadin.[14] Thus, afadin might modulate neuronal activity in the retina and cerebral cortex where gap junctions are abundant.

3.5. Cell migration

Cell migration is a fundamental cellular function implicated in many biological and pathological processes, including development, wound healing, invasion of cancer cells, and atherosclerosis. Migrating cells polarize and form a leading edge. Necl-5 accumulates at this leading edge and regulates cell migration.[64] Nectin-free afadin is also localized at leading edges and regulates directional cell migration in fibroblasts.[65,66]

3.5.1 Regulation by afadin of directional cell migration

Afadin plays a key role in persistent directional cell migration by facilitating clustering of the Necl-5–platelet-derived growth factor (PDGF) receptor–integrin $\alpha_V\beta_3$ complex, presumably in a positive feedback amplification manner.[65] In afadin-knockdown NIH3T3 cells, leading edge structures including lamellipodia, ruffles, and focal complexes are not properly formed, resulting in reduction of PDGF-induced directional cell migration. However, the knockdown of afadin does not affect random cell movement in the presence of PDGF. During directional cell migration of wild-type NIH3T3 cells, the leading edge is continuously formed toward the PDGF source and the Necl-5–PDGF receptor–integrin $\alpha_V\beta_3$ complexes accumulate there. By contrast, the leading edge transiently forms and immediately disappears in afadin-knockdown NIH3T3 cells. Thus, afadin stabilizes the Necl-5–PDGF receptor–integrin $\alpha_V\beta_3$ complexes at the leading edge and initiates the polarized intracellular signals from the leading edge in a fixed direction. In addition, the interaction of afadin with Rap1 is necessary for the localization of afadin at the leading edge, the formation of the leading edge structure, and the PDGF-induced directional cell migration.[65] The interaction of afadin with Rap1 increases the activity of Rap1, which in turn augments the interaction of afadin with Rap1.[67] Moreover, the interaction of afadin with its dilute domain-binding protein(s) like ADIP is also required for the formation of the leading edge and directional cell migration[68] (Fig. 19.1). A TJ transmembrane protein, JAM-A, also regulates cell migration in cooperation with afadin.[69] JAM-A dimerizes and interacts with afadin and PDZ-GEF2, a GEF activating Rap1. The active Rap1 increases the β1 integrin level and augments cell migration.

3.5.2 Regulation by afadin of activation and inactivation of small G proteins

Migrating cells dynamically form and disassemble leading edge structures. The spatially and temporally regulated activation and inactivation of the Rho family small G proteins, including Rac1, Cdc42, and RhoA, regulate these dynamic events in NIH3T3 cells.[66] When Rap1 locally activates Rac1 at the sites of membrane protrusions, cells spread by formation of membrane protrusions adjacent to the extracellular matrix.[70] Stimulation of the cells with PDGF induces the local and transient activation of Rap1 and Rac1 at the leading edge.[71] An active, GTP-bound form of Rap1 directly binds to nectin-free afadin through the first N-terminal Ras-associated domain of afadin and recruits afadin to the leading edge. Afadin then recruits SPA-1, a Rap1 GTPase-activating protein, to the leading edge.[66] However, afadin preferentially interacts with active Rap1 when this is present, because the binding affinity of afadin to active Rap1 is greater than that to SPA-1. In this way, inactivation of Rap1 is prevented. Active Rap1 induces the accumulation of Vav2 at peripheral ruffles, and thereby Rac1 is activated there. Collectively, the Necl-5–PDGF receptor–integrin $\alpha_V\beta_3$ complex-mediated activation and inactivation of these small G proteins, which are critically regulated by afadin, play a key role in the dynamic formation and disassembly of leading edge structures required for directional cell migration.

3.5.3 Regulation by afadin of transition form cell migration to cell adhesion

When migrating or proliferating cells collide, they form primordial cell–cell contacts. In the case of epithelial cells, mature cell–cell adhesions subsequently form including specialized cell–cell adhesion complexes, such as AJs and TJs. Simultaneously, apicobasal polarization establishes. After the formation of mature cell–cell adhesions, cells cease migration and proliferation. This process is called mesenchymal–epithelial transition (MET). During MET, cells use many of the molecular components of cell adhesion and signaling systems. For instance, integrin $\alpha_V\beta_3$ and PDGF receptor play key roles in both cell migration and formation of cell–cell adhesions (Fig. 19.2). It is conceivable that integrin $\alpha_V\beta_3$ supports the formation of AJ by enhancing nectin-mediated signaling, because most signaling molecules that are associated with integrin $\alpha_V\beta_3$ are also associated with nectins and are involved in the formation of cell–cell adhesions and regulation of cell migration. In addition, afadin, which directly interacts with nectins at AJs, is also localized at a leading edge of migrating cells and regulates directional cell migration including formation of leading-edge structures, as mentioned in Section 3.5.1. Furthermore, when cells collide, Necl-5 and nectins *trans*-interact at contact sites. This interaction enhances the nectin–afadin interaction, which in turn enhances the nectin–nectin interaction.[46] As discussed in Section 3.2, this nectin–nectin interaction further enhances the nectin–afadin interaction in the manner of a positive feedback. Thus, the Necl-5–nectin, nectin–nectin, and nectin–afadin interactions cooperatively increase the clustering of the nectin–afadin complex at the primordial cell–cell contact sites, promoting the formation of the mature nectin-based cell–cell adhesion structures.

3.6. Cell survival, proliferation, and differentiation

Embryoid bodies derived from afadin-knockout embryonic stem cells contain more apoptotic cells than those derived from wild-type embryonic stem cells.[72] Apoptosis induced by serum deprivation or Fas-ligand stimulation is more often observed in cultured afadin-knockdown fibroblasts and endothelial cells than in control cells.[67,72] Thus, afadin is involved in the cell death and/or survival-signaling cascade. Growth factors, such as PDGF and vascular endothelial growth factor (VEGF), can activate the phosphatidylinositol 3-kinase (PI3K)–Akt signaling pathway, which is critically involved in cell survival. Knockdown of afadin impaired the activation of this signaling pathway induced by PDGF and VEGF in fibroblasts and endothelial cells, respectively.[67,72] Moreover, knockdown of afadin inhibits the VEGF- or sphingosine 1-phosphate-induced proliferation and tube formation of endothelial cells on Matrigel.[67] Thus, afadin regulates cell proliferation and tubulogenesis of endothelial cells. Afadin also regulates vascular barrier function *in vitro* and *in vivo* through its interaction with $p120^{ctn}$ and ZO-1.[73] Additionally, afadin is required for postnatal angiogenesis in retina.[67] In cultured neurons, afadin controls axon branching in cooperation with Ras.[74] Thus, afadin regulates cell survival, proliferation, and differentiation.

The other example that shows the requirement for afadin and nectins in cell differentiation is spermatogenesis. In the testis, unique cell–cell junctions between Sertoli cells and spermatids play critical roles in the morphogenesis and differentiation of spermatids. The junctions depend mainly on the *trans*-, *hetero*-interaction between nectin-2 and nectin-3 expressed in Sertoli cells and spermatids, respectively.[75] The cytoplasmic region of nectin-2 in Sertoli cells associates with afadin to anchor F-actin bundles to the junctions. Disruption of the junctions by knocking out the *nectin-2* or *nectin-3* gene results in abnormal differentiation of spermatids, leading to male-specific infertility. Thus, nectins and afadin are involved in the differentiation of spermatids.

s-Afadin has been demonstrated in the nucleus and this nuclear localization is regulated by the cell cycle, transcriptional activity, and tyrosine kinase-mediated signaling.[76] Thus, s-afadin can control cellular growth by its translocation to the nucleus. Moreover, s-afadin does not stabilize E-cadherin-dependent cell adhesion, whereas l-afadin does.[77] Furthermore, both s-afadin and l-afadin regulate cell migration (unpublished observations). s-Afadin binds to nectins more efficiently than l-afadin (unpublished observations). These observations suggest that s-afadin and l-afadin have distinct roles. It is intriguing to address this issue especially in the brain, which abundantly expresses both s-afadin and l-afadin.

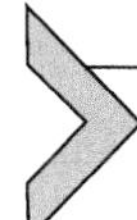

4. AFADIN/AF-6 IN LEUKEMOGENESIS AND ONCOGENESIS

AF-6 was originally identified as a fusion partner of the *MLL* gene in acute myeloid leukemia patients with the chromosomal translocation t(6;11)(q27;q23).[1] The *MLL* gene encodes a transcriptional regulator and is required for hematopoiesis.[78] Translocations of the *MLL* gene are found in 15–20% of pediatric acute myeloid leukemia cases.[79] The translocation forms fusion genes in which the truncated form of *MLL* and the partner gene are fused in frame, leading to a gain of function of *MLL*-fusion gene complexes. More than 60 fusion partners have been identified for *MLL*.[80] These fusion partners can be broadly classified into cytoplasmic proteins and nuclear proteins.[79] The cytoplasmic fusion partners, such as afadin/AF-6, AF1Q, GAS7, and GPHN, do not show transcriptional activity by themselves but promote the dimerization of the N-terminal portion of MLL.[16,79] The 103 amino acids in the first Ras-associated domain of afadin/AF-6 mediate its self-association, which activates the oncogenic potential of the MLL–AF-6 fusion protein in myeloid progenitor cells.[16] Seemingly, the oncogenic transformation by *MLL–AF-6* fusion gene does not involve recruitment of Ras to the nucleus. However, it is still unknown how the dimerization of afadin/AF-6 or other MLL-cytoplasmic fusion partners inappropriately activates the MLL target genes and leads to leukemogenesis. Furthermore, it is unknown whether afadin/AF-6 itself plays roles in normal hematopoiesis and in leukemogenesis.

Afadin also plays a role in oncogenesis of solid tumors. Loss of afadin has been reported in 15% of breast cancer cases and is associated with adverse prognosis, suggesting that *afadin/AF-6* might be a tumor suppressor gene.[81] Experimentally, downregulation of afadin increases tumorigenicity and metastasis in mice[9] and increases invasive behavior in cultured breast cancer cells.[82] The 300-kb segment of human chromosome 6q27, in which the *AF-6* gene locus is located, is commonly deleted in ovarian cancers, thereby the chromosomal segment is presumed to contain a tumor suppressor gene.[83] However, the role of the *afadin/AF-6* gene in oncogenesis of ovarian cancers is not clear. In contrast, stronger afadin expression is associated with adverse eight-year outcome of breast cancer patients.[84] Further clinical studies are required to clarify the role of afadin in oncogenesis.

5. FUNCTIONS OF Cno IN FLY

In *Drosophila,* Cno physically or genetically interacts with Ras–MAPK, JNK, Notch, and Wnt signaling pathways during morphogenesis of the eyes, bristles, and/or wings.[6,7,27,29,85] Cno is localized at AJs in embryonic epithelia and colocalized with *Drosophila* ZO-1, Pyd.[7] These two proteins interact with each other and are required for dorsal closure in the *Drosophila* embryo, indicating that the Pyd–Cno complex transduces signals at AJs for dynamic changes in cytoskeletal organization and cell adhesion during dorsal closure. Cno also regulates cell shape changes during dorsal closure[6,7,86] and links AJs with an actomyosin network during apical constriction of mesodermal cells.[20] Cno and Rap1 are colocalized at AJs, and cooperatively act during dorsal closure. In addition, they also act together in the planar cell polarity pathway.[87] Remarkably, Cno is not essential for AJ assembly or maintenance in many *Drosophila* tissues, in sharp contrast to afadin in mammals.[20]

The *Drosophila* CAM echinoid is a member of the immunoglobulin superfamily. However, its domain composition is different from that of nectins. Although echinoid is not considered to be a *Drosophila* counterpart of nectins, it is a component of AJs and binds to Cno.[22] Notably, AJs are absent between the cells when one of them lacks echinoid. Thus, echinoid regulates cell–cell adhesion in *Drosophila.*[22] Moreover, Cno forms complexes with *Drosophila* E-cadherin shg, wrapper, neurexin IV, and robo, and regulates neuron–glia interactions and axon guidance at the midline of the *Drosophila* central nervous system.[25,26] Thus, Cno regulates cell adhesion, polarization, and axon guidance in *Drosophila.*

Cno physically interacts with Pins and regulates asymmetric cell division in *Drosophila* neuroblasts.[24] *Cno* also regulates spindle orientation of the neuroblasts in cooperation with *Mud*, and Ran and Rap1 small G proteins.[23,24,30] Thus, Cno regulates cell differentiation. However, it is unknown whether afadin is involved in the regulation of asymmetric cell division and/or production of neurons and glia from progenitor cells in the mammalian brain.

6. PERSPECTIVES

In this chapter, we have described how afadin and Cno regulate a wide variety of cellular processes, such as the formation of cell–cell adhesion, establishment of apicobasal polarity, synapse formation, as well as cell migration, proliferation, and differentiation. In addition, Cno regulates cytoskeletal organization, planar cell polarity, and asymmetric cell division in neuroblasts. Despite these basic characterizations of afadin/Cno, the contribution of afadin to disease processes has not been well clarified. Considering afadin's functions in the abovementioned fundamental cellular processes, it could be involved in the pathogenesis of cancers, neuropsychiatric disorders, or other diseases. Many recent human genetic studies have identified disease-linked variants of many genes, including the *MLLT4/AF-6* gene. In addition, many high-throughput analyses are presently performed in an unbiased way. The results of these analyses will contribute to translational research on afadin. Furthermore, to facilitate the analysis, sophisticated genetically manipulated mouse models are necessary. Such mouse models would be powerful tools to promote our understanding of afadin functions *in vivo*.

The remaining fundamental questions concerning the biological functions of afadin include its specific roles in cell division of progenitor or stem cells, cell differentiation, the establishment of planar cell polarity, synaptic plasticity and neuronal circuit formation, and the formation of heterotypic cell adhesions. Moreover, a role for s-afadin in cells has not been well clarified. Furthermore, structural analysis for domains in afadin is also intriguing.

REFERENCES

1. Prasad R, Gu Y, Alder H, Nakamura T, Canaani O, Saito H, et al. Cloning of the ALL-1 fusion partner, the AF-6 gene, involved in acute myeloid leukemias with the *t*(6;11) chromosome translocation. *Cancer Res* 1993;**53**:5624–8.
2. Mandai K, Nakanishi H, Satoh A, Obaishi H, Wada M, Nishioka H, et al. Afadin: a novel actin filament-binding protein with one PDZ domain localized at cadherin-based cell-to-cell adherens junction. *J Cell Biol* 1997;**139**:517–28.
3. Takai Y, Ikeda W, Ogita H, Rikitake Y. The immunoglobulin-like cell adhesion molecule nectin and its associated protein afadin. *Annu Rev Cell Dev Biol* 2008;**24**:309–42.

4. Aoki J, Koike S, Asou H, Ise I, Suwa H, Tanaka T, et al. Mouse homolog of poliovirus receptor-related gene 2 product, mPRR2, mediates homophilic cell aggregation. *Exp Cell Res* 1997;**235**:374–84.
5. Lopez M, Aoubala M, Jordier F, Isnardon D, Gomez S, Dubreuil P. The human poliovirus receptor related 2 protein is a new hematopoietic/endothelial homophilic adhesion molecule. *Blood* 1998;**92**:4602–11.
6. Miyamoto H, Nihonmatsu I, Kondo S, Ueda R, Togashi S, Hirata K, et al. Canoe encodes a novel protein containing a GLGF/DHR motif and functions with Notch and scabrous in common developmental pathways in Drosophila. *Genes Dev* 1995; **9**:612–25.
7. Takahashi K, Matsuo T, Katsube T, Ueda R, Yamamoto D. Direct binding between two PDZ domain proteins Canoe and ZO-1 and their roles in regulation of the jun N-terminal kinase pathway in Drosophila morphogenesis. *Mech Dev* 1998;**78**: 97–111.
8. Saito S, Matsushima M, Shirahama S, Minaguchi T, Kanamori Y, Minami M, et al. Complete genomic structure DNA polymorphisms, and alternative splicing of the human AF-6 gene. *DNA Res* 1998;**5**:115–20.
9. Fournier G, Cabaud O, Josselin E, Chaix A, Adelaide J, Isnardon D, et al. Loss of AF6/afadin, a marker of poor outcome in breast cancer, induces cell migration, invasiveness and tumor growth. *Oncogene* 2011;**30**:3862–74.
10. Tanaka-Okamoto M, Hori K, Ishizaki H, Itoh Y, Onishi S, Yonemura S, et al. Involvement of afadin in barrier function and homeostasis of mouse intestinal epithelia. *J Cell Sci* 2011;**124**:2231–40.
11. Begay-Muller V, Ansieau S, Leutz A. The LIM domain protein Lmo2 binds to AF6, a translocation partner of the MLL oncogene. *FEBS Lett* 2002;**521**:36–8.
12. Kawabe H, Hata Y, Takeuchi M, Ide N, Mizoguchi A, Takai Y. nArgBP2, a novel neural member of ponsin/ArgBP2/vinexin family that interacts with synapse-associated protein 90/postsynaptic density-95-associated protein (SAPAP). *J Biol Chem* 1999;**274**: 30914–8.
13. Shao H, Kadono-Okuda K, Finlin BS, Andres DA. Biochemical characterization of the Ras-related GTPases Rit and Rin. *Arch Biochem Biophys* 1999;**371**:207–19.
14. Li X, Lynn BD, Nagy JI. The effector and scaffolding proteins AF6 and MUPP1 interact with connexin36 and localize at gap junctions that form electrical synapses in rodent brain. *Eur J Neurosci* 2012;**35**:166–81.
15. Trivier E, Ganesan TS. RYK, a catalytically inactive receptor tyrosine kinase, associates with EphB2 and EphB3 but does not interact with AF-6. *J Biol Chem* 2002;**277**: 23037–43.
16. Liedtke M, Ayton PM, Somervaille TC, Smith KS, Cleary ML. Self-association mediated by the Ras association 1 domain of AF6 activates the oncogenic potential of MLL-AF6. *Blood* 2010;**116**:63–70.
17. Zlotogora J, Zilberman Y, Tenenbaum A, Wexler MR. Cleft lip and palate, pili torti, malformed ears, partial syndactyly of fingers and toes, and mental retardation: a new syndrome? *J Med Genet* 1987;**24**:291–3.
18. Suzuki K, Hu D, Bustos T, Zlotogora J, Richieri-Costa A, Helms JA, et al. Mutations of PVRL1, encoding a cell-cell adhesion molecule/herpesvirus receptor, in cleft lip/palate-ectodermal dysplasia. *Nat Genet* 2000;**25**:427–30.
19. Brancati F, Fortugno P, Bottillo I, Lopez M, Josselin E, Boudghene-Stambouli O, et al. Mutations in PVRL4, encoding cell adhesion molecule nectin-4, cause ectodermal dysplasia-syndactyly syndrome. *Am J Hum Genet* 2010;**87**:265–73.
20. Sawyer JK, Harris NJ, Slep KC, Gaul U, Peifer M. The Drosophila afadin homologue Canoe regulates linkage of the actin cytoskeleton to adherens junctions during apical constriction. *J Cell Biol* 2009;**186**:57–73.

21. Formstecher E, Aresta S, Collura V, Hamburger A, Meil A, Trehin A, et al. Protein interaction mapping: a Drosophila case study. *Genome Res* 2005;**15**:376–84.
22. Wei SY, Escudero LM, Yu F, Chang LH, Chen LY, Ho YH, et al. Echinoid is a component of adherens junctions that cooperates with DE-Cadherin to mediate cell adhesion. *Dev Cell* 2005;**8**:493–504.
23. Carmena A, Makarova A, Speicher S. The Rap1-Rgl-Ral signaling network regulates neuroblast cortical polarity and spindle orientation. *J Cell Biol* 2011;**195**:553–62.
24. Speicher S, Fischer A, Knoblich J, Carmena A. The PDZ protein Canoe regulates the asymmetric division of Drosophila neuroblasts and muscle progenitors. *Curr Biol* 2008;**18**:831–7.
25. Slovakova J, Speicher S, Sanchez-Soriano N, Prokop A, Carmena A. The actin-binding protein Canoe/AF-6 forms a complex with Robo and is required for Slit-Robo signaling during axon pathfinding at the CNS midline. *J Neurosci* 2012;**32**:10035–44.
26. Slovakova J, Carmena A. Canoe functions at the CNS midline glia in a complex with Shotgun and Wrapper-Nrx-IV during neuron-glia interactions. *Development* 2011;**138**: 1563–71.
27. Johannes B, Preiss A. Wing vein formation in Drosophila melanogaster: hairless is involved in the cross-talk between Notch and EGF signaling pathways. *Mech Dev* 2002;**115**:3–14.
28. Matsuo T, Takahashi K, Kondo S, Kaibuchi K, Yamamoto D. Regulation of cone cell formation by Canoe and Ras in the developing Drosophila eye. *Development* 1997;**124**: 2671–80.
29. Carmena A, Speicher S, Baylies M. The PDZ protein Canoe/AF-6 links Ras-MAPK, Notch and Wingless/Wnt signaling pathways by directly interacting with Ras, Notch and Dishevelled. *PLoS One* 2006;**1**:e66.
30. Wee B, Johnston CA, Prehoda KE, Doe CQ. Canoe binds RanGTP to promote Pins (TPR)/Mud-mediated spindle orientation. *J Cell Biol* 2011;**195**:369–76.
31. Ikeda W, Nakanishi H, Miyoshi J, Mandai K, Ishizaki H, Tanaka M, et al. Afadin: a key molecule essential for structural organization of cell-cell junctions of polarized epithelia during embryogenesis. *J Cell Biol* 1999;**146**:1117–32.
32. Zhadanov AB, Provance Jr DW, Speer CA, Coffin JD, Goss D, Blixt JA, et al. Absence of the tight junctional protein AF-6 disrupts epithelial cell-cell junctions and cell polarity during mouse development. *Curr Biol* 1999;**9**:880–8.
33. Oda H, Takeichi M. Evolution: structural and functional diversity of cadherin at the adherens junction. *J Cell Biol* 2011;**193**:1137–46.
34. Mandai K, Nakanishi H, Satoh A, Takahashi K, Satoh K, Nishioka H, et al. Ponsin/ SH3P12: an l-afadin- and vinculin-binding protein localized at cell-cell and cell-matrix adherens junctions. *J Cell Biol* 1999;**144**:1001–17.
35. Meng W, Mushika Y, Ichii T, Takeichi M. Anchorage of microtubule minus ends to adherens junctions regulates epithelial cell-cell contacts. *Cell* 2008;**135**:948–59.
36. Yamamoto T, Harada N, Kano K, Taya S, Canaani E, Matsuura Y, et al. The Ras target AF-6 interacts with ZO-1 and serves as a peripheral component of tight junctions in epithelial cells. *J Cell Biol* 1997;**139**:785–95.
37. Sakisaka T, Nakanishi H, Takahashi K, Mandai K, Miyahara M, Satoh A, et al. Different behavior of l-afadin and neurabin-II during the formation and destruction of cell-cell adherens junction. *Oncogene* 1999;**18**:1609–17.
38. Tachibana K, Nakanishi H, Mandai K, Ozaki K, Ikeda W, Yamamoto Y, et al. Two cell adhesion molecules, nectin and cadherin, interact through their cytoplasmic domain-associated proteins. *J Cell Biol* 2000;**150**:1161–76.
39. Honda T, Shimizu K, Kawakatsu T, Yasumi M, Shingai T, Fukuhara A, et al. Antagonistic and agonistic effects of an extracellular fragment of nectin on formation of E-cadherin-based cell-cell adhesion. *Genes Cells* 2003;**8**:51–63.

40. Takai Y, Miyoshi J, Ikeda W, Ogita H. Nectins and nectin-like molecules: roles in contact inhibition of cell movement and proliferation. *Nat Rev Mol Cell Biol* 2008;**9**:603–15.
41. Ooshio T, Kobayashi R, Ikeda W, Miyata M, Fukumoto Y, Matsuzawa N, et al. Involvement of the interaction of afadin with ZO-1 in the formation of tight junctions in Madin-Darby canine kidney cells. *J Biol Chem* 2010;**285**:5003–12.
42. Fukuhara T, Shimizu K, Kawakatsu T, Fukuyama T, Minami Y, Honda T, et al. Activation of Cdc42 by trans interactions of the cell adhesion molecules nectins through c-Src and Cdc42-GEF FRG. *J Cell Biol* 2004;**166**:393–405.
43. Hoshino T, Sakisaka T, Baba T, Yamada T, Kimura T, Takai Y. Regulation of E-cadherin endocytosis by nectin through afadin, Rap1, and p120ctn. *J Biol Chem* 2005;**280**:24095–103.
44. Fukuyama T, Ogita H, Kawakatsu T, Fukuhara T, Yamada T, Sato T, et al. Involvement of the c-Src-Crk-C3G-Rap1 signaling in the nectin-induced activation of Cdc42 and formation of adherens junctions. *J Biol Chem* 2005;**280**:815–25.
45. Sato T, Fujita N, Yamada A, Ooshio T, Okamoto R, Irie K, et al. Regulation of the assembly and adhesion activity of E-cadherin by nectin and afadin for the formation of adherens junctions in Madin-Darby canine kidney cells. *J Biol Chem* 2006;**281**:5288–99.
46. Kurita S, Ogita H, Takai Y. Cooperative role of nectin-nectin and nectin-afadin interactions in formation of nectin-based cell-cell adhesion. *J Biol Chem* 2011;**286**:36297–303.
47. Yamada A, Irie K, Fukuhara A, Ooshio T, Takai Y. Requirement of the actin cytoskeleton for the association of nectins with other cell adhesion molecules at adherens and tight junctions in MDCK cells. *Genes Cells* 2004;**9**:843–55.
48. Fukuhara A, Irie K, Nakanishi H, Takekuni K, Kawakatsu T, Ikeda W, et al. Involvement of nectin in the localization of junctional adhesion molecule at tight junctions. *Oncogene* 2002;**21**:7642–55.
49. Komura H, Ogita H, Ikeda W, Mizoguchi A, Miyoshi J, Takai Y. Establishment of cell polarity by afadin during the formation of embryoid bodies. *Genes Cells* 2008;**13**:79–90.
50. Kuramitsu K, Ikeda W, Inoue N, Tamaru Y, Takai Y. Novel role of nectin: implication in the co-localization of JAM-A and claudin-1 at the same cell-cell adhesion membrane domain. *Genes Cells* 2008;**13**:797–805.
51. Ooshio T, Fujita N, Yamada A, Sato T, Kitagawa Y, Okamoto R, et al. Cooperative roles of Par-3 and afadin in the formation of adherens and tight junctions. *J Cell Sci* 2007;**120**:2352–65.
52. Ebnet K, Schulz CU, Meyer Zu Brickwedde MK, Pendl GG, Vestweber D. Junctional adhesion molecule interacts with the PDZ domain-containing proteins AF-6 and ZO-1. *J Biol Chem* 2000;**275**:27979–88.
53. Spacek J, Lieberman AR. Ultrastructure and three-dimensional organization of synaptic glomeruli in rat somatosensory thalamus. *J Anat* 1974;**117**:487–516.
54. Yamada A, Irie K, Deguchi-Tawarada M, Ohtsuka T, Takai Y. Nectin-dependent localization of synaptic scaffolding molecule (S-SCAM) at the puncta adherentia junctions formed between the mossy fibre terminals and the dendrites of pyramidal cells in the CA3 area of the mouse hippocampus. *Genes Cells* 2003;**8**:985–94.
55. Uchida N, Honjo Y, Johnson KR, Wheelock MJ, Takeichi M. The catenin/cadherin adhesion system is localized in synaptic junctions bordering transmitter release zones. *J Cell Biol* 1996;**135**:767–79.
56. Suzuki SC, Takeichi M. Cadherins in neuronal morphogenesis and function. *Dev Growth Differ* 2008;**50**(Suppl. 1):S119–S130.
57. Tang L, Hung CP, Schuman EM. A role for the cadherin family of cell adhesion molecules in hippocampal long-term potentiation. *Neuron* 1998;**20**:1165–75.
58. Mizoguchi A, Nakanishi H, Kimura K, Matsubara K, Ozaki-Kuroda K, Katata T, et al. Nectin: an adhesion molecule involved in formation of synapses. *J Cell Biol* 2002;**156**: 555–65.

59. Honda T, Sakisaka T, Yamada T, Kumazawa N, Hoshino T, Kajita M, et al. Involvement of nectins in the formation of puncta adherentia junctions and the mossy fiber trajectory in the mouse hippocampus. *Mol Cell Neurosci* 2006;**31**:315–25.
60. Beaudoin 3rd GM, Schofield CM, Nuwal T, Zang K, Ullian EM, Huang B, et al. Afadin, a ras/rap effector that controls cadherin function, promotes spine and excitatory synapse density in the hippocampus. *J Neurosci* 2012;**32**:99–110.
61. Majima T, Ogita H, Yamada T, Amano H, Togashi H, Sakisaka T, et al. Involvement of afadin in the formation and remodeling of synapses in the hippocampus. *Biochem Biophys Res Commun* 2009;**385**:539–44.
62. Danielson E, Zhang N, Metallo J, Kaleka K, Shin SM, Gerges N, et al. S-SCAM/MAGI-2 is an essential synaptic scaffolding molecule for the GluA2-containing maintenance pool of AMPA receptors. *J Neurosci* 2012;**32**:6967–80.
63. Xie Z, Huganir RL, Penzes P. Activity-dependent dendritic spine structural plasticity is regulated by small GTPase Rap1 and its target AF-6. *Neuron* 2005;**48**:605–18.
64. Ikeda W, Kakunaga S, Takekuni K, Shingai T, Satoh K, Morimoto K, et al. Nectin-like molecule-5/Tage4 enhances cell migration in an integrin-dependent, Nectin-3-independent manner. *J Biol Chem* 2004;**279**:18015–25.
65. Miyata M, Ogita H, Komura H, Nakata S, Okamoto R, Ozaki M, et al. Localization of nectin-free afadin at the leading edge and its involvement in directional cell movement induced by platelet-derived growth factor. *J Cell Sci* 2009;**122**:4319–29.
66. Miyata M, Rikitake Y, Takahashi M, Nagamatsu Y, Yamauchi Y, Ogita H, et al. Regulation by afadin of cyclical activation and inactivation of Rap1, Rac1, and RhoA small G proteins at leading edges of moving NIH3T3 cells. *J Biol Chem* 2009;**284**:24595–609.
67. Tawa H, Rikitake Y, Takahashi M, Amano H, Miyata M, Satomi-Kobayashi S, et al. Role of afadin in vascular endothelial growth factor and sphingosine 1-phosphate-induced angiogenesis. *Circ Res* 2010;**106**:1731–42.
68. Fukumoto Y, Kurita S, Takai Y, Ogita H. Role of scaffold protein afadin dilute domain-interacting protein (ADIP) in platelet-derived growth factor-induced cell movement by activating Rac protein through Vav2 protein. *J Biol Chem* 2011;**286**:43537–48.
69. Severson EA, Lee WY, Capaldo CT, Nusrat A, Parkos CA. Junctional adhesion molecule A interacts with Afadin and PDZ-GEF2 to activate Rap1A, regulate β1 integrin levels, and enhance cell migration. *Mol Biol Cell* 2009;**20**:1916–25.
70. Arthur WT, Quilliam LA, Cooper JA. Rap1 promotes cell spreading by localizing Rac guanine nucleotide exchange factors. *J Cell Biol* 2004;**167**:111–22.
71. Takahashi M, Rikitake Y, Nagamatsu Y, Hara T, Ikeda W, Hirata K, et al. Sequential activation of Rap1 and Rac1 small G proteins by PDGF locally at leading edges of NIH3T3 cells. *Genes Cells* 2008;**13**:549–69.
72. Kanzaki N, Ogita H, Komura H, Ozaki M, Sakamoto Y, Majima T, et al. Involvement of the nectin-afadin complex in PDGF-induced cell survival. *J Cell Sci* 2008;**121**:2008–17.
73. Birukova AA, Fu P, Wu T, Dubrovskyi O, Sarich N, Poroyko V, et al. Afadin controls p120-catenin-ZO-1 interactions leading to endothelial barrier enhancement by oxidized phospholipids. *J Cell Physiol* 2012;**227**:1883–90.
74. Iwasawa N, Negishi M, Oinuma I. R-Ras controls axon branching through afadin in cortical neurons. *Mol Biol Cell* 2012;**23**:2793–804.
75. Ozaki-Kuroda K, Nakanishi H, Ohta H, Tanaka H, Kurihara H, Mueller S, et al. Nectin couples cell-cell adhesion and the actin scaffold at heterotypic testicular junctions. *Curr Biol* 2002;**12**:1145–50.
76. Buchert M, Poon C, King JA, Baechi T, D'Abaco G, Hollande F, et al. AF6/s-afadin is a dual residency protein and localizes to a novel subnuclear compartment. *J Cell Physiol* 2007;**210**:212–23.
77. Lorger M, Moelling K. Regulation of epithelial wound closure and intercellular adhesion by interaction of AF6 with actin cytoskeleton. *J Cell Sci* 2006;**119**:3385–98.

78. Ernst P, Fisher JK, Avery W, Wade S, Foy D, Korsmeyer SJ. Definitive hematopoiesis requires the mixed-lineage leukemia gene. *Dev Cell* 2004;**6**:437–43.
79. Balgobind BV, Zwaan CM, Pieters R, Van den Heuvel-Eibrink MM. The heterogeneity of pediatric MLL-rearranged acute myeloid leukemia. *Leukemia* 2011;**25**:1239–48.
80. Meyer C, Kowarz E, Hofmann J, Renneville A, Zuna J, Trka J, et al. New insights to the MLL recombinome of acute leukemias. *Leukemia* 2009;**23**:1490–9.
81. Letessier A, Garrido-Urbani S, Ginestier C, Fournier G, Esterni B, Monville F, et al. Correlated break at PARK2/FRA6E and loss of AF-6/Afadin protein expression are associated with poor outcome in breast cancer. *Oncogene* 2007;**26**:298–307.
82. Chatterjee S, Seifried L, Feigin ME, Gibbons DL, Scuoppo C, Lin W, et al. Dysregulation of cell polarity proteins synergize with oncogenes or the microenvironment to induce invasive behavior in epithelial cells. *PLoS One* 2012;**7**:e34343.
83. Saito S, Sirahama S, Matsushima M, Suzuki M, Sagae S, Kudo R, et al. Definition of a commonly deleted region in ovarian cancers to a 300-kb segment of chromosome 6q27. *Cancer Res* 1996;**56**:5586–9.
84. Charpin C, Tavassoli F, Secq V, Giusiano S, Villeret J, Garcia S, et al. Validation of an immunohistochemical signature predictive of 8-year outcome for patients with breast carcinoma. *Int J Cancer* 2012;**131**:E236–E243.
85. Matsuo T, Takahashi K, Suzuki E, Yamamoto D. The Canoe protein is necessary in adherens junctions for development of ommatidial architecture in the Drosophila compound eye. *Cell Tissue Res* 1999;**298**:397–404.
86. Boettner B, Harjes P, Ishimaru S, Heke M, Fan HQ, Qin Y, et al. The AF-6 homolog canoe acts as a Rap1 effector during dorsal closure of the Drosophila embryo. *Genetics* 2003;**165**:159–69.
87. O'Keefe DD, Gonzalez-Nino E, Burnett M, Dylla L, Lambeth SM, Licon E, et al. Rap1 maintains adhesion between cells to affect Egfr signaling and planar cell polarity in Drosophila. *Dev Biol* 2009;**333**:143–60.

INDEX

Note: Page numbers followed by "*f*" indicate figures, and "*t*" indicate tables.

N

O

P

R

S

T

U

V

W

Z

Printed and bound by CPI Group (UK) Ltd, Croydon, CR0 4YY
08/05/2025
01864954-0002